CAUSES OF EVOLUTION

Causes of Evolution

A Paleontological Perspective

Edited by Robert M. Ross
and Warren D. Allmon

With a Foreword by Stephen Jay Gould

The University of Chicago Press
Chicago and London

Robert M. Ross is a postdoctoral fellow at the Geologisch-Paläontologisches Institut at the University of Kiel.
Warren D. Allmon is assistant professor of geology at the University of South Florida.

The University of Chicago Press, Chicago 60637
The University of Chicago Press, Ltd., London

Printed in the United States of America

99 98 97 96 95 94 93 92 91 90 54321

Library of Congress Cataloging in Publication Data

Causes of Evolution : a paleontological perspective / edited by Robert M. Ross and Warren D. Allmon ; with a foreword by Stephen Jay Gould.
p. cm.
Includes bibliographical references and index.
ISBN 0-226-72823-4. —ISBN 0-226-72824-2 (pbk.)
1. Evolution. 2. Paleontology. I. Ross, Robert M. II. Allmon, Warren D.
QH366.2.C39 1990
575—dc20 90-11049
CIP

∞ The paper used in this publication meets the minimum requirements of the American National Standard for Information Sciences—Permanence of Paper for Printed Library Materials, ANSI Z39.48-1984.

Contents

Foreword

Stephen Jay Gould

If we are forced (by some aspect of universal structure, psychic necessity, or merely the norms of Western culture) to order our world by dichotomies, we may at least choose some grand divisions that bring clarity to our subjects and insight to ourselves. Sterile dichotomies are established as debating points to make one side (usually invented) look foolish, and the other (our own) undeniably right and just. Great dichotomies make even divisions between alternatives with intellectual punch and strong empirical backing. Neither side is "right"; both are infallibly interesting, and their competing claims can only be decided by relative frequency.

The two largely orthogonal dichotomies selected by the editors (and, *mirabile dictu*, actually addressed by the authors) have been central to evolutionary biology ever since Darwin—and never resolved, if only because paleontological evidence of relative frequency is so crucial to any decision, and too few evolutionists have been sufficientliy conversant with the fossil record. Darwin's own treatment of these issues both illustrates their crucial character and continues to set an agenda for our current concerns.

1. Intrinsic-extrinsic. Natural selection, in its pure form, is a theory of trial-and-error externalism—organisms propose, and external forces (biotic and abiotic) dispose. To defend this pure form of extrinsic control on evolutionary change, Darwin denied direction-giving properties to variation and treated this internal source as raw material only. Variation, to Darwin, is copious, small in effect, and undirected—thus throwing up a dense and isotropic sphere of equipotentiality about the modal form of a population. Change occurs when the external forces of selection differentially preserve and accumulate a biased portion of this sphere. Of course, Darwin was too good a naturalist to espouse this pure form as exclusive. He identified, as "correlations of growth" for example, certain channels and constraints that might bias change from the inside. But his sense of relative frequency is unmistakable and absolutely necessary to the central logic of natural selection—external forces control evolutionary change (while internal forces supply isotropic raw material); constraints are either past adaptations imposed by

history upon current contexts or direct sequelae of current adaptations. Adaptation has both causal primacy and extrinsic status.

2. Biotic-abiotic. This second dichotomy does not strike quite so close to the heart of Darwinian logic, for both sides are permissible modes of extrinsic (that is, orthodox) control upon evolutionary change. (Since the dichotomies are orthogonal, the unorthodox intrinsic controls also have biotic and abiotic modes; nonetheless, a strict Darwinian may neglect or ignore this full taxonomy and view the biotic-abiotic contrast as a debate about styles of extrinsic change. In this sense, the biotic-abiotic dichotomy has usually been pursued as a debate *within* Darwinian theory, while the intrinsic-extrinsic dichotomy is the *locus classicus* for critics of Darwinism. This difference, no doubt, explains Allmon and Ross' revealing tabulation of positions and concerns in the first 12 years of articles from the leading paleobiological journal of America—biotic and abiotic are espoused with equal frequency, while extrinsic exceeds intrinsic by some three to one. Most scientists remain orthodox in their basic theoretical position.)

But Darwin provided special punch to the biotic-abiotic distinction by basing his most important argument for large-scale pattern in the history of life upon his contention that biotic factors maintain an overwhelming dominance in relative frequency. Darwin, in his most radical gesture during an age of unparalleled faith in expansion and domination, constructed the theory of natural selection without any statement about, or allowable prediction of, progress. In the bare bones mechanics of his theory, organisms adapt to changing local environments—and that is all. Yet Darwin, as an eminent Victorian, was not prepared to abandon his culture's most cherished view that progress, however inconstant and fitful, should characterize the history of life as it pervaded the story of culture.

Since Darwin could not justify progress by the bare bones mechanism of natural selection, he smuggled it back in with an ancillary theory of ecology—namely, the dominance of biotic over abiotic forces as agents of selection. If abiotic causes dominate, then no progress accrues—for why should Tertiary climatic warming induce more anatomical advance than similar Cambrian amelioration? But if the world is always full of life (the old principle of plentitude), and if overt struggle in biotic competition generally prevails in such a perpetual crowd, then new creatures find a place only by forcing others out ("wedging" in Darwin's favorite metaphor)—and the accumulating results of battle will impose a vector of progress on life's history. In a remarkable passage from the first edition of the *Origin* (1859: 487–8), Darwin sums up his views on gradualism, uniformity, dominance of biotic competition, and rationale for progress, with a claim for such general evenness that the summation of morphological change might even roughly measure the amount of elapsed time: "As species are produced and exterminated by slowly acting and still existing causes . . . and as the most important of all causes for organic change is one which is almost independent of altered . . . physical conditions, namely the mutual relation of organism to organism,—the improvement of one being entailing the improvement or the extermination of others; it follows that the amount of organic change in the fossils of consecutive formations probably serves as a fair measure of the lapse of actual time." Interestingly—and why not, for Darwin was so incredibly astute—the dominance of biotic competition remains the usual and strongest argument for vectors of biomechanical improvement in the history of major groups. In this volume, Jackson and McKinney attribute such vectors in bryozoans to "escalation in the defenses employed in biological interactions," while Hallam identifies the abiotic influences of mass extinction as primary controls upon the apparently nondirectional pattern of morphological evolution in Mesozoic molluscs.

Paleontology has not always been the most welcome of partners in the goodly fellowship of evolutionary subdisciplines—an odd situation indeed for the keepers of the archives of evolution's actual results. The reasons for this suspicion and low status are many—all invalid. Darwin, citing the notorious imperfection of the fossil record, viewed paleontology as more of an embarrassment than an aid. In more recent years, paleontologists have gained some ground in upgrading active suspicion to benign neglect—but then most of us would rather be reviled than ignored.

The primary rationale for such neglect has centered on the claim that all evolutionary theory must be formulated in the maximally accessible world of microevolution—and that the fossil record is therefore only a playground for the operation of mechanisms discovered elsewhere. I cannot, in this foreword, present a full defense for paleontology's necessary and equal role in the formulation of evolutionary theory (lest I usurp this book by prolixity). But two anchors of this defense are beautifully illustrated in the many excellent articles of this volume.

First, natural history is a domain of relative frequency, rarely of exclusivity. Nearly all of our most interesting debates involve alternatives equally consistent with the best of general theory. (Our situation, in this respect, is very different from the popular image of stereotypical science, where crucial experiments validate one outcome and debar others, and where theory narrowly sets a range of interpretations. We can cite some examples in this canonical mode—the exclusion of Lamarckian inheritance and the validation of Mendelism, for example. But nature's multifarious complexity and variety dictate that we usually face the world with a range of coherent possibilities, each "vying" for greater representation). In this situation, resolution is not achieved at the laboratory workbench or on the modeler's pad of paper (though these standard activities perform great service in setting plausible ranges and conditions for operation of alternatives that must then be assessed empirically). We must go to the vastness of the actual record of life's history and tabulate relative frequencies in actual occurrence (not at all a simple matter of find 'em and count 'em, given the spatial, temporal, and taxonomic inhomogeneity of nature, but still the only path to resolution). The fossil record is therefore not merely a playground for causes and processes determined elsewhere but the arbiter of basic theoretical questions on the testing ground of relative frequency.

I showed in the first part of this foreword that the two focal dichotomies of this book—intrinsic-extrinsic and biotic-abiotic—have basic implications for the deepest questions of Darwinian, indeed of all, evolutionary theory. The alternatives, in each case, represent classic exercises in testing by relative frequency: clearly extrinsic and intrinsic agents of direction exist and, just as clearly, biotic and abiotic factors regulate the history of life. But the world is a very different place, and lineages have very disparate properties when one or the other alternative dominates in relative frequency. We do need to know these relative frequencies, and we can only find out by consulting the results of life's actual history—the fossil record. Most articles in this book are exercises in the exploration of relative frequency. Schindel tries to set intrinsic versus extrinsic weights for aspects of form in the coiled gastropod shell. Tiffney and Niklas use broadest-scale empirical correlation to assess the relative importances of biotic and abiotic factors in the history of plants. I have already illustrated the importance of differing weights in the case of vectors for biomechanical improvement in bryozoans based on a dominance of biotic effects (Jackson and McKinney) versus an apparent nondirectionality correlated with abiotic control in Mesozoic molluscs (Hallam).

Second, I think that most evolutionists (in contrast with claims of the last generation for a stricter Darwinism) would now allow that not all causal processes of large-scale evolution can be apprehended in modern populations and then fully rendered by extrapolation into the vastness of time (though great differences persist as to the relative frequency of effect for added processes of the macroevolutionary realm!). So much of the macroevolutionary play is a story of differential sorting and selection among species and higher units—phenomena that cannot be resolved by extrapolating the anagenetic changes of populations (usually absent in a world dominated by stasis in any case) into the fullness of time.

Most articles in this volume work with irreducible units and levels existing in domains of space and time so broad that only the fossil record reports the evidence of their histories. Jablonski and Bottjer, for example, find that while orders tend to originate onshore, lower taxonomic units show no such bias of appearance. (Orders, of course, are abstractions, not entities—but the pattern must be saying something about substantial versus minor morphological innovation, and how could we ever know about such an important pattern except by direct consultation with the fossil record?) Valentine works with the macroevolution of clade shape; Stanley with origination and extinction rates of species properly treated as entities. Jackson and McKinney, in the midst of their largely extrapolationist account, make the powerful point that no adaptive trends are known in the unbroken anagenetic history of single monophyletic lineages; all are cumulated through numerous events of branching speciation.

The fossil record has fascinated us for its sheer phenomenology at least since Thomas Jefferson misidentified a ground sloth as a lion. We are now learning something just as precious: that paleontology is also a domain of principles working in partnership with, not in subservience to, Darwin's small world of mighty import.

Preface

This volume originated in a student-run seminar series in "Earth History and Paleobiology" held at Harvard University in the spring of 1987. The annual series was intended to foster interaction among five research groups in these fields at Harvard: invertebrate and vertebrate paleontology, paleonathropology, paleobotany/Precambrian paleontology, and sedimentology. In a group meeting to decide the fate of the 1987 series, John Barry suggested the theme that we ultimately adopted, "Biotic and Abiotic Factors in Evolution." The speakers for the series were Alfred Fischer, Anthony Hallam, Jeremy Jackson, Jennifer Kitchell, Adolf Seilacher, and Elisabeth Vrba. Roger Buick suggested that we ask speakers to submit papers for a volume. We then solicited additional papers to round out the variety of taxa, times, and points of view. In our instructions to authors, we suggested they might also address another dichotomy of cause: that of intrinsic and extrinsic factors.

We would like to thank all of the students, staff, and faculty who participated in this EHAP series for their time, assistance, and support. Although only two edited it, this volume belongs to all of us. Special thanks are due Stephen Grant, Julian Green, and Roger Buick for exploring potential publishers. We are especially grateful to Peter Williamson for virtually single-handedly managing the series in years previous to us. Without his efforts the series would not have survived to assume its present form. We also thank the Graduate School of Arts and Sciences of Harvard University for financial support of the series.

We cannot imagine this project without the constant assistance, advice, support, and friendship of our editor at the University of Chicago Press, Susan Abrams. She has skillfully guided our efforts while always leaving us great freedom. We are grateful to Geerat Vermeij and two anonymous reviewers for their comments on the volume.

All royalties from this volume are being donated to the Paleontological Society.

1

Specifying Causal Factors in Evolution: The Paleontological Contribution

Warren D. Allmon and Robert M. Ross

Introduction

Amid disparate conceptions of what comprises *the* or *a* scientific method, at least one idea elicits little diagreement: much of science consists of asking the right questions. Often the most productive path of inquiry in a complex subject comes, not from a new discovery, but from a new quesetion or from renewed attention to one among many long-standing questions. If we are clear about what we are asking, says this line of reasoning, we can be clearer about the answers. This is the context in which this volume was conceived.

It is impossible to think about causes of evolution without considering, implicitly or explicitly, the relative influence of factors intrinsic or extrinsic to organisms or whether these factors are more biologically or physically mediated. Often, however, questions about the relative roles of intrinsic and extrinsic, or biotic and abiotic, factors in evolution are not asked; if they are asked, it may be only in passing, and explicit answers are seldom given. This volume is an experiment. We wanted to see what would happen if paleobiologists were asked to identify, based on the results of their ongoing research programs, what they believed were the principal causal factors controlling evolution in the organisms under study. We asked them to comment on the relative roles of biotic and abiotic, intrinsic and extrinsic factors—not because these are the only logical or possible taxonomies of course, but because they are relatively exhaustive and so potentially interesting and relevant to anyone doing evolutionary biology. We wanted first to test the idea that explicit taxonomies of cause, by identifying and categorizing the sorts of causal factors proposed by evolutionary biologists, can lead not only to more critical examination of these factors but also to new hypotheses that fill gaps revealed in existing theories. Second, we hoped this type of methodological exercise could, in turn, contribute to better understanding of both the substantive roles that different factors play in evolutionary processes and the methodological roles that different dis-

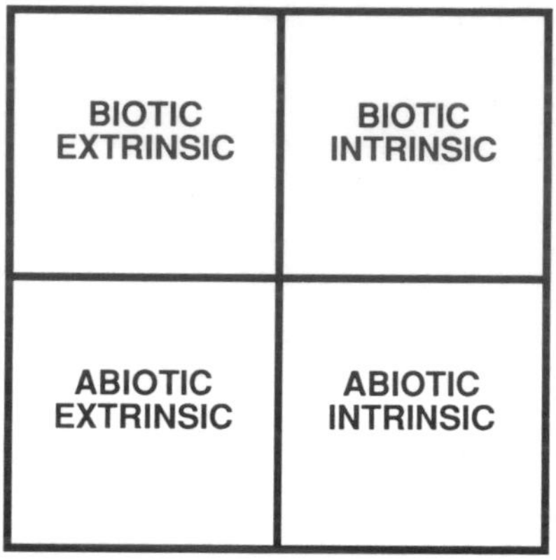

BIOTIC EXTRINSIC	BIOTIC INTRINSIC
ABIOTIC EXTRINSIC	ABIOTIC INTRINSIC

Figure 1.1 Combinations of major factors controlling evolutionary processes and patterns.

ciplines, particularly paleontology, can play in elucidating these processes. Must paleontology be content to document processes identified and studied in more detail by neontologists, or can paleontology, as some authors have argued, contribute genuinely new ideas and perspectives to evolutionary biology? If the latter, what sorts of paleontological studies are most useful? In what form should or can the paleontological contribution be?

A Causal Taxonomy

Biotic and abiotic factors correspond to biological and physical conditions, respectively. Intrinsic factors are qualities of an individual organism (or other unit at other hierarchical levels); extrinsic factors are outside and independent of that organism (or other unit). The intrinsic-extrinsic and biotic-abiotic dichotomies can be combined in a 2 × 2 table (fig. 1; cf. Vrba 1985). *Biotic intrinsic* factors are those biological factors in some way inherent or internal to an individual organism and that limit, direct, or guide the direction, rate, or mode of evolutionary change. Limitations or regularities in variation, often referred to as genetic or developmental constraints, are examples of such factors. *Abiotic intrinsic* factors are physical laws or purely physical properties of materials that affect the morphology of a developing organism. Examples include the patterns of growth of calcite crystals or constraints imposed by surface-volume relationships. *Biotic extrinsic* factors are biologically mediated processes, external to individual organisms, that affect individual organisms in whole or in part. Common examples

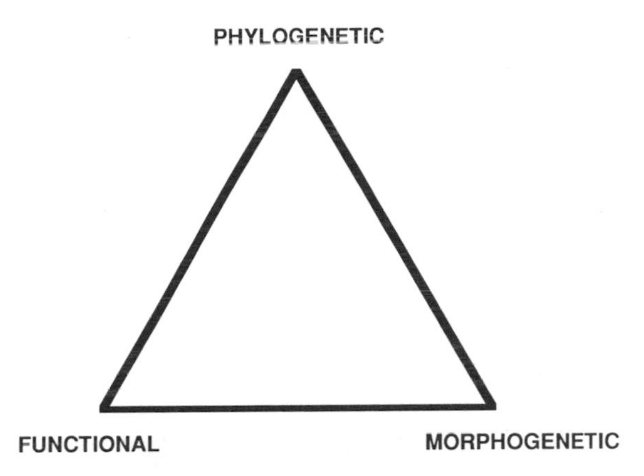

Figure 1.2. Seilacher's (1970) classification of constraints controlling morphological evolution.

include predator-prey relationships and interspecific competition. *Abiotic extrinsic* factors are purely physical phenomen external to organisms that affect them in whole or in part. Examples include climate, substrate, and wave energy.

This causal classification is similar in many ways to that proposed by Seilacher (1970; see also Raup 1972; Thomas 1979; Gould 1989). Seilacher's triangle (fig. 2) provides a threefold classification of possible factors or forces that influence the evolution or morphology. His *phylogenetic* constraints are roughly equivalent to intrinsic biotic factors in fig. 1.1; *morphogenetic* constraints are roughly equivalent to intrinsic abiotic factors; *functional* constraints include all extrinsic factors, biotic and abiotic, by which natural selection acts on organisms. The triangle was primarily intended to address problems of morphology (as reflected in the term *Konstruktions-Morphologie*), whereas the system in figure 1.1 is relevant to any sort of evolutionary change (e.g., morphological change, ecological change, diversification). The same distinction applies to the classification of causal factors suggested by Hickman (1980).

Others have discussed the significance of the abiotic-biotic and extrinsic-intrinsic dichotomies. Kingsland (1985) documented the history of debate among ecologists on the role of biotic and abiotic factors in controlling community structure. Recently Stenseth and Maynard Smith (1984) argued that evolution occurs only when stimulated by changes in the abiotic environment rather than being driven continuously by biotic interactions, as predicted by their formulation of Van Valen's (1973) Red Queen model (see also Hoffman and Kitchell 1984). Significantly, Stenseth and Maynard Smith conclude that only the paleontological record can resolve the differences between these two models.

Debates concerning the roles of intrinsic and extrinsic controls are as old as the concept of evolution (cf., Gould 1977 and this volume). The controversy has taken various guises since Lamarck's contention that evolution is governed both by internal tendencies (perhaps analogous to intrinsic constraints) and by the direct influence of the environment. Many 19th-century American paleontologists, such as E. D. Cope and Alpheus Hyatt, adopted Lamarckian ideas to explain apparent trends in the fossil record, while others, such as O. C. Marsh, explained them using the extrinsic forces driving natural selection (see Bowler 1985).

The system given in figure 1.1 has a number of problems, the first of which lies in assigning particular causal factors to one or more of the four cells. Since organisms are part and parcel of their environment (e.g., Lewontin 1982, 1983), it may be difficult in practice to determine what is biotic or abiotic or intrinsic or extrinsic. It is also clear that assigning any given evolutionary example to one or another of the four cells is an oversimplification. There is no case in which biotic forces act to the exclusion, or in the absence of, abiotic forces. Evolution is certainly not governed by one sort of factor or another but by all—or rather, not by discrete classes but by a spectrum of causes. Furthermore, it may be the interactions among individual causal factors, more than the factors themselves, that are most important in controlling evolutionary change (cf., Hutchinson 1965; Kitchell, this volume; Geary, this volume).

The taxonomy of cause given in figure 1.1 may, however, itself contain solutions to these problems. Forcing an investigator to attempt to identify causal processes according to this system may lead to better understanding of where the organism ends and the environment begins; providing a framework for tabulating the possibilities may lead to a better understanding of a continuous spectrum of causal factors. By leading to improved specification and description, it may lead to more successful identification and examination of causal interactions.

Evolutionary Causes and Paleontology

Paleontologists have in fact been considering this range of causal factors for a long time, if not always very explicitly. Whenever an explanation is given in the form of "environmental change led to . . ." or "this taxon apparently never evolved adaptations for . . . ," an author is discussing one or more of the causal factors identified in figure 1.1. What paleontologists say about causes of evolution is at best, however, only an oblique indicator of what may or may not be the case in nature. Paleontological opinion presumably has some relation to reality, but it certainly also reflects the preferences, biases, and assumptions of individual workers and, perhaps to a lesser degree, of the discipline as a whole.

Examination of this opinion can be useful in at least two ways. First, although scientific theory is not decided by vote, science operates at least in part by building and dismantling consensuses on individual problems. The current "state of the art" is, to a significant degree, a measure of what members of a given scientific community think they "know" about their corner of the natural world. Second, and more importantly, since we cannot ever completely escape bias (in the purely descriptive, textile industry sense of trend or inclination), we can only hope to acknowledge it openly. A survey of paleontological opinion on evolutionary causes may help to reveal biases and to point out implicit assumptions under which we all operate.

Toward this end we classified all papers published in the journal *Paleobiology* during the period 1975 to 1987 according to what perspective each took within the taxonomy of cause given in figure 1.1. To be included in our sample, papers had to make a reasonably extensive statement about evolutionary process; thus papers dealing strictly with descriptive paleoecology, descriptive functional morphology, taphonomy, extinction, or purely descriptive accounts of historical pattern (taxonomic diversity or morphological change) were not considered. A paper could fall into more than one category; therefore percentages are based on the total number of entries in the table (211) rather than the total number of papers classified (146). It was frequently difficult to detect exactly where an author's opinion fell in the classification, so some subjectivity in placement was unavoidable; we nevertheless believe that the data are sufficiently accurate for our purposes.

Results (fig. 1.3) suggest the following:

1. Although overall biotic and abiotic explanations seem to be used about equally, authors publishing in *Paleobiology* between 1975 and 1987 discussed extrinsic factors almost three times as often as intrinsic factors. This pattern is stronger for abiotic (approximately 6:1) than for biotic (approximately 2:1). Equal numbers of studies considered extrinsic biotic factors (e.g., Jeppson 1986; Martin 1986; Aronson 1987) as extrinsic abiotic factors (e.g., Thayer 1986; Wolfe 1987) as more important in explaining evolutionary patterns. Some authors (e.g., Bottjer and Ausich 1986) emphasize both.

	EXTRINSIC FACTORS	INTRINSIC FACTORS	TOTAL
BIOTIC FACTORS	80 (38%)	37 (18%)	117 (55%)
ABIOTIC FACTORS	80 (38%)	14 (7%)	104 (45%)
TOTAL	160 (76%)	50 (24%)	211

Figure 1.3. Frequency of sorts of evolutionary factors considered in the journal *Paleobiology* in the years 1975–87. In each cell, the top number is the frequency out of 146 papers classified, the bottom number the percentage of the total number of factors considered (211). (Some papers considered more than one factor.) Papers that did not extensively discuss forces influencing evolutionary patterns were not classified.

2. When intrinsic factors have been discussed, they have usually been biotic (e.g., Thomas 1976; Meyer and Macurda 1977; Linsley 1977; Pachut and Anstey 1979; Hansen 1980; Lauder 1981; Gould 1984a, 1984b). Intrinsic abiotic factors have been considered less often (e.g., Gould and Katz 1975; McGhee 1980; Seilacher 1979, 1984; Fortey 1983; Dafni 1986).

3. Although some papers were difficult to classify, most were not, and several authors whom we personally queried about their point of view largely confirmed our diagnoses and implied that their positions on these issues were fairly obvious. Yet in the papers themselves, few authors are explicit about the subset of potential evolutionary factors they were investigating, and most do not comment explicitly on the conceptual limitations of their program.

The data base employed here is obviously limited and probably somewhat biased. The survey reflects only the papers submitted to and accepted by *Paleobiology*, not necessarily the views of all workers who would call themselves paleobiologists (see e.g., Reif 1981). Surveys of other journals would have given different results.

Assuming, however, that this sample is representative of at least a significant proportion of the research activity in evolutionary paleontology during this period, we can explore more interesting explanations for these results. Some are more straightforward than others. The preference for studies focusing on extrinsic rather than intrinsic factors is probably best explained by the character of most paleontological data: more intrinsic than extrinsic information is lost during fossilization. Although intrinsic information is available from morphology, extrinsic information from paleoenvironmental and paleoecological reconstruction is often more abundant and easier to interpret.

The relative influence of a particular category of cause is partly a function of the evolutionary patterns being investigated. A large proportion of recent papers describe either trends in taxonomic diversification or trends in morphology. Many suggest that diversity is, at least in part, abiotically mediated (e.g., Potts 1984; Raymond 1985; Wolfe 1987); fewer have considered physical laws or physical environmental changes as explanations for long-term trends in morphology (e.g., Gilinsky 1984; McKinney 1986; Wolfe 1987). Biotic explanations are commonly advanced for diversity *or* morphological trends. Of course, these evolutionary patterns are not independent themselves; morphological trends may result from diversification at lower taxonomic levels.

The paucity of explicit statements about theoretical framework is perhaps the most striking aspect of this survey. Possible reasons for this are many and to some degree unique to each paper. In some cases authors may not feel the need to acknowledge that the data used in their study may allow exploration of a limited number of questions, either because they do not believe that implications are limited or because they think it is obvious that they are discussing only one aspect of a larger problem. Authors may assume, after long experience with a particular data set, that explication is unnecessary because the theoretical context of their conclusions is obvious or that it is clear how local conclusions extrapolate to the rest of nature. We do not suggest that every published paper should provide a full explanation of all processes conceivably involved; we do hope that plotting an exhaustive chart of possible processes can remind us where a research program has been, where it is, where it can go, and where complementary programs can be connected. A comprehensive framework makes it easier to conceive of possible processes in some regular fashion.

The Present Volume

This book is about a way of looking at evolution, especially using evidence available to paleontologists, and about the way evolution occurs. Since the papers were solicited with the understanding that they were to be framed in the context of an explicit causal taxonomy, they may be able to supplement the methodological and substantive conclusions derived from a larger literature that was not thus constrained.

METHODOLOGICAL CONCLUSIONS

1. Is this taxonomy of cause useful? Authors responded (through personal communication and in their papers) to the central theme of the volume in a number of ways. Many had been thinking about similar ideas for many years (and, indeed, that is why they were solicited to contribute), but like the authors of most *Paleobiology* papers, they had seldom articulated the differences among factors in exactly this way. Some had not been thinking in these terms before (and were solicited because they were pursuing research we thought amenable to this approach). Most of the contributors in both groups responded favorably, apparently finding the idea provocative enough to stimulate a somewhat different look at their research. This was true despite the evident difficulty many authors experienced in successfully teasing apart extrinsic from intrinsic and biotic from abiotic. A few contributors, however, found the suggested framework too vague, or an artificial distraction, or unilluminating because it covers everything and excludes nothing. As Kitchell (p. 151) states in her contribution, "Neither is wrong."

Several contributors address only a single type of causal factor. Tiffney and Niklas, for example, examine correlations between patterns of terrestrial plant diversity and a variety of abiotic factors, such as continental area, fragmentation, and topography. Schindel considers the effects of geometric constraints on gastropod shell evolution. Cronin and Ikeya discuss the differential susceptibility of eurytopic and stenotopic species to speciation but concentrate on the creation of "opportunities" for speciation by abiotically induced range fragmentation events. All of these authors include statements near the beginning or end of their papers indicating they recognize that "other factors are involved" but are consciously addressing only one. Such an approach clearly does not guarantee that the causal factor examined is the most important (as these authors acknowledge), but it may still demonstrate the advantage of applying an explicit causal taxonomy (whether it be that proposed in figure 1.1 or some other); results of considering a single or a limited range of causal factors can be put into perspective against the range of other alternatives and so can throw alternative causal hypotheses into greater relief.

2. Role of the fossil record. Results of the application of this explicit

causal taxonomy suggest a number of general conclusions about using paleontological data to study evolution. None are particularly novel but in aggregate they offer a useful perspective.

a. Evolutionary process is accessible to paleontologists, if only indirectly, but some processes are more accessible than others. Even though, as Bleiweiss puts it, "so much of nature's complexity is lost in the fossil record," considerable detail often remains. It is important to realize, however, that this detail is not uniformly distributed; the fossil record is strongly biased toward details of some processes and away from details of others. On the extrinsic side, the fossil record regularly contains information about abiotic environmental conditions, often including periodic events such as storms or volcanic eruptions. Less often it yields more or less detailed information on the biotic environment (i.e., interactions among individuals and species) and on subtle or short-term abiotic environmental changes. On the intrinsic side, the fossil record usually yields abundant data on hard-part morphology but relatively little about soft-part morphology or behavior. The temporal and geographical incompleteness of the record certainly precludes recognition of many events on "ecological" time scales (cf., Kitchell 1985; Valentine 1989; see discussion below), but not necessarily of the significance of ecological processes over longer time intervals (cf., Jackson and McKinney, this volume).

The same features of the fossil record may (e.g., Levinton, 1989) or may not (e.g., Stanley, this volume) preclude studies of factors influencing speciation. In his discussion of East African cichlid fishes, Dorit (this volume) shows that, were a good fossil record available, a dedicated paleoichthyologist would have been able to detect cichlid morphological diversification and perhaps even intraspecific geographical variation. He or she might also be able to reconstruct the sequence of fluctuating lake levels over the last 750,000 years and so suggest at least some of the abiotic factors potentially important in diversification. The same paleoichthyologist, however, would probably not be able to test hypotheses about habitat specialization and ease of dispersal between habitats. In fact, Dorit's favored explanation for much of cichlid diveristy—namely, low juvenile dispersal due to the habit of mouthbrooding—would probably not have been imagined at all had not observations of living animals been available.

Most of the authors in this volume, like most evolutionary paleontologists generally, wrestle with the problem of inferring unobservable process from observable pattern. Some make more effort to explore methodological problems and implications of this issue than others. Geary, for example, attempts to infer the causes of melanopsid snail diversification in a basin for which the geological and geochemical history is well known. Combining temporal and geographical patterns of intrinsically constrained morphological change and abiotic environmental change, she notes that while one clade underwent radiation, another did not. This "natural experiment" obviously allows neither manipulation nor direct observation of events but is nevertheless a revealing demonstration of the potential of pattern to limit procesess we consider as satisfactory explanations.

That even robust patterns do not testify to unique processes is well illustrated by Barry et al., who show that mammalian turnover events in the Neogene of Pakistan form pulses (i.e., are distributed nonuniformly through time), but they are unable to correlate these pulses with individual climatic, tectonic, or eustatic events.

b. Ecology matters and can be studied in the fossil record. As discussed in greater detail later; it is clear to some authors that interactions between an organism and its environment have important evolutionary consequences, even if these are not precisely the consequences expected 20 or 30 years ago. Ecological problems are important components of a majority of the contributions in this volume. Hallam, Heaton, Barry et al., Jablonski and Bottjer, Jackson and McKinney, and Stanley all provide empirical cases of ecological relationships from the fossil record.

SUBSTANTIVE CONCLUSIONS

1. Actual relative importance. Ideally, we would like to know, for any given evolutionary case and perhaps for the entire history of life, the actual relative importance of each of the types of causal factors specified in figure 1.1. If this is a valid classification of actual causal factors, after all, there is some actual, true proportional causal contribution, even if different factors have acted primarily through interaction rather than on their own. The contributions to this volume are only a small sample, and few of our authors make strong statements about the importance of one or another type of causal factor. Some generalizations, however, do emerge.

Hallam, for example, states that abiotic factors predominated over biotic factors in effecting evolutionary change among early Mesozoic ammonoid and bivalve faunas in Western Europe. Where biotic factors did play a significance role, Hallam believes that it was only after ecospace had been vacated by (largely abiotically caused) mass extinction. Jackson and McKinney conclude exactly the opposite: that for post-Paleozoic clonal benthos (bryozoans, corals, and coralline algae), morphological trends are the consequences of escalation in defenses in response to biotic interactions.

Schindel argues that geometric constraints in certain gastropod lineages may have substantially limited the evolutionary avenues open to them. Dorit confronts what he sees as the traditional view of diversity of cichlid fishes in the African Great Lakes, namely that speciation is primarily a passive consequence of extrinsic, environmental fragmentation. He argues instead for the greater importance of intrinsic factors, such as population structure, that make the generation of species more likely.

2. Interactions. In her paper, Kitchell assumes from the outset that interactions among different causal factors must be more important than any set of individual factors acting independently. Other authors in general behave more inductively, attempting to understand the causal factors underlying particular patterns of evolutionary change, only to decide in the end that complex patterns of interaction are probably responsible. Geary, for example, concludes that intrinsic factors establish potential morphological pathways, but that extrinsic environmental factors determine when and exactly which of those pathways are followed. Similarly, Heaton concludes that many range shifts among Pleistocene mammals in the Great Basin of the United States were due directly to climatic change, while others were effects of biotic factors, especially competition, that had themselves ultimately been due to other abiotic changes. Bleiweiss concludes that the effect of intrinsic attributes on diversity depends on the environmental setting. In complex heterogeneous habitats, he argues, clades containing mostly generalist species tend to diversify because they can take advantage of this complexity. In uniform homogeneous habitats, clades containing mostly specialist species tend to diversify because they tend to form restricted populations in such settings.

In discussing the factors that control speciation and extinction, both Stanley and Valentine explain that the effects of extrinsic factors on speciation and extinction depend on the intrinsic qualities of the organisms of study. Although he recognizes extrinsic biotic controls over certain aspects of equilibrium clade diversity, Valentine explains the waxing and waning of clades primarily through abiotically driven forces that increase or decrease adaptive opportunities and interact with intrinsic properties. Stanley considers the apparently fortuitous link between speciation and extinction rates that occurs even during episodes of taxonomic radiation. He argues that this linkage is a result of a number of factors, some extrinsic (e.g., abiotic habitat fragmentation), some intrinsic (e.g., behavioral complexity, niche breadth, dispersal ability), and some a result of both (e.g., population size/stability).

Jablonski and Bottjer struggle to identify potential individual biotic and abiotic factors responsible for the onshore-offshore evolutionary gradient they have identified but wind up favoring a complex set of possible alternative explanations. Origination of high taxonomic level morphological novelties may be due to higher environmental heterogeneity onshore (abiotic extrinsic), different population structure of near-shore species (biotic intrinsic), or differential impact of mass extinction in near-shore communities (abiotic extrinsic, mediated by biotic intrinsic). Expansion of novel morphologies offshore may be caused by their passive diffusion outward due to lower onshore extinction rates or by speciation preferentially in an offshore direction (biotic intrinsic factors in concert with abiotic extrinsic gradients). The final retreat of many of these groups into offshore habitats may result from higher rates of extinction onshore (biotic intrinsic and abiotic extrinsic factors) or from active competitive displacement of older taxa by newer (biotic intrinsic and extrinsic together).

3. Taxon specificity. The relative influence of abiotic and biotic factors is a function of the intrinsic properties of particular organisms. For example Stanley and Valentine each compare a variety of factors that may create low or high turnover clades in higher taxa. Dorit, Bleiweiss, and Geary each examine interactions of intrinsic and extrinsic factors that resulted in differential speciation within an individual lower taxon.

Clonal benthos respond to selective pressures at both the level of the modular units and the level of the colony. Jackson and McKinney (this volume) document progressive morphological trends in the level of zooidal integration and growth form and mode in bryozoans and other colonial organisms. These changes seem to be an effect of adaptation to ordinary physical processes and also escalating defense and competitive mechanisms, as suggested for solitary organisms such as molluscs, brachiopods, and crinoids (e.g., Vermeij 1977, 1987; Signor and Brett 1984). Key (1988) suggests that net improvements in efficiency of design in Ordovician trepostome bryozoans are a response to modified selective pressures at the level of the colony.

Conclusions regarding plants and animals are rather different. Tiffney and Niklas here conclude that, unlike marine invertebrates, geographical provinciality does not correlate with total diversity, suggesting that evolutionary response to geographical dispersion may be different. Their conclusion is based on the correlation of northern hemisphere land plant diversity with major physical variables, which shows that total land area, and perhaps number of uplands, are better predictors of diversity. Niklas and Kerchner (1984) have shown that early land plants improved in their biomechanical efficiency. Tiffney and Valentine (in prep.) believe that species durations decrease over the Phanerozoic in land plants, rather than increase as suggested for marine animals; many have suggested that the evolutionary response of plants to mass extinctions was different from that of animals (cf., Knoll and Rothwell 1981).

ARRANGEMENT OF PAPERS

The papers in this volume demonstrate that the range of causal factors examined, and the way in which these examinations are framed in any particular research program, are quite varied. Few of the contributions overlapped in the combination of factors considered by the approach to them. We have thus chosen to organize papers independently of process and instead according to taxonomic breadth, that is, those that consider evolutionary forces for a single taxon (class level or lower) versus those that consider patterns across higher taxa. Geary and Schindel, for example, discuss only gastropods, Bleiweiss only hummingbirds, and Dorit only fish. These authors discuss factors that account for variation in evolutionary patterns within their clade; the implication is that similar variation within other clades may have similar origins; thus the results may be generally applicable. Kitchell and Valentine, on the other hand, provide models that might apply to any organisms. Jablonski and Bottjer and Stanley similarly give examples of different sorts of animals, examining intrinsic similarities and differences; Tiffney and Niklas include all taxa of vascular plants in their analysis.

Conclusion: The Paleontological Contribution

Was the experiment a success? Did encouraging authors to classify their evolutionary results improve our ability to interpret these results? Individual readers will, of course, judge for themselves. Our overall reaction is that most of our authors earnestly tried to apply at least one of the two causal dichotomies in fig. 1.1 and that the results are decidedly mixed. There are two possible explanations for the lack of consensus on causal factors in evolution, at least as represented by this classification. One is that no general, overarching causes exist. Large-scale patterns may be just passive consequences of a multitude of smaller-scale processes. This point of view has recently been argued by Hoffman (1989).

The other is that general processes exist but are seen very differently by different researchers. The individual papers in this volume represent a variety of research styles, scales, and points of view. Some differences are therefore to be expected. More substantively, however, differences in temporal and geographical scale, taxonomic group, and theoretical approach (e.g., case study versus overall pattern) have affected the results each author presents. Common principles can be agreed to in theory, but the causes of evolution one actually sees in one's own work may depend largely on the causes of evolution one is looking for.

In 1932 J. B. S. Haldane published a book with a title similar to this one. In it, Haldane accurately reported that most paleontologists of the time despaired of understanding the patterns they saw in the fossil record according to Darwinian natural selection and so tended to limit themselves "to stating the facts of evolution, and laying down general laws which they obey, rather than attempting to discover the causes underlying those laws" (1932:29). Although Haldane briefly considers breaks in the fossil records of lineages and speculates about their significance for evolutionary tempo and mode, the book is really about the nature of heritable differences between species, the occurrence of selection in natural populations, and whether natural selection is sufficient to account for species formation. "The way to still further knowledge," Haldane writes, "lies largely in the accumulation of more facts concerning variation and selection" (1932:169).

The situation has of course dramatically changed since Haldane wrote. Paleontologists are preoccupied today as never before with the causes of evolution. Most of the major themes in contemporary paleobiology (e.g., large-scale diversity patterns, evolutionary constraints, evolutionary paleoecology, hierarchy, mass extinction) are directly related to this issue. Paleontology has succeeded in exploring virtually every area of evolutionary biology, and conflicting interpretations abound. It remains to be decided to what degree macroevolution is microevolution or paleoecology is ecology, or whether the nature of nature is more of unity or diversity of cause. The blind men disagreed but were, after all, studying the same elephant. Whether in the context of the causal classification presented here or some other, one of the chief challenges facing modern paleontology is to explore a more complete integration with biology, to see whether the causes of evolution discussed by one group of researchers are really the same as the causes discussed by another.

Acknowledgments

We are grateful to T. Cronin, M. Foote, D. Geary, J. Pojeta, G. Vermeij, and three anonymous reviewers for comments on previous drafts of the manuscript.

Literature Cited

Aronson, R. B. 1987. Predation on fossil and Recent ophiuroids. *Paleobiology* 13(2):187–92.

Bottjer, D. J., and W. I. Ausich. 1986. Phanerozoic development of tiering in soft substrata suspension-feeding communities. *Paleobiology* 12(4):400–420.

Bowler, P. J. 1985. Evolution. The history of an idea. Berkeley and Los Angeles: University of California Press.

Dafni, J. 1986. A biomechanical model for the morphogenesis of regular echinoid tests. *Paleobiology* 12(2):143–60.

Eldredge, N., and J. Cracraft. 1980. Phylogenetic patterns and the evolutionary process. New York: Columbia University Press.

Fortey, R. A. 1983. Geometrical constraints in the construction of graptolite stipes. *Paleobiology* 9(2):116–25.

Gilinsky, N. L. 1984. Does archeogastropod respiration fail in turbid water? *Paleobiology* 10:459–68.

Gould, S. J. 1977. Eternal metaphors of paleontology. In *Patterns of evolution as illustrated by the fossil record*, ed. A. Hallam, 1–26. Amsterdam: Elsevier.

Gould, S. J. 1984a. Covariance sets and ordered geographic variation in *Cerion* from Aruba, Bonaire and Curacao: a way of studying nonadaptation. *Syst. Zool.* 33:217–37.

Gould, S. J. 1984b. Morphological channeling by structural constraint: Convergence in styles of dwarfing and giantism in *Cerion*, with a description of two new fossil species and a report on the discovery of the largest *Cerion*. *Paleobiology* 10(2)1:172–94.

Gould, S. J. 1989. A developmental constraint in *Cerion*, with comments on the definition and interpretation of constraint in evolution. *Evolution* 43(3):516–39

Gould, S. J., and M. Katz. 1975. Disruption of ideal geometry in the growth of receptaculitids: a natural experiment in theoretical morphology. *Paleobiology* 1(1):1–20.

Haldane, J. B. S. 1932. The causes of evolution. London: Longmans, Green & Co.

Hansen, T. A. 1980. Influence of larval dispersal and geographic distribution on species longevity in neogastropods. *Paleobiology* 6:193–207.

Hansen, T. A. 1988. Early Tertiary radiation of marine molluscs and the long-term effects of the Cretaceous-Tertiary extinction. *Paleobiology* 14(1):37–51.

Hickman, C. S. 1980. Gastropod radulae and the assessment of form in evolutionary paleontology. *Paleobiology* 6:276–94.

Hoffman, A. 1989. Arguments on evolution. A paleontologist's perspective. New York: Oxford University Press.

Hoffman, A., and J. A. Kitchell. 1984. Evolution in a pelagic ecosystem: a paleobiologic test of models of multispecies evolution. *Paleobiology* 10(1):9–33.

Hutchinson, G. E. 1965. The ecological theater and the evolutionary play. New Haven, Conn.: Yale University Press.

Jeppson, L. 1986. A possible mechanism in convergent evolution. *Paleobiology* 12(1):80–88.

Key, M. M., Jr. 1988. Progressive macroevolutionary patterns in colonial animals. Geol. Soc. Am. (abstracts with programs) 20(7):A201.

Kingsland, S. E. 1985. Modeling nature: Episodes in the history of population ecology. Chicago: University of Chicago Press.

Kitchell, J. A. 1985. Evolutionary paleoecology: Recent contributions to evolutionary theory. *Paleobiology* 11(1):91–104.

Knoll, A. H., and G. Rothwell. 1981. Paleobotany. Perspectives in 1980. *Paleobiology* 7:7–35.

Lauder, G. V. 1981. Form and function: Structural analysis in evolutionary morphology. *Paleobiology* 7(4):430–42.

Levinton, J. A. 1989. Genetics, paleontology and macroevolution. Cambridge: Cambridge University Press.

Lewontin, R. C. 1982. The organism as the subject and object of selection. *Scientia* 118:65–82.

Lewontin, R. C. 1983. Gene, organism and environment. In *Evolution from molecules to men*, ed. D. S. Bendall, ed. 273–85. Cambridge: Cambridge University Press.

Linsley, R. 1977. Some laws of gastropod shell form. *Paleobiology* 3:196–206.

Martin, R. A. 1986. Energy, ecology and cotton rat evolution. *Paleobiology* 12(4):370–82.

McGhee, G. 1980. Shell form in the biconvex articulate Brachiopoda: a geometric analysis. *Paleobiology* 6:57–76.

McKinney, M. 1986. Ecological causation of heterochrony: a test and implications for evolutionary theory. *Paleobiology* 12(3):282–9.

Meyer, D. L., and D. B. Macurda. 1977. Adaptive radiation of the comatulid crinoids. *Paleobiology* 3:74–82.

Niklas, K. J., and V. Kerchner. 1984. Mechanical and photosynthetic constraints on the evolution of plant shape. *Paleobiology* 10(1):79–101.

Pachut, J. F., and R. L. Anstey. 1979. A developmental explanation of stability-diversity-variation hypotheses: Morphogenetic regulation in Ordovician bryozoan colonies. *Paleobiology* 5:168–87.

Potts, D. C. 1984. Generation times and the Quaternary evolution of reef-building corals. *Paleobiology* 10:48–58.

Raup, D. M. 1972. Approaches to morphologic analysis. In *Models in paleobiology*, ed. T. J. M. Schopf, 28–45. San Francisco: Freeman, Cooper & Co.

Raymond, A. 1985. Floral diversity, phytogeography and climatic amelioration during the Early Carboniferous (Dinatian). *Paleobiology* 11:293–309.

Reif, W.-E. 1981. Paleobiology today and fifty years ago: a review of two journals. *Neues Jahrbuch für Geologie und Paläontologie Mh.* 6:361–72.

Schwietzer, P., et al. 1986. Ontogeny and heterochrony in the ostracode *Cavellina* Coryell from Lower Permian rocks in Kansas. *Paleobiology* 12:290–301.

Seilacher, A. 1970. Arbeitskonzept sur Konstruktions-Morphologie. *Lethaia* 5:393–6.

Seilacher, A. 1979. Constructional morphology of sand dollars. *Paleobiology* 5(3):191–221.

Seilacher, A. 1984. Constructional morphology of bivalves: Evolutionary pathways in primary versus secondary soft-bottom dwellers. *Paleobiology* 27:207–37.

Signor, P. W., III, and C. E. Brett. 1984. The mid-Paleozoic precursor to the Mesozoic marine revolution. *Paleobiology* 10:229–45.

Stenseth, N. C., and J. Maynard Smith. 1984. Coevolution in ecosystems: Red Queen evolution or stasis? *Evolution* 39:870–80.

Thayer, C. W. 1986. Are brachiopods better than bivalves? Mechanisms of turbidity tolerance and their interaction with feeding in articulates. *Paleobiology* 12(2):161–74.

Thomas, R. D. K. 1976. Constraints of ligament growth, form and function on evolution in the Arcoida (Mollusca: Bivalvia). *Paleobiology* 2:61–83.

Thomas, R. D. K. 1979. Morphology, constructional. In *The encyclopedia of paleontology,* ed. R. W. Fairbridge and D. Jablonski, 482–7. Stroudsburg, Pa.: Dowden, Hutchinson and Ross.

Valentine, J. W. 1989. Phanerozoic marine faunas and the stability of the earth system. *Palaeoecol. Palaeoclimatol. Palaeogeograph.* 75:137–55.

Van Valen, L. 1973. A new evolutionary law. *Evol. Theory* 1:1–30.

Vermeij, G. J. 1977. The Mesozoic marine revolution: Evidence from snails, predators and grazers. *Paleobiology* 3:245–58.

Vermeij, G. J. 1987. Evolution and escalation. Princeton, N.J.: Princeton Univ. Press.

Vrba, E. S. 1985. Environment and evolution: alternative causes of the temporal distribution of evolutionary events *S. Afr. J. Sci.* 81:229–36.

Wolfe, J. A. 1987. Late Cretaceous-Cenozoic history of deciduousness and the terminal Cretaceous event. *Paleobiology* 13(2):215–26.

Part 1

PATTERNS ACROSS HIGHER TAXA

2

Onshore-Offshore Trends in Marine Invertebrate Evolution

David Jablonski and David J. Bottjer

Despite its imperfections, the fossil record has great potential for illuminating the origin of higher taxa and major adaptations. Population-level processes may be generally inaccessible to the paleontologist, but deficiencies of this scale are balanced by a wealth of data on the timing, sequence, geography, and ecology of evolutionary novelties in the fossil record. Most research has focused on the timing and sequence of novelties, a prerequisite to the study of other aspects of the evolutionary process. Here we examine one aspect of the ecology of evolutionary innovation by documenting patterns of first occurrences—at several levels within the biological hierarchy—along marine environmental gradients. We present new data on the environment of origin for novelties at the ordinal, family, and genus levels, taking an explicitly phylogenetic approach. We show that major novelties in post-Paleozoic marine organisms, associated with the origin of orders, show a pronounced bias toward onshore first appearances, whereas novelties at lower levels, such as families and genera within the orders we analyzed, show no such bias. We weigh the relative roles of biotic and abiotic factors in generating the three components of the onshore-offshore pattern and discuss the implications of the discordance of origination patterns across taxonomic levels. Although we frame no definitive explanation for the contrast, we suggest that differences in pattern imply differences in evolutionary process and offer some hypotheses for driving mechanisms.

The Time-Environment Matrix

Comparative analysis of the environmental histories of higher taxa requires a standardized paleoenvironmental framework. As discussed by Bottjer and Jablonski (1988), a nearshore-to-slope transect can be broken into five subdivisions recognized on the basis of physical sedimentary features (table 2.1). These features reflect environmental energy levels rather than geographical position relative to the strand line per se; for example, our term *inner shelf* indicates the suite of environments

Table 2.1. Paleoenvironmental scheme used in time-environment diagrams and in tables A2.1–A2.4, with most important criteria for recognition

A. *Nearshore:* Subtidal but above fair-weather wave base.
Criteria: (1) Thick laterally continuous beds of sand and (commonly) coarser sediments with parallel laminae and cross-bedding; (2) fossils disarticulated, abraded, well bedded.
B. *Inner shelf:* Below fair-weather wave base; above normal storm wave base.
Criteria: (1) Common storm beds with parallel laminae or hummocky-cross-stratification interbedded with fair-weather fine-grained beds; (2) fossils commonly in storm beds.
C. *Middle shelf:* Below normal storm wave base; below normal storm wave base.
Criteria: (1) Massive fine-grained beds, commonly rhythmic, storm beds relatively uncommon; (2) fossils unabraded.
D. *Outer shelf:* No storm influence; landward of shelf edge.
Criteria: (1) Massive fine-grained strata, commonly rhythmic on fine scale, no evidence of slumping, mass movement, or turbidites; (2) fossils unabraded.
E. *Slope and deep basin:* Beyond shelf edge.
Criteria: (1) Slumping, mass movements, turbidites, deep-sea fan facies geometries; (2) fossils from beds deposited as turbidites or by some other type of mass movement omitted.

Source: Bottjer and Jablonski 1988.

Note: Environments A and B can be combined into a more inclusive onshore category, and environments C, D, and E can be combined into an offshore category.

that fall between fair-weather wave base and normal storm wave base, rather than a specific location a short distance from shore. There is, of course, some correspondence between geographical location and environmental energy, so that our middle shelf environments will not occur shoreward of our inner shelf environments. Although our terminology is not paleogeographical in intent, we also use the more general term *onshore* for our two environments above normal storm wave base and *offshore* for our three environments below that point. This environmental scheme is extremely simplified but is therefore sufficiently general to allow incorporation of a great diversity of published sources.

The environmental subdivisions used here correspond closely to those of Sepkoski (1987, 1988). The only difference is that his zone 3, an "offshore turbulent zone" comprising oolite shoals, bioherm-rich areas, and delta-front sands, would be subsumed in our environments A and B. Otherwise, our environment C equals his 4, our D his 5, our E his 6. Previous work on Paleozoic onshore-offshore patterns had adopted an environmental continuum (e.g., Sepkoski and Sheehan 1983; Sepkoski and Miller 1985), but because placement along this continuum was based on sedimentary criteria similar to ours, the Paleozoic data matrices are comparable to our own results (and to Sepkoski 1987).

Two kinds of data are presented in this paper. The first documents environmental histories for three post-Paleozoic higher taxa in the Euroamerican region: the crinoid order Isocrinida, the bryozoan order Cheilostomata, and the bivalve superfamily Tellinacea. The three groups are chosen as dissimilar participants in the profound faunal changes that unfolded during the post-Paleozoic: the first waned through the late Phanerozoic and had a relatively archaic mode of life; the other two thrived, one clonal and epifaunal, the other aclonal and infaunal. The histories of these groups are presented in time-environment diagrams as presence-absence or contoured diversity plots for the Euramerican region (north of the Tethyan tropical region, west of the Urals, east of the allochthonous terranes of western North America), and are discussed with full documentation by Bottjer and Jablonski (1988). In these time-environment diagrams, negative data (i.e., absences from a given point in the time-environment matrix) are entered only where taphonomic control taxa, inferred to have preservational characteristics similar to those of the group of interest, have been reported (see Bottjer and Jablonski 1988).

The second kind of information involves the time and environment for first occurrences of isocrinid, cheilostome, and tellinacean taxa ranging from the ordinal to the generic level. These data were compiled globally and are superimposed on the appropriate time-environment diagrams only to put them in a historical context; they do not necessarily correspond to the particular faunal and environmental data that were used to construct the original diagrams. None of these global first occurrences, however, seriously contradicts the environmental histories constructed from Euramerican faunas alone.

Patterns at High Taxonomic Levels

Ecological patterns in the origins and fates of higher taxa have been recognized before, by both neontologists and paleontologists. Bottjer and Jablonski (1988) tabulated a variety of taxa today restricted to deep-water shelf environments or the deep sea that have been inferred on phylogenetic, functional, or paleontological grounds to have originated in—or at least to have previously inhabited—shallow-water settings. Berry (1972, 1974), Crimes (1974) and others reported onshore origination of major benthic community types and trace fossils during the Cambrian and Ordovician, and Sepkoski and Sheehan (1983; Sepkoski and Miller 1985; Miller 1988) more fully documented and analyzed the Paleozoic community-level pattern. A similar pattern was recognized in post-Paleozoic faunas by Jablonski and Bottjer (1983; Jablonski et al. 1983), who found that onshore habitats tended to support derived morphologies and community types, whereas offshore habitats contained more archaic forms and communities. In addition, Bottjer, Droser, and Jablonski (1988) demonstrated distinct onshore-offshore patterns in the trace fossils *Ophiomorpha* and *Zoophycos*.

We have now adopted a more phylogenetic approach, decomposing communities into their constituent taxa and tracking individual clades or paraclades (cf. Raup 1985) in a time-environment matrix. We constructed time-environment diagrams for three taxa (Bottjer and Jablonski 1988) and concluded that the individualistic histories of those three clades argued against community-level driving mechanisms. In an analysis of post-Paleozoic echinoderm orders, we found that 13 orders of benthic crinoids and echinoids first appeared in onshore habitats, whereas only 5 orders began offshore (Jablonski and Bottjer 1988). We found no significant environmental patterns for the seven asteroid orders recognized by Blake (1987), but this is not surprising given the scarcity of the conditions required for adequate preservation of the group (the even more problematical ophiuroids and holothuroids were omitted from the analysis). Given that marine benthic orders appear to originate preferentially (although not exclusively) in nearshore settings, we now ask whether this environmental bias is evident at all levels in the hierarchy, with ordinal originations simply an extrapolation of processes at lower levels.

Novelties within Higher Taxa

To assess the relationships among evolutionary novelties at different hierarchical levels, we have analyzed the environment of the first occurrence in the fossil record for the 15 genera and 4 families of isocrinid crinoids (21 genera and 6 families if the bourgueticrinids are included), the 11 families of tellinacean bivalves, and of six innovations within the cheilostome bryozoans.

ISOCRINIDS

The crinoid order Isocrinida (sensu Rasmussen 1978a) includes the dominant stalked crinoids of today's oceans. They first appeared in near-shore and inner-shelf environments in the early Triassic and were widely distributed across the shelf by the early Jurassic (Bottjer and Jablonski 1988 and references therein). Onshore occurrences diminished in the late Cretaceous and early Tertiary, and today the group is restricted to outer-shelf, slope, and deep-sea environments (fig. 2.1; Roux 1987). For this analysis we have adopted a slightly modified version of Rasmussen's (1978a) genus-level and familial classification, with changes based mostly on Simms' recent work (1986a, 1988a; see also Klikushin 1982a and Roux 1987 for somewhat divergent views). First occurrences of isocrinid genera are documented in Table A2.1.

In cladistic terms, the Isocrinida as traced in figure 2.1 is a paraphyletic group, with one major branch excluded from our analysis for ecological and functional reasons. The Pentacrinitidae were evidently pseudopelagic, inferred to have lived attached to floating logs, and are most commonly preserved in strata deposited under anaerobic or dysaerobic conditions hostile to most benthos (Seilacher et al. 1968; Haude 1980; Simms 1986a, 1986b). This group apparently adopted its unusual life-style at the time of its origination in the Late Triassic (Klikushin 1987a), and so we exclude it from our analysis of strictly benthic forms. Derived in turn from the Pentacrinitidae are the most successful of the post-Paleozoic crinoids, the Comatulida, a group of free-living crinoids that are afforded ordinal status by Rasmussen (1978a). Although the earliest members of this group retained a short stalk (Simms 1986a, 1988a), and the attached habit returned in at least one seemingly paedomorphic family (Klikushin 1987b; Simms 1988a), we have excluded the entire group from our analysis of Isocrinida and treat them as a separate entity because they are ecologically distinct from the sessile, or only slightly mobile relatively long-stalked isocrinids.

The phylogenetic affinities of the bourgueticrinids, treated as a distinct order by Rasmussen (1978a), are uncertain. Although some authors ally them with the order Millericrinida (e.g., Gislén 1938 but not 1924; Roux 1978, 1987; Klikushin 1982b), Rasmussen (1978b) and Simms (1986a, 1988a) derive them paedomorphically from within the clade Isocrinida + Comatulida sensu Rasmussen (1978a). Simms (1986a) presents new evidence suggesting derivation of the Bourgueticrinida directly from the isocrinids rather than the more circuitous isocrinid-pentacrinitid-comatulid-bourgueticrinid route favored by Rasmussen (1978b). (Simms [1988a] also rejects a millericrinid origin but does not choose between an isocrinid or comatulid source.) If interpreted correctly, Simms' data bring the bourgueticrinids within the purview of our analysis; cladistically they are simply another branch of the isocrinids, and as stalked benthic crinoids they are included ecologically as well, unlike the comatulids and the pentacrinitids. In light of the taxonomic uncertainties, the bourgueticrinids, which were excluded from our original time-environment analysis (fig. 2.1), are tab-

ISOCRINIDA

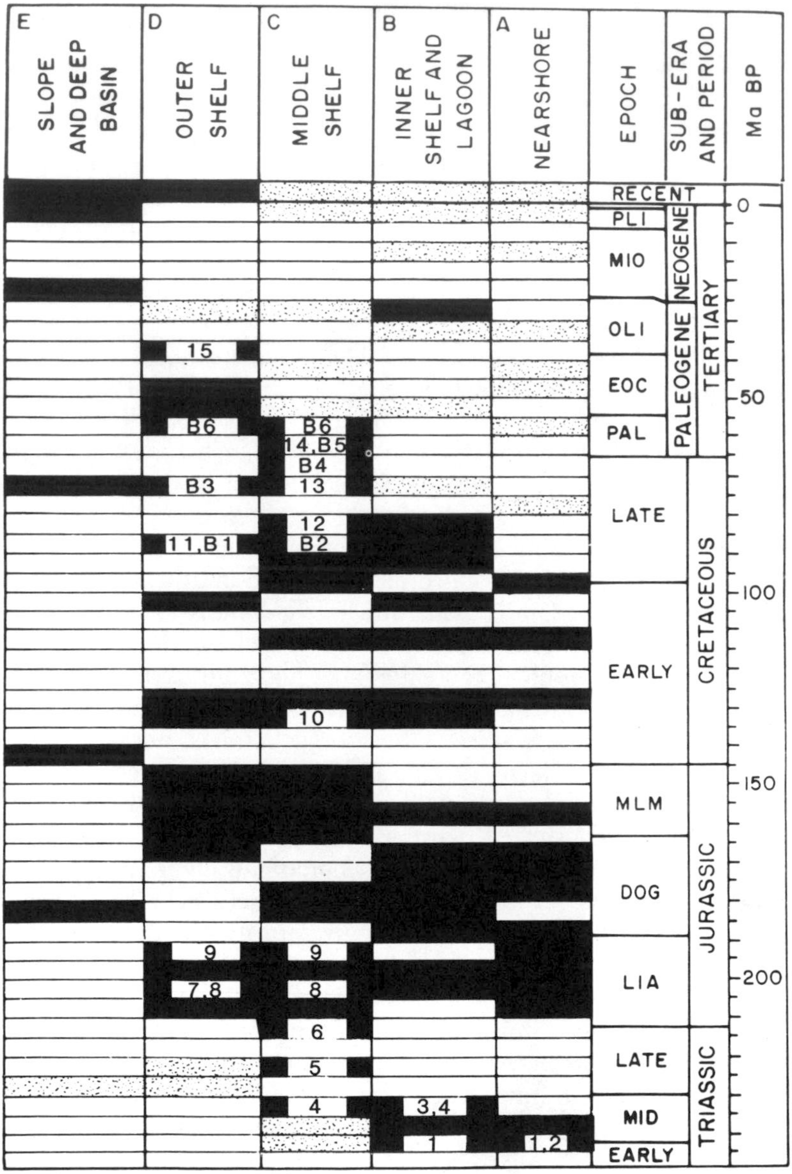

E SLOPE AND DEEP BASIN
D OUTER SHELF
C MIDDLE SHELF
B INNER SHELF AND LAGOON
A NEARSHORE
EPOCH
SUB-ERA AND PERIOD
Ma BP

RECENT
PLI
MIO
OLI
EOC
PAL
LATE
EARLY
MLM
DOG
LIA
LATE
MID
EARLY

NEOGENE
PALEOGENE
TERTIARY
CRETACEOUS
JURASSIC
TRIASSIC

0
50
100
150
200

15
B6
B6
14,B5
B4
B3
13
12
11,B1
B2
10
9
9
7,8
8
6
5
4
3,4
1
1,2

ulated separately (table A2.2), and analyses are performed both with and without them.

CHEILOSTOMES

The bryozoan order Cheilostomata is one of the youngest higher taxa of skeletonized marine invertebrates, first appearing in nearshore and inner shelf environments in the latest Jurassic (Pohowsky 1973; Taylor 1981b, 1986). The group expanded across the shelf during the early Cretaceous but remained rare and at low specific and generic diversity until the mid-Cretaceous, some 45 million years after their origination (Bottjer and Jablonski 1988). Today the cheilostomes, which constitute a strictly monophyletic clade (Cheetham and Cook 1983; see also Taylor 1988), are taxonomically diverse and occupy an extremely wide range of habitats from the intertidal to the deep sea (e.g., McKinney and Jackson 1989). The intermediate-level (generic and familial) taxonomy is in a state of uncertainty, and so we selected for analysis the following six evolutionary novelties from those surveyed by Voigt (1985) (table A2.3); even these traits, however, may have originated repeatedly, so that genealogically meaningful time-environment histories are difficult to construct (see also Lidgard 1985, 1986, on multiple origins of budding geometries).

Ovicells. Many living cheilostomes brood their young, although the protective structures may or may not be calcified. Calcified ovicells first appeared in the Late Albian of Texas (*Wilbertopora*) and Britain (*Marginaria*). Taylor (1988) argues that these first occurrences, which are in equivalent ammonite zones and thus contemporaneous at available time resolution, are close to the origin of brooding in the Cheilostomata. He builds a forceful case for the monophyletic origin of non-planktotrophic development, that is, a life cycle that includes a non-feeding, brooded larva with relatively short planktonic duration (see also Ryland 1974). Taylor (1988) further suggests that the presence of ovicells in Albian bryozoans represents the acquisition of nonplanktotrophic development and that the low larval dispersal ability of this

◀ Figure 2.1. Environmental history of benthic isocrinid crinoids in the Euramerican region (after Bottjer and Jablonski 1988, who provide full documentation), and global survey of first known occurrences of isocrinid genera (table A2.1) and bourgueticrinid genera (table A2.2). Black = presence at that time-environment coordinate; stipple = absence confirmed by taphonomioc control taxa; blank = no data. Based on the information in table A2.1, the following occurrences were added to the time-environment diagram of Bottjer and Jablonski (1988): 230–235 Ma, environment B; 225–230 Ma, B; 200–205 Ma, C; 80–85 Ma, C; 70–75 Ma, C.

mode fostered high speciation rates, thereby triggering the explosive diversification of the cheilostomes in the mid-Cretaceous.

Erect colony form. Another important innovation in cheilostome evolution was the evolution of arborescent colony form, which minimizes constraints of substratum size, probability of competitive overgrowth, and resource limitation in the slow-moving boundary layer and increases surface area devoted to feeding and reproduction (Jackson 1979; Cheetham 1986; McKinney and Jackson 1989). This growth habit has arisen repeatedly in the phylum Bryozoa (Cheetham 1986; McKinney 1986a, 1986b; see Cheetham and Thomsen 1981; Cheetham and Hayek 1983; Cheetham 1986; McKinney and Jackson 1989 for valuable analyses of biomechanical costs and constraints), but here we focus on its first recorded appearance in the Cheilostomata. The oldest known erect bryozoan colonies occur in the Late Albian, in the same ammonite zone as the oldest ovicells. McKinney and Jackson (1989: 93–94) state that the multiserial erect growth form has shown a general retreat from nearshore environments since the Paleozoic.

Cribrimorphs. Some bryozons create an arched wall of marginal spines that fuse above the zooid midline and along lateral projections of the spines, creating a discontinuous, calcified shield with an array of small, regularly spaced gaps. The result is a sievelike structure that protects the frontal membrane while permitting the passage of water necessary for tentacle eversion (Ryland 1970; Taylor 1981b). Bryozoans having this organization, most abundant and diverse in the Late Cretaceous (e.g., Ryland 1970, McKinney and Jackson 1989) are often grouped in the suborder Cribrimorpha (see review by Cheetham and Cook 1983: 147–48), although again a polyphyletic origin cannot be ruled out. Gordon (1984) regards them as a superfamily within the Ascophora (see section on Ascophorans). The cribrimorphs first appeared in the Early Cenomanian, with several genera in England and northwest Germany. Such diversity might suggest a long prior history, but no cribrimorphs are known from the Late Albian (e.g., Taylor 1986); presumably the data reflect rapid diversification or sudden polyphyletic derivation of the trait. About four genera in which the marginal spines are elongate and arched but not fused, and are thus intermediate between the primitive membranimorphs and the cribrimorph state, occur along with some of the earliest cribrimorphs in the lower Chalk Marl of England (Lange 1921; Larwood 1985); these intermediate "myagromorphs" persist at least until the Late Campanian.

Zoarial flexibility. Articulated colonies, in which an organic stalk, rootlets, or filaments impart a degree of flexibility to an otherwise rigid colony, evidently permitted the colonization of turbulent waters by erect colonies (e.g., Voigt 1985). This adaptation, which can be recognized in fossil material by a deep central hole in the tapered, cylindrical trunk of an arborescent colony form (Voigt 1985), doubtless arose repeatedly within the Cheilostomata, but the oldest known examples occur in the Cenomanian of France.

Lunulitiform colonies. Cup-shaped, radially budded colonies have also originated repeatedly in the cheilostomes (Cook and Chimonides 1983; Cook and Voigt 1986). Many of these lunulitiform species are free living and mobile, supported at or near the surface by elongated, active mandibles (see Cook and Chimonides 1983, 1986). The first records of this remarkable morphotype, from Upper Cretaceous strata of western Europe, are placed in the family Lunulitidae, a family possibly derived polyphyleticalliy from several stocks of the protean family Onychocellidae (which may itself be polyphyletic) (Voigt 1981; Cook and Chimonides 1983, 1986; Cook and Voigt 1986). The mid-Cretaceous genera *Reptolunulites* and *Pavolunulites* present morphotypes intermediate between onychocellids and lunulitids. The oldest known lunulitid with an integrated colony organization is *Lunulites plana* d'Orbigny from the Early Santonian of France, although a poorly preserved form in the Turonian or Lower Coniacian of northern Germany (E. Voigt, in litt. 1987) may represent an even earlier full-fledged lunulitid (see table A2.3).

Ascophorans. Another important novelty in cheilostome evolution was the origin of the ascus, a flexible sac that permits hydrostatic control in the presence of a solid, calcified frontal shield. (Hydrostatic control in the other cheilostome suborders is provided by a flexible frontal membrane.) The Ascophora, typified by the ascus, is the youngest cheilostome suborder but according to Cheetham and Cook (1983:196) is "probably polyphyletic" (see also Banta and Wass 1979:31; Taylor 1981b:244), so that the ascophoran diversification and rise to present dominance may be a gradal rather than a cladal phenomenon. Thus, as with the lunulitiforms, we record the oldest known appearance of the trait without assuming this is the only derivation. In fossil material, presence of an ascus is recognized by "a pore in the frontal wall (the ascopore) or by a tubular prolongation of the apertural rim (the spiramen) or by a proximal apertural sinus or slit" (Voigt 1985:334). Ambiguities remain, however, and some non-ascophorans exhibit morphological features that misleadingly suggest ascophoran anatomy, whereas some ascophorans lack definitive skeletal indicators of their soft parts (E. Voigt, in litt. 1987). Possible first records are given in table A2.3. We have several candidates for oldest known ascophoran, but although they range from latest Turonian to late (possibly early) Coniacian in age, they are all within the 85 to 90 Ma interval (following Harland et al. 1982); most are in middle or outer shelf environments, but Nowicki's *Boreasina* n.sp. may be an inner shelf occurrence.

TELLINACEA

The Tellinacea is a distinctive bivalve superfamily that originated in near-shore or inner-shelf environments in the Middle Triassic and underwent a prolific radiation comprising some 11 families and over 200 genera and subgenera (see Bottjer and Jablonski 1988). The families are reasonably distinctive, morphologically and in life habit, with each family defined by a set of derived traits in shell form, hinge and ligament morphology, and internal features of the shell reflecting soft-part anatomy (Yonge 1949; Keen 1969; Pohlo 1982; Coan 1988 advocates inclusion of Scrobiculariidae in Semelidae, however). Perhaps the most problematic grouping is the basal family, Sowerbyidae; *Rhaetidia*, questionably placed here by Keen (1969), is as likely to be the sister group to all other Tellinacea as it is to be most closely related to *Sowerbya* and may not belong in the superfamily at all (see table A2.4).

The Tellinacea has generally been regarded as strictly monophyletic (e.g., Taylor, Kennedy, & Hall, 1973; Pohlo, 1982). Allen (1985:382) suggested that "the Tellinacea possibly have a common origin with the Mesodesmatacea, a small, highly specialized superfamily paralleling the tellinacean donacids in form, habits, and habitat." The mesodesmataceans (formerly a family in the Mactracea and raised to superfamily status by Yonge and Allen 1985) date back only to the Late Cretaceous (Coniacian) (Saul 1989). Their record is so sparse that it is not clear whether, should their phylogenetic affinities lie with the Tellinacea, the taxonomic separation of the two groups is more likely to render the Tellinacea paraphyletic than to reflect a common ancestry with over 50 million years of missing record. Because the phylogenetic case is far from proven (cf. Yonge and Allen 1985), we have omitted them from our analyses. Further, the inclusion of this small group, which has a fossil record consisting only of two Eocene genera and two more from the Neogene according to Keen (1969)—and only the Late Cretaceous *Califadesma* and two Neogene genera according to Saul (1989)—would in no way alter the reported pattern. A larger phylogenetic issue is the inclusion of the family Tancrediidae. Following the *Treatise* (Keen 1969; also Cox 1965; Hallam 1976, 1987; Fürsich 1982), we have included this family in the Tellinacea, although other affinities have been suggested (e.g., Saul and Popenoe 1962; Speden 1970).

Discordance Across Hierarchical Levels

The three major groups we studied in detail all appear to begin onshore (i.e., in near-shore and/or inner-shelf environments), but novelties within the crinoids (fig. 2.1) and the cheilostomes (fig. 2.2) show no such propensity. Indeed, benthic isocrinid genera (with or without bourgueticrinids) and cheilostome novelties tend to appear first in mid-shelf environments. At the family level, the Order Isocrinida began onshore with the family Holocrinidae (*Holocrinus* and *Menocrinus*), and the Isocrinidae also appeared in the inner shelf soon thereafter (with *Isocrinius*); Simms (1988a) recognizes two additional isocrinid families, both of which began offshore in the Cretaceous—the Cainocrinidae (starting with *Nielsenicrinus*) and Isselicrinidae (with *Isselicrinus*). The two bourgueticrinid families also first appear in the middle- or outer-shelf environments (Bourgueticrinidae with *Bourgueticrinus*, and Bathycrinidae with *Monachocrinus*). The numbers are small but give no indication of an onshore bias in familial origination once the clade is underway, despite a broad occupation of environments when the later families originated.

A paleoenvironmental bias in origination might arise taphonomically if crinoids tend to disarticulate more easily in higher-energy (more onshore) environments. Because novel taxa might be more readily identified as relatively complete specimens, reported first occurrences could be skewed offshore. In addition, the pattern might reflect a potential bias in echinoderm-lagerstätten, where storm deposition of fine sediments can provide exceptional preservation (see Rosenkrantz 1971; Seilacher, Reif, & Westphal 1985; Simms 1986b). However, given our large number of nearshore and inner shelf records, the generic origination pattern is at least as likely to reflect within-habitat genus (and species?) richness, particularly as occupation of onshore habitats dwindled through the Late Mesozoic. The cheilostome data are also scanty, but the fact that only one of the six novelties in table A2.2 begins onshore is again consistent with our observations on the crinoids.

The tellinacean families do tend to originate onshore, but this clade's standing diversity gradient at the genus level (fig. 2.3) (and pre-

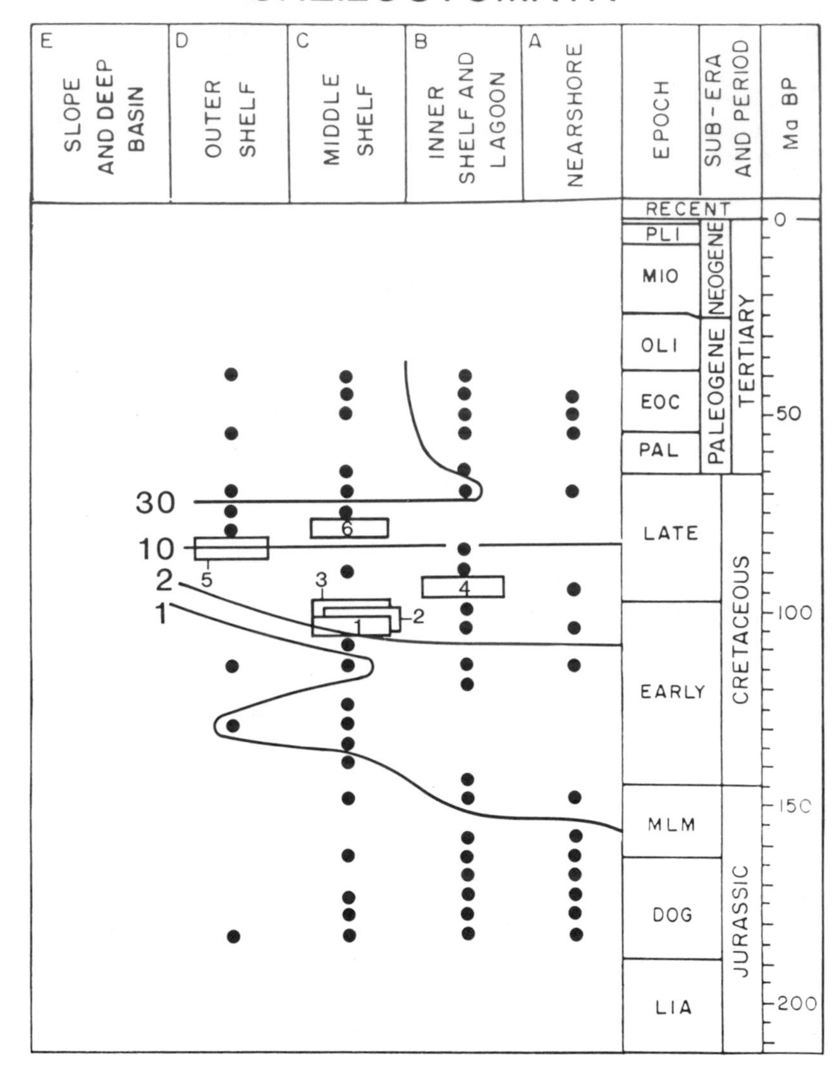

Figure 2.2. Environmental history of cheilostome bryozoans, contoured for within-habitat generic richness in the Euramerican region, and global survey of first known occurrence of five novelties within the cheilostome clade. Documentation for diversity contours provided by Bottjer and Jablonski (1988); novelties listed in table A2.3. Faunal data added to Bottjer and Jablonski (1988): 55–60 B and C (Graecen 1941; Owens and Sohl 1969; Cheetham 1977; Bernstein 1987).

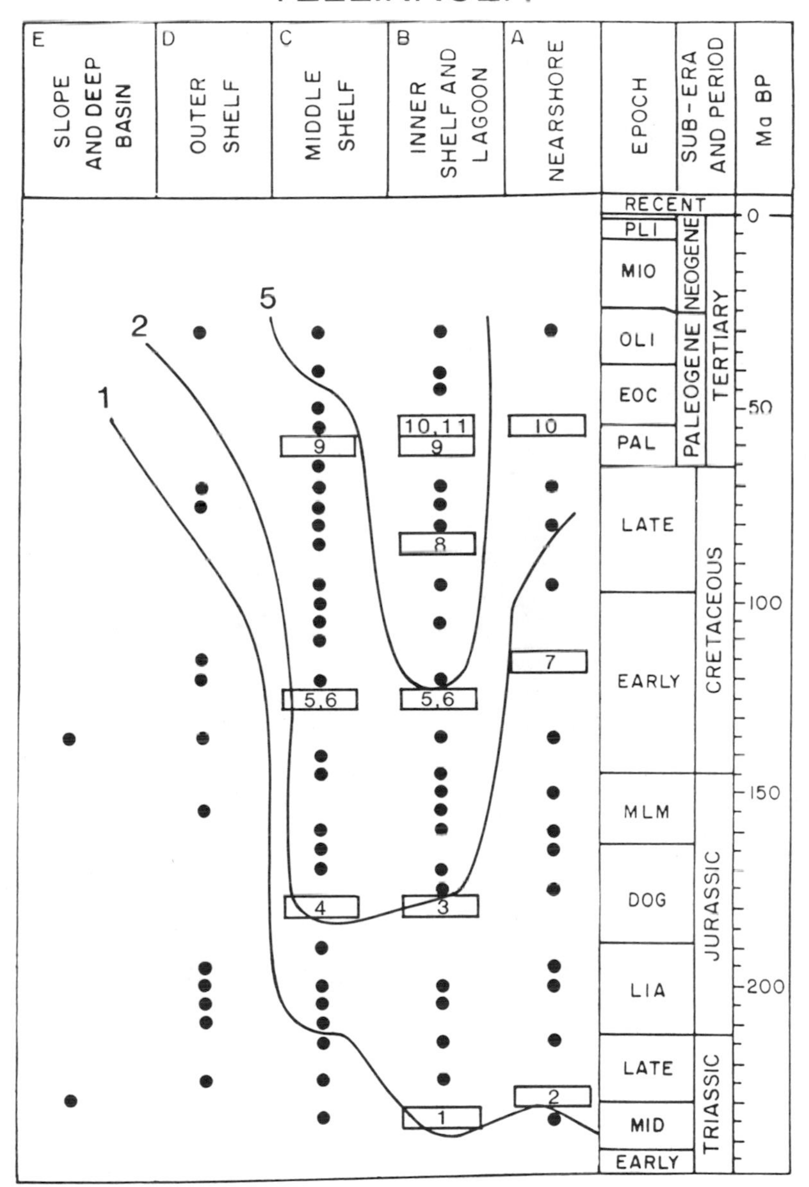

Figure 2.3. Environmental history of tellinacean bivalves, contoured for within-habitat generic richness in the Euramerican region, and global survey of first known occurrence of tellinacean families. Documentation for diversity contours provided by Bottjer and Jablonski (1988); families listed in table A2.4. Note: Occurrence 2 lies within the zero diversity field because it is Japanese, not Euramerican, and is thus outside the geographical limits of the within-habitat diversity analysis.

sumably species level) is also skewed toward onshore environments, as are the population densities and species richness of present-day tellinaceans (Yonge 1949). The origination pattern of tellinacean families suggests a dependency on within-habitat diversity at lower taxonomic levels, and the patterns for crinoid genera and families, and cheilostome novelties, are consistent with this interpretation.

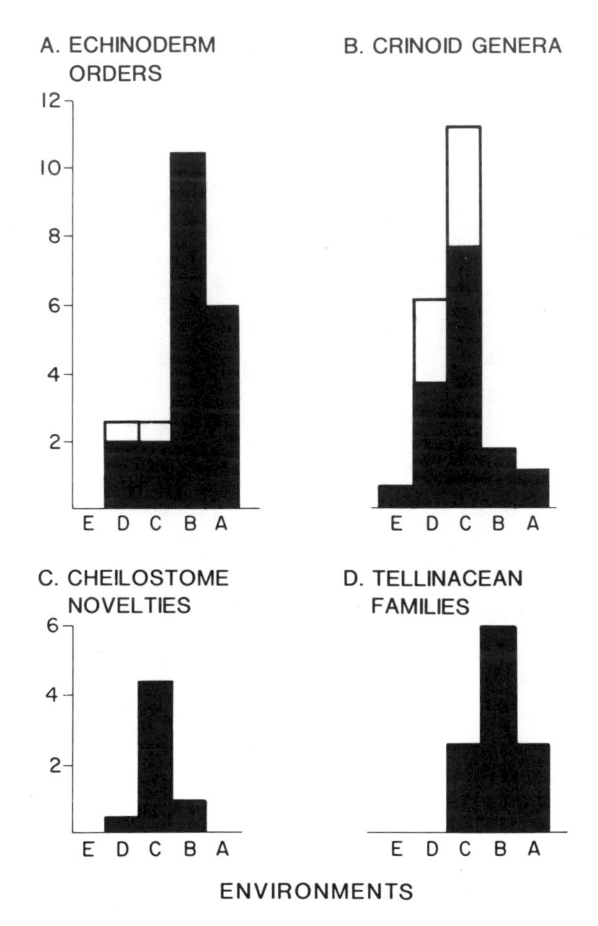

Figure 2.4. Onshore-offshore distribution of first occurrences of (**A**) post-Paleozoic crinoid and echinoid orders and (**B, C, D**) novelties within the isocrinid crinoids (genera), cheilostome bryozoans (novel morphologies), and tellinacean bivalves (families). Environments A (nearshore) through E (slope and deep basin) defined in table 2.1. For A and B, open pattern indicates effects of recognition of the bourgueticrinid crinoids as a discrete order or as part of the broader isocrinid clade, respectively. Taxa whose occurrence are only resolved to two environments are scored as 0.5 in each environment. See table 2.2. for statistical analyses.

To assess differences between ordinal first appearances and within-clade patterns, we have compared our data on within-clade novelties with those on all post-Paleozoic echinoid and crinoid orders; first occurrences that could only be resolved to two adjacent environments rather than to a single one (owing to ambiguities in stratigraphic placement or paleoenvironmental information, or to multiple occurrences) were scored as 0.5 in each environment. Environmental placement of the echinoderm orders largely follows Jablonski and Bottjer (1988), but new phylogenetic and stratigraphic information (documented in Jablonski and Bottjer 1990) prompts us to place the oldest Salenioida in nearshore environments, the oldest Pedinoida, Hemicidaroida, and Holasteroida in the inner shelf and the oldest members of Disasteroida and Echinoida in the middle shelf. The echinoids wih the poorest fossil records, the Echinothurioida and Diadematoida (see Kier, 1977; Smith 1984), may not be recorded until the Late Cretaceous, in outer-shelf environments (Smith and Wright 1990). It is unlikely that these occurrences reflect the time or place of origin of these groups, but they oppose the general pattern so that their inclusion in our analysis is conservative. We still lack adequate phylogenetic and temporal resolution to pinpoint the first appearance of the short-lived order Oligopygoida.

Environmental patterns of origination differ significantly across hierarchical levels, judging by our analyses of this unfortunately still rather small data set (table 2.2). Using the Kolmogorov-Smirnov test, we compared the frequency distribution of first occurrences of orders in our five-partition environmental transect to that of within-clade novelties documented in tables A2.1–A2.4. If all novelties are included (despite the skewed distribution of tellinacean families), and bourgueticrinids are included in the benthic Isocrinida, the frequency distributions for orders and within-clade novelties are significantly different ($p < 0.05$); the difference is not significant, however, if the bourgueticrinids are considered a distinct order ($0.10 < p < 0.20$). As noted previously, the pattern for tellinacean families apears to be constrained by the diversity gradient of their clade, and so a more appropriate comparison might be between echinoderm orders and echinoderm within-clade novelties. Differences are significant regardless of the treatment of bourgueticrinids (table 2.2). If we omit the problematic first records for Echinothurioida and Diadematoida, the three comparisons that were already significant increase to $p < 0.005$ or better, and the formerly nonsignificant comparison reaches $p < 0.05$. (See Jablonski and Bottjer forthcoming, written two years after this chapter was first submitted, for updated first occurrences of *all* post-Paleozoic invertebrate

Table 2.2. Statistical comparisons between environmental patterns of ordinal origination (benthic crinoids and echinoids; data in Jablonski and Bottjer 1988, with modifications noted in the text) and novelties within the benthic isocrinid crinoids (genera in table A2.1, plus or minus the bourgueticrinid genera listed in table A2.2), cheilostome bryozoans (innovations in table A2.3), and tellinacean bivalves (families in table A2.4).

I. Kolmogorov-Smirnov tests	Orders (*n*)	Novelties (*n*)	D_{max}	Significance
A. All novelties				
1. Bourgueticrinida as distinct order	22	32	0.29	N.S. ($D_{0.05} = 0.38$)
2. Bourgueticrinids as Isocrinida	21	38	0.40	$p < 0.05$ ($D_{0.01} = 0.44$)
B. Crinoid genera only novelties				
1. Bourgueticrinida as distinct order	22	15	0.47	$p < 0.05$ ($D_{0.01} = 0.55$)
2. Bourgueticrinids as Isocrinida	21	21	0.57	$p < 0.005$ ($D_{0.005} = 0.54$)

II. Chi-square tests	Orders (*n*)	Novelties (*n*)	X^2	Significance
A. All novelties				
1. Bourgueticrinida as distinct order				
Environments A + B	15.5	13	4.68	$p < 0.05$
Environments C + D + E	6.5	19		
2. Bourgueticrinids as Isocrinida				
Environments A + B	15.5	13	8.64	$p < 0.005$
Environments C + D + E	5.5	25		
B. Crinoid genera only novelties				
1. Bourgueticrinida as distinct order				
Environments A + B	15.5	3.5	8.22	$p < 0.005$
Environments C + D + E	6.5	11.5		
2. Bourgueticrinids as Isocrinida				
Environments A + B	15.5	3.5	13.84	$p < 0.001$
Environments C + D + E	5.5	17.5		

orders, and fuller treatment of patterns among well-preserved taxa, which do show onshore origination, and poorly preserved taxa, which show no environmental pattern.)

The vagaries of sampling, preservation, and imprecise environmental inferences can be minimized by combining data into onshore (= nearshore and inner shelf) and offshore (= middle and outer shelf plus slope and deep basin) categories. Frequencies of onshore and offshore first occurrences are significantly different for orders and within-clade novelties regardless of the treatment of bourgueticrinids and whether comparisons include all within-clade novelties or only echinoderms (chi-square tests; see table 2.2).

In addition to providing a firmer statistical basis for the claim of pervasive (though by no means universal) onshore-offshore evolutionary patterns in post-Paleozoic higher taxa, these analyses allow us to rule out one class of explanations for the observed onshore bias in the first occurrence of orders. One logical approach, which might be termed the extrapolationist argument, would attribute the onshore bias entirely to processes operating on individual organisms over ecological time scales. Onshore environments are more variable, spatially and temporally, than offshore environments, and the diversity and heterogeneity of selective forces might be expected to elicit the greatest amount of evolutionary divergence onshore. However, if this were the case, an onshore concentration of new genera and families would also be expected as this greater tendency for divergence manifested itself across the hierarchy. We do not see such a pattern for crinoid genera or families or for cheilostome novelties, and attribute the tellinacean pattern to constraints imposed by the diversity gradient of the clade. We certainly have not ruled out all mechanisms focusing on lower levels, but one of the simplest and most appealing evidently is falsified.

The origination pattern for within-clade novelties, then, suggests processes operating at lower levels: possibilities include selection and adaptation at the organismic level, differential speciation rates, or sorting processes at the level of species or lineages (e.g., Stanley 1978, 1979; Vrba and Eldredge 1984; Vrba and Gould 1986), and of course these need not be mutually exclusive. We therefore should not be surprised if wihin-clade novelties tend to follow the clade's diversity gradient: the more evolutionary experiments per unit time (be they speciation events or individual populations undergoing drift or local adaptation), the more novelties produced. However, the bias in onshore origination at the ordinal level is sufficiently pervasive, even in groups whose subsequent novelties show no such environmental bias, that simple extrapolation will not explain the higher-level pattern.

Biotic and Abiotic Factors

The data presented here, although still sparse, indicate that environmental patterns in the origin of higher taxa could not have been predicted from patterns of their constituent taxa or within-clade novelties. Clearly, data are needed on more orders and their constituents and on diversity patterns within probable sister groups near the time of an order's origin, but some hypotheses can be tested regarding driving mechanisms for the higher-level phenomenon. Three aspects of onshore-offshore patterns must be considered: preferential origin of orders onshore in the absence of a similar bias at lower taxonomic levels; environmental expansion of clades offshore; and, in some taxa such as the isocrinids, eventual retreat into an exclusively offshore distribution. Given the physical environmental gradient across which these patterns are displayed, we find it hard to imagine that the driving mechanisms are *exclusively* biotic. However, abiotic factors may exert their influence at more than one hierarchical level and time scale and may interact with biotic factors in a variety of ways.

ONSHORE ORIGINATION

The onshore origination of higher taxa is the most important aspect of the pattern—as discussed subsequently, it may well be the motor for the other components—and the most difficult to explain. One possibility is that particular physical events, such as mass extinctions (which we take to be essentially abiotic in origin; see Jablonski 1986a; Raup 1986; Flessa et al. 1987), opened ecological opportunities that somehow fostered the origin of higher taxa. However, there is little evidence for greater severity of mass extinctions onshore (few authors have looked). This is not to say that mass extinctions do not play a role in the origin of higher taxa or of evolutionary novelty. The origins of the post-Paleozoic echinoderm orders and of the within-clade novelties are not time homogeneous, but concentrated in the Triassic and Early Jurassic: 60% of these orders, plus the tellinaceans (but not the cheilostomes), originate in the first quarter of Mesozoic/Cenozoic time (see also Hallam 1987 and Erwin, Valentine, & Sepkoski 1987 for a general treatment). Orders that began later may have been slightly less likely than the early ones to start onshore, with four of the later echinoderm orders for which we have data (plus the cheilostomes) beginning onshore, and three beginning offshore, but the numbers are too small to argue for any significant temporal change in the environmental locus of origin. These data suggest, then, that while the end-Permian extinction (abetted perhaps by the end-Triassic event) may have fostered the origin of large numbers of higher taxa, it lacked the environmental selectivity required to set the stage for the observed onshore bias in origination.

Another possibility is that bathymetic gradients in abiotic factors structure biological responses over ecological and short-term evolutionary time scales to give rise to the long-term onshore bias. We have already rejected a simple extrapolationist model in which divergences are heightened in the spatial and temporal heterogeneity of onshore settings; however, environmental heterogeneity could spark innovations in more complex ways. Levinton (1988), for example, constructed a niche-packing argument, suggesting that the greater "habitat extinction rate" onshore would permit new evolutionary experiments, whereas the biotas in more offshore environments may be too densely packed and too persistent to permit much evolutionary divergence (see also Sheehan 1986). We believe that a similar argument emerges from Van Valen's (1985) general theory of origination based on invasion and interaction of adaptive zones. This is an intriguing approach to the problem, but we suspect it is vulnerable to the same lack of genus- and family-level origination bias mentioned previously.

Jablonski and Bottjer (1983; Bottjer and Jablonski 1988; see also Templeton 1986) suggested that the population structures and the kinds of isolates likely to occur in onshore settings were more likely than those offshore to undergo genetic transiliences (sensu Templeton 1980; see also Carson and Templeton 1984) and thus produce morphological innovations that could give rise to higher taxa. These environmental trends in the kinds of speciation events potentially decouple the origin of major groups from both per-taxon and overall speciation rates; it is not the number of speciation events that are important in this hypothesis but the kinds of speciation events. As Bottjer and Jablonski (1988) note, there is some evidence that the appropriate, high-dispersal modes of larval development were present in the early history of some onshore-originated groups, even when scarce or absent today, but more information is required for a critical test of this hypothesis.

M. L. McKinney (1986), invoking Gould's (1977) arguments for the ecological basis of heterochrony, attributed the onshore bias to the greater probability of progenesis in onshore, more disturbed, environments relative to offshore, more stable, environments. Selection for shorter generation times would give rise via progenesis to small-sized species, which "have greater evolutionary potential than larger ones, which are allometrically constrained by problems of similitude." However, heterochrony has been implicated in only a few of the ordinal originations documented here. Of those, Simms (1986a, 1988a, 1988b, 1988c) shows that at least some of the onshore ones occurred via neoteny rather than progenesis (e.g., the Order Isocrinida as used here), and progenetic originations are not confined to onshore habitats (e.g., the bourgueticrinids). A systematic survey is needed, but thus far we have found little evidence for progenesis as the principal avenue for the onshore origin of most orders.

A final way in which organisms' short-term responses to abiotic factors could generate an apparent onshore bias in the origin of higher taxa is through differential extinction. Onshore species tend to be geographically widespread, physiologically tolerant, and, as a consequence of these and related factors, geologically long-lived (see Jackson 1974; Jablonski 1980; Jablonski and Valentine 1981—all on post-Paleozoic mollusks; but see Sepkoski 1987 for contrary within-clade trends at the genus level). Any major innovation that happens to appear in one of these more extinction-resistant species will have a greater chance of surviving the perilous, low-diversity phase of early diversification than will an innovation of similar adaptive value that appeared in a more extinction-prone offshore species (Jablonski and Bottjer 1983; Bottjer and Jablonski 1988; see also Valentine, this volume, for a thoughtful treatment of clade dynamics). This form of upward causation (Vrba and Eldredge 1984), with processes at the organismic and species level indirectly shaping a macroevolutionary pattern, could be given a preliminary test using species duration frequencies across environmental transects in onshore-originating clades.

EXPANSION OFFSHORE

The next component of the pattern is the offshore expansion of major groups. This could have a strictly biotic origin, although played out at a high level: given an onshore origin, simple diffusion of a marine clade in the course of its diversification would inevitably carry it across the shelf (Jablonski and Bottjer 1983; Gould 1988)—in the absence of competitive, physiological, and other constraints. Even if preemptive competition is important (i.e., incumbent species tend to win most competitive interactions), offshore expansion will be the rule if onshore taxa tend to have lower extinction rates than offshore ones (see Van Valen 1985; Valentine, this volume). This passive expansion process could be played out at a level closer to that of individual organisms if successful speciation is more likely in an offshore direction than an onshore direction (for example, if broadening tolerances to fit more variable habitats is less likely than narrowing tolerances during invasion of more stable habitats) (Jablonski and Bottjer 1983; Levinton 1988). This mechanism could be tested if detailed within-clade phylogenies were placed in a time-environment matrix.

Strictly abiotic mechanisms seem unlikely to drive offshore expansions. Mass extinctions, oceanic anoxic events, or other perturbations might preferentially remove offshore taxa, thereby opening space for outward expansion, but the timing appears to be wrong. Expansions are played out over tens of millions of years (see figs. 2.1 to 2.3) and thus do not suggest a process long held in check that has abruptly been released by an extinction event. Furthermore, offshore expansion histories are individualistic; each clade evidently proceeds at its own time and its own rate, arguing against the wholesale clearing of ecospace in the wake of an offshore perturbation. One possible exception is the expansion of bivalves in carbonate environments following extinction of dominant brachiopod and stromatoporoid inhabitants in the Late Devonian extinction (Miller 1988). However, none of the post-Paleozoic mass extinctions (e.g., Sepkoski 1986) seem to preface the expansion histories documented here. Oceanic anoxic events, probably the result of a shallowing of the oxygen-minimum zone with subsequent onlap of anoxic bottom waters, would logically affect offshore habitats more than onshore ones. However, the timing of the major anoxic events—restricted to the *falciferum* zone of the early Toarcian (Jenkyns 1988); to the *plenus* and *labiatus* zones at the Cenomanian-Turonian boundary (Schlanger et al. 1987); and to broader intervals in the Aptian-Albian and Coniancian-Santonian (Jenkyns 1980)—again does not correspond closely to the offshore expansions known to us.

RETREAT OFFSHORE

The final component of the pattern is the retreat offshore exhibited by some higher taxa (e.g., fig. 2.1). In theory this could be driven abiotically, if onshore representatives are more vulnerable to background or mass extinction than offshore ones. Any lag in reoccupation of vacated habitats could result in preemption by surviving onshore clades—assuming preemptive niche occupation is a major force in evolution; see Valentine (1980; Walker and Valentine 1985; Valentine and Walker 1987) for an attractive macroevolutionary model based on this concept (and also Van Valen's [1985] theory of adaptive zones; also Gilinsky and Bambach 1987), and Hallam (1987) on "pre-emptive rather than displacive" competition in Lower Jurassic faunas. Although mass extinctions have almost certainly mediated a number of major faunal replacements (Benton 1983; 1987; Jablonski 1986a, 1986b, 1989), the retreats that we have studied do not correspond well with mass extinction events. For example, the last nearshore isocrinids disappeared well before the end-Cretaceous extinction; the retreat dragged on for some tens of millions of years, and the last middle shelf occurrence in the main body of the retreat shown in figure 2.1 came about 10 my after the end-Cretaceous event, with an inner-shelf straggler as late as Late Oligocene. Of course, the isocrinids lost only one genus in the end-Cretaceous event (Rasmussen 1978a, 1979). This fact is probably significant in itself, but a more persuasive test of the role of mass extinctions in driving offshore retreats would entail environmental histories for groups that suffered more severely. To give a Paleozoic example of apparent bathymetric selectivity in mass extinction, the deep-water rugosan corals weathered the Frasnian-Fammenian extinctions of the late Devonian better than did the more diverse shallow-water rugosans (Sorauf and Pedder 1986). However, a strictly bathymetric effect could be confounded with disruption of tropical carbonate platforms or reefal communities (see McGhee 1982 on the ambiguity of cold- versus deep-water survivorship).

Environmental selectivity in the opposite sense appears to be the rule in Cambro-Ordovician trilobite turnover events. Shelf-edge and slope trilobites evidently enjoyed greater survivorship than did cratonic taxa, subsequently giving rise to replacements for the post-extinction platform faunas (Fortey 1983; Westrop and Ludvigsen 1987; and references therein). We have no explanation for this offshore-onshore pattern and why trilobite patterns should contrast so strongly with the other taxa we have discussed, but note that both of the cited papers suggest a biogeographical component—preferential extinction of shelf endemics—to the extinctions. Perhaps the selectivity reflects bathymetric trends in trilobite biogeography rather than bathymetric trends in severity of the extinction perturbation per se (see Jablonski 1986a, 1986b, 1989, for reviews of the positive relation between clades' geographical ranges and their survivorship during mass extinctions).

Turning to the interplay of biotic and abiotic factors, the simple model of niche-preemption under background extinction discussed previously could account for both expansion and retreat patterns, particularly if onshore species are more extinction resistant than offshore ones (see Valentine, this volume; see also Van Valen 1985, who places a greater emphasis on clade interactions but reaches similar conclusions regarding the consequences of differential turnover rates among clades). This macroevolutionary displacement process is difficult to test, but a demonstration of the required dynamics would be a major first step: Do two clades thought to be potential interactors exhibit the predicted complementary environmental histories? Do they both exhibit a bathymetric gradient in species durations? Is the displacement accomplished piecemeal, in a stepwise fashion? Unfortunately, there is no reason for the interaction to be restricted to a single pair of clades. A single epifaunal suspension-feeding clade in theory could be preempted in different parts of its geographical, temporal, or environmental range by sponges, bryozoans, bivalves, gastropods, polychaetes, and arthropods. Valentine (this volume) has argued that the process could take place under a set of reasonable assumptions and reasonable parameters, but a closely documented example would be a staggering task.

Other sorts of potential clade interactions are more strictly biotic. Expansion of predators, bioturbators, or competitors capable of active displacement rather than passive preemption could drive older clades offshore (see, for example, Vermeij 1986, 1987, who argues for deep water as a refuge for species with poorly developed competitive and defensive capacities; Meyer and Macurda [1977]; Oji [1985], Aronson [1987], Schneider [1988] on predation-mediated crinoid retreat). Discussing bryozoan growth forms rather than clades, McKinney and Jackson (1989: 94) attribute the offshore retreat of multiserial erect forms to "the dramatic escalation in effectiveness of the grazing, browsing, and gouging machinery of potential predators upon bryozoans throughout the Mesozoic." As with the preemption argument, these interaction-driven processes are plausible but difficult to demonstrate. Complementary time-environment patterns of bioturbators and immobile epibenthos (cf. Thayer 1983), of predators and potential prey, or of predation damage and prey retreat, would help to support such hypotheses. All of these mechanisms ultimately reduce to our initial problem: the mechanism for the onshore origination of major groups.

Conclusions

Major abiotic perturbations evidently play little role in shaping the onshore-offshore evolutionary patterns documented here. Abiotic factors come into play not as a primary driving mechanism but as the template across which differential rates and patterns of biotic processes are arrayed. In the short term, bathymetric gradients in abiotic factors such as temperature, salinity, turbidity, disturbance, and spatial heterogeneity influence a host of biotic factors such as population densities and persistence, life history strategies, genetic population structures. In the long term, those abiotic gradients impose differences in speciation and extinction rates (and possibly modes), and taxonomic durations.

Biotic factors, then, as shaped by the environmental gradient, must underlie these onshore-offshore patterns, but a bewildering array of alternative mechanisms remains. We have been able to rule out only a few: an extrapolationist hypothesis based on local, short-term, selective forces and a macroevolutionary hypothesis based on ecological determinants of heterochrony. The onshore origin of higher taxa is still the central question, and may be all that is required to explain the subsequent expansion offshore of new taxa and perhaps even the eventual retreat into deep water exhibited by some groups. However, the nature of interactions among clades and major adaptive types and the level at which those interactions are manifested are still poorly understood (e.g., Vermeij 1987; Benton 1987), and plenty of complex scenarios are plausible. The alternatives are approaching the realm of testability for the first time.

Our most striking result is the discordance between the environmental patterns for first appearances of major groups and of novelties within major groups: higher taxa preferentially first appear in onshore environments, despite the absence of such a bias for contained lower taxa or novelties. This is strong evidence for a hierarchical view of evolution. Lower-level novelties seem diversity-dependent in their origination: the more evolutionary experiments per unit time, whether within species or at speciation events, the more frequent the production of a new genus, family, or other within-clade novelty (see Stanley 1978, 1979, for the argument from speciation rates). The tellinacean families are the exception that proves this rule. In contrast to the crinoid and cheilostomes, tellinacean novelties appear onshore, but as with crinoids and cheilostomes, tellinacean novelties first appear where tellinaceans are already most diverse and abundant. Anstey (1986) records a comparably discordant pattern in Ordovician bryozoans in North America: major novelties (e.g., the oldest known ctenostome) first appeared in the clastic wedges of the Reedsville-Lorraine Province of the basinal margins, while the highest standing diversity and the most generic originations occurred in the more offshore Cincinnati Province.

Valentine (1969, 1973, 1980, 1986, etc.), Van Valen (1971, 1973, 1975, etc.), and Bambach (1983, 1985) have fruitfully argued that taxa of high rank tend to correspond to ground plans, discrete from one another and with broad functional and ecological identities. Our data support and extend this view. Higher taxa (orders and the large and distinct superfamily Tellinacea) do not conform to a pattern predictable from lower levels, and to us the different pattern implies a different process. We are not suggesting that orders arise in single jumps by massive macromutations or that systematists have developed a taxonomy so objective that ordinal rank is strictly comparable across phyla. However, some aspect of the onshore environment fosters or enhances survivorship or morphologies sufficiently distinct or with sufficient long-term evolutionary productivity that a disproportionate fraction of the clades or paraclades afforded high taxonomic rank will be traceable to onshore species. The evolutionary origins and fates of those major evolutionary branches are not simply a smooth extrapolation of evolutionary dynamics at the lowermost levels in the hierarchy.

Acknowledgments

Many people have generously assisted us with advice, information, and inaccessible references. We are grateful to A. H. Cheetham, B. Clavel, K. W.

Flessa, A. Hallam, H. Hess, S. R. A. Kelly, M. LaBarbera, S. Lidgard, M. R. A. Listokin, F. K. McKinney, N. J. Morris, M. J. Nowicki, A. Oravecz-Scheffer, D. B. Rowley, LouElla R. Saul, J. J. Sepkoski, Jr., M. J. Simms, A. B. Smith, S. Suter, P. D. Taylor, E. Voigt, S. E. Walker, and A. M. Ziegler for all their help. We also thank the staffs at the following libraries, which gave us essential access to the literature: Hancock Library, University of Southern California (especially Suzanne Henderson); Geology and Geophysics Library, University of California, Los Angeles (especially Michael Noga); Field Museum of Natural History Library; British Museum (Natural History) Libraries; and the John Crerar Library, University of Chicago (especially Kathleen Zar). Mary L. Droser performed herculean tasks of data management. We thank Susan M. Kidwell for, as always, an insightful review of the manuscript; we also thank Scott Lidgard, Geerat J. Vermeij, and the editors of this volume for their comments. We acknowledge the Donors of the Petroleum Research Fund, administered by the American Chemical Society, for support of this research. Additional support was provided by the Visiting Scientist Program of the Field Museum of Natural History (to DB) and by the National Science Foundation (Grants EAR 85-08970 and EAR 85-19941 to DB; Ear 84-177011 and INT 86-2045 to DJ).

Literature Cited

Allen, J. A. 1985. Recent Bivalvia: Their form and evolution. In ed. E. R. Trueman and M. R. Clarke, 337–403. *The Mollusca*, vol. 10, *Evolution*, Orlando, Fla.: Academic Press.

Anstey, R. L. 1986. Bryozoan provinces and patterns of generic evolution and extinction in the Late Ordovician of North America. *Lethaia* 19:33–51.

Aronson, R. B. 1987. Predation on fossil and Recent ophiuroids. *Paleobiology* 13:187–92.

Bambach, R. K. 1983. Ecospace utilization and guilds in marine communities through the Phanerozoic. In *Biotic interactions in Recent and fossil benthic communities*, ed. M. J. S. Tevesz and P. L. McCall, 719–46. New York: Plenum.

———. 1985. Classes and adaptive variety: the ecology of diversification in marine faunas through the Phanerozoic. In *Phanerozoic diversity patterns*, ed. J. W. Valentine, 191–25. Princeton, N.J.: Princeton University Press.

Banta, W. C., and R. E. Wass. 1979. Catenicellid cheilostome Bryozoa. I. Frontal walls. *Australian J. Zool. Suppl. Ser.* 68:.

Benton, M. J. 1983. Large-scale replacements in the history of life. *Nature* 302:16–17.

———. 1987. Progress and competition in macroevolution. *Biol. Rev.* 62:305–38.

Berry, W. B. N. 1972. Early Ordovician bathyurid province lithofacies, biofacies, and correlations—their relationship to a proto-Atlantic Ocean. *Lethaia* 5:69–84.

Berry, W. B. N. 1974. Types of Early Paleozoic faunal replacements in North America: Their relationship to environmental change. *J. Geol.* 82:371–82.

Blake, D. B. 1987. A classification and phylogeny of post-Palaeozoic sea stars (Asteroidea: Echinodermata). *J. Natural Hist.* 21:481–528.

Bottjer, D. J., M. L. Droser, and D. Jablonski, 1988. Palaeoenvironmental trends in the history of trace fossils. *Nature* 333:252–5.

Bottjer, D. J., and D. Jablonski. 1988. Paleoenvironmental patterns in the evolution of post-Paleozoic benthic marine invertebrates. *Palaios* 3:540–60.

Carson, H. L., and A. R. Templeton. 1984. Genetic revolutions in relation to speciation phenomena: the founding of new populations. *Ann. Rev. Ecol. Syst.* 15:91–131.

Cheetham, A. H. 1986. Branching, biomechanics, and bryozoan evolution. *Proc. R. Soc. London* B 228:151–71.

Cheetham, A. H., and P. L. Cook. 1983. General features of the Class Gymnolaemata. In *Treatise on invertebrate paleontology,* part G, *Bryozoa,* vol. 1, rev., ed. R. A. Robison, 138–207. Boulder, Colo.: Geological Society of America; Lawrence, Kans.: University of Kansas.

Cheetham, A. H., and L. A. C. Hayek. 1983. Geometric consequences of branching growth in adeoniform Bryozoa. *Paleobiology* 9:240–60.

Cheetham, A. H., and E. Thomsen. 1981. Functional morphology of arborescent animals: Strength and design of cheilostome bryozoan skeletons. *Paleobiology* 7:355–83.

Coan, E. V. 1988. Recent eastern Pacific species of the bivalve genus *Semele. Veliger* 31:1–42.

Cook, P. L., and P. J. Chimonides. 1983. A short history of the lunulite Bryozoa. *Bull. Marine Sci.* 33:566–81.

———. 1986. Recent and fossil Lunulitidae (Bryozoa, Cheilostomata). 6. *Lunulites sensu lato* and the genus *Lunularia* from Australasia. *J. Natural Hist.* 20:681–705.

Cook, P. L., and E. Voigt. 1986. *Pseudolunulites* gen. nov., a new kind of lunulitiform cheilostome from the Upper Oligocene of northern Germany. *Verh. naturwiss. Ver. Hamburg N.F.* 28:107–27.

Cox, L. R. 1985. Jurassic Bivalvia and Gastropoda from Tanganyika and Kenya. *Bull. Br. Mus. (Nat. Hist.) Geol. Suppl.* 1.

Crimes, T. P. 1974. Colonisation of the early ocean floor. *Nature* 248:328–30.

Erwin, D. H., J. W. Valentine, and J. J. Sepkoski, Jr. 1987. A comparative study of diversification events: the early Paleozoic versus the Mesozoic. *Evolution* 41:1177–86.

Flessa, K. W., et al. 1987. Causes and consequences of extinction. In *Patterns and processes in the history of life,* ed. D. M. Raup and D. Jablonski, 235–57. Berlin: Springer-Verlag.

Fortey, R. A. 1983. Cambrian-Ordovician trilobites from the boundary beds in western Newfoundland and their phylogenetic significance. *Spec. Pap. Palaeontol.* 30:179–211.

Fürsich, F. T. 1982. Upper Jurassic bivalves from Milne Land, East Greenland. *Gronlands Geol. Unders. Bull. 144.*

Gilinsky, N. L., and R. K. Bambach, 1987. Asymmetrical patterns of origination and extinction. *Paleobiology* 13:427–45.

Gilsén, T. 1924. Echinoderm studies. *Zool. Bidrag Uppsala* 9:1–330.

———. 1938. A revision of the Recent Bathycrinidae. *Lunds Univ. Arsskr. N. S. Avd.* 2 34(10):1–30.

Gordon, D. P. 1984. The marine fauna of New Zealand: Bryozoa: Gymnolaemata from the Kermadec Ridge. *N.Z. Oceanogr. Inst. Mem. 91.*

Gould, S. J. 1977. *Ontogeny and phylogeny.* Cambridge, Mass.: Harvard University Press.

———. 1988. Trends as changes in variance: a new slant on progress and directionality in evolution. *J. Paleontol.* 62:319–29.

Hallam, A. 1976. Stratigraphic distribution and ecology of European Jurasic bivalves. *Lethaia* 9:245–59.

———. 1987. Radiations and extinctions in relation to environmental change in the marine Lower Jurassic of northwest Europe. *Paleobiology* 13:152–68.

Harland, W. B., A. V. Cox, P. G. Llewellyn, C. A. G. Pickton, A. G. Smith, and R. Walters. 1982. A *geologic time scale.* Cambridge: Cambridge University Press.

Haude, R. 1980. Constructional morphology of the stems of Pentacrinitidae, and way of life of *Seirocrinus.* In: *Echinoderms: Present and Past,* ed. M. Jangoux, 17–23. Rotterdam: Balkema.

Jablonski, D. 1980. Apparent versus real effects of transgressions and regressions. *Paleobiology* 6:397–407.

———. 1986a. Causes and consequences of mass extinctions: a comparative approach. In *Dynamics of extinction.* ed. D. K. Elliott, 183–229. New York: John Wiley & Sons.

———. 1986b. Evolutionary consequences of mass extinctions. In *Patterns and processes in the history of life.* ed. D. M. Raup and D. Jablonski, 313–29. Berlin: Springer-Verlag.

———. 1989. The biology of mass extinction: a paleontological view. *Phil. Trans. R. Soc. London* 325:357–368.

Jablonski, D., and D. J. Bottjer. 1983. Soft-bottom epifaunal suspension-feeding assemblages in the Late Cretaceous: Implications for the evolution of benthic paleocommunities. In *Biotic interactions in living and fossil benthic communities,* ed. M. J. S. Tevesz and P. L. McCall, 747–812. New York: Plenum.

———. 1988. Onshore-offshore evolutionary patterns in post-Paleozoic echinoderms: a preliminary analysis. In *Echinoderm biology, Proceedings of the 6th International Echinoderm Conference.* ed. R. D. Burke, P. V. Mladenov, P. Lambert, and R. L. Parsley, 81–90. Rotterdam: Balkema.

———. Forthcoming. The origin and diversification of major groups: Environmental patterns and macroevolutionary lags. In *Major evolutionary ra-*

diations, Syst. Assoc. Spec. volume 42, ed. P. D. Taylor and G. P. Larwood, 17–57. Oxford: Oxford University Press.

Jablonski, D., J. J. Sepkoski, Jr., D. J. Bottjer, and P. M. Sheehan. 1983. Onshore-offshore patterns in the evolution of Phanerozoic shelf communities. *Science* 222:1123–5.

Jablonski, D., and J. W. Valentine. 1981. Onshore-offshore gradients in eastern Pacific shelf faunas and their paleobiogeographic significance. In *Evolution today*, ed. G. G. E. Scudder and J. L. Reveal, 441–53. Pittsburgh: Carnegie-Mellon University.

Jackson, J. B. C. 1974. Biogeographic consequences of eurytopy and stenotopy among marine bivalves and their evolutionary significance. *Am. Nat.* 108:541–60.

———. 1979. Morphological strategies of sessile animals. In *Biology and systematics of colonial organisms*, Syst. Assoc. Spec. vol. 11, ed. G. Larwood and B. R. Rosen, 499–555. London: Academic Press.

Jenkyns, H. C. 1980. Cretaceous anoxic events: from continents to oceans. *J. Geol. Soc. London* 137:171–88.

———. 1988. The Early Toarcian (Jurassic) anoxic event: Stratigraphic, sedimentary, and geochemical evidence. *Am. J. Sci.* 288:101–51.

Keen, A. M. 1969. Superfamily Tellinacea de Blainville, 1814. In *Treatise on invertebrate paleontology. Mollusca 6, Bivalvia*, ed. R. C. Moore, N613–43. Boulder, Col.: Geological Society of America; Lawrence, Kans.: University of Kansas.

Kier, P. M. 1977. The poor fossil record of the regular echinoid. *Paleobiology* 3:168–74.

Klikushin, V. G. 1982a. Taxonomic survey of fossil isocrinids with a list of the species found in the USSR. *Géobios* 15:299–325.

———. 1982b. Cretaceous and Paleogene Bourgueticrinina. (Echinodermata, Crinoidea) of the USSR. *Géobios* 15:811–43.

———. 1987a. Distribution of crinoids remains in the Triassic of the U.S.S.R. *N. Jb. Geol. Paläont. Abh.* 173:321–38.

———. 1987b. Thiolliericrinid crinoids from the Lower Cretaceous of Crimea. *Géobios* 20:625–65.

Lang, W. D. 1921. *Catalogue of the fossil Bryozoa (Polyzoa) in the Department of Geology, British Museum (Natural History); the Cretaceous Bryozoa (Polyzoa), vol. III, The cribrimorphs*, part *I*. London: British Museum (Natural History).

Larwood, G. P. 1985. Form and evolution of Cretaceous myagromorph Bryozoa. In *Bryozoa: Ordovician to Recent*, ed. C. Nielsen and G. P. Larwood, 169–74. Fredensborg, Denmark: Olsen & Olsen.

Levinton, J. 1988. *Genetics, paleontology, and macroevolution*. New York: Cambridge University Press.

Lidgard, S. 1985. Zooid and colony growth in encrusting cheilostome bryozoans. *Palaeontology* 28:255–91.

———. 1986. Ontogeny in animal colonies: a persistent trend in the bryozoan fossil record. *Science* 232:230–2.

McGhee, G. R., Jr. 1982. The Frasnian-Famennian extinction event: a preliminary analysis of Appalachian marine ecosystems. *Geol. Soc. Am. Spec. Pap.* 190:491–500.

McKinney, F. K. 1986a. Historical record of erect bryozoan growth forms. *Proc. R. Soc. London* B228:133–49.

———. 1986b. Evolution of erect marine bryozoan faunas: Repeated success of unilaminate species. *Am. Nat.* 128:795–809.

McKinney, F. K., and J. B. C. Jackson. 1989. *Bryozoan evolution.* Boston: Unwin Hyman.

McKinney, M. L. 1986. Ecological causation of heterochrony: a test and implications for evolutionary theory. *Paleobiology* 12:282–9.

Meyer, D. L., and D. B. Macurda. 1977. Adaptive radiation of the comatulid crinoids. *Paleobiology* 12:282–9.

Miller, A. I. 1988. Spatio-temporal transitions in Paleozoic Bivalvia: an analysis of North American fossil assemblages. *Hist. Biol.* 1:251–73.

Oji, T. 1985. Early Cretaceous *Isocrinus* from northeast Japan. *Palaeontology* 28:629–42.

Pohowsky, R. A. 1973. A Jurassic cheilostome from England. In *Living and fossil Bryozoa*, ed. G. P. Larwood, 447–56. London: Academic Press.

Pohlo, R. 1982. Evolution of the Tellinacea (Bivalvia). *J. Moll. Stud.* 48:245–56.

Rasmussen, H. W. 1978a. Articulata. In *Treatise on invertebrate paleontology,* part T. *Echinodermata* 2, *Crinoidea*, vol. 3, ed. R. C. Moore and C. Teichert, T813–T1027. Boulder, Colo.: Geological Society of America; Lawrence, Kans.: University of Kansas.

———. 1978b. Evolution of articulate crinoids. In *Treatise on invertebrate paleontology,* part T. *Echinodermata* 2, *Crinoidea*, vol 1, ed. R. C. Moore and C. Teichert, T302–16. Boulder, Colo.: Geological Society of America; and Lawrence, Kans.: University of Kansas.

———. 1979. Crinoids, asteroids and ophiuroids in relation to the boundary. In *Cretaceous-Tertiary boundary events symposium. I. The Maastrichtian and Danian of Denmark*, ed. T. Birkelund and R. G. Bromley, 65–71. Copenhagen: University of Copenhagen.

Raup, D. M. 1985. Mathematical models of cladogenesis. *Paleobiology* 11:42–52.

———. 1986. Biological extinction in Earth history. *Science* 231:1528–33.

Rosenkrantz, D. 1971. Fossil-Lagerstätten. *N. Jb. Geol. Paläont. Abh.* 138:221–58.

Roux, M. 1978. Ontogénèse, variabilité et évolution morphofonctionelle du pédoncule et du calice chez les Millercrinida (Echinodermes, Crinoides). *Géobios* 11:213–41.

———. 1987. Evolutionary ecology and biogeography of Recent stalked crinoids as a model for the fossil record. In *Echinoderm studies*, vol. 2, ed. M. Jangoux and J. M. Lawrence, 1–53. Rotterdam: Balkema.

Ryland, J. S. 1970. *Bryozoans.* London: Hutchinson University Library.

———. 1974. Behaviour, settlement and metamorphosis of bryozoan larvae: a review. *Thal. Jugosl.* 10:239–62.

Saul, L. R. 1989. California Late Cretaceous donaciform bivalves. *Veliger* 32:188–208.

Saul, L. R., and W. P. Popenoe. 1962. *Meekia*, enigmatic Cretaceous pelecypod genus. *Univ. Calif. Geol. Sci.* 40:289–344.

Schlanger, S. O., M. A. Arthur, H. C. Jenkyns, and P. A. Scholle. 1987. The Cenomanian-Turonian Oceanic Anoxic Event. I. Stratigraphy and distribution of organic carbon-rich beds and the marine delta-^{13}C excursion. In *Marine petroleum source rocks*, Geolog. Soc. Lond. Spec. Pub. 26, ed. J. Brooks and A. J. Fleet, 371–99.

Schneider, J. A. 1988. Frequency of arm regeneration of comatulid crinoids in relation to life habit. In *Echinoderm biology, Proceedings of the 6th International Echinoderm Conference*, ed. R. D. Burke, P. V. Mladenov, P. Lambert, and R. L. Parsley, 531–8. Rotterdam: Balkema.

Seilacher, A., G. Drozdzewski, and R. Haude. 1968. Form and function of the stem in a pseudoplanktonic crinoid. *Palaeontology* 11:275–282.

Seilacher, A., W.-E. Reif, and F. Westphal, 1985. Sedimentological, ecological and temporal patterns of fossil Lagerstätten. *Phil. Trans. R. Soc. London* B311:5–23.

Sepkoski, J. J., Jr. 1986. Phanerozoic overview of mass extinction. In *Patterns and processes in the history of life*, ed. D. M. Raup and D. Jablonski, 277–95. Berlin: Springer-Verlag.

———. 1987. Environmental trends in extinction during the Paleozoic. *Science* 235:64–66.

———. 1988. Alpha, beta, or gamma: Where does all the diversity go? *Paleobiology* 14:221–34.

Sepkoski, J. J., Jr., and A. I. Miller. 1985. Evolutionary faunas and the distribution of Paleozoic benthic communities in space and time. In *Phanerozoic diversity patterns*, ed. J. W. Valentine, 153–90. Princeton, N.J.: Princeton University Press.

Sepkoski, J. J., Jr., and P. M. Sheehan. 1983. Diversification, faunal change, and community replacement during the Ordovician radiation. In *Biotic interactions in Recent and fossil benthic communities*, ed. M. J. S. Tevesz and P. L. McCall, 673–718. New York: Plenum.

Sheehan, P. M. 1986. Macroevolution and low diversity faunas (abstract). *Geolog. Soc. Am. Progr.* 18:324.

Simms, M. J. 1986a. The taxonomy and palaeobiology of British Lower Jurassic crinoids. Ph.D. diss., University of Birmingham, Birmingham, England.

———. 1986b. Contrasting lifestyles in Lower Jurassic crinoids: a comparison of benthic and pseudopelagic Isocrinida. *Palaeontology* 29:475–93.

———. 1988a. The phylogeny of post-Palaeozoic crinoids. In *Echinoderm phylogeny and evolutionary biology*, ed. C. R. C. Paul and A. B. Smith, 269–84. Oxford: Clarendon Press.

———. 1988b. The role of heterochrony in the evolution of post-Palaeozoic crinoids. In *Echinoderm biology, Proceedings of the 6th International Echi-*

noderm Conference, ed. R. D. Burke, P. V. Mladenov, P. Lambert, and R. L. Parsley, 97–102. Rotterdam: Balkema.

———. 1988c. Patterns of evolution among Lower Jurassic crinoids. *Hist. Biol.* 1:17–44.

———. in press. Crinoids across the Triassic-Jurassic boundary. *Cahiers Scientifique de l'Institut Catholique de Lyon.*

Smith, A. B. 1984. *Echinoid palaeobiology.* London: Allen and Unwin.

Smith, A. B., and C. W. Wright. 1990. *British Cretaceous echinoids*, part 2. *Echinothurioida, Diadematoida and Stirodona.* London: Palaeontographical Society.

Sorauf, J. E., and A. E. H. Pedder. 1986. Late Devonian rugose corals and the Frasnian-Famennian boundary. *Can. J. Earth Sci.* 23:1265–87.

Speden, I. G. 1970. The type Fox Hills Formation, Cretaceous (Maestrichtian), South Dakota, part 2, Systematics of the Bivalvia. *Peabody Mus. Natural Hist. Yale Univ. Bull.* 33.

Stanley, S. M. 1978. Chronospecies' longevities, the origin of genera, and the punctuational model of evolution. *Paleobiology* 4:26–40.

———. 1979. *Macroevolution: Pattern and process.* San Francisco: W. H. Freeman.

Taylor, J. D., W. J. Kennedy, and A. Hall. 1973. The shell structure and mineralogy of the Bivalvia. II. Lucinacea–Clavagellacea. Conclusions. *Bull. Br. Mus. (Nat. Hist.) Zool.* 22:255–94.

Taylor, P. D. 1981a. Bryozoa from the Jurassic Portland Beds of England. *Palaeontology* 24:863–75.

———. 1981b. Functional morphology and evolutionary significance of differing modes of tentacle eversion in marine bryozoans. In *Recent and fossil Bryozoa*, ed. G. P. Larwood and C. Nielsen, 235–47. Fredensborg, Denmark: Olsen & Olsen.

———. 1986. *Charixa* Lang and *Spinicharixa* gen. nov., cheilostome byrozoans from the Lower Cretaceous. *Bull. Br. Mus. (Nat. Hist.) Geol.* 40:197–222.

———. 1988. Major radiation of cheilostome bryozoans: Triggered by the evolution of a new larval type? *Hist. Biol.* 1:45–64.

Templeton, A. R. 1980. The theory of speciation via the founder principle. *Genetics* 94:1011–38.

———. 1986. The relation between speciation mechanisms and macroevolutionary pattern. In *Evolutionary processes and theory*, ed. S. Karlins and E. Nevo, 497–512. Orlando, Fla: Academic Press.

Thayer, C. W. 1983. Sediment-mediated biological disturbance and the evolution of marine benthos. In *Biotic interactions in Recent and fossil benthic communities*, ed. M. J. S. Tevesz and P. L. McCall, 400–595. New York: Plenum.

Valentine, J. W. 1969. Patterns of taxonomic and ecological structure of the shelf benthos during Phanerozoic time. *Palaeontology* 12:684–709.

———. 1973. *Evolutionary paleoecology of the marine biosphere.* Englewood Cliffs, N.J.: Prentice-Hall.

———. 1980. Determinants of diversity in higher taxonomic categories. *Paleobiology* 6:440–50.

———. 1986. Fossil record of the origin of Baupläne and its implications. In *Patterns and Processes in the History of Life*, ed. D. M. Raup and D. Jablonski, 209–22. Berlin: Springer-Verlag.

Valentine, J. W., and T. D. Walker. 1987. Extinctions in a model taxonomic hierarchy. *Paleobiology* 13:193–207.

Van Valen, L. M. 1971. Adaptive zones and the orders of mammals. *Evolution* 25:420–8.

———. 1973. A new evolutionary law. *Evol. Theory* 1:1–30.

———. 1985. A theory of origination and extinction. *Evol. Theory* 7:133–42.

Vermeij, G. J. 1986. Survival during biotic crises: the properties and evolutionary significance of refuges. In *Dynamics of extinction*, ed. D. K. Elliott, 231–46. New York: John Wiley & Sons.

———. 1987. *Evolution and escalation*. Princeton, N.J.: Princeton University Press.

Voigt, E. 1981. Répartition et utilisation stratigraphique des Bryozoaires du Crétacé moyen (Aptien-Coniacien). *Cret. Res.* 2:439–62.

———. 1985. The Bryozoa of the Cretaceous-Tertiary boundary. In *Bryozoa: Ordovician to Recent*, ed. C. Nielsen and G. P. Larwood, 329–42. Fredensborg, Denmark: Olsen & Olsen.

Vrba, E. S., and N. Eldredge. 1984. Individuals, hierarchies and processes: Towards a more complete evolutionary theory. *Paleobiology* 10:146–71.

Vrba, E. S., and S. J. Gould. 1986. The hierarchical expansion of sorting and selection: Sorting and selection cannot be equated. *Paleobiology* 12:217–28.

Walker, T. D., and J. W. Valentine. 1985. Equilibrium models of evolutionary species diversity and the number of empty niches. *Am. Nat.* 124:887–99.

Westrop, S. R., and R. Ludvigsen. 1987. Biogeographic control of trilobite mass extinction at an Upper Cambrian "biomere" boundary. *Paleobiology* 13:84–99.

Yonge, C. M. 1949. On the structure and adaptations of the Tellinacea, deposit-feeding Eulamellibranchia. *Phil. Trans. R. Soc. London* B234:29–76.

Yonge, C. M., and J. A. Allen. 1985. On significant criteria in establishment of superfamilies in the Bivalvia: the creation of the superfamily Mesodesmatacea. *J. Moll. Stud.* 51:345–9.

Appendixes

A2.1. Time and environment of first occurrence for genera of the crinoid Order Isocrinida (sensu Rasmussen 1978)

---

1. *Holocrinus*. 240–245 Ma, A and B
   Taxon: *Holocrinus smithi* (Clark)

Age: Mid-Spathian
Unit: Thaynes Fm., Paris (Bear Lake County), Idaho

Taxon: "*Isocrinus* sp." [= *H. smithi* according to Klikushin 1987a]
Age: Mid-Spathian
Unit: Virgin Fm., Utah and Nevada, many localities

References: Kummel (1957); Carr and Paull (1983); Hagdorn (1986), Klikushin (1987a)

2. *Moenocrinus*. 240–245 Ma, A.
   Taxon: *M. deecki* Hildebrand
   Age: Late Early Anisian
   Unit: Konglomeratbänke des Unteren Wellenkalk, north Württemberg, SW Germany
   References: Schwarz (1975); Hagdorn (1983, 1985, 1986)

3. *Laevigatocrinus*. 230–235 Ma, B.
   Taxon: *L. laevigatus* (Münster)
   Age: Late Ladinian-Early Carnian
   Unit: Cassian Formation, several localities, northern Italy
   References: Ogilvie (1893); Zardini (1973); Urlichs (1977); Fürsich and Wendt (1977); Wendt and Fürsich (1979); Klikushin (1979); Simms (in press)
   Note: This species also recorded by Klikushin (1987) from the Carnian of the Koryak Mountains, Far Eastern U.S.S.R.

4. *Isocrinus*. 230–235 Ma, B and C
   Taxon: *I. tyrolensis* (Laube) (includes *scipio* Bather, *candelabrum* Bather, and *anulatus* Leonardi and Lovo; Simms in press)
   Age: Late Ladinian-Carnian
   Units: Cassian Formation, several localities, northern Italy. Also recorded from the Carnian Veszprem Formation, Bakony, Hungary; mid-Carnian Raibl Formation, northern Italy; mid-Carnian Tor Formation, western Austria; Carnian of northern Afghanistan; and Carnian of Koryak Mountains and Primorye, Far Eastern U.S.S.R.
   References: Cassian references cited above; Wöhrmann (1889); Bather (1909); Angermeier et al. (1963); Jerz (1966); Balogh (1981); Galacz et al. (1985); Oravecz-Scheffer (1987); Klikushin (1983a, 1987a)
   Notes: (1) Includes *Terocrinus* Klikushin 1982a and *Tyrolecrinus* Klikushin 1983a (Jäger 1985; Simms 1986a, in press). (2) Simms (in press) does not accept the record of *I. tyrolensis* from the Late Anisian of south China (Kristan-Tollmann and Tollmann 1983).

5. *Singularicrinus*. 220–225 Ma, C
   Taxon: *S. singularis* Klikushin
   Age: Early Norian
   Unit: Unnamed, Bolshoj Tkatch Hill, Laba River, northwest Caucasus, U.S.S.R.

References: Rostovstev et al. (1966); Efimova (1975); Beznosov and Efimova (1979); Klikushin (1982a)
Note: Possibly a continuation of the *Laevigatocrinus* lineage (Simms in press).

6. *Chladocrinus*. 210–225 Ma, C
Taxon: *C. psilonoti* (Quenstedt)
Age: Earliest Hettangian (pre-*planorbis* Subzone)
Unit: Preplanorbis Beds, Dorset, Somerset, Gloucestershire, Worcestershire, South Wales, United Kingdom
References: Hallam (1960); Wobber (1965, 1968a, 1968b); Donovan et al. (1979); Donovan and Kelloway (1984); Simms (1986a, 1986b, 1988).

7. *Balanocrinus*. 200–205 Ma, D
Taxon: *B. quiaiosensis* de Loriol
Age: Late Sinemurian, *oxynotum* subzone
Unit: Lower Lias Clays, pyritic mudstone facies; Gloucestershire and Worcestershire, Great Britain
References: Cope et al. (1980); Simms (1986a, 1986b, 1988)

8. *Hispidocrinus*. 200–205 Ma, C/D
Taxon: "*Pentacrinus*" *scalaris* Goldfuss
Age: Late Sinemurian, *stellare* Subzone
Unit: Redcar Mudstone Formation (= Upper Calcareous Shales of previous authors), Robin Hood's Bay, Yorkshire, Great Britain
References: Buckman (1915); Hemingway (1974); Powell (1984); Simms (1986a, 1986b, 1988)

9. *Chariocrinus*. 190–195 Ma, C or D
Taxon: *C. wuerttembergicus* (Oppel)
Age: Late Early Toarcian, *fibulatum* Subzone
Unit: Upper Lias Clay Formation, Cheltenham (Gloucestershire), Great Britain
References: Davies (1969); Cope et al. (1980); Simms (1986a, 1986b, 1988)

10. *Nielsenocrinus*. 130–135 Ma, C
Taxon: *N. chavannesi* (de Loriol)
Age: Early Hauterivian
Unit: Marnes d'Hauterive. Romainmotier (Vaud), Switzerland
References: Bashang (1921); Custer (1928); Hess (1975)

11. *Austinocrinus*. 85–90 Ma, D
Taxon: *A. albaticus* Klikushin (1985a)
Age: Early Coniacian, *Inoceramus wandereri* Zone
Unit: *wandereri* Zone, Kujbyshevo, Belbeck Valley, Crimea, U.S.S.R.
References: Naidin (1981); Klikushin (1985a, 1985c)

12. *Isselicrinus* (includes *Buchicrinus* Klikushin 1977 and *Praeisselicrinus* Klikushin 1977), 80–85 Ma, C

Taxon: *I. atabekjani* Klikushin (1973)
Age: Early (Klikushin 1983b) or Late (Atabekjan 1979) Campanian
Unit: Unnamed, Maly Balkhan Ridge, Turkmenia, U.S.S.R.
References: Atabekjan (1979); Klikushin (1983b, 1985b)
Note: Of uncertain age within the Senonian is *I. tibiensis* (Dupuy de Lome and Revilla 1956), Tibi, Prov. Alicante, Spain, B or C (LeClerc and Azema 1976).

13. *Doreckicrinus*. 70–75 Ma, C
Taxon: *D. indentatus* Klikushin (1985b)
Age: Late Campanian
Unit: Unnamed, Tuarkyr, NW Turkmenia, U.S.S.R.
References: Vinogradov (1968); Nalivkin (1973); Klikushin (1983b, 1985b)

14. *Cainocrinus*. 60–65 Ma, C
Taxon: *C.? aff. C. tintinnabulus* Forbes (Rasmussen 1972)
Age: Early Danian
Unit: Pulawy Beds and Sochaczew Beds, boring at Sochaczew, central Poland
References: Pozaryska (1965); Krach (1981)

15. *Tauriniocrinus*. 35–40 Ma, D or E
Taxon: *T. gastaldii* (Michelotti)
Age: Early Oligocene
Unit: Pteropod marls, road cut between Bruzzi and Crenna, Cassianelle, and Cairo Montenotte (Liguria), NW Italy
References: Rovereto (1939); Franceschetti (1967); Lorenz (1968, 1984); Desio (1973)
Note: Congeneric with the Recent genus *Metacrinus* according to Klikushin (1982a).

A2.2. Time and environment of first appearance for genera of the crinoid Order Bourgueticrinida (sensu Rasmussen 1978)

B1. *Bourgeticrinus*. 85–90 Ma, D (and C?)
Taxon: *Bourgueticrinus* sp.
Age: Late Early Turonian
Unit: Unnamed, Belbeck Valley, Crimea, U.S.S.R.
References: Klikushin (1982b, 1985c)

Taxon: *B. fischeri* (Geinitz)
Age: Turonian
Unit: Upper Chalk, Kent; Middle Chalk and lower part of Upper Chalk, Yorkshire; lower part of Sussex White Chalk Formation, Isle of Wight; Plänerkalk of Strehlen, NW Germany.
References: Rasmussen (1961); Peake and Hancock (1961); Neale (1974); Wood and Smith (1978); Ernest and Schmid (1979); Ernst et al. (1984); Schulz et al (1984); Mortimore (1986); Robinson (1986)

Note: Supposed Early Cretaceous bourgueticrinids cited by Klikushin (1975) are almost certainly the comatulid family Thiollierierinidae (see Klikushin 1987b).

B2. *Monachocrinus*. 85–90 Ma, C
Taxon: *M.? aff. M. regnelli* Rasmussen
Age: Latest Santonian
Unit: *Marsupites* Zone, Gleidingen Brickworks, NW Germany
References: Ernst (1968, 1975); Rasmussen (1975)

B3. *Democrinus*. 70–75 Ma, D
Taxon: *D. gilseni* Rasmussen
Age: Early Maastrichtian
Unit: White Chalk, Mons Klint, Denmark, and Rügen, East Germany
References: Rasmussen (1961); Nestler (1965); Hakansson et al. (1974)

B4. *Dunnicrinus*. 65–70 Ma, C
Taxon: *Dunnicrinus mississippiensis* Moore
Age: Late Maastrichtian
Unit: Upper Prairie Bluff Formation, NE Mississippi.
References: Stephenson and Monroe (1940); Moore (1967); Russell et al. (1983)

B5. *Bathycrinus*. 60–65 Ma, C
Taxon: *B. windi* Rasmussen.
Age: Late Danian
Unit: Bryozoan Chalk, Nyvang Gaard, Denmark
References: Rasmussen (1961); Berthelsen (1962); Asgaard (1968); Cheetham (1971); Thomsen (1976); Hakansson and Thomsen (1979)

B6. *Conocrinus*. 55–60 Ma, C and D
Taxon: *Conocrinus* sp.
Age: Early Thanetian
Unit: Sables Inférieurs and Calcaire Detrique, Bearn, Pyrenees, France
References: Roux and Plaziat (1978; see also Gaemers 1978 for Late Thanetian occurrences)

A2.3. Time and environment of first appearance of selected evolutionary novelties within the bryozoan Order Cheilostomata

1. Ovicells. 100–105 Ma, C
Taxon: *Marginaria* sp. of Taylor (1988)
Age: Late Albian, *inflatum* Zone
Unit: Red Chalk (Hunstanton Red Rock), A Beds, Norfolk, England
References: Rastall (1930); Larwood (1961); Rawson et al. (1978)

Taxon: *Wilbertopora mutabilis* Cheetham
Age: Late Albian, *Adkinsites bravoensis* Zone (approximate correlative of *inflatum* Zone)

Unit: Kiamichi Formation, Washita Group, Texas
References: Cheetham (1954, 1975); Scott et al. (1978); Taylor (1986, 1988)

2. Erect colony form. 100–105 Ma, C
Taxon: New Genus, new species (BMNH D38164, Taylor 1986:199; P. D. Taylor, in litt., August 1986)
Age: Late Albian, *inflatum* Zone
Unit: Cowstones at the base of the Foxmould, Dorset, England
References: Tresise (1960); Ager and Smith (1965); Garrison et al. (1987)

3. Cribrimorphs. 95–100 Ma, C
Taxa: *Andriopora* sp. and 3 or 4 indet. genera
Age: Earliest Cenomanian, *carcitanense* Subzone (= lower *mantelli* Zone) and perhaps *saxbii* Subzone (= upper *mantelli* Zone)
Unit: Untercenomanen Rotkalk (Klippenfazies sensu Kahrs 1926), Mülheim-Broich, NW Germany
References: Hancock et al. (1972); Wiedmann and Schneider (1979); Voigt (1981) and pers. comm., July 1988

Taxa: *Andriopora mockleri* Lang, *Ctenopora pecten* Lang, *Octopora auricula* Lang
Age: Early Cenomanian, upper *carcitenense* Subzone and *saxbii* Subzone
Unit: Lower Chalk Marl, Cambridge, England; 5 to 20 ft. above base
References: Lang (1921); Worssam and Taylor (1969); Kennedy and Garrison (1975); Kennedy and Hancock (1976); Voigt (1981); Weaver (1982); Wright and Kennedy (1984); Larwood (1985)
Note: The single corroded colony of *Otopora* sp. reported by Dzik (1975) from the "Lower (or Middle) Cenomanian" of Korzkiew, Poland, may also be one of the oldest cribrimorphs, but because several units of Cenomanian age are exposed at this locality (Maryanska 1968; Marcinowski 1974; Tarkowska and Liszka 1982), we are unable to pinpoint this record.

4. Articulated colony form. 90–95 Ma, B
Taxon: *Cellarinidra clavata* (d'Orbigny)
Age: Mid-Cenomanian
Unit: Sables et Grès du Mans a *Scaphites equalis*, Le Mans (Maine), France; also in Mid- and/or Late Cenomanian of Ile Madame and Port-des Barques (Charente Maritime), France
References: Juignet (1968, 1980); Moreau (1976, 1977); Juignet et al. (1978); Voigt (1981); Moreau et al. (1983)

5. Ascophorans. 85–90 Ma, C or D
Taxon: *Rotiporina culverensis* Brydone, *R. altonensis* Brydone
Age: Latest Turonian, *H. planus* Zone

Unit: Lewes Chalk Member, White Chalk Formation, Isle of Wight and Alton, Hampshire, respectively
References: Brydone (1929–36); Voigt (1981, 1985); Mortimore (1986); Mortimore and Pomerol (1987); Pomerol et al. (1987)
Notes: (1) The affinities of these species are not entirely certain (E. Voigt, in litt., 1987; P. D. Taylor, pers. comm., March 1989). If they prove not to be ascophoran, then the following occurrence is the oldest known ascophoran.
(2) These Turonian species are probably not congeneric with the Maastrichtian type species of *Rotiporina* (E. Voigt, pers. comm., July 1988).

Taxon: *Boreasina* n.sp. of M. J. Nowicki, unpublished
Age: Early Coniacian–Santonian
Unit: Bioclastic limestones, Saintonge and Periogord Blanc, N Aquitaine, France
References: M. J. Nowicki (in litt., March 1988); Séronie-Vivien (1972); Kennedy (1984)
Note: The species, now considered by Voigt (1990) to be the oldest undoubted ascophoran, is recorded throughout the Santonian chalky facies of Saintonge and Angoumas, which would also be 85 to 90 Ma, C; two questionable specimens have been found in the Lower Coniacian, which would probably be 85 to 90 Ma, B (M. J. Nowicki, in litt., March 1988).
General notes: *Porina cenomana* Lecointre is a cyclostome (Voigt, 1990). *Dacryoporella reussi* (Lang), accepted as oldest ascophoran by Larwood et al. (1967) is excluded from the group by Cheetham (1971) and Voigt and Hilmer (1983).

6. Lunulitiform organization. 85–90 Ma, C
Taxon: *Pavolunulites* sp. (or *Lunulites* sp.) (poorly preserved)
Age: Turonian
Unit: Turonian Plänerkalk, Halle, Westfalia, NW Germany
References: Ernst and Schmid (1979); Ernst et al. (1984); Wood et al. (1984); E. Voigt (in litt., 1987)
Note: Oldest named species is *Lunulites plana*, Early Santonian Chateau Member, Craie de Villedieu (Voigt 1981; Cook and Chimonides 1983, 1986; Jarvis et al. 1982; Jarvis and Gale 1984), also 85–90 Ma, C.

A2.4. Time and environment of first appearance for families of the bivalve Superfamily Tellinacea

---

1. Sowerbyidea. 235–240 Ma? B
Taxon: *Rhaetidia praenuntia* (Stoppani)
Age: Late Anisian–Early Carnian?
Unit: Calcare di Esino, Esino, Lombardy, N Italy
References: Ronchetti (1959); Assereto and Casati (1965); Casati and Gnaccolini (1967)

Note: This species also reported by Conti (1954) from the Late Norian of Val Solda, Lago di Lugano, N Italy.

Taxon: *R. salomoni* Bittner
Age: Early Ladinian
Unit: Calcare della Marmolada, western Dolomites, NE Italy
References: Ronchetti (1959); Desio (1973); Gaetani et al. (1981)
Notes: (1) Slightly younger is *R.* aff. *R. zitteli* Bittner of Reed (1927), reportedly from the Carnian of Yunnan, China (Norian according to Kobayashi and Tamura (1984) and the Committee on Chinese Lamellibranch Fossils, 1976); poor preservation hinders generic placement of this material.
(2) *R. timorensis* Krumbeck (1924), from the Late Triassic of Timor, is almost certainly not this genus, and probably not tellinacean.
(3) Keen (1969) places *Rhaetidia* in the Sowerbyidae with a query. We share her uncertainty: *Rhaetidia* resembles *Sowerbya* in external form but unlike *Sowerbya* lacks cardinal teeth and has an entire pallial line (Bittner 1895); we also note a gap of about 50 my of richly fossiliferous deposits between *Rhaetidia* and the next sowerbyid. *Rhaetidia* may be the sister group to the rest of the Tellinacea and would then be a separate plesion; alternatively, it may not be tellinacean or even heterodont in affinities (N. J. Morris, pers. comm., July 1988). If Sowerbyidae is restricted by exclusion of *Rhaetidia*, then the family would begin with *Sowerbya triangularis* in the Late Toarcian (*thouarsense* Zone), 185–190 Ma, B (Hallam 1987 and in litt., 1986)—which does not change the results discussed here.
(4) Keen (1969) lists two "Permo-Triassic" genera from the Parana Basin of South America as "doubtful Tellinacea," but further study has shown these to be Crassatellacea and Pholadomyacea (Runnegar and Newell 1971; N. J. Morris, pers. comm., July 1988) of Late Permian (probably Guadalupian) age (Fulfaro and Landim 1976; Gama et al. 1982; Petri and Fulfaro 1983)

2. Tancrediidae. 225–230 Ma, A
Taxon: *Sakawanella triadica* Ichikawa (1950)
Age: Early Carnian
Unit: *Oxytoma-Mytilus* bed, lower Kochigatani Group, Tosa Province, Kochi Prefecture, Shikoku Island, Japan
References: Ichikawa (1950); Kobayashi and Ichikawa (1950)

3. Quenstedtiidae. 185–190 Ma, B
Taxon: *Quenstedtia laevigata* (Phillips)
Age: Late Toarcian (*thouarsense* Zone)
Unit: Cephalopod Bed
References: Davies (1969); Hallam (1987, in litt., 1988)

4. Unicardiopsidae. 185–190 Ma, C
Taxon: *Unicardiopsis incertum* (Phillips)
Age: Latest Toarcian

Unit: Blea Wyck Sands. Yorkshire, England
References: Dean (1954); Hemingway (1974); Knox (1984)
Note: Unicardiopsidae is distinguished from the lucinacean Mactromyidae only on the basis of hinge morphology, which is rarely preserved in small Jurassic bivalves (Chavan 1962; A. Hallam, in litt., 1988), so that earliest occurrences are uncertain.

5. Tellinidea. 125–130 Ma, B and C
Taxon: *Linearia subhercynica* (Maas)
Age: Early Hauterivian
Unit: Hilskonglomerat, "Sonderfazies"
References: Wollemann (1900: Michael and Pape (1971); Hillmer (1971); Kemper (1973); Michael (1974)
Note: Maas' (1895) material from Gersdorfer Berg is probably Barremian (Kollmann 1982)

Taxon: *Linearia subconcentrica* (d'Orbigny)
Age: Early Hauterivian
Unit: Calcaire à Spatangues, many localities in Paris Basin, and Marne d'Hauterive, Morteau (Doubs), France
References: Rittener (1902); Gillet (1921a, 1924–5); Burri (1956); Mégnien and Mégnien (1980); Rat et al. (1987)
Note: This species recorded as high as Lower Aptian Couche Rouge of Wassy (Gilbert 1921b).

Taxon: *"Tellina" carteroni* d'Orbigny
Age: Early Hauterivian
Unit: Calcaire à Spatangues: many localities in Paris Basin
References: Gillet (1921, 1924–5); Mégnien and Mégnien (1980); Rat et al. (1987)
Note: Most records of *Tellina carteroni* d'Orbigny are probably icanotiids (Casey 1961); however, the lateral dentition reported for the type material (Stoliczka 1870:123; Gillet 1924:136) excludes at least some specimens from Icanotiidae.

Other candidates: *Tellina sobralensis* Sharpe (1849) from the Early Cretaceous of Portugal is of uncertain affinities but is probably not tellinacean.
Tellina? (Arcopagia) n.sp. Harbort (1905) and *Tellina (Lagvignon) ovalis* Harbort (1905) are insufficiently known for taxonomic assignment.

6. Icanotiidae. 125–130 Ma, B and C
Taxon: *Scittilia* cf. *S. nasuta* Casey (1961) (= *Tellina carteroni* d'Orbigny of de Loriol (1861; Pictet and Campiche 1864–67)
Age: Early Hauterivian
Unit: Marne d'Hauterive, Sainte Croix (Vaud), Varappe (Mont-Saleve), NW Switzerland
References: Rittener (1902); Gillet (1924); Baschang (1921); Casey (1961); Haefeli et al. (1965)

Notes: (1) *Scittila japonica* Hayami (1965) evidently occurs in the upper part of the Hanoura Formation and therefore is probably Barremian in age (Hayami 1966).
(2) Slightly younger is *Scittila minuta* Lan (in Ma et al. 1976), from the Mangang Formation, late Early Cretaceous (Stratigraphic Workers of Yunnan 1978) of Yunnan Province, China.

Taxon: *Tancretella* sp. of Day (1967)
Age: Hauterivian
Unit: Battle Camp Formation or Gilbert River Formation, Laura Basin, N Queensland
References: Day (1967); de Keyser and Lucas (1968); Day et al. (1983)

7. Donacidae. 115–120 Ma, A
Taxon: *Protodonax minutissimus* (Whitfield)
Age: Early Aptian
Unit: Abeih Formation (Couches à Gastéropodes of earlier authors), "Olive Locality," Abeih, Lebanon
References: Vokes (1946); Saint-Marc (1970, 1981); Dubertret (1975); Walley (1983)
Note: This species recorded as high as Late Aptian Seend Ironsands of Seend, England (Casey 1961; Casey and Bristow 1964).

8. Solecurtidae. 80–85 Ma, B
Taxon: *Protagelus albertinus* (d'Orbigny)
Age: Early Campanian
Unit: Silakkudi Formation, Ariyalur Group, Comarapolliam, Tiruchirapalli District, Tamil Nadu State, India
References: Blanford (1862); Stolickza (1870–71); Rasheed and Ravindran (1978); Sundaram and Rao (1979); Tapaswi (1979); Chiplonkar (1985)
Note: *Tagelus cocholguei* Stinnesbeck (1986), Maastrichtian Quiriquina Formation of central Chile, is probably congeneric. Other possible Cretaceous solecurtids, all Southern Hemisphere in distribution, include *Solecurtus gratus* Wilckens (1905), Upper Senonian of southern Chile (see also Pérez and Reyes 1978); and *Solecurtus (Azor) woodsi* Rennie (1930), Campanian of South Africa.

9. Psammobiidae. 60–65 Ma, B and C
Taxon: *Gari (Gobraeus) debilis* (Deshayes)
Age: Danian
Unit: Unnamed, Luzankova, N Ukraine; Sochaczew Beds, Bochotnica, central Poland: *Echinanthus carinatus* Zone, eastern flank of Caspian Sea
References: Pozaryska (1965); Makarenko (1970); Panteleev (1974); Krach (1981)
Note: This species recorded as high as Thanetian (Farchad 1936).

Taxon: *Adansonella duponti* (Cossmann)
Age: Danian (NP2-NP3; Aubry 1986)

Unit: Calcaire grossier de Mons, Mons, Belgium
Refererences: Pozaryska (1965); Glibert and van de Poel (1973)
Note: This species also recorded from Hückelhovener Schichten, Well Sophia Jacoba 6, NW Germany (Anderson 1974, 1986; Aubry 1986 dates this as NP5). Slightly older, but much less well preserved: *Gobraeus* (?) *dejaeri* (Vincent) (Glibert and van de Poel 1973), Danian (NP2-NP3; Aubry 1986) Pondingue de Ciply, Ciply, Belgium.
Note: *Rhectomyax*, oldest psammobiid in the *Treatise*, is actually the oldest known member of the leptonacean family Kelliidae (Saul 1988).

10. Scrobiculariidae. 55–60 Ma, A and B
Taxon: *Scrobiculabra condamini* (Morris)
Age: Lat Thanetian (NP9, Aubry 1986)
Unit: Woolwich Shell Beds, Woolwich and Reading Formation, Thames Valley, England
References: Morris (1854); Prestwich (1854); Hester (1965); Tracey (1986); Bateman and Moffat (1986)

11. Semelidae. 55–60 Ma, B
Taxon: *Semele langdoniana* Aldrich
Age: Late Thanetian (NP9, Siesser et al. 1985)
Unit: Bells Landing Member, Tuscahoma Formation, Bells Landing, Alabama
References: Palmer and Brann (1965); Mancini and Oliver (1981); Gibson et al., (1982); Siesser (1983)

A2.5. References for Tables 2.2 to 2.5 and Figures

Ager, D. V. and W. E. Smith. 1965. The coast of South Devon and Dorset between Branscombe and Burton Bradstock. *Geol. Assoc. Guides*, 23.

Anderson, H.-J. 1974. Die Fauna der paläocänen Hückelhovener Schichten aus dem Schacht Sophia Jacoba 6 (Erkelenzer Horst, Niederrheinische Bucht, Teil 2, Bivalvia: Heterodonta und Anomalodesmata. *Geol. Palaeont.* 8:159–92.

Anderson, H.-J. 1986. The northwest German Tertiary Basin—Paleocene. In *Nordwestdeutschland im Tertiär*, ed. H. Tobien, 650–9. Berlin: Gebrüder Borntraeger.

Angermeier, H.-A., A. Pöschl, and H.-J. Schneider. 1963. Die Gliederung der Raibler Schichten und die Ausbildung ihrer Liegendgrenze in der "Tirolischen Einheit" des östlichen Chiemgauer Alpen. *Mitt. Bayer. Staatssamml. Paläont. Hist. Geol.* 3:83–105.

Asgaard, U., 1968. Brachiopod palaeoecology in Middle Danian limestones at Fakse, Denmark. *Lethaia* 1:103–21.

Assereto, R., and P. Casati. 1965. Revisione della stratigrafia Permo-Triassico delle Val Camonica (Lombardia). *Riv. Ital. Paleont. Stratigr.* 71:999–1097.

Atabekjan, A. A. 1979. Correlation of the Campanian Stage in Kopetdag and Western Europe. In *Aspekte der Kreide Europas*, ed. J. Wiedmann, 511–26. Stuttgart: E. Schweizerbart'sche Verlagsbuchhandlung (Nägele u. Obermiller).

Aubry, M.-P. 1986. Paleogene calcareous nannoplankton biostratigraphy of northwestern Europe. *Palaeogeogr. Palaeoclimatol. Palaeoecol.* 55:267–334.

Balogh, K. 1981. Correlation of the Hungarian Triassic. *Acta Geol. Acad. Sci. Hungaricae* 24:3–48.

Baschang, J. H. 1921. Beiträge zur Kenntnis der Bryozoen-Horizonte in der untern-Kreide des Westschweizerischen und französischen Jura. *Mem. Soc. Paleont. Suisse* 45:1–78.

Bather, F. A. 1909. Triassic echinoderms of Bakony. *Resultate der Wissenschaftlichen Erforschung des Balatonsees* 1(1), 1:1–288.

Bateman, R. M., and A. J. Moffat. 1986. Petrography of the Woodwich & Reading Formation (Late Palaeocene) of the Chiltern Hills, southern England. *Tertiary Res.* 8:75–103.

Bernstein, M. R. 1987. Paleontologic and biostratigraphic survey of the Vincentown Formation (Paleocene) along the valley of Big Timber Creek in southern New Jersey. *Northeastern Geol.* 9:133–44.

Berthelsen, O. 1962. Cheilostome Bryozoa in the Danian deposits of east Denmark. *Danmarks Geol. Unders. II. Raekke,* vol. 83.

Beznosov, N. V., and N. A. Efimova. 1979. Stratigraphy of the Triassic in the northwestern Caucasus. *Sovetskaia Geologiia* 1979 10:52–63. (In Russian).

Bittner, A. 1895. Lamellibranchiaten der alpinen Trias. I. Revision der Lamellibranchiaten von St. Cassien. *Abh. Kaiserlich-Koniglichen Geol. Riechsanst. Wien* 18(1):236.

Blanford, H. F. 1862. On the Cretaceous and other rocks of the South Arcot and Trichinopoly Districts, Madras. *Mem. Indian Geol. Surv.* 4(1):1–222.

Brydone, R. M. 1929–36. Further notes on new or imperfectly known Chalk Polyzoa. London: Dulau & Co.

Buckman, S. S. 1915. A palaeontological classification of the Jurassic rocks of the Whitby District, with a zonal table of Lias ammonites. In *The geology of the country between Whitby and Scarborough*, 2d. ed., Mem. Geol. Surv. Great Britain, ed. C. Fox-Strangeways and G. Borrow, 59–108.

Burri, F. 1956. Die Rhynchonelliden der unteren Kreide (Valanginien-Barremien) im westschweizerischen Juragebirge. *Eclogae Geol. Helv.* 49:599–701.

Carr, T. R., and R. K. Paull. 1983. Early Triassic stratigraphy and paleogeography of the Cordilleran miogeosyncline. In *Mesozoic paleogeography of the west-central United States*, ed. M. W. Reynolods and E. D. Dolly, 39–56. Denver, Colo.: S. E. P. M. Rocky Mountain Symposium.

Casati, P., and M. Gnaccolini. 1967. Geologia delle Alpi Orobie occidentale. *Riv. Ital. Paleont. Stratigr.* 73:25–144.

Casey, R. 1961. The stratigraphical palaeontology of the Lower Greensand. *Palaeontology* 3:487–621.

Casey, R., and C. R. Bristow. 1964. Notes on some ferruginous strata in Buckinghamshire and Wiltshire. *Geol. Mag.* 101:116–28.

Chavan, A. 1962. Essai critique de classification des Ungulinidae. *Inst. R. Sci. Nat. Belgique Bull.*, vol. 38, no. 23.

Cheetham, A. H. 1954. A new Early Cretaceous cheilostome bryozoan from Texas. *J. Paleontol.* 28:177–84.

Cheetham, A. H. 1971. Functional morphology and biofacies distribution of cheilostome Bryozoa in the Danian Stage (Paleocene) of southern Scandinavia. *Smithsonian Contrib. Paleobiol.*, vol. 6.

Cheetham, A. H. 1975. Taxonomic significance of autozooid size and shape in some early multiserial cheilostomes from the Gulf Coast of the USA. *Doc. Lab. Géol. Faci. Sci. Lyon Hors Série* 3:547–64.

Cheetham, A. H. 1977. Notes on Vincentown Bryozoa. *International Bryozoology Association Field Trip Guidebook*. Washington, D.C.

Chiplonkar, G. W. 1985. Attempts at litho- and biostratigraphic subdivision of the Upper Cretaceous rocks of South India—a review. *Q. J. Geol. Min. Met. Soc. India* 57:1–32.

Committee on Chinese Lamellibranch Fossils, Nanjing Institute of Geology and Palaeontology. 1976. *Fossil Lamellibranchiata of China*. Beijing: Science Press, Academia Sinica.

Conti, S. 1954. Stratigrafia e paleontologia della Val Solda (Lago di Lagano). *Mem. Descr. Carta Geol. d'Italia*, vol. 30.

Cook, P. L., and P. J. Chimonides. 1983. A short history of the lunulite Bryozoa. *Bull. Marine Sci.* 33:566–81.

Cook, P. L., and P. J. Chimonides. 1986. Recent and fossil Lunulitidae (Bryozoa, Cheilostomata). 6. *Lunulites sensu lato* and the genus *Lunularia* from Australasia. *J. Nat. Hist.* 20:681–705.

Cope, J. W. C., T. A. Getty, M. K. Howarth, N. Morton, and H. S. Torrens. 1980. A correlation of Jurassic rocks in the British Isles, part 1. Introduction and Lower Jurassic. *Geol. Soc. London Spec. Rept. 14*.

Custer, W. 1928. Étude géologique du Pied du Jura vaudois. *Mat. Carte Géol. Suisse* N.S. 59:1–72.

Davies, D. K. 1969. Shelf sedimentation: an example from the Jurassic of Britain. *J. Sed. Pet.* 39:1344–70.

Day, R. W. 1967. Marine lower Cretaceous fossils from the Minmi Member, Blythesdale Formation, Roma-Wallumbilla area. *Geol. Surv. Queensland Publ.* 335.

Day, R. W. W. G. Whitaker, C. G. Murray, I. H. Wilson, and K. G. Grimes. 1983. Queensland geology. *Geol. Surv. Queensland Publ.* 383.

Dean, W. T. 1954. Notes on part of the Upper Lias succession at Blea Wyke. Yorkshire. *Proc. Yorkshire Geol. Soc.* 29:161–79.

de Keyser, F., and K. G. Lucas. 1968. Geology of the Hodgkinson and Laura Basins, north Queensland. *Bur. Min. Res. Australia Bull.* 84.

de Loriol, P. 1861. *Description des animaux invertébrés fossiles contenus dans l'étage Néocomien moyen du Mont-Salève*. Geneva.

Desio, A., ed. 1973. *Geologia dell'Italia*. Torino: Unione Tipografico-Editrice Torinese.

Donovan, D. T., A. Horton, and H. C. Ivimey-Cook. 1979. The transgression of the Lower Lias over the northern flank of the London Platform. *J. Geol. Soc. London* 136:165–73.

Donovan, D. T., and G. A. Kellaway. 1984. *Geology of the Bristol district: the Lower Jurassic rocks*. Memoirs of the Geological Survey of Great Britain. London: H.M.S.O. *Mem. British Geological Survey*.

Dubertret, L. 1975. Introduction à la carte géologique a 1/50,000e du Liban. *Notes et Mémoires sur le Moyen-Orient* 13:343–402.

Dupuy de Lome, E., and J. de la Revilla. 1956. Dos especies fosiles nuevas en las Provincias de Valencia y Alicante. *Notas y Comunicaciones del Instituto Geologico y Minero de Espana* 43:5–9.

Dzik, J. 1975. The origin and early phylogeny of the cheilostomatous Bryozoa. *Acta Palaeont. Polonica* 20:395–423.

Efimova, N. A. 1975. Foraminifera and deposition of the Kodzinski Series of the north-western Caucasus (R. Tkatch). *Trudy VINIGRI* 171:47–61.

Ernst, G. 1968. Die Oberkreide-Aufschlüsse im Raume Braunschweig-Hannover und ihre stratigraphische Gliederung mit Echinodermen und Belemniten. Teil Die jüngere Oberkreide (Stanton-Maastricht). *Beihefte zu den Berichten der Naturhistorischen Gesellschaft zu Hannover* 5:235–84.

Ernst, G. 1975. Stratigraphie, Fauna und Sedimentologie der Oberkreide von Misburg und Höver bei Hannover. *Mitt. Geol.-Paläont. Inst. Univ. Hamburg* 44:69–97.

Ernst, G., and F. Schmid, 1979. Multistratigraphische Untersuchungen in der Oberkreide des Raumes Braunschweig-Hannover. In Aspekte der Kreide Europas, ed. J. Weidmann, 11–46. Stuttgart: E. Schweizerbart'sche Verlagsbuchhandlung (Nägele u. Obermiller).

Ernst, G., C. J. Wood, and H. Hilbrecht. 1984. The Cenomanian-Turonian boundary problem in NW-Germany with comments on the north-south correlation to the Regensburg area. *Bull. Soc. Geol. Denmark* 33:103–13.

Farchad, H. 1936.Etude du Thanétien (Landennien marin) du bassin de Paris. *Mém. Soc. Géol. France* N.S. 30:1–101.

Franeschetti, B. 1967. Stui geologici sulla regione ad ovest di Ocada (Provincia di Alessandri). *Mem. Soc. Geol. Italiana* 6:379–420.

Fulfaro, V. J., P. M. B. Landim. 1976. Stratigraphic sequences of the intracratonic Paraná Basin. *Newsl. Stratigr.* 4:150–68.

Fürsich, F. T., and J. Wendt. 1977. Biostratinomy and paleoecology of the Cassian Formation (Triassic) of the Southern Alps. *Palaeogeogr. Palaeoclimatol. Palaeoecol.* 22:257–323.

Gaemers, P. A. M. 1978. Biostratigraphy, palaeoecology and palaeogeography of the mainly marine Ager Formation (Upper Paleocene-Lower Eocene) in the Tremp Basin, central-south Pyrenees, Spain. *Leidse Geol. Meded.* 51:151–231.

Gaetani, M., E. Fois, F. Jadoul, and A. Nicora. 1981. Nature and evolution of Middle Triassic carbonate buildups in the Dolomites (Italy). *Mar. Geol.* 44:25–57.

Galacz, A., F. Horvath, and A. Vörös. 1985. Sedimentary and structural evolution of the Bakony Mountains (Transdanubian Central Range, Hungary): Paleogeographic implications. *Acta Geol. Hungarica* 28:85–100.

Gama, E., Jr., A. N. Bandeira, Jr., and A. B. França. 1982. Distrubuçao espacial e temporal das unidades litoestratigráficas Paleozóicas na parte central da Bacia do Paraná. *Rev. Bras. Geocienc.* 12:578–89.

Garrison, R. E., W. J. Kennedy, and T. J. Palmer, 1987. Early lithification and hardgrounds in Upper Albian and Cenomanian calcarenites, southwest England. *Proc. Geol. Assoc.* 98:103–40.

Gibson, T. G., E. A. Mancini, and L. M. Bybell. 1982. Paleocene of Middle Eocene stratigraphy of Alabama. *Trans. Gulf Coast Assoc. Geol. Soc.* 32:449–58.

Gillet, S. 1921a. Etude de la faune de lamellibranches du Calcaire à Spatangues (Hautérivien supérieur). *Bull. Soc. Sci. Hist. Nat. Yonne, Auxerre* 75:45–108.

Gillet, S. 1921b. Etude du Barrémien supérieur de Wassy (Haute-Marne). *Bull. Soc. Géol. France 4e Série* 21:3–47.

Gillet, S. 1924–5. Etudes sur les Lamellibranches néocomiens. *Soc. Géol. France Mém. N.S.* 3:1–224, 225–339.

Glibert, M., and L. van de Poel. 1973. Les Bivalvia du Danien et du Montian de la Belgique. *Inst. R. Sci. Nat. Belgique Mém., 175.*

Graecen, K. F. 1941. The stratigraphy, fauna, and correlation of the Vincentown Formation. *N.J. Dept. Conserv. Devel. Geol. Ser. Bull.*, vol. 52.

Haefeli, C., W. Maync, H. J. Oertli, and R. F. Rutsch. 1965. Die Typus-Profile des Valanginien und Hauterivien. *Bull. Ver. Schweiz. Petrol.-Geol. und -Ing.* 31:41–75.

Hagdorn, H. 1983. *Holocrinus doreckae* n.sp. aus dem Oberen Muschelkalk und die Entwicklung von Sollbruchstellen im Stiel der Isocrinida. *N. Jb. Geol. Paläont. Mh.* 1983:345–68.

Hagdorn, H. 1985. Immigration of crinoids into the German Muschelkalk Basin. In *Sedimentary and evolutionary cycles*, ed. U. Bayer and A. Seilacher, 23–54. Berlin: Springer-Verlag.

Hagdorn, H. 1986. *Isocrinus? dubius* (Goldfuss, 1831) aus dem Unteren Muschelkalk (Trias, Anis). *Z. Geol. Wiss. Berlin* 14:705–27.

Hakansson, E., R. Bromley, and K. Perch-Nielsen. 1974. Maastrichtian chalk of north-west Europe—a pelagic shelf sediment. *Spec. Publ. Internat. Assoc. Seidmentol.* 1:211–33.

Hakansson, E., and E. Thomsen. 1979. Distribution and types of bryozoan communities at the boundary in Denmark. In *Cretaceous/tertiary boundary events symposium.* I. *The Maastrichtian and Danian of Denmark*, ed T. Berkelund and R. G. Bromley, 78–91. Copenhagen: University of Copenhagen.

Hallam, A. 1960. A sedimentary and faunal study of the Blue Lias of Dorset and Glamorgan. *Phil.Trans. R. Soc. London* B243:1–44.

Hallam, A. 1987. Radiations and extinctions in relation to environmental change in the marine Lower Jurassic of northwest Europe. *Paleobiology* 13:152–68.

Hancock, J. M., W. J. Kennedy, and H. Klaumann. 1972. Ammonites from the transgressive Cretaceous on the Rhenish Massif, Germany. *Palaeontology* 15:445–9.

Harbort, E. 1905. Die Fauna der Schaumburg Lippe'schen Kreidemulde. *Abhandlungen Königlich Preussischen Geologischen Landesanstalt und Bergakademie* N.F. 45.

Hayami, I. 1965. Lower Cretaceous marine pelecypods of Japan, part II. *Mem. Fac. Sci. Kyuhsu Univ. Ser. D.* 17:73–150.

Hayami, I. 1966. Lower Cretaceous marine pelecypods of Japan, part III. *Mem. Fac. Sci. Kyushu Univ. Ser. D* 17:151–249.

Hemingway, J. E. 1974. Jurassic. In *The geology and mineral resources of Yorkshire*, ed. D. H. Rayner and J. E. Hemingway, 161–223. Leeds: Yorkshire Geological Society.

Hess, H. 1975. Die fossilen Echinodermen des Schweizer Juras. *Veröffentlichungen aus dem Naturhistorischen Museum Basel*, vol. 8.

Hester, S. W. 1965. Stratigraphy and palaeogeography of the Woolwich and Reading Beds. *Bull. Geol. Surv. Gt. Brit.* 23:117–23.

Hillmer, G. 1971. Bryozoen (Cyclostomata) aus dem Unter-Hauterivium von Nordwest-deutschland. *Mitt. Geol-Paläont. Inst. Univ. Hamburg* 40:5–106.

Ichikawa, K. 1950. *Sakawanella*, a new genus, and some other pelecypods from the Upper Traissic Kochigatani Group in the Sakawa Basin, Shikoku, Japan. *J. Fac. Sci. Univ. Tokyo Sect.* 2 7:245–56.

Jäger, M. 1985. Die Crinoiden aus dem Pliensbachium (mittlerer Lias) von Rottorf am Klei unde Empelde (Südniedersachsen). *Ber. naturhist. Ges. Hannover* 128:71–151.

Jarvis, I., and A. Gale. 1984. The Late Cretaceous transgression in the SW Anglo-Paris Basin: Stratigraphy of the Craie de Villedieu Formation. *Cret. Res.* 5:195–224.

Jarvis, I., A. Gale, and C. Clayton. 1982. Litho- and biostratigraphical observations on the type sections of the Craie de Villedieu Formation (Upper Cretaceous, western France). *Newsl. Stratigr.* 11:64–82.

Jerz, H. 1966. Untersuchungen über Stoffbestand, Bildungsbedingungen und Paläogeographie der Raibler Schichten zwischen Lech und Inn (Nördliche Kalkalpen). *Geologica Bavarica* 56:3–102.

Juignet, P. 1968. Facies littoraux du Cénomanien des environs du Mans (Sarthe). *Bull. Bur. Rech. Géol. Min. section* IV 1968(4):5–20.

Juignet, P. 1980. Cénomanian. *Bur. Rech. Géol. Min. Mém.* 109:130–8.

Juignet, P., W. J. Kennedy, and A. Lebert. 1978. Le Cénomanien du Maine: Formations sédimentaires et faunes d'ammonites du stratotype. *Géol. Médit.* 5:87–100.

Kahrs, E. 1927. Zur Paläogeographie der Oberkreide in Rheinland-Westfalen. *N. Jb. Geol. Paläont. Mineral. Geol. Paläont. Beilage-Bände* 58B:627–87.

Keen, A. M. 1969. Superfamily Tellinacea de Blainville, 1814. In *Treatise on invertebrate paleontology*, part N, *Mollusca* 6, vol. 2, ed. R. C. Moore, N613–43. Boulder, Colo.: Geological Society of America; Lawrence, Kan.: University of Kansas.

Kemper, E. 1973. The Valanginian and Hauterivian stages in northwest Germany. In *The boreal lower Cretaceous*, Geol. J. Spec. Issue 5, ed. R. Casey and P. F. Rawson, 327–43.

Kennedy, W. J. 1984. Systematic palaeontology and stratigraphic distribution of the ammonite faunas of the French Coniacian. *Spec. Pap. Palaeontol.* 31.

Kennedy, W. J., and R. E. Garrison. 1975. Morphology and genesis of nodular chalks and hardgrounds in the Upper Cretaceous of southern England. *Sedimontology* 22:311–86.

Kennedy, W. J., and J. M. Hancock. 1976. The mid-Cretaceous of the United Kingdom. *Ann. Mus. Hist. Nat. Nice* 4:V.1–V.72.

Klikushin, V. G. 1973. Upper Cretaceous isocrinids (Crinoidea) from Maly Balkham Ridge. *Novye isseledovania v geologii Leningrad* 5:41–50. (In Russian.)

———. 1975. Mechanics of the column in the Bourgueticrinidae. *Paleontol. J.* 9(1):121–4.

———. 1977. Sea lilies of the genus *Isselicrinus*. *Paleontol. J.* 11(1):82–89.

———. 1979. Sea lilies of the genera *Balanocrinus* and *Laevigatocrinus*. *Paleontolo. J.* 13(3):346–54.

———. 1982a. Taxonomic survey of fossil isocrinids with a list of the species found in the U.S.S.R. *Géobios* 15:299–325.

———. 1982b. Cretaceous and Paleogene Bourgueticrinina (Echinodermata, Crinoidea) of the U.S.S.R. *Géobios* 15:811–43.

———. 1983a. The Triassic crinoids of northern Afghanistan. *Paleontol. J.* 17(2):82–90.

———. 1983b. Distribution of crinoidal remains in the Upper Cretaceous of the U.S.S.R. *Cret. Res.* 4:101–6.

———. 1985a. Crinoids of the genus *Austinocrinus* Loriol in the U.S.S.R. *Palaeontographica* 190:159–92.

———. 1985b. New Late Cretaceous and Paleogene sea-lillies from the over-Caspian region. *Paleont. Sb. Lvov* 22:44–50.

———. 1985c. Turonian, Coniacian and Santonian deposits of the Belbeck River, Crimea. *Biul. Mosk. Obschch. Ispyt. Prirody Otd. Geol.* 60(2):69–82. (In Russian.)

———. 1987a. Distribution of crinoidal remains in Triassic of the U.S.S.R. *N. Jb. Geol. Paläont. Abh.* 173:321–38.

———. 1987b. Thiolliericrinid crinoids from the Lower Cretaceous of Crimea. *Géobios* 20:625–65.

Knox, R. W. O'B. 1984. Lithostratigraphy and depositional history of the late Toarcian sequence of Ravenscar, Yorkshire. *Proc. Yorkshire Geol. Soc.* 45:99–108.

Kobayashi, T., and K. Ichikawa. 1950. On the Upper Triassic Kochigatani Series in the Sakawa Basin, in the province of Tosa (Kochi Prefecture),

Shikoku Island, Japan, and its pelecypod-faunas. *J. Fac. Sci. Univ. Tokyo Sect.* 2 7:179–206.

Kobayashi, T., and M. Tamura. 1984. The Triassic Bivalvia of Malaysia, Thailand and adjacent areas. *Geol. Palaeont. Southeast Asia* 25:201–27.

Kollmann, H. A. 1982. Gastropoden-Faunen aus der höheren Unterkreide Nordwestdeutschlands. *Geol. Jb.* A65:517–51.

Krach, W. 1981. Paleocene fauna and stratigraphy of the Middle Vistula River, Poland. *Studia Geol. Polonica,* vol. 71.

Kristan-Tollmann, E., and A. Tollmann. 1983. Überregionale Züge der Tethys in Schichtfolge und Fauna am Beispiel der Trias zwischen Europa und Fernost, speziell China. *Schriftenreihe erdwiss. Komm. österr. Akad. Wiss.* 5:177–230.

Krumbeck, L. 1924. Die Brachiopoden, Lamellibranchiaten und Gastropoden der Trias von Timor II. Palaontologischer Teil. *Palaeontologie von Timor* 13:1–275.

Kummel, B. 1957. Paleoecology of Lower Triassic formations of southeastern Idaho and adjacent areas. *Geol. Soc. Am. Mem.* 67(2):437–68.

Lang, W. D. 1921. *Catalogue of the fossil Bryozoa (Polyzoa) in the Department of Geology, British Museum (Natural History). The Cretaceous Bryozoa (Polyzoa). Volume III. The cribrimorphs.—Part I.* London: British Museum (Natural History).

Larwood, G. P. 1961. The Lower Cretaceous deposits of Norfolk. *Trans. Norfolk and Norwich Naturalists Soc.* 19:280–92.

Larwood, G. P. 1985. Form and evolution of Cretaceous myagromorph Bryozoa. In *Bryozoa: Ordovician to Recent,* ed. C. Nielsen and G. P. Larwood, 169–74. Fredensborg, Denmark: Olsen & Olsen.

Larwood, G. P., A. W. Medd, D. E. Owen, and R. Tavener-Smith. 1967. Bryozoa. In *The fossil record,* ed. W. B. Harland et al., 379–95. London: Geological Society.

LeClerc, J., and J. Azema. 1976. Le Crétacé dans la région d'Agost (Province d'Alicante—Espagne) et ses accidents sédimentaires. *Cuad. Geol. Univ. Granada* 7:35–51.

Lorenz, C. R. 1968 (1969). Contribution à l'étude stratigraphique de l'Oligocène et du Miocène inférieur des confins Liguro-Piemontais (Italie). *Atti Inst. Geol. Univ. Genova* 6:255–888.

Lorenz, C. R. (1986). Evolution stratigraphique et structurale des Alpes Ligures depuis l'Eocène supérieur. *Mem. Soc. Geol. Italiana* 28:211–28.

Ma Qihong, Chen Jinhua, Lan Xiu, Gu Zhiwei, and Chen Chuzhen. 1976. Mesozoic bivalves from Yunnan Province, southwest China. In *Mesozoic fossils of Yunnan Province,* Nanjing Institute of Geology and Palaeontology, Academia Sinica. Beijing: Science Press.

Maas, G. 1895. Die untere Kreide des subhercynen Quadersandsteingebirges. *Z. Deutsch. Geol. Gesell.* 47:227–302.

Makarenko, D. E. 1970. *Early Paleocene molluscs of the northern Ukraine.* Kiev: Naukova Dumka. (In Russian.)

Mancini, E. A., and G. E. Oliver. 1981. Planktic foraminifers from the Tuscahoma Sand (upper Paleocene) of southwest Alabama. *Micropaleontology* 27:204–25.

Marcinowski, R. 1974. The transgressive Cretaceous (Upper Albian through Turonian) deposits of the Polish Jura Chain. *Acta Geol. Polonica* 24:117–217.

Maryanska, T. 1968. On a new subspecies of *Multicrescis variabilis* d'Orb., a bryozoan from the Cenomianian of Korzkiew near Cracow. *Prace Muzeum Ziemi, Warszawa* 12:169–76.

Mégnien, C., and F. Mégnien, eds. 1980. Synthèse géologique du Bassin de Paris. III. Lexique des noms de formation. *Bur. Rech. Géol. Min. Mém.*, vol. 103.

Michael, E. 1974. Zur Palökologie und Faunenführung im westlichen Bereich des norddeutschen Unterkreide-Meeres. *Geol. Jb.* A19:1–68.

Michael, E., and H.-G. Pape. 1971. Eine bemerkenswerte Bio- und Lithofazies an der Basis des Unte-Hauterivium Nordwestdeutschlands. *Mitt. Geol. Inst. Tech. Univ. Hannover* 10:43–107.

Moore, R. C. 1967. Unique stalked crinoids from Upper Cretaceous of Mississippi. *Univ. Kan. Paleontol. Contrib. pap.* 17.

Moreau, P. 1976. Cadre stratigraphique et rythmes sédimentaires du Cénomanien nord-aquitain (Région de Rochefort). *Bull. Soc. Géol. France Sér.* 7 18:747–755.

Moreau, P. 1977. Les environements sédimentaires marins dans le Cénomanien du nord du Bassin de Paris. *Bull. Soc. Géol. France Sér.* 7 19:281–8.

Moreau, P., I. H. Francis, and W. J. Kennedy. 1983. Cenomanian ammonites from northern Aquitaine. *Cret. Res.* 4:317–39.

Morris, J. 1854. Descriptions of some new species of shells from the "Woolwich and Reading Series." *Q. J. Geol. Soc.* 10:157–62.

Mortimore, R. N. 1986. Stratigraphy of the Upper Cretaceous White Chalk of Sussex. *Proc. Geol. Assoc.* 97:97–139.

Mortimore, R. N., and B. Pomerol. 1987. Correlation of the Upper Cretaceous White Chalk (Turonian to Campanian) in the Anglo-Paris Basin. *Proc. Geol. Assoc.* 98:97–143.

Naidin, D. P. 1981. The Russian Platform and the Crimea. In *Aspects of Mid-Cretaceous regional geology*, ed. R. A. Reyment & P. Bengtson, 29–68. London: Academic Press.

Nalivkin, D. V. 1973. *Geology of the U.S.S.R.* Edinburgh: Oliver & Boyd.

Neale, J. W. 1974. Cretaceous. In *The geology and mineral resources of Yorkshire*, ed. D. H. Rayner and J. E. Hemingway, 225–43. Leeds: Yorkshire Geological Society.

Nestler, H. 1965. Die Rekonstruktion des Lebensraumes der Rügener Schreibkereide-Fauna (Unter-Maastricht) mit Hilfe der Paläoökologie und Paläobiologie. *Geologie Beiheft*, vol. 49.

Ogilvie, M. M. 1893. Contributions to the geology of the Wengen and St. Cassian strata in southern Tyrol. *Q. J. Geol. Soc. London* 49:1–78.

Oravacez-Scheffer, A. 1987. Triassic foraminifers of the Transdanubian Central Range. *Geol. Hungarica Ser. Paleont.* 50:75–331.

Owens, J. P., and N. F. Sohl. 1969. Shelf and deltaic paleoenvironments in the Cretaceous-Tertiary of the New Jersey Coastal Plain. In *Geology of selected areas in New Jersey and eastern Pennsylvania and guidebook of excursions*, ed. S. Subitzky, 214–78. New Brunswick, N.J.: Rutgers University Press.

Palmer, K. V. W., and D. C. Brann. 1965. Catalogue of the Paleocene and Eocene Mollusca of the southern and eastern United States. Part I. Pelecypoda, Amphineura, Pteropoda, Scaphopoda, and Cephalopoda. *Bull. Amer. Paleont.* 48:1–466.

Panteleev, G. S. 1974. *Stratigraphy and bivalve mollusks of Danian and Paleocene deposits of Zakaspia*. Moscow: Nauka.

Peake, N. B., and J. M. Hancock. 1961. The Upper Cretaceous of Norfolk. *Trans. Norfolk Norwich Nat. Soc.* 19:293–339.

Pérez, d'A., E., and R. Reyes B. 1978. Las Trigonias del Cretácico superior de Chile y su valor cronoestratigráfico. *Bol. Inst. Inv. Geol. Chile*, vol. 34.

Petri, S., and V. J. Fulfaro. 1983. *Geologia do Brasil*. Sao Paulo: Universidade de Sao Paolo.

Pictet, F.-J., and G. Campiche. 1864–7. Description des fossiles du terrain crétacé des environs de Sainte-Croix, 3me partie. Classe des Mollusques Acéphales. *Mat. pour la Paléont. Suisse*, vol. 4.

Pomerol, B., H. W. Bailey, C. Monciardini, and R. N. Mortimore. 1987. Lithostratigraphy and biostratigraphy of the Lewes and Seaford Chalks: a link across the Anglo-Paris Basin at the Turonian-Senonian boundary. *Cret. Res.* 8:289–304.

Powell, J. H. 1984. Lithostratigraphical nomenclature of the Lias Group in the Yorkshire Basin. *Proc. Yorkshire Geol. Soc.* 45:51–57.

Pozaryska, K. 1965. Foraminifera and biostratigraphy of the Danian and Montian in Poland. *Palaeont. Polonica*, vol. 14.

Prestwich, J. 1854. On the structure of the strata between the London Clay and the Chalk in the London and Hampshire Tertiary systems. II. The Woolwich and Reading Series. *Q. J. Geol. Soc. London* 10:75–157.

Rasheed, D. A., and C. N. Ravindran. 1978. Foraminiferal biostratigraphic studies of the Ariyalur Group of Tiruchirapalli Cretaceous rocks of Tamil Nadu State. In *Proceedings of the VII Indian Colloquium on micropalaeontology and stratigraphy*, ed. D. A. Rasheed, 321–36. Madras: Department of Geology, University of Madras.

Rasmussen, H. W. 1961. A monograph on the Cretaceous Crinoidea. *Biol. Skr. Kong. Danske Vidensk. Selsk.* vol. 12, no. 1.

Rasmussen, H. W. 1972. Lower Tertiary Crinoidea, Asteroidea and Ophiuroidea from northern Europe and Greenland. *Biol. Skr. Kong. Danske Vidensk. Selsk.* 19:1–83.

Rasmussen, H. W. 1975. Neue Crinoiden aus den Oberkreide bei Hannover. *Berichte Naturhist. Ges. Hannover* 119:279–83.

Rastall, R. H. 1930. The petrography of the Hunstanton Red Rock. *Geol. Mag.* 57:436—57.

Rat, P., B. David, F. Magniez-Janin, and O. Pernet. 1987. Le golfe du Crétacé inférieur sur le sud'est du Bassin parisien: milieu (échinides, foramnifères) et évolution de la transgression. *Mém. Géol. Univ. Dijon* 11:15–29.

Rawson, P. F., D. Curry, F. C. Dilley, J. M. Hancock, W. J. Kennedy, J. W. Neale, C. J. Wood, and B. C. Worssam. 1978. A correlation of Cretaceous rocks in the British Isles. *Geol. Soc. London Spec. Rept.* 9.

Reed, F. R. C. 1927. Palaeozoic and Mesozoic fossils from Yunnan. *Palaeontologia Indica N.S* 10(1):1–331.

Rennie, J. L. V. 1930. New Lamellibranchia and Gastropoda from the Upper Cretaceous of Pondoland. *Ann. South Afr. Mus.* 28:159–260.

Rittener, T. 1902. Etude géologique de la Côte-aux-Fées et des environs de Sainte-Croix et Baulmes. *Mat. Carte Géol. Suisse N.S.* 13:1–116.

Robinson, R. N. 1986. Stratigraphy of the Chalk Group of the North Downs, southeast England. *Proc. Geol. Assoc.* 97:141–70.

Ronchetti, C. R. 1959. Il Trias in Lombardia (Studi geologici e paleontologici). I. Lamellibranchi ladinici del gruppo delle Grigne. *Riv. Ital. Paleont. Stratigr.* 65:271–357.

Rostovstev, K. O., G. M. Aladatov, and N. R. Azarian. 1966. Triassic of the Caucasus and Precaucasus. *Akademiia Nauk SSSR Izvestiia Seriia Geologicheskia* 1966(3):88–100. (In Russian.)

Roux, M., and J.-C. Plaziat. 1978. Inventaire des Crinoides et interprétation paléobathymétrique de gisements du Paléogène pyrénéen franco-espagnol. *Bull. Soc. Géol. France Sér.* 7 20:299–308.

Roverto, G. 1939. Liguria geologica. *Mem. Soc. Geol. Italiana* vol. 2.

Runnegar, B., and N. D. Newell. 1971. Caspian-like relict molluscan fauna in the South American Permian. *Bull. Am. Mus. Nat. Hist.* 146:1–66.

Russell, E. E., D. M. Keady, E. A. Mancini, and C. C. Smith. 1983. Upper Cretaceous lithostratigraphy and biostratigraphy in northeast Mississippi, southwest Tennessee and northwest Alabama, shelf chalks and coastal clastics. *Soc. Econ. Paleontol. Mineral. Gulf Coast Section Guidebook.*

Saint-Marc, P. 1970. Contribution à la connaissance du Crétacé basal au Liban. *Rev. Micropaléont.* 12:224–33.

Saint-Marc, P. 1981. Lebanon. In *Aspects of Mid-Cretaceous regional geology,* ed. R. A. Reyment and P. Bengtson, 103–31. London: Academic Press.

Saul, L. R. 1988. *Rhectomyax* Stewart, 1930 (Bivalvia: Kelliidae): familial reassignment. *J. Paleontol.* 62:481.

Schwarz, H.-U. 1975. Sedimentary structures and facies analysis of shallow marine carbonates (Lower Muschelkalk, Middle Triassic, southwestern Germany). *Contrib. Sedimentol.,* vol. 3.

Scott, R. W., D. Fee, R. Magee, and N. Laali. 1978. Epeiric depositional models for the Lower Cretaceous Washita Group, north-central Texas. *Bur. Econ. Geol. Univ. Texas Rept. Inv.,* vol. 94.

Séronie-Vivien, M. 1972. *Contribution à l'étude de Sénonien en Aquitaine septentrionale, les stratotypes français* 2. Paris: C. N. R. S.

Sharpe, D. 1849. On the Secondary District of Portugal which lies on the north of the Tagus. *Q. J. Geol. Soc. London* 6:135–201.

Siesser, W. G. 1983. Paleogene calcareous nannoplankton biostratigraphy: Mississippi, Alabama and Tennessee. *Miss. Dept. Nat. Res. Bur. Geol. Bull.* 125.

Siesser, W. G., B. G. Fitzgerald, and D. J. Kronman. 1985. Correlation of Gulf Coast provincial Paleogene stages with European standard stages. *Geol. Soc. Am. Bull.* 96:827–31.

Simms, M. J. 1986a. The taxonomy and palaeobiology of British Lower Jurassic crinoids. Ph.D. diss., University of Birmingham, Birmingham, England.

Simms, M. J. 1986b. Contrasting lifestyles in Lower Jurassic crinoids: a comparison of benthic and pseudopelagic Isocrinida. *Palaeontology* 29:475–93.

Simms, M. J. 1988. Patterns of evolution among lower Jurassic crinoids. *Hist. Biol.* 1:17–44.

Simms, M. J. In press. Crinoids across the Triassic-Jurassic boundary. *Cahiers Scientifiques de l'Institut Catholique de Lyon.*

Stephenson, L. W., and W. H. Monroe. 1940. The Upper Cretaceous deposits. *Miss. Geol. Surv. Bull.* 40.

Stinnesbeck, W. 1986. Zu den faunistischen und palökologischen Verhältnissen in der Quiriquina Formation (Maastrichtium) Zentral-Chiles. *Palaeontographica* A194:99–237.

Stoliczka, F. 1870–1. Cretaceous fauna of southern India, vol. 3. The Pelecypoda, with a review of all known genera of this class, fossil and Recent. *Geol. Surv. India Palaeont. Indica Ser.* 6, vol. 3.

Stratigraphic Workers of Yunnan. 1978. Regional stratigraphy of the southwestern region: Yunnan part. Beijing: Publishing House of Geology.

Sundaram, R., and P. S. Rao. 1979. Lithostratigraphic classification of Uttatur and Trichinopoly Groups of Upper Cretaceous rocks of Tiruchirapalli District, Tamilnadu. *Geol. Surv. India Misc. Publ.* 45:111–19.

Tapaswi, P. M. 1979. New veneroid bivalve taxa from Trichinopoly Cretaceous. *Biovigyanam* 5:93–95.

Tarkowski, R., and S. Liszka. 1982. Foraminifera and age of the Korzkiew sands near Cracow. *Ann. Soc. Geol. Polon.* 52:231–238.

Taylor, P. D. 1986. *Charixa* Lang and *Spinicharixa* gen. nov., cheilostome bryozoans from the Lower Cretaceous. *Bull. Br. Mus. Nat. Hist. Geol.* 40:197–222.

Taylor, P. D. 1988. Major radiation of cheilostome bryozoans: Triggered by the evolution of a new larval type? *Hist. Biol.* 1:45–64.

Thomsen, E. 1976. Depositional environment and development of Danian bryozoan biomicrite mounds (Karlby Klint, Denmark). *Sedimentology* 23:485–509.

Tracey, S. 1986. Lower Tertiary strata exposed in a temporary excavation at Well Hall, Eltham, south east London. *Tertiary Res.* 7:107–23.

Tresise, G. R. 1960. Aspects of the lithology of the Wessex Upper Greensand. *Proc. Geol. Assoc.* 71:316–39.

Urlichs, M. 1977. Der Alterstellung der Pachycardienstuffe und der Unteren

Cassianer Schichten in der Dolomiten (Italien). *Mitt. Bayer. Staatssl. Paläont. Hist. Geol.* 17:15–25.

Vinogradov, A. P., ed. 1968. *Atlas of the lithological-palaeogeographical maps of the U.S.S.R.*, vol. III. *Triassic, Jurassic and Cretaceous.* Moscow: Ministry of Geology of U.S.S.R., Academy of Sciences of U.S.S.R.

Voigt, E. 1981. Répartition et utilisation stratigraphique des Bryozoaires du Crétacé moyen (Aptien-Coniacien). *Cret. Res.* 2:439–62.

Voigt, E. 1985. The Bryozoa of the Cretaceous-Tertiary boundary. In *Bryozoa: Ordovician to Recent*, ed. C. Nielsen and G. P. Larwood, 329–42. Fredensborg, Denmark: Olsen & Olsen.

Voigt, E. 1990. Mono– or polyphyletic evolution of cheilostomatous bryozoan divisions? *Proc. Internat. Bryozool. Assoc. Mtg.* (in press).

Voigt, E., and G. Hillmer. 1983. Oberkretazische Hippothoidae (Bryozoa Cheilostomata) aus dem Campanium von Schweden und Maastrichtium de Niederlande. *Mitt. geol.-paläont. Inst. Univ. Hamburg* 54:169–208.

Vokes, H. E. 1946. Contributions to the paleontology of the Lebanon Mountains, Republic of Lebanon, part 3, The pelecypod fauna of the "Olive Locality" (Aptian) at Abeih. *Bull. Am. Mus. Nat. Hist.* 87:139–216.

Walley, C. D. 1983. A revision of the Lower Cretaceous stratigraphy of Lebanon. *Geol. Rundschau* 72:377–88.

Weaver, P. P. E. 1982. Ostracoda from the British Lower Chalk and Plenus Marls. London: Palaeontographical Society.

Wendt, J., and F. T. Fürisch. 1979. Facies analysis and paleogeography of the Cassian Formation, Triassic, Southern Alps. *Riv. Ital. Paleont. Stratigr.* 85:1003–28.

Wiedmann, J., and H. L. Schneider. 1979. Cephalopoden und Alter der Cenoman-Transgression von Mülheim-Broich, SW-Westfalen. In *Aspekte der Kreide Europas*, ed. J. Wiedmann, 645–80. Stuttgart: E. Schweizerbart'sche Verlagsbucchandlung.

Wilckens, O. 1907 (privately issued 1905). Die Lamellibranchiaten, Gastropoden etc. der oberen Kreide Südpatagoniens. *Ber. Naturf. Ges. Feiburg im Breigau* 15:97–166.

Wobber, F. J. 1965. Sedimentology of the Lias (Lower Jurassic) of South Wales. *J. Sed. Pet.* 35:683–703.

Wobber, F. J. 1968a. Microsedimentary analysis of the Lias in South Wales. *Sed. Geol.* 2:13–49.

Wobber, F. J. 1968b. A faunal analysis of the Lias (Lower Jurassic) of South Wales (Great Britian). *Palaeogeogr. Palaeoclimatol. Palaeoecol.* 5:269–308.

Wöhrmann, S. v. 1889. Die Fauna der sogennanten Cardita- und Raibler-Schichten in den Nordtiroler und bayerischen Alpen. *Jb. K.K. Geol. Reichsanstalt Wien* 39:181–258.

Wollemann, A. 1900. Die Bivalven und Gastropoden des deutschen und holländischen Neocoms. *Abhandlungen der Königlich Preussischen geologischen Landesanstalt* N.F., vol. 31.

Wood, C. J., G. Ernst, and G. Rasemann. 1984. The Turonian-Coniacian stage boundary in Lower Saxony (Germany) and adjacent areas: the

Salzgitter-Salder Quarry as a proposed international standard. *Bull. Geol. Soc. Denmark* 33:225–38.

Wood, C. J., and E. G. Smith. 1978. Lithostratigraphical classification of the Chalk in north Yorkshire, Humberside and Lincolnshire. *Proc. Yorkshire Geol. Soc.* 42:263–87.

Worssam, B. C., and J. H. Taylor. 1969. *Geology of the country around Cambridge*. London: HMSO.

Wright, C. W., and W. J. Kennedy. 1984. *The Ammonoidea of the Lower Chalk*. I. London: Palaeontographical Society, 1–126.

Zardini, R. 1973. Fossili di Cortini. Atlante degli Echinodermi Cassiani (Trias medio-superiore) della region dolomitica attorno a Cortina d'Ampezzo. Cortina d'Ampezzo: Foto Ghedina.

3

Continental Area, Dispersion, Latitudinal Distribution, and Topographic Variety: A Test of Correlation with Terrestrial Plant Diversity

Bruce H. Tiffney and Karl J. Niklas

Current ecological and macroevolutionary theory postulates two broad categories of factors that influence the Phanerozoic diversification of terrestrial life: heterogeneity of the physical environment and biotic interaction. Both find support in neontologic and paleontologic data.

Modern ecological studies often stress the influence of total land area, geographical diversity, and latitudinal distribution of land on biotic diversity. The importance of these three features is borne out by empirical data, although the exact mechanisms by which each influences diversity remain under debate. Land area has been recognized as a potentially important control of diversity (Preston 1962; Johnson and Raven 1970) through "island biogeography" and general species/area effects (e.g., MacArthur and Wilson 1967). The dispersion of geographical features affects allopatric speciation at different scales. Topographical diversity within a single land mass provides a range of habitats ("patchiness") within one area, as well as creating topographical barriers to biotic exchange (e.g., MacArthur and Wilson 1967; Bridgewater 1988). On a greater scale, the separation of continents is the basis for identifying biotic provinces. Latitudinal gradients of diversity have been observed in both terrestrial and marine systems (Fischer 1960; McCoy and Conner 1980) and presumably relate to climate. Global climatic gradients combine with continental positions to create barriers and differing environments. Climatic fluctuations and varied topography have been theorized to act as a "species pump" in the Quaternary, accounting for species diversity in the extant tropics (e.g. Haffer 1969; Brown, Sheppard, and Turner 1974).

The effect of geographical features on diversity in the fossil record has not been widely explored. Valentine demonstrated the strong positive correlation of increasing continental fragmentation and increasing

Phanerozoic marine invertebrate diversity (Valentine 1973; Valentine and Moores 1970, 1972), while Flessa and Imbrie (1973) independently summarized patterns of diversification and developed a model of species-area relationships on a geological scale. Kurtén (1967) suggested that continental dispersion and terrestrial vertebrate diversity are positively correlated, and Rich et al. (1986) identified a strong correlation between pulses of seafloor spreading and the diversity histories of various planktonic lineages. In a more general paper on "Biological Diversification and its Causes," Cracraft (1985) implicated geographical complexity as a prime factor in driving diversification.

In contrast to this emphasis on the importance of physical factors, it can be argued that the present biotic diversity reflects a synergistic biological "snowballing," in which increasingly complex organisms both create a greater diversity of organism-organism interactions, and become increasingly able to modify the original physical environment. In the ecological time frame, this explanation conforms to observations of "coevolution" in the broadest sense, wherein organisms create resources or environments that enable other organisms to become so specialized as to become distinct species (e.g., the various concepts expressed in Futuyma and Slatkin 1983). In the paleontological perspective, observations suggest that the increasing structural and chemical complexity of plants directly affected the environment in which animals and other plants evolved (e.g., Tiffney 1985; Wing and Tiffney 1987). For example, the appearance of secondary wood in the Devonian resulted in a multilayer community, altered moisture and sunlight at the forest floor, and created a vertical zonation of resources that influenced subsequent animal evolution, of which the present high diversity in tropical rain forests is the end result.

The relative importance of biotic and physical influences on diversity has also been a cause of debate in "macroevolutionary" theory, resulting two models of evolution: the *Red Queen's Hypothesis* (Van Valen 1973) and the *Stationary Hypothesis* (Stenseth and Maynard Smith 1984). Put succinctly, "the Red Queen model predicts that evolution is largely autonomous; biotic interactions within an ecosystem represent a sufficient driving force for evolution to continue even in the absence of any changes in the abiotic environment. The Stationary model, by contrast, predicts that evolution will cease in the absence of changes in abiotic parameters." (Hoffman and Kitchell 1984:10). Tests of the two models (Hoffman and Kitchell 1984; Wei and Kennett 1986) based on diversity data and climatic change yield inconclusive results. It is worth noting here that, while the question of the relative importance of biotic and abiotic stimuli in evolution is a significant one, the

Red Queen Hypothesis was initially erected to explain patterns of extinction (Van Valen 1973) and, as originally constructed, cannot explain the increased rates of variation necessary for origination and diversification (Tiffney in prep.).

The sum of published observations suggests that overall Phanerozoic diversity is influenced by both the physical environment and by biotic interactions. Paleobotanical observation and resulting theory suggest some trends in land plant diversity may have been driven by biotic interactions (Niklas 1986), whereas others result from physical stimuli (Tiffney 1981; Knoll 1984, 1986). It seems clear that arguments that one or the other is all important are probably fruitless, although individual diversity changes in the fossil record may often appear to correlate with major patterns of physical or of biological change. Biological interactions must have played a role in terrestrial diversification. As evidence, within-community diversity (unaffected by changing total land area, topography, or land dispersion) has increased in terrestrial plant communities from the Devonian to the present (Niklas, Tiffney, and Knoll 1980; Knoll 1984). However, it is difficult to quantify (and thus to empirically test) the factors involved in these biotic interactions. Too often such interpretations have to rest on narrative scenarios of "how organisms interacted," as pieced together from their morphological fossil record, together with analogy to the ecologies of living counterparts.

By contrast, correlation of physical variables with diversity through geological time may be quantified. Recent geological advances make it possible to obtain a first approximation of the numbers of land masses, land area, upland ranges, and so on in the geological past. In this paper we have quantified changes in continental area, continental dispersion, topography, and latitudinal distribution for 12 stages from the Early Devonian to the Middle Miocene of the Northern Hemisphere. These data are statistically compared with changes in vascular land plant diversity for the same time intervals. We emphasize that this study is a first approximation. New diversity data (e.g. Lidgard and Crane 1988) will alter the numbers used here. The quantification of geographical features is approximate and is at the scale of geographical and temporal resolution offered by the available maps (see the section on Errors and Biases). Finally, our analyses do not account for the possible biological effects of climatic change, which may in some instances vary directly as a function of changing geography but in others may not. Like buying a personal computer, many might caution us to wait for a better and more robust product (data set) before essaying this study, but such a prospect will always lie before us. Hence, we choose to precede the angels in order to establish a model succeeding workers can improve upon.

Methods and Materials

DATA GATHERING

Data on plant diversity are taken from the compilation of over 18,000 plant species records assembled for studies on patterns of terrestrial plant diversification (Niklas 1978; Knoll, Niklas, and Tiffney 1979; Niklas, Tiffney, and Knoll 1980, 1983, 1985). Each record involves one or more reports of the occurrence of a species, providing a point or duration for the species' existence. Data were collected from the primary literature and from secondary compilations. Care was taken to account for synonyms and multiple organs of one plant ("organ genera"), since both can inflate recorded diversity over the actual number of organisms within a flora. Niklas, Tiffney and Knoll (1980) provide an extensive discussion of inherent caveats attending this data base.

The plant data are largely Northern Hemisphere in origin, which directly reflects the primary literature, which is predominantly from North America and Eurasia. This geographical bias requires that the diversity data be regressed on land masses *presently* in the Northern Hemisphere, including North America, Central America, Greenland, Eurasia including Indomalaysia, and the Japanese islands. For example, in measuring geographic diversity in the earliest Jurassic, our data reflect the conditions on the above mentioned land masses or their parts, and not necessarily on land masses resident in the earliest Jurassic Northern Hemisphere. We also exclude those portions of Africa and South America presently in the Northern Hemisphere, as well as Arabia and the subcontinent of India, as each is minimally represented in our data set.

Paleogeographical data were measured or counted from the reconstructions of Ziegler et al. (1979) and Scotese et al. (1979) for the Paleozoic and Parrish, Ziegler, and Scotese (1982) and Ziegler, Scotese, and Barrett (1983) for the Mesozoic and Cenozoic. The reconstruction for the Early Cretaceous came from Ziegler et al. (1987). In total, we analyzed 12 maps. All are Mollweide equal-area projection, permitting direct measurement of area from each map.

Measurements of the area of land masses and topographical features were compiled using the graphics program "SECTION" (Niklas and Boyd 1987). Numbers of islands and upland ranges were counted directly from the maps. Physical features spanning two hemispheres were counted as occurring in the Northern Hemisphere. The distinction between continents and islands was qualitative; units the size of Greenland and smaller are considered as islands; those larger, as continents. Similarly, the distinction between "small" versus "large" uplands was qualitative, as two visually distinct size classes existed. Upland, lowland, and total land area were calculated as percentages of the total hemispheric surface area.

Data on the latitudinal distribution of land in the Northern Hemisphere were assembled from Parrish (1985), who presented latitudinal distribution of land area for 11 time periods relevant to this study. For some, the latitudinal bands included land masses extraneous to those we wished to sample. In these cases, SECTION was used to determine the percent of land in each latitudinal band contributed by the Northern Hemisphere as here defined. In the period from the Early Devonian to the Early Triassic, some portions of the land presently in the Northern Hemisphere lay between the equator and 20° south latitude. Since latitudinal distribution of land was of interest here, rather than with its dispersion (and in order to have a consistent sample size for statistical purposes), we combined the area of these land masses in southern latitudes with the area of land found in the corresponding northern latitudinal band.

Data on geography and diversity are presented in tables 3.1, 3.2, and 3.3. Data for the latitudinal distribution of area through time are presented in table 3.4.

DATA ANALYSIS

We performed uni- and multivariate statistical analyses. Univariate, linear regressions consistently used 12 time periods as the independent

Table 3.1. Data on plant diversity given as total number of species, number of pteridophytic species (spore-bearing), number of seed plant species, percent of total flora comprised by pteridophytes, percent comprised by seed plants

Age	Total Species no.	Pteridophyte Species no.	Seed Plant Species no.	Pteridophyte (%)	Seed Plant (%)
Early Devonian	23	23	0	100	0
Early Carboniferous	80	72	8	90	10
Late Carboniferous	210	180	30	85.7	14.3
Late Permian	220	130	90	59	41
Early Triassic	185	132	53	71.4	28.6
Early Jurassic	223	43	180	19.3	80.7
Late Jurassic	228	47	181	20.6	79.4
Early Cretaceous	250	84	166	33.6	66.4
Middle Cretaceous	263	82	181	31.2	68.8
Late Cretaceous	310	78	232	25.2	74.8
Middle Eocene	412	80	332	19.6	80.4
Middle Miocene	528	37	491	7.1	92.9

Table 3.2. Geographical data on percent of shallow sea, lowland, and upland area from the Devonian to the Middle Miocene in the Northern Hemisphere as defined in this paper

Time	A	B	C	D	E
	% Upland	% Lowland	A + B	% Shallow Sea	A + B + D
Early Devonian	0.95	4.7	5.65	7.2	12.9
Early Carboniferous	0.92	6.3	7.22	3.9	11.1
Late Carboniferous	1.2	9.4	10.6	9.9	20.5
Late Permian	2.1	7.0	9.1	4.1	13.2
Early Triassic	1.6	11.3	12.9	6.5	19.4
Early Jurassic	3.1	11.9	15.0	5.4	20.4
Late Jurassic	2.9	9.4	12.3	2.6	14.9
Early Cretaceous	1.3	10.2	11.5	n.a.	n.a.
Middle Cretaceous	3.2	7.9	11.1	7.3	18.4
Late Cretaceous	2.3	10.3	12.6	5.4	18.0
Middle Eocene	3.2	10.6	13.8	4.1	17.9
Middle Miocene	4.7	10.8	15.5	5.1	20.6

Note: Columns A through E, percentage area figured from total area (sea plus land) of the hemisphere. Column A, percent of surface area of the hemisphere occupied by uplands. Column B, percent of surface area of hemisphere occupied by lowlands. Column C, sum of percent of hemisphere occupied by dry land (A + B). Column D, percent of hemisphere occupied by shallow seas. Column E, percent of hemisphere occupied by presumed continental crust (A + B + D).

variable; significance levels of r-values are indicated accordingly. Over 80 pairwise comparisons were made between geographical features and plant data; thus the potential for interdependence among variables is high. Accordingly, r-values that are significant in single pairwise comparisons become suspect, as legitimate probabilities cannot be assigned when many tests of significance are made. Rather, a correlation can only be considered as "significant" or "not significant" (Manley 1986). Only p-values of 0.01 or greater were considered as "likely to be significant" in this study, but attention is called to values as low as $p = 0.05$ for the sake of discussion.

We used multivariate statistical procedures to test the relationships among geographical factors and between geographical factors and plant diversity, and performed partial correlation analyses after correcting for time. Partial correlation posits a situation where none of the measured variables are assumed to be functionally dependent on the other variables. Canonical correlation attempts to relate independent variables (geography and age) in the order of their decreasing contribution to explaining the observed pattern in the dependent variables (plant diversity). Since multivariate statistics work best with a reasonably limited suite of variables, we selected geographical variables that showed the best potential for significance, based on the simple univariate regressions and their biogeographical significance (total land area, total num-

Table 3.3. Data on numbers of geographical features (continents, islands, uplands on islands) from the Early Devonian to the Middle Miocene for the Northern Hemisphere as defined here (North America, Eurasia, Greenland) of the present day

Age	Continents	Island	Total Land	Uplands on Conts.	Uplands on Islands	Total Uplands	Small Uplands	Large Uplands
Early Devonian	3	5	8	2	7	9	8	1
Early Carboniferous	2	13	15	6	8	14	14	0
Late Carboniferous	3	8	11	8	7	15	14	1
Late Permian	4	6	10	12	3	15	12	3
Early Triassic	3	5	8	16	2	18	18	0
Early Jurassic	2	4	6	14	1	15	13	2
Late Jurassic	2	8	10	11	2	13	10	3
Early Cretaceous	3	4	7	10	2	12	9	3
Middle Cretaceous	3	8	11	12	1	13	11	2
Late Cretaceous	3	7	10	10	3	13	10	3
Middle Eocene	3	6	9	12	5	17	14	3
Middle Miocene	2	7	9	11	7	18	16	2

Table 3.4. Latitudinal distribution of land area in the Northern Hemisphere (as defined in this paper) by 10° increments.

Age	Degrees Latitude										
	20–10	10–0	0–10	10–20	20–30	30–40	40–50	50–60	60–70	70–80	80–90
Early Devonian	0	1.93	7.8	8.9	7.6	3.4	2.4	1.2	0	0	0
Early Carboniferous	1.07	1.6	6.5	7.0	6.5	5.8	1.0	1.9	2.4	1.4	0
Late Carboniferous	0	7.0	8.0	8.1	6.4	4.4	3.8	3.8	2.2	0.3	0
Late Permian	0	2.75	6.8	7.4	8.6	6.6	4.6	5.2	5.0	2.8	0.4
Early Triassic	0	3.98	5.6	8.4	9.5	13.2	11.2	6.1	3.4	1.4	0.2
Early Jurassic	0	0	1.76	4.6	8.8	12.4	14.7	12.4	7.6	4.7	1.1
Late Jurassic	0	0	0.66	2.16	12.5	14.8	12.5	8.4	4.0	1.8	0.7
Middle Cretaceous	0	0	2.09	2.99	6.4	11.5	13.7	10.2	7.4	4.4	0.8
Late Cretaceous	0	0	1.42	2.05	8.22	8.4	8.4	12.0	9.5	7.3	1.1
Middle Eocene	0	0	2.16	1.88	5.15	11.4	12.5	14.0	14.7	6.7	0
Middle Miocene	0	0	4.98	4.1	7.6	12.1	15.1	15.1	13.8	6.6	0.1

Note: Figures are for 10^6 km^2. Data from Parrish (1985), slightly modified. In time periods where some "Northern Hemisphere" land is in the Southern Hemisphere, the Southern Hemisphere area was combined with the corresponding Northern Hemisphere latitudinal band for calculations.

ber of uplands), together with measures of latitudinal distribution (standard deviation, skewness and kurtosis of land distribution), geological age, and diversity (total, pteridophyte, and seed plant diversity). The multivariate tests were run using the area data from Parrish (1985) in order to include the measure of latitudinal distribution of land. These data do not include a measurement for the Early Cretaceous, thus $n = 11$. Significant levels of r-values are indicated accordingly.

The relation of geological age to diversity and geographical variables was first tested with univariate linear regressions. Multivariate hypothesis tests were then run with time set as the independent variable and the cited geographical and diversity features as the dependent variables. Several competing statistics were evaluated for consistency (e.g., Wilks' criterion, Roy's maximum root criterion).

Univariate, linear regressions were run on hand-held, preprogrammed calculators (Texas Instrument TI-55; Sharp EL-512). The canonical correlation, partial correlation, and multivariate hypothesis tests were all run on SAS using a Cornell University mainframe computer.

Univariate regressions are run with area data measured by SECTION and computed as percent of hemispheric surface; however, the multivariate statistics and the univariate tests involving time were based on direct area measurements of Parrish (1985) and are presented in square kilometers. These two measurements of area were necessary, since Parrish (1985) provides only total land mass area without making the distinction between upland and lowland. To convert our percentage readings to square kilometers would have increased the probability of mathematical error. While actual area and percent area are different forms of measure, they compare closely: a regression provides $r = +0.952$ with $n = 11$, which is significant above the 0.1% level.

Errors and Biases

Our data represent a first approximation, since several layers of estimation underlie the apparent numerical precision. Further, the number of paleogeographical reconstructions restrict our sample size to 11 or 12, depending on the test involved. Although new data on plant diversity are unlikely to greatly influence the overall botanical trends observed in our data set, they could alter patterns of correlation with geographic features in cases where correlation is presently marginal. For this reason, the occurrence of levels of significance beyond the normally accepted 1% level is noted in order to indicate where new plant data might have an effect.

The primary source of possible inferential error, however, comes from the paleogeographical data. We make three primary assessments of paleogeographical data: area, counted features, and latitudinal dispersion.

AREAL MEASUREMENTS

We measured the total area of shallow seas (inferred continental shelf), lowlands, and uplands from existing paleogeographical maps and derive from lowland + upland area a measure of total land area. However, paleogeographical maps convey a greater sense of exactitude than their creators intended. Ziegler et al. (1979) review the evidence that goes into determining continental outlines and relations, and caution that these are, at best, highly informed estimates. This is compounded by errors made in digitizing the continents in order to obtain specific areas. Because the available paleogeographical maps are small (and simple enlargement would have added errors of optical distortion), some discrepancy between the published map and the electronic image will occur. To minimize systematic error, the images were digitized by one person.

COUNTED FEATURES

Continents are large features and are fairly accurately recorded in making paleogeographical maps. This is not the case for islands and uplands, since both are inferred from depositional and biotic patterns in paleocartography. Islands are presumably recognized by virtue of being distant from continents or associated with island-arc features. However, there is no guarantee that an island recorded on a map was not a peninsula, with consequent effects upon the relationship of its biota to that of the mainland. Islands may also have existed in the past for which evidence has subsequently been either obliterated or obscured by amalgamation with continents. Finally, given the large scale of the maps, only larger islands are indicated.

Uplands were plotted by Ziegler et al. (1979) on the basis of presence of andesitic volcanism and along identified colliding plate boundaries. Given that other sources of uplift exist, their reconstruction would seem to be a minimum estimate. This method of "upland identification" provides no accurate indication of the absolute height of uplands and only the broadest indication of their areal extent. Thus, the difference in biogeographical diversity and evolutionary potential that exists between a "low" upland and a major mountain range with its several altitudinal zones is obscured. Our estimate of numbers of uplifts will, as with islands, omit ranges that are too small to show on the scale of the paleogeographical maps.

LATITUDINAL DISTRIBUTION OF LAND IN THE NORTHERN HEMISPHERE

The estimate of the amount of land in 10° latitudinal bands rests upon two assumptions: (1) the correct measurement of area in those cases where we had to recalculate data from Parrish (1985) and (2) the correct paleolatitudinal placement of the continents. Mesozoic and Cenozoic maps are fairly accurate, due largely to the availability of dated oceanic crust. However, earliest Mesozoic and older oceanic crust has been subducted, removing one of the primary sources of positional data, and necessitating the estimation of continental positions through other, less accurate, means. Thus, Paleozoic maps have a far greater probability of error in the position of land masses than do Mesozoic or Cenozoic maps. Of particular relevance are changing notions of the placement of the Chinese land mass in the later Paleozoic. In the maps used here (Ziegler et al. 1979; Scotese et al. 1979), China is located at approximately 30° north latitude in the later Paleozoic, with a northerly drift. Ziegler et al. (1979) note that Permian gradients of invertebrate diversity suggest a more southerly location than in their reconstruction, and more recent estimations (Scotese, Raymond, pers. comm.) place it further south in the later Paleozoic than indicated in Ziegler et al. (1979). However, for consistency, we have used the published maps. Our computations on diversity and latitude suggest at present that an alteration of China's position would not significantly alter the final conclusion.

Tests and Results

GEOGRAPHICAL CORRELATIONS

The relationship of diversity to geographical features cannot be evaluated without an estimate of the degree of covariance in geographical features. Fifty-five univariate regressions were run; those significant at the 5% level or better are summarized in table 3.5. Each regression is identified by a number in order to expedite discussion.

A group of correlations in table 3.5 displays a logical association: (1) Total number of land masses/ number of islands; islands form the most numerous class of land masses. (2) Total land area/lowland area; most land area is lowland, hence its area parallels total land area. (3) Total number uplands/number small uplands; small uplands are more common than large ranges, hence they dominate the total number of uplands. (7) Total land area/total number of continents; this suggests that the differentiation of a new continent involves the creation of new surface area. (8) Total number of uplands/uplands on continents; most islands have one central upland, but continents have room for several

Table 3.5. Regression coefficients for the association of geographical features of the Northern Hemisphere (see tables 3.2 to 3.4 for data)

Simple Univariate Regressions	
1. Number of land masses/number of islands	r = +0.968***
2. Total land area/lowland area	r = +0.940***
3. Number of uplands/number small uplands	r = +0.915***
4. Total land area/number uplands on continents	r = +0.769**
5. Lowland area/number uplands on continents	r = +0.766**
6. Total land area/upland area	r = +0.763**
7. Total land area/total number continents	r = +0.736**
8. Total number uplands/number uplands on continents	r = +0.699*
9. Total number uplands/total land area	r = +0.668*
10. Total number uplands/total lowland area	r = +0.638*
Partial Correlation (corrected for the effect of time)	
11. Total land area/total number uplands	r = +0.858***
12. Standard deviation of the distribution of Northern Hemisphere land masses/kurtosis of distribution of same	r = −0.841**

Note: Items are listed by type of statistical test and in descending order of significance. *** = 0.1% significance level; ** = 1% significance level; * = 5% significance level.

uplands; thus, continental uplands dominate total uplands. These correlations display an internal consistency and suggest that total area, lowland area, and total number of continents are effectively similar measures of geography. Similarly, total land mass number measures island number, and total number of uplands tracks both number of small uplands and perhaps more importantly, numbers of uplands on continents. Note that correlations 1, 2, and 3 are at the 0.1% level, but 7 and 8 are at the 1% level.

The remaining correlations are marginal but form an interesting and coherent group. Regressions 4, 5, 6, 9, and 10 have in common measures of total land area (including lowland area, noted as synonymous with total area) and upland area or number. It is likely that these correlations are a further reflection of the correlation between the numbers of uplands and numbers of uplands on continents (8). Correlations 4 to 7 are at the 1% level and 8 to 10 are at the 5% level and are therefore all suspect because of type I error.

Two results from the partial correlation analysis (table 3.5) are significant: total land area/total number uplands (11) is significant at the 0.1% level, after correcting for the effect of time. This bears out univariate correlations and indicates that total area and total number of uplands are highly correlated. The second correlation (12) is specific to the measurement of latitudinal distribution of land in the Northern Hemisphere (see figure 3.1, a presentation of these data as a series of histograms through time). The correlation compares the absolute lati-

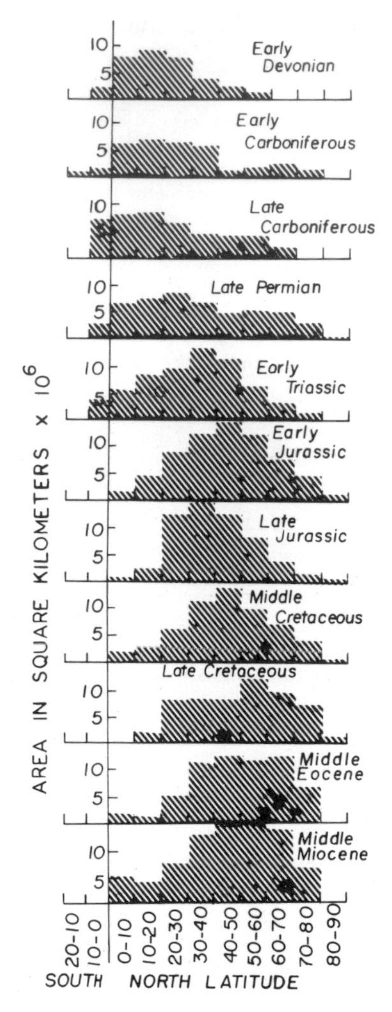

Figure 3.1. Latitudinal distribution of absolute area of land (lowland and upland) of Northern Hemisphere land masses from the Devonian to the Middle Miocene in $km^2 \times 10^6$ (vertical axis). Note that Northern Hemisphere as defined here includes Eurasia, North America, Greenland, and associated small land masses but not those portions of Africa, South America, or India presently in the Northern Hemisphere. Addition of the latter land masses would increase the amount of land in lower latitudes in the later Mesozoic and Cenozoic but not change the increase in amount of land at high latitudes.

tudinal spread of Northern Hemisphere land masses (standard deviation) and the latitudinal location of the bulk of this land (kurtosis). The negative sign indicates a displacement towards the North Pole and indicates that land has simultaneously been concentrated at high latitudes as the latitudinal spread of Northern Hemisphere continents has decreased.

DIVERSITY AND GEOGRAPHICAL FACTORS

Two features stand out in the univariate correlations of diversity with geography (table 3.6). First, in the six correlations significant to the 5% level or better, pteridophyte diversity is not present. This indicates that the history of vascular land plant diversity is essentially that of seed plant diversity. However, this would not have been true in the later Paleozoic, when pteridophytes dominated the terrestrial flora, suggesting the advisability of a separate regression of Devonian-Permian pteridophyte diversity against geographical variables. Second, total diversity correlates most strongly with area; upland area at the 0.1% level and total area at the 1% level. The correlation of total land plant diversity with the total number of uplands is only significant at the 5% level. Hence, area remains a significant factor. The apparent importance of numbers of uplands and their correlation with area (table 3.5) appears to be reinforced.

To permit evaluation of the response of pteridophytes to geographical factors in the basence of the overwhelming dominance of seed plants, regressions of pteridophyte diversity from the Early Devonian to the Early Triassic were run against geographical variables. Total pteri-

Table 3.6. Regression coefficients for the association of physical features with vascular plant diversity (see tables 3.1 to 3.4 for data)

Simple Univariate Regressions	
1. Number seed plant species/upland area	r = +0.897***
2. Total number land plant species/upland area	r = +0.837***
3. Total number land plant species/total land area	r = +0.806**
4. Number seed plant species/total land area	r = +0.768**
5. Total number land plant species/lowland area	r = +0.641*
6. Total number land plant species/total number uplands	r = +0.615*
Partial Correlation (corrected for the effect of time)	
7. Total number land plant species/total number uplands	r = +0.788**

Note: Items are listed by type of statistical test and in descending order of significance. *** = 0.1% significance level; ** = 1% significance level; * = 5% significance level.

dophyte species number correlated most significantly with number of uplands (r = 0.770) and total area of lowlands (+0.786). Both values are suggestive but significant only at the 20% level (n = 5).

DIVERSITY AND LATITUDINAL DISTRIBUTION OF LAND

Three aspects of diversity correlated with measurements of land distribution in the Northern Hemisphere (table 3.7).

1. Seed plant numbers/skewness of land distribution. This could indicate that seed plant numbers increased as the land masses moved toward more northerly latitudes, spanning a wider range of climatic zones. However, diversity may also have evolved over long periods of time, independent of latitudinal position.

2. Pteridophyte species number/standard deviation of land mass distribution. This indicates that pteridophyte diversity is greatest when land masses are more equitably distributed among latitudes. Given the trend of land distribution during the period under consideration, this is equivalent to stating that pteridophyte diversity and land area in low latitudes are directly proportional.

3. Pteridophyte species number/kurtosis (concentration of land). As kurtosis decreases, the diversity of pteridophytes increases, suggesting that as more land occurs in more latitudes (and thus more land in lower latitudes), pteridophyte diversity increases.

CANONICAL CORRELATION

A canonical correlation of land plant diversity and selected geographical features (total land area, total number uplands, standard deviation, skewness and kurtosis of the latitudinal distribution of land and geological age) yields a p-value of +0.1212; F-tests reveal this is not significantly different from zero, suggesting that the collective behavior of the biological variables, and of the geological variables, shows no signifi-

Table 3.7. Regression coefficients for the association of the latitudinal distribution of land area in the Northern Hemisphere with vascular plant diversity (see tables 3.1 and 3.4 for data)

Simple Univariate Regressions	
1. Seed plant species number/skewness	r = −0.832**
2. Pteridophyte species number/standard deviation	r = +0.692*
3. Pteridophyte species number/kurtosis	r = −0.599*

Note: Standard deviation, skewness, and kurtosis reflect the latitudinal distribution of land masses visualized as a bar graph (see figures 3.1, 3.2). Listed in descending order of significance. *** = 0.1% significance level; ** = 1% significance level; * = 5% significance level.

Table 3.8. Regression coefficients for the association of absolute age with total diversity and selected geographical features

1. Total species number/time	r = −0.903***
2. Skewness of continental distribution/time	r = +0.879***
3. Total land area/time	r = +0.875***

Note: *** = 0.1% significance level.

cant correlation. This could be interpreted in two ways. It could imply that the other correlations are, by this measure, spurious, and that it is impossible to separate the influence of specific geographical features on changes in diversity. Alternatively, two or more of the variables may have antagonistic trends that result in countercorrelations.

GEOLOGIC AGE VERSUS VARIABLES

Unvariate simple regressions of total species number, total land area, standard deviation of the latitudinal distribution of land area in the Northern Hemisphere (a measure of absolute spread of land masses), skewness of this distribution of land masses (displacement of preponderance of area toward equator or pole), kurtosis of land mass distribution (concentration of land along its distribution), and numbers of uplands against geological age were performed. Three correlations were significant at the 0.1% level (table 3.8): geological age correlates with total species number, with total land area, and with skewness. This indicates that, as time has progressed, more species appeared, more land area was present in the Northern Hemisphere, and that land masses increasingly shifted toward more northerly latitudes.

Multivariate hypothesis tests of the effect of age upon the dependent variables (area, number of uplands, standard deviation, kurtosis and skewness of Northern Hemisphere land, species number) yield a probability > F = 0.0052, which is significant [F (6, 4) = 21.59], based on all F-test criteria. This indicates that geological age is probably the single most important factor explaining the observed variation in biological and geological parameters.

Discussion

AREA AND DIVERSITY

Statistical analyses (see table 3.6) suggest a strong correlation between area and diversity. Upland area correlates with total plant diversity and seed plant diversity at the 0.1% level of significance, and total area correlates with total and seed plant diversity at the 1.0% level. Total number of uplands correlates with diversity, significant at the 5% level. The virtual absence of other correlated geographical features suggests the primacy of area in explaining variations in diversity. This agrees with Flessa's (1975) finding that continental area is significantly correlated with the extant diversity of mammalian orders, families, and genera.

These correlations of diversity and area may be influenced by the biases of the available volume and surface exposure of sedimentary rocks, which previously have been noted to closely correlate with recorded marine and terrestrial diversity (Raup 1972, 1979; Niklas, Tiffney and Knoll 1980). In the present case, the absolute dominance of this bias is questionable, as upland area is generally assumed to be poorly represented in the fossil record, yet the strongest correlation is between upland area and diversity. Further, increases in within-community diversity in marine (Bambach 1977) and land plant (Niklas, Tiffney, and Knoll 1980) assemblages suggest that Phanerozoic diversity trends are not solely a function of available collecting sites.

Is it possible, however, that changes in area reflect changes in other geographical features with which area may be linked? (see table 3.5). Three pairs of geographical features correlate at the 0.1% level of significance (nos. 1, 2, 3); none links area to another geographical feature. At the 1.0% level, area is linked to the total number of continents (no. 7), and to the total number of uplands on continents (nos. 4 and 5). The former correlation seems unimportant in the present case; the total numbers of continents are generally very low (two to three); if area were to be correlated with geographical separation, the significant correlation would be between area and total numbers of land masses or area and numbers of islands (which dominate total numbers of land masses). The correlation with numbers of uplands on continents is more suggestive, since it parallels the correlation of total area and lowland area with total numbers of uplands (significant at the 5% level; nos. 9, 10). Again, these correlations have 1% or 5% levels of confidence, which may not be significant in view of the number of variables considered. However, the partial correlation (no. 11) indicates that there are strong relationships between total land area and total number of uplands through time (significant at the 0.1% level). Flessa (1975) hypothesized that mammalian diversity and total area were correlated because habitat patchiness increased as total continental area increased. It is possible that plant diversity correlates with total land area at least in part because total land area correlates with numbers of uplands, which form environmental patches.

AREA AND TOPOGRAPHY

Both simple univariate statistics and multivariate statistics demonstrate a correlation of land area with numbers of uplands. This indicates that numbers of mountains correlate with land area and raises an unresolved group of questions. Do larger areas increase the potential of recognizing mountains in the fossil record? Do mountains somehow reflect tectonic processes that generate area? Do larger areas provide a statistically greater chance for the formation of mountains? In this context, the independence of upland area and upland number is counterintuitive and might possess some geological significance. However, in the biological context, the correlation of numbers of uplands and land area makes it impossible to distinguish the hypothesis that vascular plant diversity is a function of area from the hypothesis that diversity is a function of terrestrial topographical variability. Both may be important.

TOPOGRAPHY AND DIVERSITY

Although topography does not directly correlate with vascular plant diversity, it appears to correlate strongly with area, which in turn correlates with diversity. Therefore, the hypothesis that topographical differentiation influences biotic diversity cannot be disentangled from area effects.

GEOGRAPHICAL DISPERSION AND DIVERSITY

Perhaps the interesting result concerns the effect of dispersion of land masses on diversity. There is no significant correlation (at the 5% level or greater) between land plant diversity and numbers of continents, islands, or total land masses in the Northern Hemisphere. This is not consistent with the trends observed by Kurtén (1967, 1969), on which he based the hypothesis that the diversity of reptiles and mammals was influenced by the number of separate land masses present when each group was radiating. The plant pattern also contrasts with the pattern in the marine record (Valentine and Moores 1970), in which invertebrate diversity is positively correlated with the dispersion of the continents. This contrast could result from one or more factors. It could be due to a bias in collecting the data, although we are unable at this time to identify such a bias. It might result from the particular Mesozoic and Cenozoic history of the Northern Hemisphere, in which the continents have never been as isolated as they have become in the Southern Hemisphere; this alternative could be tested with Southern Hemisphere fossil data. It could also be that land mass dispersion does stimulate plant diversification, but that this effect is overwhelmed by the stimulation of diversity by other factors. Finally, it is possible that the different biologies of plants versus animals influence their response to geographical separation, as further discussed in the summary.

LATITUDINAL DISTRIBUTION OF LAND AND DIVERSITY

The present increase in biological diversity from the poles to the equator (e.g., Schall and Pianka 1978) leads to the prediction that the latitudinal distribution of land in the past should affect patterns of diversity. Four statistical correlations are significant in this regard; however their interpretation is complex.

1. The negative correlation of standard deviation of the latitudinal distribution of land with the degree of kurtosis in that distribution (see Table 3.5, no. 12) indicates that the latitudinal spread of land in the Northern Hemisphere has decreased toward the present, while the land has simultaneously been displaced toward the North Pole, a conclusion apparent from visual inspection of sequential paleogeographical maps and presented graphically in figures 3.1 and 3.2.

2. The three diversity correlations with the latitudinal data are not particularly strong; that with seed plant diversity is at the 1% level (see table 3.7, no. 1), the two with pteridophyte diversity are at the 5% level (see table 3.7, nos. 2 and 3).

3. Seed plant diversity negatively correlates with skewness; thus seed plant diversity increases as continents are displaced toward the North Pole.

4. By contrast, pteridophyte diversity correlates positively with standard deviation and inversely with kurtosis, indicating that pteridophyte diversity increases with increased equatorial land mass area.

The pteridophyte data make sense, as pteridophytes are inherently more tropical than temperate, and their diversity would be expected to decrease as land moved out of the lower latitudes. The seed plant data could be interpreted to indicate that greater diversity occurs as land occupies a greater latitudinal, and thus presumably climatic, range. However, perhaps more than any other correlation, this one is beset with difficulties. First, latitudinal diversity gradients are correlated with sharp climatic gradients such as exist in the present glacial period. Such steep climatic gradients may also have been operative in the later Carboniferous and Permian but were otherwise not a common feature in the time span under investigation. It is even possible that the Mesozoic equator hosted a markedly lower terrestrial diversity than the coeval temperate zone (Ziegler et al. 1987), which would thoroughly confound a statistical test based on the assumption that the present is a

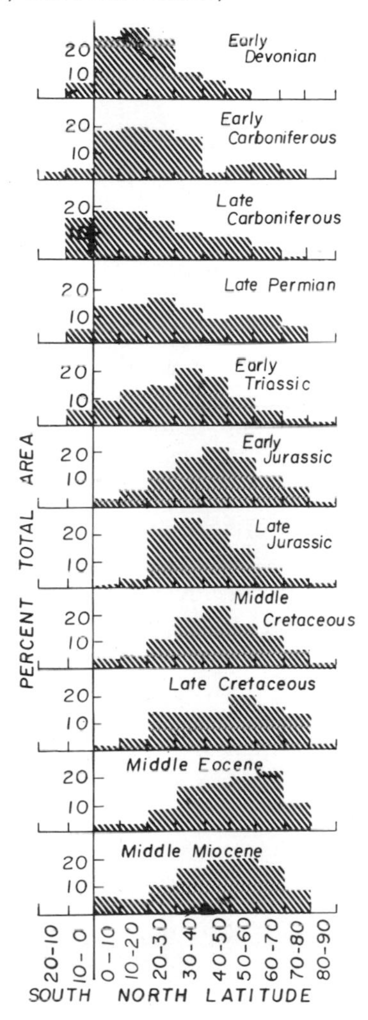

Figure 3.2. Latitudinal distribution by 10° latitudinal increments of present-day Northern Hemisphere land masses (see fig. 3.1) from the Devonian to the Middle Miocene. The vertical axis is percent of total provided by each latitudinal band.

correct model for the past. Second, while we can provide a test for the relation of total diversity to the latitudinal distribution of land through time, it does not track the latitudinal distribution of diversity in the past. For these reasons, we suspect that the relation between seed plant diversity and increased land area in higher latitudes may not be related, but that while land was displaced poleward, diversity rose for other, unconnected, reasons.

TIME, GEOGRAPHICAL FEATURES AND DIVERSITY

The correlation of time with total land area in the Northern Hemisphere (significant at the 0.1% level) indicates a pattern of increasing land area at higher latitudes through time. This pattern is more pronounced in the Northern Hemisphere than globally, as the correlation of time with total world land area is significant only at the 5% level ($r = -0.651$; $n = 11$). Figure 1 of Parrish (1985) corroborates this pattern, displaying a Paleozoic rise in total land area, followed by fluctuation around 140 $km^2 \times 10^6$ in the Mesozoic and Cenozoic. Cogley (1981) presents a similar conclusion. The correlation of increasing skewness of continental distribution with time is evident in figures 3.1 and 3.2 and by visual inspection of sequential global paleographical maps.

Time also correlates with total diversity at the 0.1% level (see Table 3.8). On the one hand, this correlation could indicate that synergistic interactions among organisms were significant factors in driving diversification and that the longer biological systems existed, the more complex they became. Alternatively, diversity could be dependent on area, and thus diversity could track time as a side effect of tracking increasing Northern Hemisphere area. If so, it is curious that time does not correlate with total number of uplands, which also generally correlate with area (see Table 3.5). Contrary to the suggestion made in our discussion of area and diversity, this might indicate that area was more important than topographical differentiation in influencing diversity.

Summary

The most important correlate with the history of plant diversity is geological age. However, age also correlates with total land area, and total land area with total number of uplands. These associations prevent us from unequivocally associating diversity with any particular geographical factor. Further, the present analyses do not support or refute the possibility that diversification had a biological basis. Nevertheless, certain correlations between diversity and physical features are worthy of mention.

In perhaps the most clear-cut result, land plants show no correlation with increasing numbers of separate land masses. This is at variance with the pattern observed with marine invertebrates (Valentine and Moores 1970) and for reptiles and mammals by Kurtén (1967, 1969). At present, we believe that this is not a function of the mode of sample collection or analysis. It might reflect different modes or rates of response to geographical change in marine invertebrates and possibly terrestrial vertebrates, as compared to terrestrial plants. In light of the significance that is accorded allopatric speciation in evolutionary theory, this seems a counterintuitive solution. However, the relative frequency of allopatric speciation in different major taxa has not been explored in a comparative manner.

More importantly, there are other differences in evolutionary patterns between invertebrates and vertebrates on the one hand, and land plants on the other (e.g., Traverse 1988). Valentine, Tiffney, and Sepkoski (submitted) have observed that average species duration increases in more derived clades of marine invertebrates while it decreases in derived clades of land plants. Further, the history of marine invertebrates witnesses a shrinkage in the total number of *bauplanes* over the Phanerozoic, while in post-Devonian land plants, the analogous diversity of growth forms appears to increase. In an attempt to separate the hypothesis that the difference in response to changing geography is a plant-animal distinction from the hypothesis that it is a marine-terrestrial distinction, we are presently analyzing the response of terrestrial vertebrate diversity (data supplied by J. Sepkoski, University of Chicago) to the same range of geographical variables treated here.

A strong correlation exists between diversity and total land area, but this is associated with a pronounced correlation between land area and total number of uplands. The total data set indicates that area may be the more important factor, but the two physical variables cannot clearly be untwined at present. The two variables could be separated by comparing a time series of (1) a land mass of changing size that either lacks uplands or possesses a fixed number of uplands or (2) a land mass of constant size upon which upland number changes against the resident plant diversity. The mode of assembly of our present data set does not permit this kind of geographical partitioning of the diversity data. We plan to re-collect the data in order to permit such a study, among others. There is also an outside, and we suspect unlikely, possibility that area and diversity may covary in a similar manner but for different reasons, both having to do with the passage of time (absolute age).

The comparison of diversity with the latitudinal distribution of land was suggestive but inconclusive. Pteridophyte species numbers apparently increase as equatorial land area increases, but the correlation of increasing seed plant diversity with the poleward displacement of continental area is suspected to be spurious. A proper test of the effect of latitudinal distribution of land area on past diversity could be achieved by (1) following diversity on one land mass through time as it changes in its latitudinal placement (the same effect might be achieved by looking at one stationary land mass affected by changing climate) or (2) by plotting diversity data by paleolatitude over several time intervals and correlating it with the latitudinal distribution of land. Again, the construction of our data set does not permit geographical separation of the data. The second generation of diversity data will make such studies possible. Data from the living taxa leave little doubt as to the reality of the equator-pole diversity gradient. The latitudinal distribution of land probably has profoundly affected diversity in times of steep global temperature gradients. However, such gradients are so scattered in the history of land plant evolution that they may have had little effect on the overall trend of land plant diversity, which we suspect was influenced in a more consistent and long-term manner by other geographical features.

A variable that we did not explore is climate, which may vary as a result of changing geographical features or independent of them. Although paleotemperatures from earlier geological times are probably imprecise, it would be valuable to correlate estimated dominant global temperature with the geographical features presented in this paper, as well as with the diversity of pteridophytes and seed plants. If it is possible to quantify, it would also be valuable to compare changing global climatic gradients with total diversity.

Our correlations raise some geological questions. First, is the observation of the increase of Northern Hemisphere terrestrial area through time significant? If it is, is it a predictable function of some geological process or an artifact of the method of assembly of paleographical maps, possibly reflecting the uncertainty of placement of Paleozoic land masses? If the former, does it suggest any predictions for the future of the Earth? Second, why is there but one significant correlation between upland area and measures of total land area, yet four correlations between numbers of uplands and total and lowland area? Is there a scaling factor involved in the relation of mountains and continents? Do collisions between plates result in mountains that accumulate in number as a continent ages? The latter seems unlikely in light of rates of erosion and the existence of mountain-building mechanisms other than continental collision, but the question requires exploration.

In sum, the dispersion of land masses is found not to be correlated with diversity. Area and possibly numbers of uplands do correlate with diversity. The diversity data set did not permit evaluation of the effect of latitudinal distribution of land on diversity. The latter question, and the teasing apart of the effects of area versus numbers of uplands, awaits the creation of a new, more flexible, diversity data set. The role of physical geography in stimulating land plant diversity cannot be clearly identified in this study.

The need for a more flexible data set is paralleled by the need for a data set that embodies a greater degree of biologically important information. We have treated land plants here as two categories separated on reproduction—pteridophytic or seed. A data set arranged so as to permit recognition of plant habit, pollination mode, and habitat would permit further fine tuning of our understanding of patterns of diversity change.

Two hypotheses have been offered to explain trends in Phanerozoic diversity. One hypothesis envisions an "internal drive" to biotic diversification; the more species that exist, the more opportunities for speciation. The other hypothesis argues that diversity change is primarily driven by fluctuations in the abiotic environment. Paleogeographical evidence clearly indicates that the geographical distribution, total area, and physiognomy of land has changed through the Phanerozoic. These observations do not demonstrate the primacy of the latter hypothesis, however, because whether external forces change or not, other observations make it clear that species create and challenge their own environments in geological and ecological time. The two hypotheses are polarized expressions of the human desire to find clear answers to messy, multifaceted problems. Indeed, philosophically, separation of physical from biological stimuli is particularly difficult with terrestrial plants, given their ability to directly influence and modify the physical environment (e.g., changes in daily temperature and moisture fluctuation in an open and closed canopy community). Both physical and biological stimuli must have influenced the course of evolution. Further studies of terrestrial plant and animal diversity and of patterns of physical change will probably result in recognition of time periods when either biological or physical factors were dominant in shaping evolution, but we think it unlikely that all of evolution will be dominated by one or the other. However, perhaps the most important thing that can come out of such inquiry will be a growing sense of how different groups of organisms respond to changes in their biological or physical environment or both.

Acknowledgments

We thank Judith Parrish (University of Arizona), Anne Raymond (Texas A&M University), and Christopher Scotese (University of Texas) for suggesting sources of data for this work and for their discussions of sources of error in paleocartography. We also thank Dr. Gail Rubin of Cornell University for advice on matters multivariate. The comments of two reviewers and the editors of this volume helped hone the written presentation. Research partially supported by NSF grant BSR-8796251 to BHT.

Literature Cited

Bambach, R. K. 1977. Species richness in marine benthic habitats through the Phanerozoic. *Paleobiology* 3:152–67.

Bridgewater, P. B. 1988. Biodiversity and landscape. *Earth-Science Rev.* 25:486–91.

Brown, K. S., Jr., P. M. Sheppard, and J. R. G. Turner. 1974. Quaternary refugia in tropical America: Evidence from race formation in *Heliconius* butterflies. *Proc. R. Entomolog. Soc. London, B.* 187:369–78.

Cogley, J. G. 1981. Late Phanerozoic extent of dry land. *Nature* 291:56–58.

Cracraft, J. 1985. Biological diversification and its causes. *Ann. Missouri Botanical Garden* 72:794–822.

Fischer, A. G. 1960. Latitudinal variations in organic diversity. *Evolution* 14:64–81.

Flessa, K. W. 1975. Area, continental drift and mammalian diversity. *Paleobiology* 1:189–94.

Flessa, K. W. and J. Imbrie. 1973. Evolutionary pulsations: Evidence from Phanerozoic diversity patterns. In *Implications of continental drift to the earth sciences*, vol. 1, ed. D. H. Tarling and S. K. Runcorn, 247–85. London: Academic Press.

Futuyma, D. J., and M. Slatkin. 1983. *Coevolution*. Sunderland Mass.: Sinauer Associates.

Haffner, J. 1969. Speciation in Amazonian forest birds. *Science* 165: 131–7.

Hoffman, A., and J. Kitchell. 1984. Evolution in a pelagic ecosystem: a paleobiologic test of models of multispecies evolution. *Paleobiology* 10:9–33.

Johnson, M. P., and P. H. Raven. 1970. Natural regulation of plant species diversity. In *Evolutionary biology.* vol. 4, ed. T. Dobzhansky, M. K. Hecht, and W. C. Steere, 127–62. New York: Appleton-Century-Crofts.

Knoll, A. H. 1984. Patterns of extinction in the fossil record of vascular plants. In *Extinctions*, ed. M. Nitecki, 21–67. Chicago: University of Chicago Press.

Knoll, A. H. 1986. Patterns of change in plant communities through geological times. In *Community ecology,* ed. J. Diamond and T. J. Case, 122–41. New York: Harper and Row.

Knoll, A. H., K. J. Niklas, and B. H. Tiffney. 1979. Phanerozoic land-plant diversity in North America. *Science* 206:1400–1402.

Kurtén, B. 1967. Continental drift and the paleogeography of reptiles and mammals. *Commentationes Biologicae Societas scientiarum fennica* 31: 1–8.

Kurtén, B. 1969. Continental drift and evolution. *Sci. Am.* 220:54–64.

Lidgard, S., and P. R. Crane. 1988. Quantitative analyses of the early angiosperm radiation. *Nature* 331:344–6.

Manley, B. F. J. 1986. *Multivariate statistical techniques: a primer.* London: Chapman and Hall.

McCoy, E. D., and E. F. Connor. 1980. Latitudinal gradients in the species diversity of North American mammals. *Evolution* 34: 193–203.

MacArthur, R. H., and E. O. Wilson. 1967. *The theory of island biogeography.* Princeton, N.J.: Princeton University Press.
Niklas, K. J. 1978. Coupled evolutionary rates and the fossil record. *Brittonia* 30:373–94.
Niklas, K. J. 1986. Large-scale changes in animal and plant terrestrial communities. In *Patterns and processes in the history of life*, ed. D. M. Raup and D. Jablonski, 383–405. Berlin: Springer-Verlag.
Niklas, K. J., and S. P. Boyd. 1987. Computer program for three-dimensional reconstructions and numerical analyses of plant organs from serial sections. *Am. J. Botany* 74:1595–9.
Niklas, K. J., B. H. Tiffney, and A. H. Knoll. 1980. Apparent changes in the diversity of fossil plants. In *Evolutionary biology.* vol. 12, ed. M. K. Hecht, W. C. Steere, and B. Wallace, 1–89. New York: Plenum.
Niklas, K. J., B. H. Tiffney, and A. H. Knoll. 1983. Patterns in land plant diversification. *Nature* 303:614–6.
Niklas, K. J., B. H. Tiffney, and A. H. Knoll. 1985. Patterns in vascular land plant diversification: An analysis at the species level. In *Phanerozoic diversity patterns: profiles in macroevolution*, ed. J. W. Valentine, 97–128. Princeton, N.J.: Princeton University Press.
Parrish, J. T. 1985. Latitudinal distribution of land and shelf and absorbed solar radiation during the Phanerozoic. *U.S. Geolog. Surv. Open-File Rept.* 85–31:1–21.
Parrish, J. T., A. M. Ziegler, and C. R. Scotese. 1982. Rainfall patterns and the distribution of coals and evaporites in the Mesozoic and Cenozoic. *Palaeogeogr. Palaeoclimatol. Palaeoecol.* 40:67–101.
Preston, F. W., 1962. The canonical distribution of commoness and rarity. *Ecology* 43:185–215; 410–32.
Raup, D. M. 1972. Taxonomic diversity during the Phanerozoic. *Science* 177:1065–71.
Raup, D. M. 1979. Biases in the fossil record of species and genera. *Bull. Carnegie Mus. Natural Hist.* 13:85–91.
Rich, J. E., G. L. Johnson, J. E. Jones, and J. Campsie. 1986. A significant correlation between fluctuations in seafloor spreading rates and evolutionary pulsations. *Paleoceanography* 1:85–95.
Schall, J. J., and E. R. Pianka. 1978. Geographical trends in numbers of species. *Science* 201:679–86.
Scotese, C. R., R. K. Bambach, C. Barton, R. van der Voo, and A. M. Ziegler. 1979. Paleozoic base maps. *J. Geol.* 87:217–77.
Stenseth, N. C., and J. Maynard Smith. 1984. Coevolution in ecosystems: Red Queen evolution or stasis? *Evolution* 38:870–80.
Tiffney, B. H. 1981. Diversity and major events in the evolution of land plants. In *Paleobotany, paleoecology and evolution*, ed. K. J. Niklas, 193–230. New York: Praeger Press.
Tiffney, B. H. 1985. Geological factors and the evolution of plants. In *Geological factors and the evolution of plants*, ed. B. H. Tiffney, 1–10. New Haven, Conn: Yale University Press.

Tiffney, B. H. In preparation. The Red Queen reigns over extinction, but not speciation.

Traverse, A. 1988. Plant evolution dances to a different beat. Plant and animal evolutionary mechanisms compared. *Historical Biol.* 1:277–301.

Valentine, J. W. 1973. Plates and Provinciality, a theoretical history of environmental discontinuities. In *Organisms and continents through time*, Paleontolog. Assoc. Spec. Pap. Paleontol. 12. ed. N. F. Hughes, 79–92.

Valentine, J. W., and E. M. Moores. 1970. Plate tectonic regulation of biotic diversity and sea level: a model. *Nature* 228:657–9.

Valentine, J. W., and E. M. Moores. 1972. Global tectonics and the fossil record. *J. Geol.* 80:167–84.

Valentine, J. W., B. H. Tiffney and J. J. Sepkoski, Jr. Submitted. Evolutionary Dynamics on Land and Sea: A Comparative Approach. *Palaios*.

Van Valen, L. 1973. A new evolutionary law. *Evolutionary Theory* 1:1–30.

Wei, K.-Y., and J. P. Kennett. 1986. Taxonomic evolution of Neogene planktonic foraminifera and paleoceanographic relations. *Paleoceanography* 1:67–84.

Wing, S. L., and B. H. Tiffney. 1987. The reciprocal interaction of angiosperm evolution and tetrapod herbivory. *Rev. Palaeobotany Palynology* 50:179–210.

Ziegler, A. M., C. R. Scotese, W. S. McKerrow, M. E. Johnson, and R. K. Bambach. 1979. Paleozoic paleogeography. *Ann. Rev. Earth Planetary Sciences* 7:473–502.

Ziegler, A. M., C. R. Scotese and S. F. Barrett. 1983. Mesozoic and Cenozoic paleogeographic maps. In *Tidal friction and the earth's rotation II*, ed. P. Brosche and J. Sundermann, 240–52 Berlin: Springer-Verlag.

Ziegler, A. M., A. L. Raymond, T. C. Gierlowski, M. A. Horrell, D. B. Rowley, and A. L. Lottes. 1987. Coal, climate and terrestrial productivity: the Present and Early Cretaceous compared. In *Coal and coal-bearing strata: Recent advances*, ed. A. C. Scott, 25–49. London: Geolog. Soc. Special Publ. 32.

4

The General Correlation Between Rate of Speciation and Rate of Extinction: Fortuitous Causal Linkages

Steven M. Stanley

An empirical rule of evolution states that rates of speciation and extinction are correlated in the animal world. Higher taxa that enjoy relatively high rates of speciation also suffer relatively high rates of extinction (Stanley 1979). This correlation might have been suspected before 1979, but paleontologists were neither documenting nor investigating it because, in the tradition of Simpson (1944, 1953), they were studying rates of evolution and extinction largely at higher taxonomic levels.

Originally, I suggested that two factors contribute to the empirical correlation: two biological variables, complexity of stereotypic behavior and capacity for geographic dispersal, influence speciation and extinction similarly. Additional factors that contribute to the correlation have since come to light. It seems quite extraordinary that several mutual controls fortuitously link rate of speciation to rate of extinction. My purpose in this article is to provide a current view of this linkage.

THE NEED FOR AN EXPLANATION

On first consideration, it might seem that a correlation between rates of speciation and extinction should require no special explanation other than one based on the requirements for niche partitioning in a crowded ecosystem: within a diverse higher taxon, extinction may be necessary for most types of new species to arise, so that rate of extinction dictates rate of speciation (see chapter 5). The empirical correlation that I documented did not reflect ecological crowding, however. This correlation was between rate of speciation in adaptive radiation and typical rate of extinction for a higher taxon (fig. 4.1). This ruled out control of speciation by extinction. Even taxa that typically speciate rampantly in adaptive radiation are generally unable to escape high rates of extinction while they are diversifying.

Differences between taxa can be so great that even some taxa undergoing radiation do not speciate as rapidly as other groups that are only

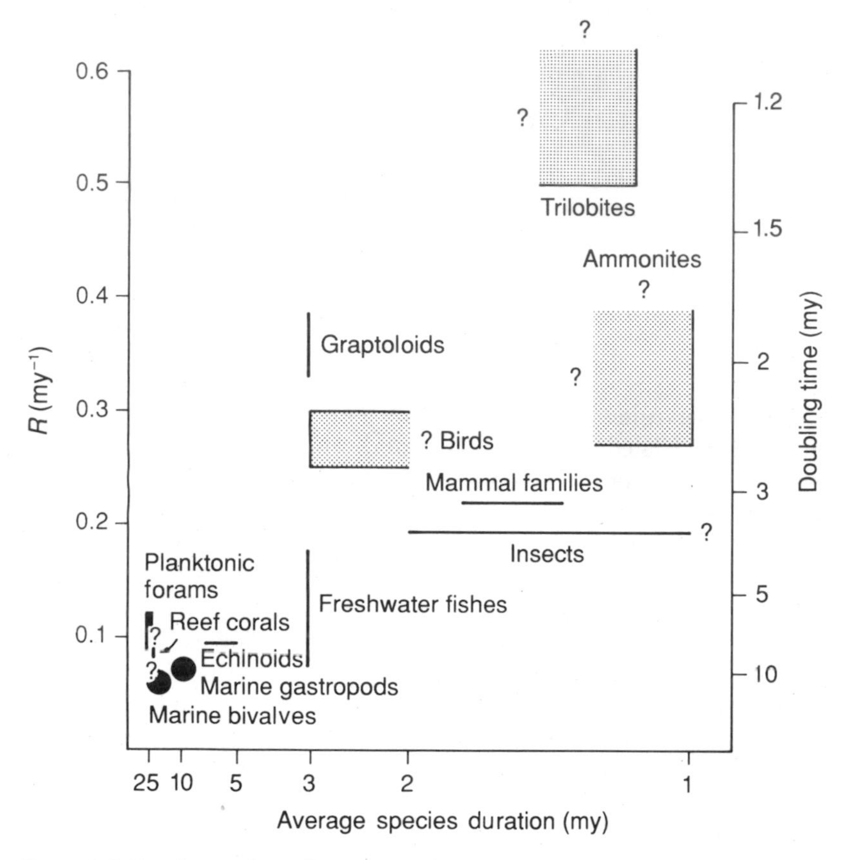

Figure 4.1 Plot showing how, for a variety of animal taxa, rate of speciation in adaptive radiation is correlated with rate of extinction. Mean rate of extinction (E) is the inverse of mean species duration. Average species duration is plotted on an inverse scale here, as a surrogate for E, because it is not a precise mean and because pseudoextinction has not been factored out; thus, the position of a group along the horizontal axis is only a rough estimate of mean rate of extinction. R is measured rate of exponential increase in number of species for a sizeable taxon early in adaptive radiation; this is plotted as a surrogate for rate of speciation because the latter (S) equals R plus E (there is an autocorrelation between S and E). Even though numbers on the graph are not precise, there are such great differences between taxa with high rates and taxa with low rates that the trend of the plotted data must be generally accurate. Small pulses of radiation (ones yielding only a few species) can occur within various taxa at higher values of R than are shown here; the rates plotted here represent radiations that have yielded dozens of species. (From Stanley 1979.)

maintaining approximately constant diversities. For example, as I will describe later, currently radiating clades of siphonate burrowing bivalve mollusks are experiencing much lower rates of speciation (and also extinction) than are clades of nonsiphonate burrowing bivalves that are *not* radiating. This particular difference appears to result from differences between the two groups in the population sizes of individual species—one of the controls to which I ascribe the general linkage between rates of speciation and extinction.

It is, of course, true that a taxon characterized by a high rate of extinction could not radiate if it speciated at a low rate. More generally, any trait that accelerated extinction and suppressed speciation would not only fail to produce adaptive radiation, it would disappear quickly from the world along with any taxon that possessed it. Thus, it is inevitable that the lower right-hand region of figure 4.1 is vacant. On the other hand, a taxon favored with a low rate of extinction would also benefit from having a high rate of speciation in adaptive radiation, being thus blessed with the best of both worlds. In other words, simple arithmetic does not prevent taxa from invading the upper left-hand region of figure 4.1. Rather, causal linkage between rate of speciation and rate of extinction must explain the rarity of taxa having high rates of speciation and low rates of extinction. (I have previously labeled exceptions *supertaxa* and will discuss these later in this paper.) In fact, the linkage must constantly be strained by species selection, as I have defined the process (Stanley 1975, 1979). Species selection, by this broad definition, favors the kinds of species that survive for relatively long intervals and also the kinds of species that speciate at relatively high rates.

Interestingly, a correlation between rate of speciation and rate of extinction has also been documented for vascular land plants (fig. 4.2), although not all of the data employed to establish this correlation represent taxa in the early stages of adaptive radiation (Niklas, Tiffney, and Knoll 1983).

Before examining the mutual controls that link rate of speciation to rate of extinction, I will digress briefly to examine the nature of speciation.

THE GEOGRAPHY OF SPECIATION

Speciation, being a creative process that produces new, often quite distinctive, biological entities, is more problematical than the simpler destructive process of extinction. The geographical pattern of speciation is a controversial aspect of the process that one must examine before addressing the factors that govern rates of speciation.

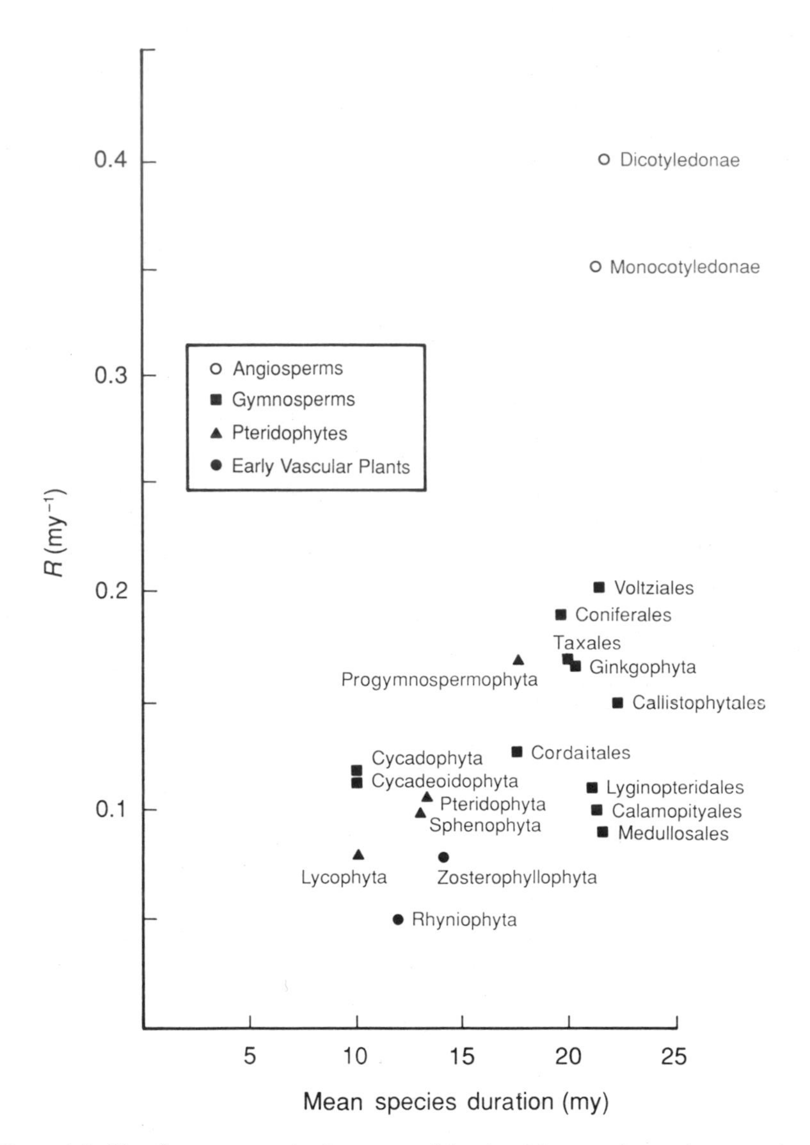

Figure 4.2. Plot, for suprageneric plant taxa, of fractional increase in species per unit time (*R*) against mean species duration. This plot displays a positive correlation that resembles the one shown for animal taxa in figure 4.1, although in this graph, the scale for the abscissa is arithmetic. Taxa belonging to more specialized, modern plant groups have experienced higher mean rates of both diversification and extinction. (After Niklas et al. 1983).

In recent years, a controversy has arisen surrounding the set of concepts known as vicariance biogeography. The central theme of vicariance biogeography is that most taxa diversify by the large-scale fragmentation of their geographical ranges. In the simile of Croizat (1964:204), the process is like the fracturing of a framed piece of glass by a series of blows. At issue is whether this mechanism of diversification dominates in nature or whether diversification by dispersal is more common. This latter mechanism entails the establishment of a small, isolated colony that evolves into a new species beyond the original geographical range of the species to which it once belonged.

In the vicariance scenario, broad dispersal occurs early in the history of a clade, and taxonomic diversification by geographical fragmentation occurs much later; many new subtaxa form via single episodes of extensive fragmentation. In the vicariance pattern, during diversification the geographical range of the descendant clade expands little or not at all beyond the range of the parent group. In contrast, diversification by dispersal entails the sporadic expansion of the parent taxon's range from a center of dispersal, via speciation.

Following Croizat (1958), vicariance biogeographers have viewed an emphasis on diversification by dispcral as a major flaw of the neo-Darwinian paradigm. That diversification by dispersal is important in nature can hardly be doubted, however, following studies such as those of Mayr (1954), Wilson (1961), and Carson (1970) for island faunas. The work of Carson is particularly compelling, in that his chromosomal studies suggest that many species of *Drosophila* in Hawaii have been founded by single gravid females that have arrived at previously unoccupied terrane.

There is no question that evolution has also frequently followed the vicariance pattern in the course of earth history. Strangely, however, proponents of vicariance biogeography have focused upon spatial fragmentation on a vast scale, with plate tectonics playing a major role. On such a scale, most speciation events would entail large, widespread populations rather than small, localized ones. The problem with this notion is that plate movements are much too slow to account for the high rates of speciation for which we have evidence. If an average species survives for roughly 5 million years, then within a global ecosystem not experiencing decline, on the order of a million new species must evolve every million years. Plate tectonic movements are so sluggish, however, that in the course of a million years they have a trivial effect on biogeographical distributions on a global scale: They cannot be producing most speciation.

Even so, there is no question that vicariance can produce much of the diversification that occurs on a much smaller spatial scale, at which geologically rapid fragmentation of environments can produce allopatric speciation during intervals of time measured in tens of thousands of years or less and within regions measured in kilometers or tens of kilometers. Later in this paper I will review some examples of this kind of phenomenon, including ones resulting from environmental changes during the recent ice age. The nature of speciation here is presumably quite similar to that of speciation entailing small populations that become spatially isolated by dispersal.

Causal Factors

If, indeed, the correlation between rate of speciation and rate of extinction reflects common causal factors, speculation about the dominant controls of these rates is constrained. At least some of the dominant controls must tend to elevate or depress both kinds of rates simultaneously. The general correlation of the two kinds of rates is a remarkable empirical rule of macroevolution, but remarkable at a more fundamental level is the fact that not one, but at least five, factors appear to influence the two kinds of rate simultaneously and in the same direction. I will describe the five identified to date.

BEHAVIORAL COMPLEXITY

Another general rule of nature is that higher taxa that exhibit advanced stereotypic behavior tend to experience high rates of extinction (Stanley 1979:265–7, fig. 9–2). Examples are most groups of mammals, trilobites, and ammonoids, which have mean species durations of 3 million years or less (Stanley 1979; Ward and Signor 1983). These groups can be contrasted to such behaviorally simple taxa as foraminifera, corals, and bivalves, in which mean species duration is 10 million years or more (Stanley 1979; Buzas and Culver 1984; Foster 1986). It is true that we lack detailed information on the habits of trilobites and ammonoids, but comparisons with living arthropods and cephalopods leave little doubt that these extinct taxa were behaviorally advanced in comparison to the single-celled foraminifera, the acephalic bivalves, and the sedentary corals. I would include in this general category sexually selected behavior and attendant morphologies and what West-Eberhard (1983) has termed *social selection*. Complex stereotypic behavior can be looked upon as a form of ecological specialization, narrowing certain dimensions of the niche hyperspace. This specialization renders species vulnerable to environmental change; on the average, it must accelerate rate of extinction.

We can predict that advanced stereotypic behavior should also accelerate rate of speciation. Relevant here is the conclusion of Mayr (1963:95) that "ethological barriers to random mating constitute the largest and most important class of isolating mechanisms in animals." This is indeed likely to be true for animals with complex behavior such as vertebrates (Wilson et al. 1975), but it cannot apply to very simple creatures. For marine animals that employ external fertilization, for example, timing of reproduction may be the only "behavioral" variable that can serve as an isolating mechanism. The facts appear to bear out these predictions. Behaviorally advanced taxa such as mammals, birds, trilobites, and ammonoids have generally speciated much more rapidly than creatures such as foraminifera, corals, and bivalve mollusks characterized by simple behavior (Stanley 1979).

Although most plants exhibit exceedingly simple behavior—tropism, for example—or none at all, species-specific pollinating mechanisms must play a role in the extinction and speciation of angiosperms similar to the role of advanced stereotypic behavior in animals. Clearly, the extinction of an exclusive pollinator must result in the extinction of the host plant; hence, higher taxa in which many species are highly specialized with respect to pollinators should tend to experience relatively high rates of extinction. Such taxa should enjoy relatively high rates of speciation as well, simply because a switch to the new pollinator will serve as a ready isolating mechanism—one less probable for species with multiple pollinators and impossible for wind-pollinated forms.

NICHE BREADTH

It seems evident that rate of speciation and rate of extinction are both negatively correlated with niche breadth. This idea seems first to have been proposed for marine organisms, in which eurytopic species of unstable near-shore or estuarine environments have been contrasted to stenotopic normal marine species (Slobodkin and Sanders 1969; Jackson 1974). Speciation may be promoted in stenotopic species by their tendency to be deployed in the form of unstable, patchy populations that are at the mercy of environmental fluctuations. An additional factor, however, is the greater opportunity that stenotopic taxa have for the invasion of new niches via speciation; eurytopic species are afforded few possibilities for niche partitioning. Stenotopic species should also suffer relatively high rates of extinction. Like behavioral specialists, discussed in the previous section, they are especially vulnerable to environmental change. Wilson (1985), for example, found that for the ant fauna preserved in Hispaniolan amber about 20 million years old, the

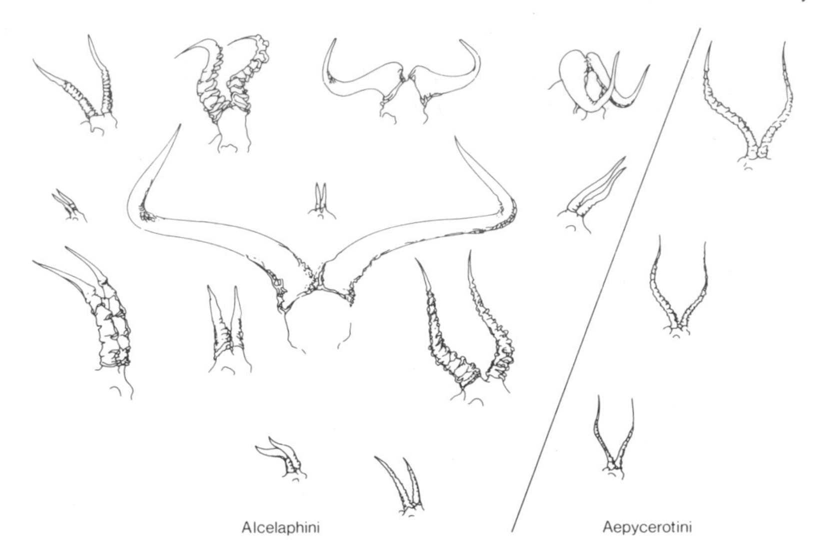

Figure 4.3. Diversity of horn and frontlet morphologies for species of two African bovid tribes (drawn to scale). Both fossil and Recent examples are illustrated. The impalas (right of diagonal line) display little diversity. (From Vrba 1984a.)

incidence of extinction was relatively high for genera highly specialized in predatory habits, social parasitism, or nest site preference.

I favor the ideas of the preceding paragraph, but Vrba (1980, 1984b) introduced a different hypothesis as to why stenotopic species should speciate and suffer extinction at high rates. In her view, "stenotopes should more frequently be subject to directional selection pressure at the levels of genotypes and phenotypes, and consequently to population divergence, speciation and extinction" (Vrba 1984b:326). This idea was conceived in part to explain the fact that alcelaphine bovids (the blesbuck-hartebeest-wildbeest group), which are trophic specialists, have speciated at much higher rates than the aepycerotine bovids (impalas), which are trophic generalists. Three competing or supplementary hypotheses can be erected, however, in light of other factors addressed in the present chapter:

1. The alcelephines exhibit a wide variety of horn morphologies (fig. 4.3); inasmuch as horn morphology seems to be an important aspect of mate recognition in such animals (Vrba 1984a), it seems possible that specialized species-specific behavior promotes reproductive isolation in this group, as described in the preceding section.

2. As outlined in the foregoing paragraph, trophic specialization among acephalines may facilitate speciation that entails niche partitioning; in contrast the great trophic generalization of the impala (there is only one living species) may inhibit further speciation.

3. Although many alcephaline species are relatively mobile, these species migrate en masse, which would seem to permit fragmentation; all alcephaline species except the blue wildebeest are vastly less abundant and less densely distributed than impalas (Vrba 1984a)—a condition that, as I will describe in the following section, may promote speciation. Perhaps tests can be devised to compare the relative merits of these hypotheses.

POPULATION SIZE AND STABILITY

From simple numerical considerations, we can predict that a species with a small total population size would be relatively vulnerable to extinction. The same should be true of a species with a highly unstable total population—one experiencing fluctuations that frequently bring it to low numerical levels. Indeed, studies have confirmed this prediction for both marine bivalve mollusks and planktonic foraminifera of Neogene age (Stanley 1986; Stanley et al. 1988 for pub.).

The significance of the size and stability of populations is revealed by differential survivorship for several ecologically discrete sets of bivalve species during the past 2 million years along the west coast of North America and in waters bordering Japan (Stanley 1986). The patterns here are as follows:

1. Burrowing species with siphons experience much greater mean geological longevity than nonsiphonate taxa (fig. 4.4A). The latter, being sluggish burrowers that live in shallowly buried life positions, are highly vulnerable to predation. Predation is the dominant factor limiting population sizes of marine bivalves (Stanley 1973), and it appears that for this reason nonsiphonate species are typified by relatively small, unstable populations.

2. For bivalves, as for other taxa, species of small body size are more populous than species of large body size (fig. 4.4B). Siphonate burrowers offer excellent opportunities to study the effects of this disparity in that they represent an ecologically coherent group that is also rich in number of species. As would be expected, it turns out that siphonate burrowing species of large body size exhibit much lower survivorship than ones of small body size.

3. Scallops (Pectinidae) are unusual among epifaunal bivalves (ones that live *on* rather than *within* the substratum) in that many of their species inhabit exposed subtidal settings, where predators abound. Apparently it is the scallops' unusual ability to swim, if only awkwardly, that gives them the necessary edge against predation. Even so, many extant species are rare, as attested to by their value to shell collectors. Again, the prediction based on population size is borne out: survivorship is much lower for scallop species than for other epifaunal bivalves

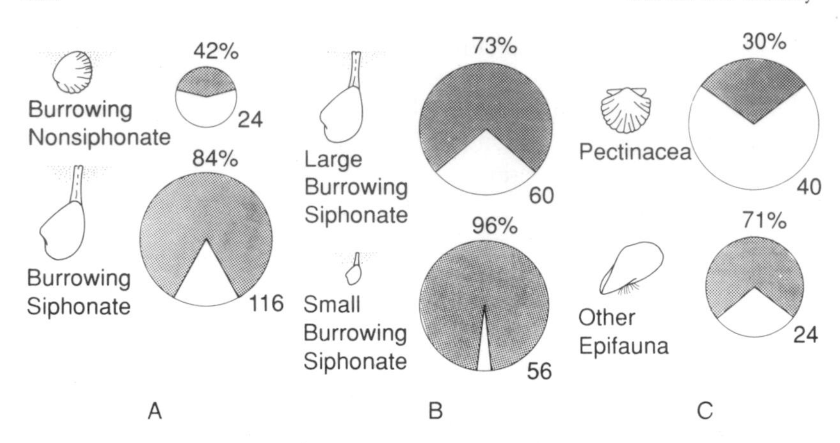

Figure 4.4. Differences in rate of extinction between groups of marine bivalve mollusks, as illustrated by differential survivorship for species that constituted the faunas of Japan and California 2 million years ago. Differences between pairs of groups are statistically significant and are observed for each of the two geographical areas considered independently. Percentages represented the fraction of species that have survived to the Recent. Number at lower right of a pie diagram gives sample size. (From Stanley 1986.)

(fig. 4.4C). Most of the latter experience less intense predation because they live in the intertidal zone or in bays or lagoons where salinities are reduced—or they live partly bored into or nestled within hard substrata. Interestingly, whereas mean species duration for scallops has been perhaps 3 million years during Neogene time, it was approximately 20 million years during Jurassic and Cretaceous time. During this earlier interval, predation on the seafloor was almost certainly less intense than it is today (Stanley 1974, 1977; Vermeij 1977).

There is also evidence that rate of speciation is relatively high for bivalve taxa with small, unstable populations (Stanley 1986). During an early stage of adaptive radiation, net rate of increase in number of species is approximately exponential. It equals rate of speciation minus rate of extinction. Calculations reveal that during adaptive radiations of siphonate burrowing bivalves, rate of speciation has been in the neighborhood of 0.12 per million years. Rate of extinction for *non*siphonate burrowing bivalves has been almost twice as high as this (fig. 4.5). The implication is that for a typical clade of the latter simply to maintain constant diversity, rate of speciation must be higher than it is within a typical radiating clade of siphonate bivalves. This may seem strange, when nonsiphonate groups have been suffering high rates of extinction. The answer seems to be that heavy predation not only accelerates rate of extinction, by reducing and destabilizing populations, but it also accelerates rate of speciation. I suggest that the explanation is what I have termed the *fission effect:* a destructive agent such as predation can elevate rates of extinction within a victimized taxon but, in fragmenting and destabilizing populations, it also elevates the rate of speciation (Stanley 1986). A vicariance pattern is implied, but one in which the isolated populations that become new species may be quite small and one in which the cause of fragmentation is biological rather than physical.

As illustrated in figure 4.5, more severe predation pressure can be expected to suppress speciation by preventing isolates from expanding into new species. While suppressing speciation, such extremely heavy predation, sustainable only by predators with alternative food resources, must also accelerate extinction. This double-edged sword will ultimately lead to the demise of the victimized clade. Thus, I am invoking only moderately heavy predation as a factor that promotes both speciation and extinction, not devastatingly heavy predation.

Other environmental factors that diminish and destabilize populations to a moderate degree should also simultaneously accelerate rates

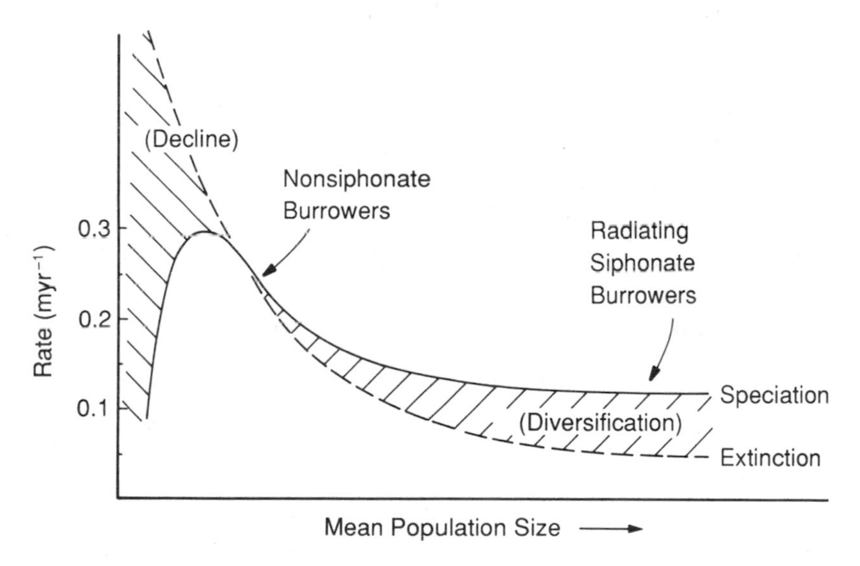

Figure 4.5. A model depicting the fission effect for burrowing bivalves of Neogene age. Rates of speciation and extinction for nonsiphonate taxa, which typically are not radiating, and for radiating siphonate taxa are empirically based. Rates of speciation and extinction are apparently high for nonsiphonate burrowers because heavy predation fragments and diminishes their populations. Presumably, as illustrated on the left side of the graph, still more severe predation would suppress rate of speciation and increase rate of extinction. (From Stanley 1986.)

of speciation and extinction. Among these is high variability of food supply in space and time. This kind of control appears to have produced a relatively high rate of evolutionary turnover in the globorotaliid clade of planktonic foraminifera during the past 20 million years (Stanley et al., 1988). Most extant globorotaliid species live relatively deep in the oceanic water column, flourishing only where upwelling brings cool water into the photic zone, where food is plentiful. The populations that constitute entire species are not only patchy and unstable, but apparently small, in aggregate. Rate of speciation and rate of extinction have been higher for the globorotaliids than for the other large Neogene clade of planktonic foraminifera, the globigerinid group (fig. 4.6). Globigerinids tend to occupy shallower waters, where stable nutrition appears to foster great abundance. Not only is food relatively plentiful here, for heterotrophy, but high light levels sustain symbiotic algae, which many globigerinids possess but deeper-dwelling globorotaliids do not.

While nutritional stability of the sort just described can be viewed as a biotic variable influencing rates of speciation and extinction, the physical environment is the more fundamental control here; it governs the availability of food resources.

DISPERSAL ABILITY

Dispersal ability has been widely viewed as inversely correlated with rate of speciation and rate of extinction. This idea has been developed especially well for benthic marine organisms, some of which have planktotrophic larvae (ones that feed while in plankton) and some of which have larvae that are nonplanktonic or that are planktonic but short lived because they do not feed (see, for example, Hansen 1978, 1980; Jablonski 1982). Great potential for dispersal has been seen as favoring the survival of a species by promoting geographical expansion. At the same time, effective dispersal has been considered to reduce the probability of speciation by reducing the likelihood of geographical isolation.

Data on the effects of dispersal ability are sparse, but the predicted relationships have been reported for gastropods of Paleogene age (Hansen 1978, 1980) and Cretaceous age (Jablonski 1982). On the other hand, late Neogene bivalves of western North America fail to show the predicted relationship for extinction: Narrowly distributed species have not, in general, survived for shorter intervals than widely distributed species. Furthermore, species lacking planktotrophic larvae have not suffered relatively high rates of extinction (Stanley 1986). Here, geographical range is, at most, of second-order importance in determining probability of extinction; as noted earlier, population size is of first-

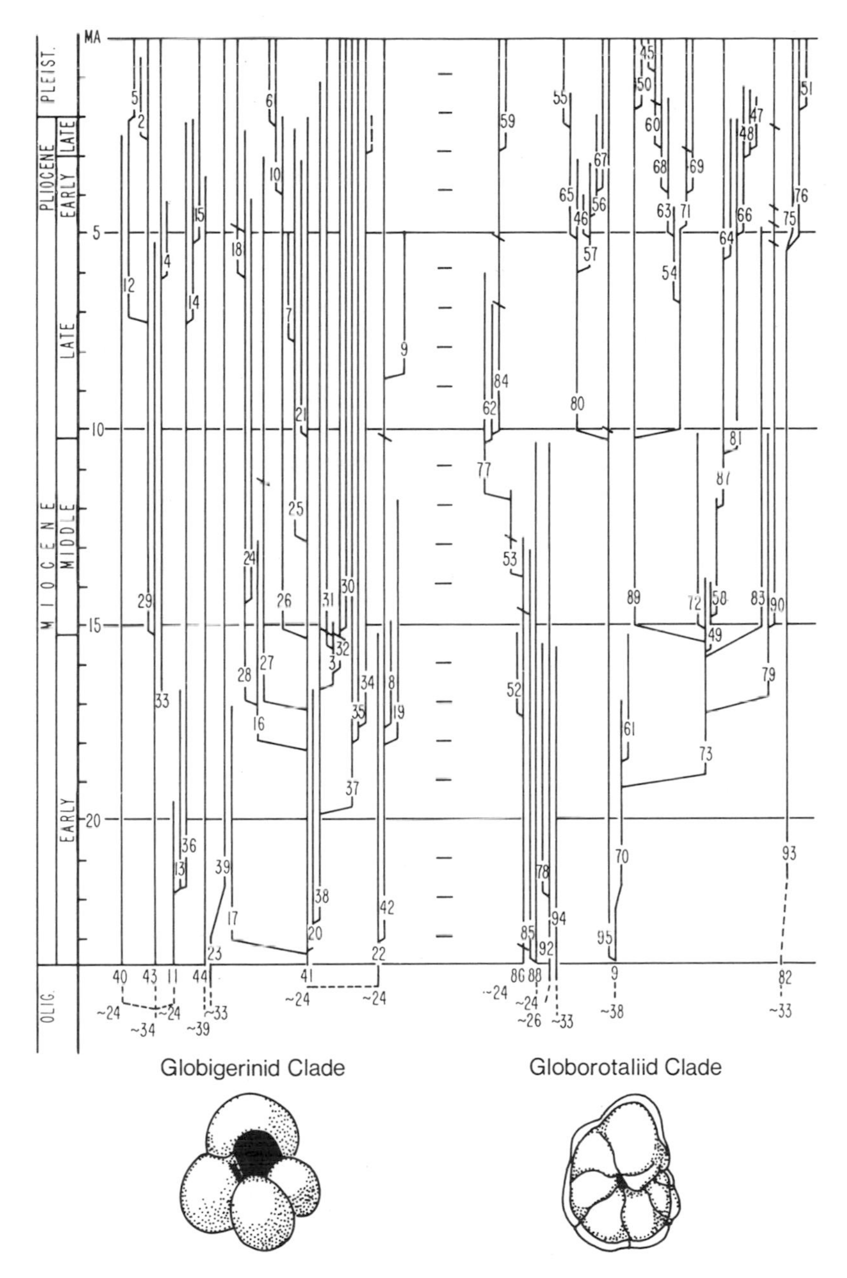

Figure 4.6. Phylogenies for the two major clades of Neogene planktonic forminifera. Each vertical bar represents a single lineage. Diagonal lines connecting lineages represent speciation events. Dashes extending into the Oligocene indicate earlier histories for several lineages. Rates of speciation and extinction have been higher in the globorotaliid clade than in the globigerinid clade. See Stanley et al. (1988) for the identities of lineages, estimates of actual rates, and statistical tests.

order importance. In the Discussion section I will suggest why the relative importance of population size and dispersal ability varies from situation to situation.

Actually, there is evidence that the relationship between dispersal ability and rate of speciation is more complex than has been envisioned by some workers. Levin and Wilson (1976) documented higher rates of both extinction and speciation for herbs than for woody plants and attributed this contrast to the small and unstable effective population sizes of herbs and to the high powers of dispersal of these plants. These writers hypothesized that for herbs, which occupy unstable, patchy habitats, effective dispersal promotes speciation because it produces numerous small isolated populations. Perhaps the same principle holds for many animals and animallike taxa. For example, the patchy, unstable populations within such groups as scallops and globorotaliid foraminifera (described earlier) may cause the highly effective dispersal mechanisms that characterize these taxa to promote speciation by generating isolates. If this is the case, we can recognize an important principle as to how dispersal affects speciation: *For species that are characterized by stable, relatively continuous geographical distributions, effective dispersal will retard rate of speciation by opposing the formation of isolates. (Rate of extinction will also be low because the total population will be large, widespread, and stable.) On the other hand, for species characterized by patchy or unstable populations, effective dispersal will promote speciation by generating isolates. (Rate of extinction will also be high because of the instability.)* Note that in both scenarios, the linkage between rate of speciation and rate of extinction remains.

Deterioration of the environment is frequently patchy. This kind of change and other kinds of habitat fragmentation are haphazard with regard to the distribution of species. Cracraft (1982) has suggested that this kind of serendipitous environmental control may govern most rates of speciation and extinction on a larger scale, with plate tectonic events serving as the driving force. As noted earlier, one problem here is that plate tectonic events are much too infrequent to account for the large majority of speciation events. A second problem is that this model, which relies on geographical fragmentation on a vast scale, predicts that those species that have the broadest geographical ranges should speciate at the highest rates (fragment most). This relationship has never been documented, whereas, as noted previously, the reverse relationship has been observed.

HABITAT FRAGMENTATION

There is evidence that during the recent ice age, environmental deterioration on a small scale has accelerated both extinction and speciation in South America and Africa. The speciation appears to have resulted from a repeated vicariance pattern of fragmentation. This amounts to a kind of fission effect, in which climatic change rather than predation has been the agent. In South America and Africa, climatic change has impinged on animal life during the past 3 million years primarily by altering the distribution patterns of vegetation. In South America, birds, reptiles, and insects adapted to forest habitats have experienced accelerated extinction and speciation at times when forests have shrunk and fragmented (review by Haffer 1981). Similarly, in Africa, antelopes experienced a high rate of species turnover after 3 million years ago, when climates and vegetation patterns apparently began to experience severe fluctuations (Vrba 1985). We can envision that under some circumstances such differential rates of speciation and extinction would follow taxonomic lines. Thus, repeated deterioration of habitats of a particular type has the potential to produce an evolutionary pattern in which rates of speciation and extinction are correlated.

Discussion

Many questions remain about the macroevolutionary roles of the variables that link rate of speciation to rate of extinction. Why should different variables prevail as determinants of rates under different circumstances? How may the linkage be strained, which is to say, what accounts for the existence of supertaxa? And finally, how large a role does mass extinction play in determining patterns of extinction? (The agents of mass extinction certainly do not accelerate speciation while a biotic crisis is in progress.) These are the final questions that I will address.

THE RELATIVE IMPORTANCE OF CONTROLS

The five identified factors that influence rate of speciation and rate of extinction must vary in relative significance from taxon to taxon and from situation to situation. Generalizations about the relative importance of the controlling factors are not easily made. I will, however, explore how two of the identified variables, geographical range and population size, may vary in relative importance in influencing rates of extinction.

Recognizing that the extinction of a species represents the diminution of both geographical range and population size to zero, what we are asking is which of these two variables plays the greater role in determining probability of extinction under particular circumstances. In general, the two variables are correlated: population size tends to increase with geographical range. The correlation is sometimes a weak one, however. Among siphonate bivalve species, for example, species of large body size are, on the average, somewhat more broadly distributed along the west coast of North America than are species of small body size; even so, the mean population densities of large forms are orders of magnitude lower than those of small forms, so that the latter have much larger total populations (Stanley 1986).

Relevant to our question are the quantitative relationships between probability of extinction and geographical range and population size. For population size, Diamond (1984) noted that if deaths are independent events, the probability that a population of n individuals will become extinct during an interval of time equals p^n where p is the probability of death for any individual during the interval. Thus, an arithmetic plot of probability of extinction against population size takes the form of a hollow curve (fig. 4.7A). Certainly some deaths are interconnected (catastrophic), which as Diamond noted will flatten the curve somewhat, but empirical plots are nonetheless hollow (fig. 4.7B).

Empirical plots that reflect the effect of area but not population size may not be available at present. For island faunas, plots of local extinction rates versus island size produce hollow curves (Diamond 1984), but a positive correlation between area and population size is likely, and this would mean that the effect of population size could account for the relationship. The question to be resolved is how for a particular population size the probability of fatal environmental deterioration during an interval of time varies with geographical area. The situation here parallels that described in the previous paragraph for population size. Let n be the number of areas of unit size and p be the probability of one area becoming uninhabitable. Then, assuming that each unit behaves independently and is equally vulnerable, the probability of total extinction by environmental deterioration will be p^n. In other words, a hollow curve like that for population size will obtain. Again, the assumption that units will behave independently is unrealistic (mass extinctions, for example, strike across broad areas), but this will only flatten the curve to some degree: It will resemble the hollow curves that relate probability of extinction to population size (fig. 4.7).

The hollow curve relationships have important implications. When we find no strong relationship between rate of extinction and either population area or population size for the species of a taxon, it may be because we are considering a range of population areas or population sizes that is positioned entirely within the asymptotic segment of the curve. For example, as figure 4.8 illustrates, I suggest that in my analysis of species survivorship for West Coast bivalves (Stanley 1986) the

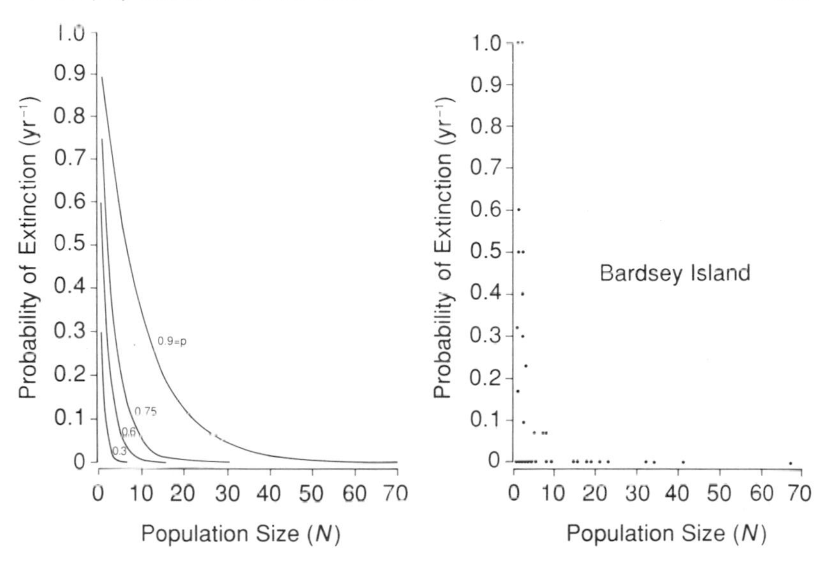

Figure 4.7. Probability of extinction as a function of population size. **A,** Predicted relationship, assuming that deaths are independent events, with a probability per unit time of p; probability of extinction is then p^n, as illustrated for various values of p by the family of curves (p values shown beside curves). **B,** Observed relationship for local extinctions of bird species on Bardsey Island, off the west coast of Britain, from 1954 to 1969; here, even though deaths have undoubtedly not all been independent events, a hollow curve still obtains. (From Diamond 1984.)

approximately twofold difference between taxa having "small" and "large" mean geographical ranges for component species has had a trivial effect on survivorship because these mean ranges fall relatively close to one another along the gently sloping right-hand tail of the curve. A set of species that all occupy nearly the entire Indo-Pacific region might well exhibit significantly greater mean longevity because their mean geographical range is vastly larger. In fact, the Indo-Pacific molluscan fauna as a whole, which includes a number of very widespread species, has apparently experienced a remarkably great mean species longevity during Neogene time (fig. 4.9).

Of course, different taxa will be characterized by different hollow curves. The range of values over which population size or population area will have strong effects will thus vary from group to group. It appears that for benthic foraminifera, probability of extinction for species fails to correlate strongly with total population size or geographical range (Martin Buzas, pers. comm., 1987). If this is true, it probably reflects the fact that for this group nearly all species are positioned far

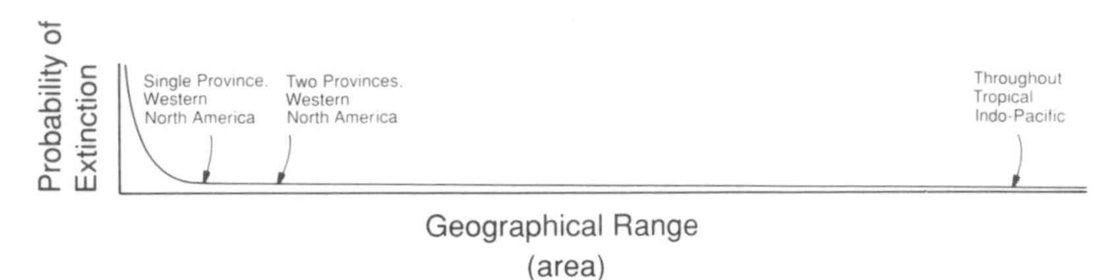

Figure 4.8. Hypothetical probability of extinction for species occupying regions of the Pacific. A species found throughout two contiguous provinces along the west coast of the Americas has only a slightly lower probability of extinction than a similar species found throughout only one province. On the other hand, a similar species ranging throughout the vast tropical Indo-Pacific realm has a substantially lower probability of extinction; although the west coast species are positioned along the tail of the curve, the tropical Indo-Pacific species occupies a much more distal position, which significantly reduces its probability of extinction.

out on the tail of each of the two hollow curves. Even relatively rare species of foraminifera, being minute, are extremely abundant, and most foraminiferan species are very widespread (many are cosmopolitan).

SUPERTAXA

Earlier I noted the existence of a few taxa that radiate at extremely high rates relative to other similar taxa, because they somehow successfully strain the linkage between rate of speciation and rate of extinction. The Muridae, the family of rodents that includes the Old World rats and mice, appears to represent one of these groups, which I have termed *supertaxa*. The murids may owe this status to traits that I have focused upon in this article. The murids are very small mammals with weak powers of dispersal; this presumably enhances their probability of speciation. At the same time, they are herbivores that frequently have such enormous populations, despite having narrow geographical ranges, that the extinction of a typical species would seem to be relatively improbable. An unusual combination of high rate of speciation and low rate of extinction may explain why the murids have radiated at a more rapid rate than any other large group of Neogene mammals (Stanley 1979:105–7).

A trait that enhances rate of speciation without linkage to rate of extinction can also produce supertaxon status. The cichlid fishes appear to be such a supertaxon (Stanley 1979:281). The cichlids are renowned for their rampant speciation in the large African rift valley lakes and other youthful bodies of water. Their tendency to radiate rapidly appears to result from their exceptional morphogenetic flexibility (see Dorit, this volume). As Liem (1973) has shown, the cichlids' special pharyngeal jaw apparatus allows them readily to evolve many different feeding mechanisms, which permit invasion of a variety of niches. This condition has no evident bearing on rate of extinction. The cichlids seem to speciate at extraordinary rates while experiencing rates of extinction that are not unusually great. This, it seems, is sufficient to make them a supertaxon.

The cheilostome bryozoans may represent another example. Among benthic marine taxa, their rate of adaptive radiation in Late Cretaceous time was exceptionally high, yielding an R-value of 0.12 to 0.15 my^{-1} (Stanley 1979:243–4). Taylor (1988) attributed this to a high rate of speciation for clades with nonplanktotrophic larvae: The cheilostome radiation followed rapidly upon the evolution of brooding structures, and brooding species constitute well over 90% of the recent fauna.

Taylor proposed that dispersal of these faunas by infrequent rafting of colonies has favored the isolation of small populations and has thus promoted speciation—but has also produced populations typically as widespread as those of planktotrophic species, so that the two groups may not have differed in rate of extinction.

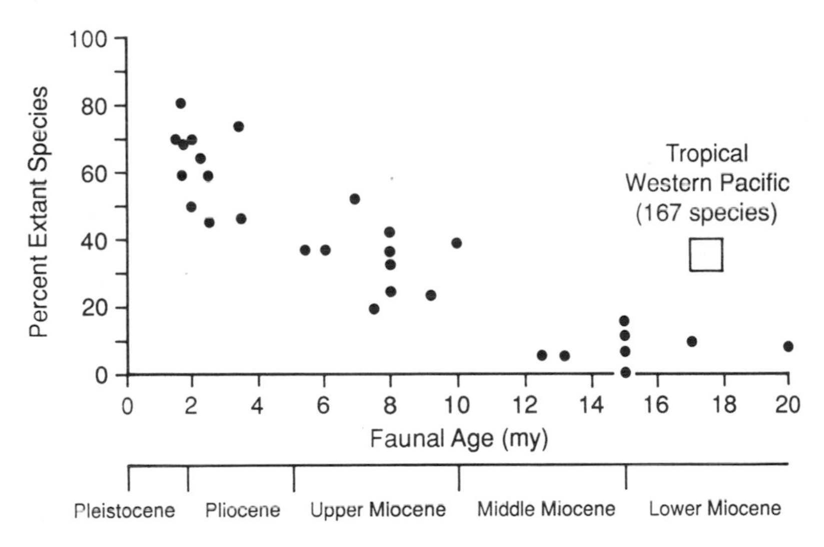

Figure 4.9. High survivorship of tropical western Pacific gastropod species from the lower Miocene (square) in comparison survivorship for gastropod species constituting fossil faunas of California and Japan (dots). The latter are, on the average, much less widely distributed. (From Stanley 1979.)

I previously noted that traits resembling those of supertaxa may accrue to species having biogeographical configurations like those of shallow-water benthic marine species of the western Pacific (Stanley 1979:298–9). In this region, even species with excellent capacities for dispersal, enormous geographical ranges, and demonstrably low rates of extinction may speciate at relatively high rates because their habitats are patchy (island-dependent). Vermeij (1987) elaborated upon this idea and conducted tests that seem to confirm it. He found that rate of speciation in tropical Pacific gastropods seems to increase with dispersal ability.

The apparent examples of supertaxa that I have cited appear to be quite unusual. The existence of these exceptions, while interesting, does not overturn the general rule of evolution that rates of speciation and extinction tend to be correlated among taxa.

THE IMPORTANCE OF CATASTROPHIC EXTINCTION

Finally, I want to point out that the normal rate of species turnover in a higher taxon may strongly influence the degree to which mass extinction alters the history of that higher taxon by eliminating species. Bearing on this question is the issue of whether the severe pulses of extinction that we refer to as mass extinctions entail an accentuation of normal or "background" rates of extinction or, alternatively, whether the relative impact of mass extinction on a particular higher taxon has little relationship to the background rate of extinction for that taxon.

I have shown that if mass extinction represents an intensification of background rates, then higher taxa with normally high rates of turnover are especially vulnerable to mass extinction (Stanley 1979). This conclusion was based on the fact that if a crisis interval raises rate of extinction or lowers rate of speciation by a certain percentage—or does both—then taxa with higher background rates than others will suffer more catastrophic declines in diversity (fig. 4.10) (see also chapter 5). Some support for this proposition comes from the observation that three invertebrate groups characterized by unusually high rates of species turnover—the trilobites, graptolites, and ammonoids—all were also plagued by multiple mass extinctions, including a final one that was terminal. Looking at the other side of the coin, it is difficult to imagine that certain taxa with very low rates of background extinction, such as the benthic foraminifera or marine Bivalvia, could disappear entirely in a mass extinction. Indeed, these latter taxa appear never to have been struck as hard by mass extinctions as were the trilobites, graptolites, and ammonoids, repeatedly.

On the other hand, there is evidence that the impact of some mass extinctions on certain taxa has departed from normal background pat-

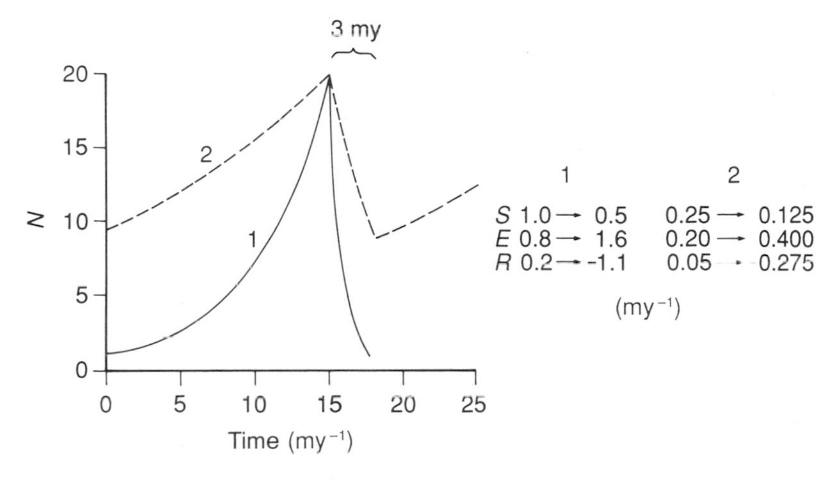

Figure 4.10. Evolutionary instability associated with high rates of evolutionary turnover (high rates of S and E). **A,** Hypothetical example in which two radiating taxa are characterized by values of S, E, and R that are proportional (four times as large in taxon 1 as in taxon 2). After 15 million years of radiation, when each taxon contains 20 species, an interval of severe environmental deterioration doubles rates of extinction and cuts rates of speciation in half. The result for the taxon characterized by high initial rates is a precipitous decline, whereas the taxon characterized by low initial rates declines only modestly, recovering after 3 million years, when environmental conditions return to normal. (From Stanley 1979.)

terns. Subtaxa that normally experience relatively low rates of extinction were preferentially victimized within the Bryozoa during the late Ordovician global crisis (Anstey 1978); within the Bivalvia (Stanley 1982, 1986) and the planktonic foraminifera (Stanley et al., 1988) during heavy Plio-Pleistocene marine extinction; and within the Mollusca during the Cretaceous global crisis (Jablonski 1986). These departures from normal patterns have been taken to indicate that the causes of the various crises differed from normal agents of extinction for the affected groups.

If the pattern described in the preceding paragraph prevails—if the impact of mass extinction on particular taxa is largely independent of the normal rates of extinction within these taxa—then an interesting contrast emerges. Taxa that normally experience low rates of background extinction will suffer a relatively large percentage of their total losses during mass extinctions. Taxa that normally experience high evolutionary turnover will suffer a smaller percentage of their total losses during mass extinctions. The two major groups of planktonic foraminifera of Neogene age provide an example of this kind of contrast. The globorotaliid clade has experienced significantly higher rates of evolutionary turnover than the globigerinid clade during noncrisis intervals.

During the climatic deterioration of Plio-Pleistocene time, however, the globigerinid clade suffered a much more severe increase in the rate of extinction. The result of this pattern is that a much higher fraction of total Neogene extinctions in the globigerinid clade occurred during the crisis interval: 16 of 26 (62%) versus 13 of 30 (43%) for the globorotaliid clade (see fig. 4.5).

The pattern that I have just described has consequences for the role of species selection in large-scale evolution. Taxa such as the globigerinids, which experience slow turnover but suffer heavily during crisis intervals, must be relatively weakly affected by the differential extinction component of species selection during most of their history. They will, however, be subject to pronounced *catastrophic* species selection. Analogous to catastrophic selection at the population level, catastrophic species selection is essentially mass extinction that selects for certain biological traits. Among globigerinids, for example, spiny taxa that harbor symbiotic algae have been struck especially hard by mass extinction (Keller 1986). Planktonic foraminiferan species of very small proportions were the principal survivors of the terminal Cretaceous mass extinction (Gerta Keller, pers. comm., 1986).

Conclusions

A notable empirical rule states that rates of speciation and rates of extinction are strongly correlated in the animal world: High rates of both processes have characterized some taxa and low rates have characterized others. This association is not arithmetically necessary. In fact, species selection tends to produce taxa that are characterized by high rates of speciation and low rates of extinction, yet only a few "supertaxa" exhibit this favored combination of traits when compared with other related taxa.

The general correlation between rate of speciation and rate of extinction results from the fact that the factors that operate as the dominant controls of the two rates fortuitously affect each in the same way (positively or negatively). On the average, rates of speciation and extinction increase with complexity of stereotypic behavior, decrease with niche breadth, and decrease with size and stability of the populations that constitute species. The situation for dispersal ability appears to be more complex: For species characterized by unstable, patchy population, effective dispersal promotes speciation and probability of extinction is also relatively high. In contrast, for groups of species whose populations are stable and continuously distributed, effective dispersal is associated with low rates of speciation and extinction. Under some circumstances, habitat fragmentation also tends to promote speciation and extinction simultaneously for certain kinds of species. Of these controls, three are intrinsic biological traits: behavioral complexity, niche breadth, and dispersal ability. One control, habitat fragmentation, is usually abiotic in ultimate causation. Size and stability of populations, on the other hand, are influenced by both biological and physical factors.

Whether any one of the five identified controlling variables plays a primary role in producing disparate rates of speciation and extinction among two or more biological groups under certain circumstances depends in part upon the degree to which that particular control differs between the two groups in question. A plot of the probability of extinction for a species against population size or biogeographical area takes the form of a hollow curve. Whether population size or biogeographical area is a significant variable in producing disparate rates of speciation and extinction between two taxa depends upon where the values for the two taxa are positioned on the hollow curve. If, for example, both taxa are represented by values that lie close together on the asymptotic tail of the curve for a variable, this variable will have only a weak differential effect on rates for the two subtaxa.

The linkage between rate of speciation and rate of extinction can be strained by the presence of a trait that favors speciation without affecting rate of extinction as profoundly. Weak dispersal may be such a trait for murid rodents, promoting speciation while failing to accelerate extinction, because population density in this group tends to be so enormous as to offset the effects of narrow geographical range. Morphogenetic flexibility appears to be an example of such a trait for the cichlid fishes.

For taxa that are characterized by low rates of species turnover under normal circumstances, pulses of extinction during biotic crises account for an unusually large percentage of total species extinctions. These pulses may exert a relatively strong control over the composition of such taxa through the operation of catastrophic species selection.

Literature Cited

Anstey, R. L. 1978. Taxonomic survivorship and morphologic complexity in Paleozoic bryozoan genera. *Paleobiology* 4:407–18.

Buzas, M. A., and S. J. Culver. 1984. Species duration and evolution: Benthic foraminifera on the Atlantic continental margin of North America. *Science* 225:829–30.

Carson, H. L. 1970. Chromosome tracers of the origin of species. *Science* 168:1414–8.

Cracraft, J. 1982. A nonequilibrium theory for the rate-control of speciation

and extinction and the origin of macroevolutionary patterns. *Syst. Zool.* 31: 348–65.
Croizat, L. 1958. *Panbiogeography,* vols. 1, 2a, 2b. Published by the author, Caracas.
Croizat, L. 1964. *Space, time, form: the biological synthesis.* Published by the author, Caracas.
Diamond, J. M. 1984. "Normal" extinctions of isolated populations. In *Extinctions,* ed. M. H. Nitecki, 191–246. Chicago: University of Chicago Press. © 1984 by The University of Chicago.
Foster, A. B. 1986. Neogene paleontology in the northern Dominican Republic. 3. The family Poritidae (Anthozoa: Scleractinia). *Bull. Am. Paleontol.* 90:45–123.
Haffer, J. 1981. Aspects of Neotropical bird speciation during the Cenozoic. In *Vicariance biogeography. a critique,* ed. G. Nelson and D. E. Rosen, 371–405. New York: Columbia University Press.
Hansen, T. 1978. Larval dispersal and species longevity in lower Tertiary gastropods. *Science* 199:885–7.
Hansen, T. A. 1980. Influence of larval dispersal and geographic distribution on species longevity in neogastropods. *Paleobiology* 6:193–209.
Jablonski, D. 1982. Evolutionary rates and modes in Late Cretaceous gastropods: Rate of larval ecology. *Third North Am. Paleontol. Conv. Proc.* 1:257–62.
Jablonski, D. 1986. Background and mass extinctions: the alteration of macroevolutionary regimes. *Science* 231:129–33.
Jackson, J. B. C. 1974. Biogeographic consequences of eurytopy and stenotopy among marine bivalves and their biogeographic significance. *Am. Nat.* 104:541–60.
Keller, G. 1986. Stepwise mass extinctions and impact events: late Eocene to early Oligocene. *Marine Micropaleontol.* 10:267–93.
Levin, D. A., and A. C. Wilson. 1976. Rates of evolution in seed plants: Net increase in diversity of chromosome numbers and species numbers through time. *Proc. Nat. Acad. Sci. U.S.A.* 73:2086–90.
Liem, K. F. 1973. Evolutionary strategies and morphological innovations: Cichlid pharyngeal jaws. *Syst. Zool.* 22:425–41.
Mayr, E. 1954. Change of genetic environment and evolution. In *Evolution as a process,* ed. J. Huxley, A. C. Hardy, and E. B. Ford, 157–80. London: Allen & Unwin.
Mayr, E. 1963. *Animal species and evolution.* Cambridge, Mass.: Harvard University Press.
Niklas, K. J., B. H. Tiffney, and A. H. Knoll. 1983. Patterns in vascular land plant diversification. *Nature* 303:614–16.
Simpson, G. G. 1944. *Tempo and mode in evolution.* New York: Columbia University Press.
Simpson, G. G. 1953. *The major features of evolution.* New York: Columbia University Press.
Slobodkin, L. B., and H. L. Sanders. 1969. On the contribution of environ-

mental predictability to species diversity. *Brookhaven Sym. Biology* 22:82–95.
Stanley, S. M. 1973. Effects of competition on rates of evolution, with special reference to bivalve mollusks and mammals. *Syst. Zool.* 22:486–506.
Stanley, S. M. 1974. What has happened to the articulate brachiopods? (abstract with programs) *Geol. Soc. Am.* 6:966–7.
Stanley, S. M. 1975. A theory of evolution above the species level. *Proc. Nat. Acad. Sci., U.S.A.* 72:646–50.
Stanley, S. M. 1977. Trends, rates, and patterns of evolution in the Bivalvia. In *Patterns of Evolution* ed. A. Hallam, 209–50. Amsterdam: Elsevier.
Stanley, S. M. 1979. *Macroevolution: pattern and process.* San Francisco: W. H. Freeman and Co.
Stanley, S. M. 1982. Species selection involving alternative character states: an approach to macroevolutionary analysis. *Third North Am. Paleontol. Conv. Proc.* 2:505–10.
Stanley, S. M. 1986. Population size, extinction, and speciation: the fission effect in Neogene Bivalvia. *Paleobiology* 12:89–110.
Stanley, S. M., K. L. Wetmore, and J. P. Kennett. Macroevolutionary differences between the two major clades of Neogene Planktonic Foraminifera. *Paleobiology* 14:235–49.
Taylor, P. D. 1988. Major radiation of cheilostone bryozoan, triggered by the evolution of a new larval type? *Hist. Biol.* 1:45–64.
Vermeij, G. J. 1977. The Mesozoic marine revolution: Evidence from snails, predators, and grazers. *Paleobiology* 3:245–58.
Vermeij, G. J. 1987. The dispersal barrier in the tropical Pacific: Implications for molluscan speciation and extinction. *Evolution* 41:1046–58.
Vrba, E. S. 1980. Evolution, species and fossils: How does life evolve? *South Afr. J. Sci.* 76:61–84.
Vrba, E. S. 1984a. Evolutionary pattern and process in the sister-group Alcelaphini-Aepycerotini (Mammalia: Bovidae). In *Living fossils*, ed. N. Eldredge and S. M. Stanley, 62–79. New York: Springer Verlag.
Vrba, E. S. 1984b. What is species selection? *Syst. Zool.* 33:318–28.
Vrba, E. S. 1985. African Bovidae: Evolutionary events since the Miocene. *South Afr. J. Sci.* 81:263–6.
Ward, P. D., and P. W. Signor. 1983. Evolutionary tempo in Jurassic and Cretaceous ammonites. *Paleobiology* 9:183–97.
West-Eberhard, M. J. 1983. Sexual selection, social competition and speciation. *Quart. Rev. Biol.* 58:155–83.
Wilson, A. C., G. L. Bush, S. M. Case, and M.-C. King. 1975. Social structuring of mammalian populations and rate of chromosomal evolution. *Proc. Nat. Acad. Sci.* 72:5061–5.
Wilson, E. O. 1961. The nature of the taxon cycle in the Melanesian ant fauna. *Am. Nat.* 95:169–93.
Wilson, E. O. 1985. Invasion and extinction in the West Indian ant fauna: Evidence from the Dominican amber. *Science* 229:265–7.

5

The Macroevolution of Clade Shape

James W. Valentine

From the perspective of geological time, the Phanerozoic fossil record is characterized by incessant change. Among the marine faunas of the shallow sea, each million years registers a significant loss, gain, or turnover at the level of species. At higher taxonomic levels there is less volatility, yet over many millions of years the numerically important animal clades expand and contract at the family level, for example, and dominance shifts from one group to another; the age of trilobites gives way to the age of brachiopods.

The purpose of this paper is to assess the nature of the processes that drive these changing patterns of clade representation in marine benthic faunas above the species level and to ask whether they are biotically or abiotically regulated. To this end it is necessary to consider the relation between microevolutionary and macroevolutionary processes and the nature of evolution within the taxonomic hierarchy.

Evolution and the Taxonomic Hierarchy

SPECIES AS UNITS IN MACROEVOLUTION

Microevolution is chiefly concerned with the origin of heritable variation, the differential perpetuation of heritable features of phenotypes within species, and with the rise of those impediments to gene flow that permit populations to differentiate and/or new species to originate. Macroevolution has been treated as the differential perpetuation of heritable features of species within clades and sometimes as involving the origin of new clades. Thus the difference between microevolution and macroevolution may be treated as a difference in the levels of the hierarchy that are involved (e.g., Salthe 1975, 1985; Stanley 1979; Gould 1982; Vrba and Eldredge 1984; Eldredge and Salthe 1984; and Eldredge 1985)—a difference in which level furnishes the unit "individuals" (individual organisms or species) and which level furnishes the collection of these units that evolves (species or clades). Not only are species in macroevolution treated as analogues of individuals in microevolution, but in fact species may be regarded as "individuals" in logic as well, a view championed by Ghiselin (1974a, 1974b, 1987; see also Hull 1976 and Mayr 1987).

Commonly, in hierarchies, properties appearing at a higher level cannot exist at a lower level (emergent properties; for examples within the ecological hierarchy see Valentine 1968, 1973 and Salthe 1983). Species, for example, exhibit mortality rates and dispersion patterns, qualities that individuals simply cannot possess. Salt (1979) has objected to this use of the term *emergent* precisely because the properties are collective ones and can be predicted to occur; however, predictability of the properties is not the issue here, but rather the principle of their nonappearance at the lower, unit level and their appearance in the collective. So far as we know in biological hierarchies, there are no properties of such collectives that are not derived from the properties of the units that compose them—whatever levels we examine; if there were, vitalism would be verified. There are also "group" properties that happen to be present in every unit in a collection, such as taxonomic characters possessed by every individual in a species. These group properties, however, are not emergent (Salt 1979; Vrba and Eldredge 1984). The most general term for the differential perpetuation of species' lineages is *sorting* (Vrba and Gould 1986). This term subsumes selection on species-level emergent properties ("species selection"; Eldredge and Gould 1972; Stanely 1975); differential speciations or extinctions based on properties of individual organisms ("effect sorting," Vrba 1980; Vrba and Eldredge 1984); and the results of processes and events that cause speciation and extinction at random with respect to the properties of species (or clades), from wherever they may stem.

For macroevolution to occur, the sorting must involve heritable features. Group properties are heritable by definition, but the heritability of emergent properties is not so clear, although it must occur. For one example, there is evidence that developmental type is a phylogenetic feature at the family level in Brachyura and other Crustacean groups (see Hines 1986 and references therein), presumably a group property. However, developmental type is also known to vary among species within clades as small as genera (see Jablonski and Lutz 1983). The loss of feeding structures in planktotrophic larvae may be virtually irreversible via microevolution (Strathmann 1978). To the extent that this is the case, trends toward increasing numbers of species with planktotrophic development within clades must be owing to the inheritance of this developmental type by proliferating daughter species—to macroevolution. For another example, Jablonski (1987) has argued that the geographical range of species may be heritable at the species level. Geographical range is certainly an emergent species property, based upon the properties of individuals (perhaps including developmental type) but displayed at the species level; if heritable, it should be subject to macroevolution.

HIGHER TAXA AS UNITS IN MACROEVOLUTION

Nested hierarchies possess the organizing power to reduce large and complicated collections of entities from level to level while usefully summarizing the information contained in the progressively larger collections; as Simon (1962) has pointed out, hierarchies may form the only useful architecture of complexity. Nowhere is this organizing power better displayed than in the Linnean taxonomic hierarchy. This construction begins with units—species—none of which can be identical in principle and then proceeds to erect a series of collectives—higher taxa—on successive levels that contain ever more disparate units, and that are difficult to match as to level. Nevertheless, the hierarchy is rich in information from base to apex and provides life science with an indispensible vocabulary.

The field of macroevolution has exploited part of the taxonomic hierarchy by formally studying processes that occur when species are considered to be individuals and to serve as units in the evolution of clades. However, it is difficult to employ species as units of study in broad-scale macroevolutionary research, principally because the fossil record of species is so spotty. The fact that many groups fossilize only under exceptional circumstances is not troublesome in this regard, because many important groups—classes and orders of Mollusca, Arthropoda, Echinodermata, and Brachiopoda, for marine invertebrate examples—have mineralized skeletons that fossilize regularly; if their records were adequate, we could simply study macroevolution among these rather different taxa. However, it is well established that the marine sedimentary record contains more gaps than records (Darwin 1859; Barrell 1917; Schindel 1980; Sadler 1981), and many of the marine rocks that we do have do not contain fossils. Perhaps better than 1 out of 10 durably skeletonized species of marine invertebrates that has lived during Phanerozoic time has been described as fossil (Signor 1985). For diversity studies, the usual remedy for this spottiness is to study the record of families, which must be many times more complete than the record of species.

One drawback to using taxa at this higher level is that important shifts in taxa at lower levels, especially at the species level where diversification and extinction operate, may be reflected only weakly (Valentine 1968, 1974). On the other hand, the less volatile behavior of higher taxa means that in extrapolating trends or conditions across temporal gaps in the record we are less likely to be confused by the fortuitous sampling of short-lived shifts or unusual preservations. The best available data indicating Phanerozoic diversity trends among fossil marine clades are compiled with the genus or family as the unit and order, class, or phylum as the collective (e.g., Sepkoski 1981 and Sepkoski and Hulver 1985).

When studying macroevolution, one solution to the spotty record of species is to employ families as proxies for species, assuming that the behaviors exhibited on the family level mirror species behavior. This assumption has been implicit in a number of diversity studies (e.g., Sepkoski et al. 1981). The interpretation of species behavior from data on higher-level taxa would represent an important step in evolutionary studies and should certainly be pursued. However, computer simulations of diversity changes in hierarchical systems based on genealogical trees indicate that a relatively narrow range of diversity behaviors on the family level may result from a wide range of diversity behaviors on the species level; the species behavior may be effectively masked if only the family behavior is consulted (Valentine and Walker 1986; Endler and Valentine, in prep.). Perhaps data on generic diversities, intermediate as they are in spottiness and presumably in volatility between those of the familial and specific levels, will eventually help in reconstructing species behaviors.

Another solution to the poor species record is simply to study genera, families, or other higher taxa as macroevolutionary units in their own right. This is also an important step to take in macroevolutionary studies, although there have been strong methodological objections raised to this approach. Here I briefly review some of these objections and try to meet them before proceeding to examine family diversity data from a broad-scale macroevolutionary standpoint.

Monophyletic families (and other supraspecific taxa) are conceded to be individuals in logic, just as are species (Ghiselin 1974a) and, by analogy with macroevolution when species are units, families may participate as units in macroevolutionary processes if they reproduce and if there is heritable variation among them. Some workers hold that families (or other supraspecific taxa) simply cannot reproduce (see Wiley 1981; Eldredge and Salthe 1984; and Vrba and Eldredge 1984). This is a matter of definition: taxa are being considered as valid entities only if they are strictly monophyletic. A strictly monophyletic taxon must consist only of the species descending from a single, founding species that is itself a member of the taxon and must consist of *all* of the descendant species. If a family gives rise to another family, one of the species that is a descendant of the founder of the parent family must, obviously, serve as founder to the daughter family, and this species and other members of the descendant family are thus not part of the parent family, which therefore cannot be monophyletic; it is *paraphyletic*. An aversion to the use of paraphyletic taxa is widespread among phylogenetic systematists. For a review of arguments from the standpoint of cladistics see Wiley (1981).

Whatever the utility of cladistics in dealing with living taxa, there are grave difficulties in applying such a system to the classification of fossil taxa. This problem has long been recognized and widely discussed, and my remarks are hardly original. Nevertheless, the problem is particularly acute for taxonomic diversity studies and for macroevolutionary studies in general and must be addressed at least briefly. A basic difficulty lies in the insistence on strictly monophyletic taxa in cladistics and the widespread use (and I believe utility) of paraphyletic taxa in paleontology.

As an example of paraphyly among fossil families, consider a phylogenetic tree hypothesized for Cretaceous ammonoids (Arkell 1957, fig. 151). There are 49 families recognized, of which 23, or 47%, are believed to give rise to other families and are therefore paraphyletic. For the Jurassic, the same source recognizes 47 ammonoid families, of which 18, or 38%, are paraphyletic. The compendium of Phanerozoic marine families of Sepkoski (1982 and emendations) contains over 3,300 marine families that occur as fossils, of which about 2,800 are invertebrates and, of these, about 2,600 belong to groups, like ammonoids, that are durably skeletonized and that make up the bulk of the invertebrate fossil record. These are the families that provide the more reliable data for curves of appearances, disappearances, and for curves and spindle diagrams that depict standing diversities and the waxing and waning of individual clades (see for examples papers in Valentine 1985). If between, say, a third and a half of these families are paraphyletic, and such taxa are inadmissible, then studies of the various curves and spindle diagrams must be so flawed as to be meaningless.

Recasting these fossil families so as to produce strictly monophyletic clades without loss would clearly be the best solution, but the resultant problems would appear to be most severe. Some of the suggested solutions, which include special conventions for dealing with ancestors, have been reviewed by Wiley (1981 and references therein, especially Patterson and Rosen 1977). As cladistic procedures ordinarily require the products of evolutionary branching to be sister groups, those groups with long fossil histories involving much branching can entail a proliferation of higher and higher categories to embrace successively older branches unless some additional convention is employed. One convention for handling fossil taxa is to place them in a new type of category, the *plesion* (Paterson and Rosen 1977), which may be unranked or may be of variable rank (Wiley 1981). By definition, there is no place for such a category within the Linnean hierarchy.

Many fossil families are composed of morphologically distinctive clusters of species that are inferred to be descended from common ancestors. The only consistent difference between the monophyletic and paraphyletic examples of these families is that a species of the latter produced a daughter that was ancestral to another family. The monophyletic families, by contrast, were terminal clusters within their higher clades. If a daughter family persists after its parent family has disappeared, that paraphyletic parent is not actually extinct but "pseudoextinct" (Van Valen 1973), since as a strictly monophyletic taxon it continues to exist. When paraphyletic taxa are ruled out, then durations of many families will be extended forward in time. This practice is certainly in keeping with the overt aims of cladistic taxonomy, which emphasize the treelike aspect of the history of life, but it is followed at the expense of some of the organizing power inherent in hierarchical architectures. In this case, cladistic practice largely undoes precisely what the taxonomic hierarchy otherwise permits, that is, the comparative study of clusters of species, descended from common ancestors, that are distinctive at levels judged to represent genera or families or some such higher taxa, whether they happen to be composed of living species or to have produced descendants or not.

A case in point has been presented by Patterson and Smith (1987), who examined the effects of recasting some of the marine family data (which contained paraphyletic families) that were used by Raup and Sepkoski (1984, 1986) in studying extinction intensities that were believed to display periodic peaks. Echinoderm and fish families from mid-Permian and later times were revised to include only strictly monophyletic families. One clear result was that the monophyletic families were far less sensitive to extinction events that were the original family data. This was largely because the monophyletic families simply did not go extinct in such numbers as did the paraphyletic ones, whenever there were in fact large numbers of species lost (see the comments of Sepkoski 1987b). The removal of clusters of species belonging to paraphyletic families no longer registered on the family extinction curves. Clearly, study of paraphyletic families is preferable if one wishes the data to reflect more closely the pace of species extinction or is interested in the presence and fate of the species clusters indicated by the use of paraphyletic families. A possible alternative is to study genera, but the completeness of the generic record is less adequate than that of families, and at any rate the same problems with cladistic practices would surface. One must balance data with aims. Whether or not paraphyletic families display an extinction periodicity is, of course, another issue.

If we simply admit paraphyletic taxa as permissible members of phylogenetic trees, we may proceed to consider macroevolutionary processes on higher levels. Beyond the formal strictures of cladistics, some workers believe that it is inappropriate to employ taxa on levels above that of species as units in evolution. One argument is that higher taxa, though units of history, are not units of evolution, since they are composed of "individual evolutionary units which have the potential to evolve independently of each other" (Wiley 1981:75; see also Wiley 1980). Adopting this notion would invalidate much of macroevolution. However, there is no reason for evolutionary units to be interdependent or even interbreed if they reproduce and display heritable differences on which sorting may occur. Indeed, in asexual clades, interbreeding does not occur; although this raises difficulties with the species concept it certainly does not preclude the evolution of such clades.

Another perceived problem with the evolution of supraspecific taxa is that they are considered not actually to reproduce—families do not found families, for example; species found families—and thus do not meet a basic criterion of evolution (Vrba and Eldredge 1984; Eldredge and Salthe 1984). If paraphyletic taxa are permitted, however, it is perfectly permissible to consider that higher taxa can reproduce and evolve without contravening logic or biology. The argument can be put as follows, using the family level as an example. Each family emanates from another family. When examined more closely, of course, the whole parental family is not responsible for the daughter family, which has arisen from only one genus. The genus is not the lowest founding unit either, but within that genus, some particular species has given rise to the founding species of the founding genus of the daughter family. But the entire parent species cannot found the daughter species (and genus and family) but rather some population of that species, which may be a very small sample, possibly as small as two genetic individuals in most invertebrate groups. Indeed, the individuals themselves merely contribute samples of their genomes as contained in their gametes. The fact that a new family may be founded by as few as a couple of gemetes does not prevent our identifying (in theory at least) given individuals as the parents and should not prevent our considering a particular species, genus, or family as parental. Of course, the significance of the exercise changes when one considers families as producing families rather than species producing families, but one is then merely addressing different questions, not biologically forbidden ones. Here, macroevolution is considered to occur within a taxon on any level of the Linnean hierarchy (as the collective) in terms of taxa on any lower level above that of the individual (as furnishing the units). For example, when a class is used as the collective, species, genera, families, or orders may be used as units.

Macroevolution of Clade Shape

SPECIES AS UNITS

Clade shape means the shape of spindle diagrams or curves that reflect the standing diversity of clades through time, a shape that results from an interplay between rates of appearance and disappearance of the taxa employed as units of the clade (fig. 5.1). As clades appear when their founding species originate and disappear when the last remaining species becomes extinct, it is useful to consider aspects of the dynamics of species turnover within clades before examining the family dynamics internal to major invertebrate groups. As Stanley (1979 and this volume) has established, species origination and extinction rates tend to be closely similar. Obviously they cannot be very far apart for very long; a clade with a large excess of speciation over extinction would soon "fill" the biosphere, whereas a clade with a large excess in extinction would quickly disappear. Stanley has suggested that during episodes of extensive adaptive radiation these rates may be linked fortuitously, since many factors that foster genetic isolation of populations and thus speciation also produce restricted species distributions and thus may promote extinction. Another interpretation for background rates of speciation and extinction is suggested below.

A computer model of species origination and extinction in a finite adaptive region partly analogous to the adaptive zone of Simpson (1944, 1953) has been explored for insight into turnover dynamics (Walker and Valentine 1984; Valentine and Walker 1987). The adaptive region, shaped as a torus, is subdivided into adaptive addresses analogous to "niches," each of which can potentially accommodate one and only one species. A hypothetical clade is assigned an intrinsic rate of speciation (S) and of extinction (E) and permitted to originate as a founder species at an arbitrary address. The program then proceeds via a series of steps, each scaled to equal a million years; at each step, any species present must either attempt to produce a daughter species, become extinct, or continue unchanged (see Raup et al. 1973). The probabilities assigned for each of these alternative events are drawn from fossil data, but the result for any given species is a matter of

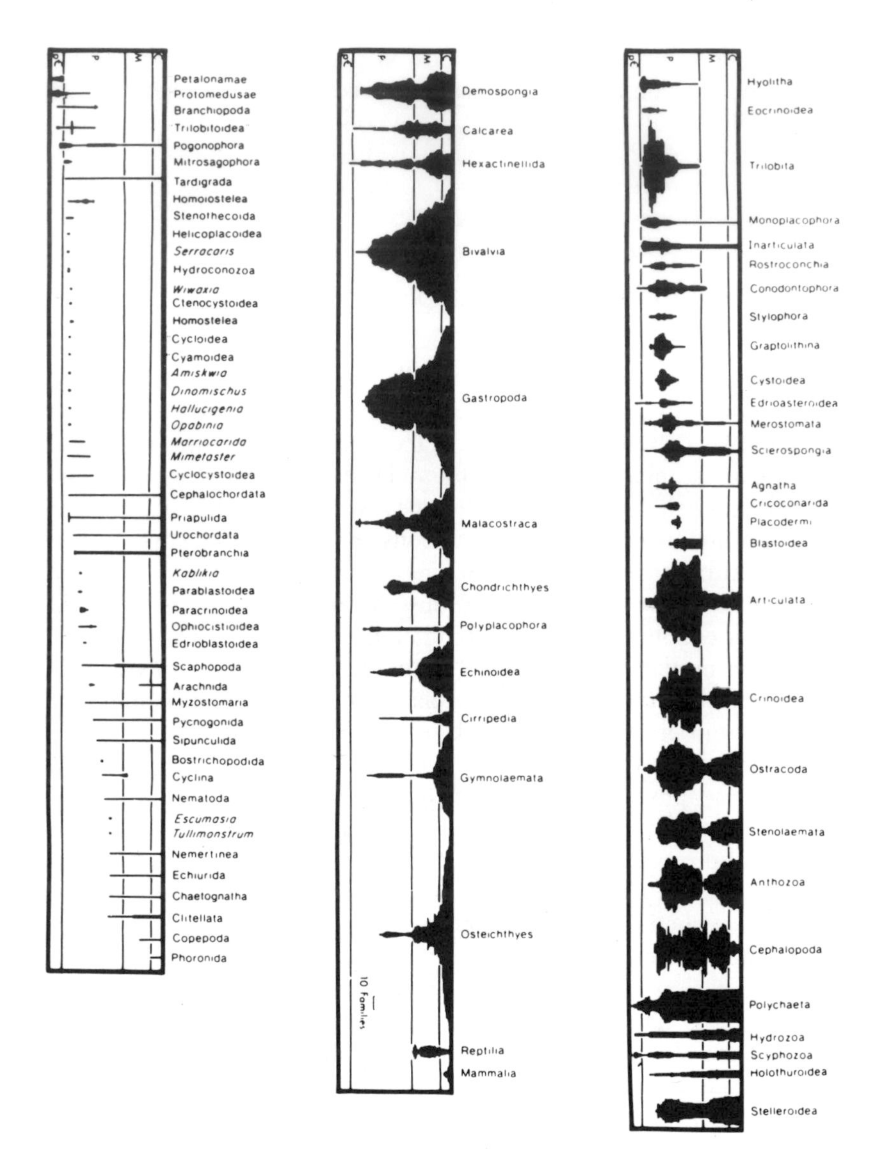

Figure 5.1. Sepkoski's spindle diagrams representing the family diversity of various marine taxa through geologic time. Cenozoic, C; Mesozoic, M; Paleozoic, P; late Precambrian ("Vendian" or "Ediacaran"), pC. (From Bambach, 1985.)

chance. Daughter species must originate on an adaptive space adjoining the space occupied by the parent, which is selected at random. If the selected space is preoccupied, the speciation attempt fails and the parent simply continues unchanged; thus there is a difference between the number of attempted speciations and those that are successful. During early steps, successful speciations significantly exceed extinctions and the adaptive space becomes increasingly densely occupied. However, the rate of successful speciation then slows, owing to an increasing frequency of encounters of attempted daughters with occupied adaptive addresses. The per species background extinction rate remains constant, and the falling rate of successful speciation comes eventually to coincide with the extinction rate and an equilibrium diversity is then established. The rates of *S* and *E* are now linked. Standing species diversity fluctuates about the equilibrium level, owing to the stochastic variables in the model. If the per species extinction rate were not constant but increased with increasing diversity as is sometimes postulated (MacArthur and Wilson 1967), the equilibrium diversity would be lower than in the model with constant background extinction. As this model creates a logistic curve of species diversities, the proportion of adaptive spaces that is unoccupied at equilibrium is expressed by *E*/*S* (Walker and Valentine 1984), which therefore indicates the equilibrium occupation density of species. A given density may be achieved equally by lineages with high levels of *S* and *E*, and thus a high rate of species turnover, or by those lineages with low levels of *S* and *E* and a low rate of species turnover. Plausible occupation densities suggested by estimates of *S* and *E* provided by the fossil record (see Stanley 1979; Walker 1984) range from about 50% to 80%; perhaps the average major clade of Phanerozoic marine invertebrates had an occupation density of about 66% at an average time (Walker and Valentine 1984).

If two separate clades are founded within the same adaptive zone model, the one with the higher occupation density (lower *E*/*S* ratio) will eventually exclude the other, no matter what their relative turnover rates may be (Walker 1984), other things being equal. That is, without considering such factors as ecological competition or interspecies interactions of any other sort, one clade will displace another simply by possessing a superior combination of the ability to maintain itself for a long time on the resources of a given adaptive region and the ability to occupy adaptive spaces (to utilize resources) more rapidly once they are made available via extinction (see Stanley 1979 and this volume). It is the superior combination of these two factors, and not superiority in either of them, that determines which clade will prevail. In this simple case, the "losing" clade will become extinct, but in life clades may usually occupy successfully another adaptive zone or some niches that do not overlap with any clades that possess potentially higher occupation densities there.

If two clades in a single adaptive zone have identical E/S ratios and therefore the same potential occupation densities, they may coexist indefinitely but not eternally. Their relative densities will fluctuate stochastically, and sooner or later one of them will happen to fluctuate to zero density; the "smaller" the adaptive zone, the sooner such extinction is likely to occur. This situation illustrates a fundamental asymmetry in clade dynamics: the upper limit of diversity permitted by resources may be tested repeatedly, but the lower limit may be tested only once; extinction is forever. Therefore in the long run, the clades with superior extinction resistance—low E—are more likely to endure than are those with high E, no matter that the origination rates (S) of the latter are correspondingly high (Valentine 1989). So, in fact, clades with low turnover rates are more likely to persist, other things being equal, than clades with high turnover rates. A number of workers have been led to this conclusion from quite different approaches (for example Sepkoski 1979; Stanley 1979; Raup 1983; and Van Valen 1985b).

In the preceding discussion it has been tacitly assumed that the adaptive zones are homogeneous, but in nature they are clearly not, and it is interesting to consider the case of two clades in an adaptive zone divided into subzones. The adaptive barriers or, "valleys," separating these subzones must be more formidable than the barriers separating species' niches within subzones, so that the probability of a speciation event penetrating into a different subzone is lower than of speciation within a subzone. If the two clades have equal E/S ratios in all subzones, then the clade with the higher turnover rate is the more likely to be eliminated from a given subzone, as discussed previously.

Consider a wave of externally imposed extinction that sweeps an adaptive zone, eliminating from some subzone all but a single species belonging (as it usually would) to the clade with the lower turnover rate. Reinvasion of the subzone across the defining barrier would usually require considerable time despite the high speciation rate of the eliminated clade, and the surviving clade would be able to diversify despite its lower speciation rate, reducing the chances of reinvasion further. Once the surviving clade had achieved a significant occupation density, even successful penetration into the subzone by a species of the high-turnover clade would usually not result in a lengthy stay there, as random extinctions would tend to eliminate it when it was still at a low diversity, and the inherent advantage of the low-turnover clade should eventually cause it to prevail in any event. Indeed, a low-turnover clade may be expected to prevail in the foregoing situation, even if it has a somewhat higher *E/S* ratio than the high-turnover clade; the advantage in occupation density necessary to favor the high-turnover clade depends upon the differences in clade turnover rates and the barrier strength of the subzone. Additionally, the low-turnover clade is the more likely to have survived at higher densities any extinction wave in neighboring subzones and thus to form the source stock of invading species.

Very high levels of extinction, however, may sometimes be more favorable to high-turnover clades, especially if the extinction is random with respect to the clade properties that are sorted during "background" extinctions. When adaptive zones and subzones are deprived of *all* of their occupants by extinction, there should be a premium on clades most likely to reinvade (see Sepkoski 1979). Thus, the clades first rebounding from mass extinctions might commonly be different from those that eventually increase during background extinctions (see below).

It should be stressed that the simplifying assumptions and subterfuges employed in the model do not appear to invalidate the advantages and disadvantages imputed to clades with various *E/S* ratios. In particular, it is not necessary for diversity to reach a stable equilibrium level or for an equilibrium level to remain constant for relations between lower- and higher-turnover clades outlined here to obtain. Indeed, the data that we have on the changing structure of the Phanerozoic biosphere indicate that the absolute level of equilibrium diversity accommodation has varied significantly, and the important fluctuations in apparent extinction rates imply that equilibrium was not always maintained or approximated. The advantages and disadvantages accruing from different turnover rates, however, are displayed as well during differential departures from or rises toward equilibrium as during stochastic fluctuations about an equilibrium level. The similarity of E and S rates within clades may arise, not from their definitional equality at equilibrium, but from the relative openness of adaptive space for high and low turnover clades. For example, a clade with low extinction resistance will lose many species during an extinction wave, creating many opportunities for speciation in adaptive space to which it has evolutionary access; thus a relatively high speciation rate would be expected. A clade with high extinction resistance would not be presented with so many speciation "opportunities" and would have a correspondingly low speciation rate—all without any necessary approximation of an equilibrium diversity. The reference made in the model to an equilibrium level is a useful conceptual strategy but may or may not conform to historical reality for any given time in any given adaptive zone.

FAMILIES AS UNITS

Insofar as clade shapes are concerned, the principles of macroevolution using species or families as units appear to be similar, for each uses lower taxa as the units and higher taxa as collectives for evaluating their differential perpetuation. However, the possible causes of the differentials become multiplied as taxa at increasingly higher levels are used as units, principally because there are increasingly more lower levels at which processes may occur that give rise to effect sorting. It is expected that in proceeding to higher levels, sorting would ordinarily occur increasingly by effects and less and less by selection at the unit level. Macroevolution need not involve selection in this restricted sense; so long as sorting occurs on characteristics that are expressed at the family level, the families may be considered as units in macroevolution.

Van Valen (1985a) and Van Valen and Maiorana (1985) have suggested that taxonomic extinction rates within clades are essentially constant and that changes in the macroevolutionary behavior of clades are therefore owing chiefly to changes in origination rates. Gilinsky and Bambach (1987) have examined the temporal trends of origination and extinction of families within 99 higher taxa of marine organisms in rather long time intervals (20 million years) in order to avoid the complications of extinctions with short-term effects. They found that family extinction rates do tend to be nearly constant throughout most of the duration of the higher taxa. This supported Van Valen's suggestion and is in keeping with the assumption of stochastically constant species' extinction rates within higher taxa by Walker and Valentine (1984; also Valentine and Walker 1987). Gilinsky and Bambach found further that family origination rates tended to decline throughout the duration of the higher taxa and thus that the long-term fates of the taxa were regulated chiefly by origination rates. They speculated that this lowering of originations might be owing to progressive stabilization of coadapted gene complexes that progressively constrained the evolution of novel families within the higher taxa. It may be conjectured that the declining origination rates partly stemmed first, from a "filling" of the adaptive subzones to which families of the higher taxa may have access and second, from preemption of these subzones by other, lower-turnover clades as time passed (see Sepkoski 1979).

There is evidence of an overall Phanerozoic trend toward declining turnover rates at the family level (Raup and Sepkoski 1982). Figure 5.2 depicts the rates of family extinction across the Phanerozoic. Discount-

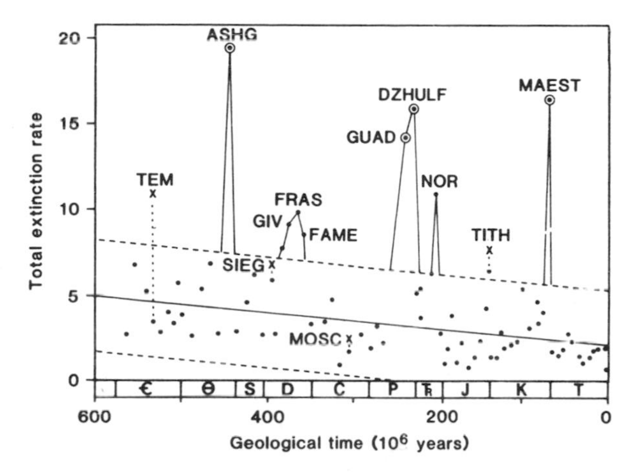

Figure 5.2. Extinction rate of marine invertebrate and vertebrate families per million years. The five most intensive extinctions are the late Ordovician (ASHG), early upper Devonian (FRAS), end-Permian (DZHULF), middle late Triassic (NOR) and end-Cretaceous (MAEST). The regression line is of background extinction intensities, with a 95% confidence interval indicated by dashed lines. (From Raup and Sepkoski 1982.)

ing the high rates suffered during the five mass extinction events, extinction rates clearly decrease, which is brought out by the regression line on the background (nonprincipal mass extinction) rates. Family origination rates must also have declined over this eon (Van Valen 1985a; Gilinsky and Bambach 1987), and indeed there is evidence that the per family rate of diversification among Metazoa in general was nearly five times higher during early Paleozoic diversification than during Mesozoic diversification (Sepkoski 1984). One implication of this trend in family turnover rates is that dominant Cambrian clades should have more volatile diversities than Paleozoic clades, and Paleozoic clades more volatile diversities than Mesozoic-Cenozoic ones. This expectation seems to be met; trilobites, the dominant Cambrian fossil clade, had high speciation and extinction rates and display volatile Cambrian diversities (fig. 5.1; Stanley 1979; this volume); articulate brachiopods and Paleozoic crinoids are dominant clades of the post-Cambrian Paleozoic and are somewhat volatile (fig. 5.1), and brachiopods have intermediate speciation and extinction rates (Walker 1984; crinoid species rates have not been documented); whereas bivalves and gastropods, rising to dominance in the Mesozoic and Cenozoic, have relatively low turnover rates (Stanley 1979; this volume) and exhibit little volatility but rather steady success in increasing their representation (fig. 5.1).

Data on generic extinction rates for an increasing number of marine invertebrate clades are available through the efforts of J. J. Sepkoski, Jr. (unpubl. data base). Some of these data have been summarized by Raup and Boyajian (1988). Though at the generic level and based on mean per stage extinctions, these data are in general agreement with the species-level data mentioned previously and with the family behavior summarized in figure 5.1. Trilobite genera have relatively high extinction rates; brachiopod and crinoid genera have intermediate rates; and bivalve and gastropod rates are lowest among these groups. Other interesting contrasts are between Paleozoic corals (high to intermediate rates) and Mesozoic-Cenozoic corals (low rates), and the essentially Paleozoic crinoids (intermediate) and essentially Mesozoic-Cenozoic echinoids (low rates). The outstanding exception to these trends is among the Cephalopoda. The chiefly Paleozoic nautilids have an appropriate intermediate extinction rate, but the chiefly Mesozoic ammonoids have the highest recorded generic extinction rate of any invertebrate clade, nearly half again as high as the trilobites. Although the possibility cannot be discounted that some of this high ammonoid rate may be a taxonomic artifact, there may be other explanations as well.

The domination of the earliest Paleozoic Evolutionary Fauna by families of high-turnover clades may be ascribed to the relative rapidity with which such clades were able to fill the open adaptive zones and subzones of those times during a general Metazoan radiation (Sepkoski 1979). During one other period of the Phanerozoic the marine biosphere appears to have been as nearly empty of metazoan species as during the early Cambrian—the Early Triassic, following the great Permian-Triassic extinctions (see Erwin, Valentine, and Sepkoski 1987). If the open adaptive space of the early Cambrian favored high-turnover clades, then the open adaptive space of the early Triassic might do so as well. And indeed Van Valen (1984) has analyzed Phanerozoic family background extinction rates to suggest that while they tend to decline steadily during the Paleozoic, they rebound to reach higher levels following the Permian-Triassic extinction and subsequently decline again to a low in the Neogene. Clearly ammonoids could most rapidly exploit the resources of early Mesozoic time, and their abundance during the early phase of the "Modern Fauna" is thus not anomalous. As many ammonoids were nektobenthonic, it may be that adaptation to this mode of life, distinctive among major invertebrate clades, is also associated with their unusually volatile macroevolutionary pattern.

Major patterns of clade dominance through the Phanerozoic may therefore reflect an interplay between clade dynamics and environmental opportunity. The three Evolutionary Faunas of Sepkoski (1981), then, are chiefly associations of major clades with similar turnover rates, with perturbations arising from mass extinctions. The general rise to dominance of low-turnover clades means that the average taxon of those present becomes progressively longer lived; this is in accord with the finding of Boyajian (1986) that background extinctions tend to remove younger families preferentially and therefore that the fossil invertebrate fauna is dominated by younger families earlier and older families later in Phanerozoic time.

Major patterns of environmental distribution of Paleozoic marine faunas can be interpreted to derive from interplays between clade dynamics and patterns of the intensity of environmental fluctuations. Bretsky (1968) has noted that extinction rates were higher in the offshore communities dominated by Paleozoic-style faunas than in onshore communities dominated by Modern elements, and Sepkoski and Miller (1985) have shown that the major evolutionary faunas of the Phanerozoic, elements of which are present throughout the Paleozoic, are each most abundant inshore at first and then gradually spread into offshore environments, progressively displacing the elements of the older faunas from shallow environments (see also Sepkoski 1987a; Jablonski and Bottjer, this volume). The transition between the Cambrian and later Paleozoic-style faunas required tens of millions of years; that between the Paleozoic and the elements of the Modern Fauna took hundreds of millions of years. A plausible interpretation of these trends is that the more extinction-resistant clades were able to persist in the rigorous inshore environments and to expand, whereas the less extinction-resistant clades increasingly found their inshore resources sequestered by low-turnover clades and became restricted to the less rigorous offshore environments where their presence created the relatively high-turnover fauna.

A possible bias in the fossil record that might affect observed extinction rates has been examined by Pease (1985, 1988 and references therein). He has pointed out that as the fossil record is incomplete, the observed durations of taxa must be shorter than the actual durations, and therefore the observed extinction rates are higher than the actual ones; when the record is poorest, this effect should be greatest. If the fossil record is poorest in the earliest Phanerozoic and improves monotonically to the present, the observed extinction rates would decrease steadily for this reason, even if actual extinction rates were constant. However, such a monotonic bias of representation within taxa has never been demonstrated and indeed seems highly unlikely. As noted previously, the observed extinction rates within major taxa are commonly rather constant throughout the Phanerozoic (Gilinsky and Bambach 1987), and it is the replacement of taxa with high extinction rates by those with low rates that accounts for much at least of the general decline in extinction rates; this is not a pattern that can be assigned to bias in the record.

There is little doubt that biases in the record affect our estimates of evolutionary rates, but they do not appear to account for the clade patterns discussed here. However, the biases exist, environmental factors fluctuate more rapidly in some environments than in others, the environment changes in a somewhat mosaic fashion, some lineages have biological properties that permit them to pursue high-turnover strategies more successfully than others, some clades may be the sole occupants of large adaptive zones, and processes of origination and extinction have stochastic components. Therefore, although macroadaptive trends in clade dynamics may be expressed by most dominant clades, some atypical clades and contrary examples are not unexpected. Mass extinctions, if their effects are random with respect to background clade dynamics, may, by extinguishing subclades with particular dynamic properties, interrupt or entirely reset the course of clade dominance (see Jablonski 1986).

Biotic or Abiotic Factors as Principal Determinants of Clade Shapes

The model investigated here has involved environmental changes that increase or decrease adaptive opportunities interacting with intrinsic clade properties; the results of the interaction appear as taxonomic origination and extinction rates. The opportunities may be biotic or abiotic, and the model itself gives few clues as to which might be the more important in producing the clade shape changes that are actually observed across the course of Phanerozoic time. It is interesting that Van Valen (especially 1985a, 1985b; see also Van Valen and Maiorana 1985) has found in studies of the patterns of origination and extinction of taxa many of the patterns and trends that appear in the present model, even though he assumes that biotic factors comprise the major driving element in these patterns, following his famous Red Queen hypothesis (Van Valen 1973). The present model does not at all require that biotic factors be important, aside from the assumption that resource utilization in an adaptive space may be precluded by an occupying taxon, and it is consistent with the stationary model of Stenseth and Maynard Smith (1984) as well.

Great increases in family diversities cannot be accommodated without the occupation of extensive and varied tracts of abiotic adaptive space. Generally, increasing diversities seem to be accompanied by the rise of intricate biological interdependencies that permit the further amplification of diversity beyond the purely physical base, for organisms themselves are exploited as habitats and of course as food, and populations may become limited by predators or competitors. Furthermore, the activities of organisms alter the physical environment, profoundly in some instances, and the feedback effects, many of which are poorly understood and controversial, further complicate the assignment of macroevolutionary determinants to one factor or the other. Nevertheless, rapid expansion of clades, such as the trilobites during the Cambrian and the crinoids and articulate brachiopods during the Ordovician, suggests the relatively rapid filling of adaptive space that was already present but largely uninhabited, particularly since these groups, especially the latter two, display little variety in trophic level so far as we know. Massive rapid familial extinctions also suggest a chiefly abiotic basis, although again amplification of an extinction through the disruption of biological dependencies would seem likely to occur. Certainly in the most current hypotheses, involving climate change, extraterrestrial bombardments, or sea level change, mass extinctions are driven by abiotic factors.

During the stretches of geological time when clade diversities are changing more slowly, the intrinsic "background" rates of diversification or extinction that ultimately determine clade success or failure are almost certainly based upon adaptations evolved at some lower level and appearing as effects at the level of families and of still higher taxa. Thus an assessment as to whether the origins of those adaptations are biotic or abiotic should be made at those lower levels. The generic level is the lowest for which Phanerozoic data are available. Raup and Boyajian (1988) have analyzed the stratigraphic records of 19,897 fossil genera and have shown that, when normalized for turnover rates and thus volatility, most classes and orders display a concordance both in the timing and in the rate intensity of generic extinction. As these investigators conclude, such congruence indicates that the stresses responsible for the extinction rates are effective across a broad range of environments throughout the Phanerozoic and thus strongly implicates abiotic factors.

Furthermore, if the background clade shape changes were being driven by biotic factors, we should expect to find that adaptations to biotic factors tend to increase in effectiveness in the more successful clades and that these clades should display such adaptations in the greatest number or in the greatest development, the most extreme expressions. Vermeij (1987) has recently examined those (chiefly morphological) adaptations most likely to be responses to biotic challenges, and he has concluded that there is widespread evidence that those sorts of adaptations have indeed been enhanced over geological time. Later organisms, for example, tend to be better armored, to burrow more deeply, and or to have enhanced locomotory abilities than did their predecessors, presumably in order to cope better with predators (or perhaps competitors in some instances). However, over the same time, predators have become more powerful or otherwise more adept at piercing or breaking shells and have their own enhanced locomotory capabilities. The "escalated" antipredatory adaptations have not necessarily retarded the extinction rates of their possessors, because these forms inhabit environments in which predators have become more capable (Vermeij 1987, especially chapter 13). There is little relation between obvious biotic adaptations and extinction resistance; if anything, the relationship is weakly negative, and highly escalated forms are somewhat more likely to suffer extinction (Vermeij 1987: chapter 14).

On the other hand, the major biotic associations of the Phanerozoic have appeared and developed, as Sepkoski and Miller (1985) have put it, on an environmental template, and the subsequent trends of sorting among clades occur along major environmental gradients (Sepkoski and Miller 1985; Sepkoski 1987a). The progress of changing clade dominance within the marine realm and over Phanerozoic time is amenable to explanation in terms of clade-dynamic responses to parameters defining the gradients in the physical environment. To be sure, there may usually be some amplifications or modifications as biotic interactions are worked out among evolving lineages, and these would be expected to be most important where biotic interactions are most intense, as on reefs. However, for families it appears that for massive changes in diversity and for lower-level diversity changes and tradeoffs associated with the waxing and waning of marine clades, the chief driving factors are abiotic.

Macroevolutionary battles evidently must be fought with armies developed by microevolutionary processes. It has been said of Wellington's famous victory at Waterloo that it was won on the playing fields of Eton; however true that may be, it does not relieve the necessity for studying the actual battlefield to understand what occurred. The fields where the battles of clades are worked out are adaptive zones, and although the results may indeed flow from the character of the individual organisms involved, it is in the zones that the waxing and waning of clades are decided, and it is to the factors operative on that level that these conclusions are meant to apply.

Acknowledgments

The manuscript was critically reviewed by Professors Bruce H. Tiffney, University of California, Santa Barbara, and Elisabeth S. Vrba, Yale University, for which many thanks are due. This paper was put into final form during tenure as Bateman Visiting Scholar at the Department of Geology and Geophysics, Yale University; the generous hospitality of the Department is gratefully acknowledged. Research on which this paper is based is supported by NSF grants EAR 84-1711 and 87-21192.

Literature Cited

Arkell, W. J. 1957. Introduction to Mesozoic Ammonoidea. In *Treatise on invertebrate paleontology*, part L, ed. R. C. Moore, 81–129. New York: Geological Society of America.

Bambach, R. K. 1985. Classes and adaptive variety: The ecology of the diversification of marine faunas through the Phanerozoic. In *Phanerozoic diversity patterns: Profiles in macroevolution*, ed. V. W. Valentine, pp. 191–253. Princeton, N.J.: Princeton University Press.

Barrell, J. 1917. Rhythms and the measurements of geologic time. *Geol. Soc. Amer. Bull.* 28:745–904.

Boyajian, G. F. 1986. Phanerozoic trends in background extinctions: Consequence of an aging fauna. *Geology* 14:955–8.

Bretsky, P. W., Jr. 1968. Evolution of Paleozoic marine invertebrate communities. *Science* 159:1231–3.

Darwin, C. R. 1859. *On the origin of species*. London: Murray.

Eldredge, N. 1985. Unfinished synthesis. New York and Oxford: Oxford University Press.

Eldredge, N., and S. J. Gould. 1972. Punctuated equilibria: an alternative to phyletic gradualism. *Models in paleobiology*, ed. T. J. M. Schopf, 82–115. San Francisco: W. H. Freeman.

Eldredge, N., and S. N. Salthe. 1984. Hierarchy and evolution. In *Oxford surveys in evolutionary biology*, vol. 1, ed. R. Daiokins and M. Ridley, 184–200. Oxford: Oxford University Press.

Erwin, D. H., J. W. Valentine, and J. J. Sepkoski, Jr. 1987. A comparative study of diversification events: the early Paleozoic versus the Mesozoic. *Evolution* 41:1177–86.

Ghiselin, M. T. 1974a. A radical solution to the species problem. *Syst. Zool.* 25:536–44.

Ghiselin, M. T. 1974b. *The economy of nature and the evolution of sex*. Berkeley and Los Angeles: University of California Press.

Ghiselin, M. T. 1987. Species concepts, individuality and objectivity. *Biol. Phil.* 2:127–43.

Gilinsky, N. L., and R. K. Bambach. 1987. Asymmetrical patterns of origination and extinction in higher taxa. *Paleobiology* 13:427–45.

Gould, S. J. 1982. The meaning of punctuated equilibrium and its role in validating a hierarchical approach to macroevolution. In *Perspectives on evolution*, ed. R. Milkman, 83–104. Sunderland, Mass.: Sinauer Associates.

Hines, A. H. 1986. Larval patterns in the life histories of brachyuran crabs (Crustacea, Decapoda, Brachyura). *Bull. Marine Sci.* 39:444–66.

Hull, D. L. 1976. Are species really individuals? *Syst. Zool.* 25:174–91.

Jablonski, D. 1986. Background and mass extinctions: the alternation of macroevolutionary regimes. *Science* 231:129–33.

Jablonski, D. 1987. Heritability at the species level: Analysis of geographic ranges of Cretaceous mollusks. *Science* 238:360–3.

Jablonski, D., and R. A. Lutz. 1983. Larval ecology of benthic marine invertebrates: Paleobiological implications. *Biol. Rev.* 58:21–89.

MacArthur, R. H., and E. O. Wilson. 1967. *The theory of island biogeography.* Princeton, N.J.: Princeton University Press.

Mayr, E. 1987. The ontological status of biological species. *Biol. Phil.* 2:145–66.

Patterson, C., and D. E. Rosen. 1977. Review of ichthyodectiform and other Mesozoic teleost fishes and the theory and practice of classifying fossils. *Bull. Am. Mus. Natural Hist.* 158:81–172.

Patterson, C., and A. B. Smith. 1987. Is the periodicity of extinctions a taxonomic artefact? *Nature* 330:248–51.

Pease, C. M. 1985. Biases in the durations and diversities of fossil taxa. *Paleobiology* 11:272–92.

Pease, C. M. 1988. Biases in the per-taxon origination and extinction rates of fossil taxa. *J. Theor. Biol.* 130:9–30.

Raup, D. M. 1983. On the early origins of major biologic groups. *Paleobiology* 9:107–15.

Raup, D. M., and G. E. Boyajian. 1988. Patterns of generic extinction in the fossil record. *Paleobiology* 14:109–25.

Raup, D. M., and J. J. Sepkoski, Jr. 1982. Mass extinctions in the marine fossil record. *Science* 215:1501–3.

Raup, D. M., and J. J. Sepkoski, Jr. 1984. Periodicity of extinctions in the geologic past. *Proc. Nat. Acad. Sci. U.S.A.* 81:801–5.

Raup, D. M., and J. J. Sepkoski, Jr. 1986. Periodic extinction of families and genera. *Science* 231:833–6.

Raup, D. M., S. J. Gould, T. J. M. Schopf, and D. S. Simberloff. 1973. Stochastic models of phylogeny and the evolution of diversity. *J. Geol.* 81:525–42.

Sadler, P. M. 1981. Sediment accumulation rates and the completeness of stratigraphic sections. *J. Geol.* 89:569–84.

Salt, G. W. 1979. A comment on the use of the term *emergent properties*. *Am. Nat.* 113:145–8.

Salthe, S. N. 1975. Problems of macroevolution (molecular evolution, phenotype definition, and canalization) as seen from a hierarchical viewpoint. *Am. Zool.* 15:295–314.

Salthe, S. N. 1983. An extensional definition of functional individuals. *Am. Nat.* 121:139–44.

Salthe, S. N. 1985. *Evolving hierarchical systems.* New York: Columbia University Press.

Schindel, D. E. 1980. Microstratigraphic sampling and the limits of paleontologic resolution. *Paleobiology* 6:408–26.

Sepkoski, J. J., Jr. 1979. A kinetic model of Phanerozoic taxonomic diversity. II. Early Phanerozoic families and multiple equilibria. *Paleobiology* 5:222–51.

Sepkoski, J. J., Jr. 1981. A factor analytic description of the marine fossil record. *Paleobiology* 7:36–53.

Sepkoski, J. J., Jr. 1982. A compendium of fossil marine families. *Milwaukee Public Mus. Contrib. Biol. Geol.* 51:1–125.

Sepkoski, J. J., Jr. 1984. A kinetic model of Phanerozoic taxonomic diversity. III. Post-Paleozoic families and mass extinctions. *Paleobiology* 10:246–67.

Sepkoski, J. J., Jr. 1987a. Environmental trends in extinction during the Paleozoic. *Science* 235:64–66.

Sepkoski, J. J., Jr. 1987b. Reply [to Patterson and Smith, 1987]. *Nature* 330:251–2.

Sepkoski, J. J., Jr., and M. L. Hulver. 1985. An atlas of Phanerozoic clade diversity diagrams. In *Phanerozoic diversity patterns: Profiles in macroevolution*, ed. J. W. Valentine, 11–35. Princeton, N.J.: Princeton University Press.

Sepkoski, J. J., Jr., and A. I. Miller. 1985. Evolutionary faunas and the distribution of Paleozoic marine communities in space and time. In *Phanerozoic diversity patterns: Profiles in macroevolution*, ed. J. W. Valentine, 153–90. Princeton, N.J.: Princeton University Press.

Sepkoski, J. J., Jr., R. K. Bambach, D. M. Raup, and J. W. Valentine. 1981. Phanerozoic marine diversity and the fossil record. *Nature* 293:435–7.

Signor, P. W., III. 1985. Real and apparent trends in species richness through time. In *Phanerozoic diversity patterns: Profiles in macroevolution*, ed. J. W. Valentine, 129–50. Princeton, N.J.: Princeton University Press.

Simon, H. A. 1962. The architecture of complexity. *Proc. Am. Philosoph. Soc.* 106:467–82.

Simpson, G. G. 1944. *Tempo and mode in evolution.* New York: Columbia University Press.

Simpson, G. G. 1953. *The major features of evolution.* New York: Columbia University Press.

Stanley, S. M. 1975. A theory of evolution above the species level. *Proc. Natl. Acad. Sci. U.S.A.* 72:646–50.

Stanley, S. M. 1979. *Macroevolution: pattern and process.* San Francisco: W. H. Freeman.

Stenseth, N. C., and J. Maynard Smith. 1984. Coevolution in ecosystems: Red Queen evolution or stasis. *Evolution* 38:870–80.

Strathmann, R. R. 1978. The evolution and loss of feeding larval stages of marine invertebrates. *Evolution* 32:894–906.

Valentine, J. W. 1968. The evolution of ecological units above the population level. *Paleontol.* 42:253–67.

Valentine, J. W. 1973. *Evolutionary paleoecology of the marine biosphere.* Englewood Cliffs, N.J.: Prentice Hall.

Valentine, J. W. 1974. Temporal bias in extinctions among taxonomic categories. *J. Paleontol.* 48:549–52.

Valentine, J. W., ed. 1985. *Phanerozoic diversity patterns: Profiles in macroevolution.* Princeton, N.J.: Princeton University Press.

Valentine, J. W. 1989. Phanerozoic marine faunas and the stability of the earth system. *Global and Planetary Change* 1:137–55.

Valentine, J. W., and T. D. Walker. 1986. Diversity trends within a model taxonomic hierarchy. *Physica D* 22:31–42.

Valentine, J. W., and T. D. Walker. 1987. Extinctions in a model taxonomic hierarchy. *Paleobiology* 13:193–207.

Van Valen, L. M. 1973. A new evolutionary law. *Evol. Theory* 1:1–30.

Van Valen, L. M. 1984. A resetting of Phanerozoic community evolution. *Nature* 307:50–52.

Van Valen, L. M. 1985a. How constant is evolution? *Evol. Theory* 7:93–106.

Van Valen, L. M. 1985b. A theory of origination and extinction. *Evol. Theory* 7:133–42.

Van Valen, L. M., and V. C. Maiorana. 1985. Patterns of origination. *Evol. Theory* 7:107–25.

Vermeij, G. J. 1987. Evolution and escalation, an ecological history of life. Princeton, N.J.: Princeton University Press.

Vrba, E. S. 1980. Evolution, species and fossils: how does life evolve? *S. Afr. J. Sci.* 76:61–84.

Vrba, E. S., and N. Eldredge. 1984. Individuals, hierarchies and processes: towards a more complete evolutionary theory. *Paleobiology* 10:146–71.

Vrba, E. S., and S. J. Gould. 1986. The hierarchical expansion of sorting and selection: Sorting and selection cannot be equated. *Paleobiology* 12:217–28.

Walker, T. D. 1984. The evolution of diversity in an adaptive mosaic. Ph.D. diss., University of California, Santa Barbara.

Walker, T. D., and J. W. Valentine. 1984. Equilibrium models of evolutionary species diversity and the number of empty niches. *Am. Nat.* 124:887–99.

Wiley, E. O. 1980. Is the evolutionary species fiction?—A consideration of classes, individuals, and historical entities. *Syst. Zool.* 29:76–80.

Wiley, E. O. 1981. *Phylogenetics: the theory and practice of Phylogenetic systematics.* New York: Wiley.

6

The Reciprocal Interaction of Organism and Effective Environment: Learning More About '*And*'

Jennifer A. Kitchell

The central question posed to contributors of this volume was, Is evolution affected more by the characteristics of organisms themselves *or* by their environment? The response to this query, of longstanding tradition, is in the category of: "Neither is wrong." The inability to answer this seemingly simple question one way or the other derives from the fallacy elegantly captured by Eddington's remark: "We used to think that if we knew one, we knew two, because one and one are two. We are beginning to realize that we need to learn a great deal more about '*and*'."

The purpose of this paper is to show, by implementing a relationship between organism and environment, that coupling the two (the "and" of Eddington) is unlike "driving" the one by the other. The coupling introduces consequences of hierarchy such that the "collective behavior of the whole is qualitatively different from that of the sum of the individual parts" and introduces consequences of nonlinear feedback such that "playing the game changes the rules" (Gleick 1987).

Why, if evolution is affected not by (1), the characteristics of organisms nor by the other (1), the environment of organisms, but by the (+) of their interaction, do most analyses fail to recognize this fundamental unity of organism and environment? This practice is in the Darwinian tradition of viewing the environment as autonomous, its flux a consequence of independent processes (e.g., cosmology, planetary formation, tectonics), and the species as "driven" (Levins and Lewontin 1985). However, "if the evolution of organisms is only a transformation of the evolution of the environment," these authors conclude, "then in a deep sense organisms really are irrelevant, and the study of evolution is nothing but a combination of molecular biology and geology."

Do biological systems participate in their own evolution, or do they, in the words of Campbell (1982), "get evolved"? That this vision—of an independent driver and a dependent driven, and a separated abiotic and biotic environment—is erroneous is attested to by an obvious fact: organisms determine a *relevant* environment from the environment-at-large. Whales and diatoms, although in physical contact, do not experience the same environment (Williams 1966). Instead, the effective environment is selected, transformed, and modulated by the activity, behavior, and characters of the organism itself. Organisms thereby *interactively* determine over time the selection regime experienced by the species. Such activity "sets the stage for its own evolution" (Levins and Lewontin 1985).

The Tyranny of Intuition

What happens when organism and 'environment' are at the same time subject and object? Intuition, or what seems rational, works reasonably well for simple systems but often impedes our understanding of complex systems. Moreover, even simple systems do not necessarily possess simple dynamics (e.g., Lorenz 1964; May 1976; Carr and Kitchell 1980; Kitchell and Carr 1985; Campbell et al. 1985). The specific question of interest is, *What are the potential evolutionary dynamics of a system coupled so that the evolution of one is the environment of the other and vice versa?*

Why this question is not a simple one is made evident in the following distinction between what Levins and Lewontin (1985) termed the *constructional* versus the *passive* approach:

> With the view that the organism is a passive object of autonomous forces, evolutionary change can be represented as two simultaneous differential equation systems. The first describes the way in which organism O evolves in response to environment E. . . . The second is the law of autonomous change of the environment as some function only of environmental variables. . . . A constructionist view that breaks down the alienation between the object-organism and the subject-environment must be written as a pair of *coupled* differential equations. . . . The coupled differential equations . . . are not easy to solve, but they represent the minimum structure of a correct theory of the evolution of such systems. It is not only that they are difficult to solve, but that they pose a conceptual complication, for there is no longer a neat separation between cause (the environment) and effect (the organism). There is, rather, a continuous process in which an organism evolves to solve an instantaneous problem that was set by the organism itself, and in evolving changes the problem slightly.

We have constructed such a coupled, simultaneous model, whose details are more fully described below and especially in DeAngelis,

Kitchell, & Post (1985, in press). We selected the organism-'environment' interaction known to have the most retrievable information content in the fossil record (see review by Kitchell 1986) and whose modern-day interaction is also well known from field and experimental studies (see Kitchell et al. 1981). Our interest in developing the model, beyond its utility in paleobiology, was directed toward exploring general features of this "*and*" of Eddington, thereby expanding our intuition when faced with complex, coupled, nonlinear systems.

As will be shown in the following discussion, the nonintuitive aspects of the dynamics are that even simple unidirectional, continual change in a single variable elicits a wide spectrum of behaviors, including stasis, abrupt change, divergence, and gradual change. Such behaviors are not only counter to intuitive expectations but counter to expectations of the general literature on species interactions and their predicted patterns of evolutionary change.

Biotic Filter: Hierarchy Not Dichotomy

Abiotic and biotic factors are not either-or ends of a dichotomy. Any focus on biotic interactions should not be interpreted to support the opinion that abiotic effects are unimportant. The importance of abiotic effects are so well established as to be beyond question. Nevertheless, a recognition of the hierarchical structure of evolutionary processes and patterns takes "pride of place" off a specific level; where one looks determines in large part what one sees.

Most abiotic effects on the evolutionary process are "filtered through the ecological hierarchy" (Eldredge and Salthe 1984). We showed previously (Kitchell et al. 1986) that even a physical disaster the magnitude of that associated with the close Cretaceous extinction had its impact mediated by biological factors operating at the individual level. A cascade hypothesis, of biotic control with an abiotic trigger, has also been proposed as a solution to the problem of the late Pleistocene extinctions (Owen-Smith 1987).

My interest, however, is not in arguing for the supremacy or significance of strong ecological interactions as an evolutionary mechanism but in showing that even to begin to understand the evolutionary *potential* of strong ecological interactions one must first understand their behavior. Without such understanding there can be no real progress on the issue.

An example may suffice. Imagine being asked to consider a future time in earth history when the earth gets increasingly colder, a trend accelerated by the albedo effect, until most life on earth becomes extinct. I would not find the latter phenomenon particularly interesting. Instead, my reply might be, "Elementary—the abiotic theater closed, and the biotic play ended its run." Physiological limits provide a sufficient explanation of process and result. What is more interesting, largely because it is less easy to come to understand, are the behaviors of complex, interdependent, nonlinear interactions between players and context.

Limitations of the Hypothesis of Escalation

An expectation of *progressive* adaptation has been advanced recently by Vermeij (1987) under the term escalation, "the adaptive improvement of species in response to the increased hazards of their biological surroundings." The escalation hypothesis posits that although "modern organisms may be no better adapted to their biological surroundings than ancient ones were to theirs . . . the biological surroundings have themselves become more rigorous within a given habitat." There are two components: the 'gap' between organism and 'hazard' (as posed by the biological surroundings) and the absolute 'level' of the hazard. Vermeij concludes that "escalation occurs when the hazards become more severe and the aptations to these hazards become better expressed." The level rises, and the gap narrows.

Escalation is introduced as a substitute term for coevolution, with the following reasoning: "Adaptive reciprocity is the essential property of coevolution, whereas adaptation to enemies is the cornerstone of escalation. If adaptive reciprocity involves enemies, as may be the case in interactions between predators and retaliating prey, then escalation has a strong coevolutionary component." Coevolution implies a two-way interaction, whereas the focus of escalation is on the one-way response, particularly of prey to predators. The escalation hypothesis equates narrowing of the adaptive gap with progress. Progress from the point of view of a predator is expressed by increased predation; progress from the point of view of a prey is expressed by decreased predation. Both are measures of the effectiveness of coping with each other as 'environment.'

There is substantial evidence (well summarized by Vermeij 1987) that predation has played a role in evolution. I see several problems, however, with the formulation of the hypothesis of escalation that limit its utility, applicability, and interpretability. First, it emphasizes the expectation of progressive or linear trends in aptations and hazards and omits any explicit consideration of the consequences of tradeoffs and feedback. This failing derives in part from a concern expressed by Kohn

(1987) that the escalation model is "described only verbally . . . [and] might have been made clearer and more precise were it augmented by a graphical or mathematical presentation."

By contrast, the model we have developed (and describe subsequently) entails both a mathematical formulation and graphical presentation of predictions. It also couples the organism and its biological 'environment', thereby introducing the potential for feedback, and explicitly involves, as subroutines of the model, age-structured populations and tradeoffs between defense, growth, and reproduction. The increased realism of our model would be moot, however, if the predictions remained the same as those of the verbally expressed escalation model. They do not.

One's understanding of a process dictates both the type of data one gathers to evaluate a hypothesis and the outcome of the evaluation. The limited categories of data presented as empirical validation of the escalation hypothesis underscore a second problem. For example, only two types of data are collated by Vermeij (1987) to assess the dynamics of escalation between drilling gastropods (as predators) and bivalves (as prey): the percent of drilled to undrilled bivalves in an assemblage (a measure of predation intensity) and percent incompletely drilled to completely drilled bivalves (a measure of disturbance or prey escape). These data focus on the score where successful predation is a point for the predator and unsuccessful predation a point for the prey.

This score is a necessary but not sufficient measure of the potential consequences of the independent and interdependent evolution between predator and prey. There is no information gathered on tradeoffs between aptations or the data requisite to determine whether or not predation (and prey escape) was selective and according to what criteria (e.g., Kitchell et al. 1981 for the same predator-prey interaction). A simple analogy to sports teams (where team refers to Chicago Bears and not a specific roster) will illustrate the problem. If the only data provided are the scores of two teams in play with one another over a period of time, it is not possible to determine—from such data—whether or not the level of play or skill of the players is changing over time. It may be or it may not be. Such questions require more and different information.

Second, only one score has been examined: that of the organism on its opposing biological 'hazard' but not of the consequences of the interaction back on the organism. The appropriate 'score' involves more than a measure of mortality. All defense and no reproductive 'offense' makes for an unrealistic organism.

In the model we have developed, survival depends in part on growth rate, in part on the allometries of prey biomass and shell growth; these in turn depend on energy allocation, as does reproductive effort. Specific coupling pathways and specific tradeoffs are made part of the model. To approximate reality as well, predation is explicitly selective rather than random.

A third problem I have with the notion of progressive adaptation embedded in the escalation hypothesis is its suggestion that some trait is absolutely better, vis-à-vis the species interaction. By this I mean, by way of example, that a trait such as a thickened aperture in gastropods is considered 'good' (suggesting it would be selected for) because it lessens the likelihood of successful predation. Again, our modeling shows that this is not the necessary case. Even when the mode and strategy of predation remain fixed, the value of a prey trait is not fixed: the consequences of its expression may vary from being highly positive (to the prey's overall reproductive effort, dependent jointly on the probability of survival and expected amount of energy available for reproduction) to highly negative. Subsequent examples will illustrate how what's best can become what's worst. These examples underscore the supervenience of fitness, meaning that it "cannot be identified with any single . . . property" (Sober 1984).

The theoretical development we have pursued is aimed at determining better *how* the process might actually work and in providing templates for the interpretation of data that bear on this "eternal" question. As theory improves, so too will its predictions, necessitating new data and a reinterpretation of old data. The motivating suspicion behind this theoretical development is that a next step is long overdue and has to be taken in order to improve our collective intuition. The processes and patterns are likely not as simple and linear as expected.

The Behavior of Coupled Organism *And* "Environment"

The simultaneous solution of coupled equations of organism and organism, each the biotic environment of the other, involves this "*and*." A central issue in contemporary *theory* is whether or not in such coupled systems, such as coupled predator-prey interactions, there is only one class of behavior—that of flight and pursuit, leading to unidirectional linear expectations of an 'arms race.' The complex causality inherent in nonlinear, coupled systems removes this question from analysis by intuition. Instead, mathematical models are used to develop and expand intuitive expectations. Ironically, however, even most of the general theoretical models have assumed that the analogy of linear escalation is correct. This concept of 'perpetual' evolution has also been applied to the fossil record. Gould (1977), in summarizing Jerison's (1973) work on the evolution of brain size in mammalian herbivores and carnivores, does just that by invoking the arms race as explanation and justification: "As the brains of herbivores grew larger (presumably under intense selective pressure imposed by their carnivorous predators), the carnivores also evolved bigger brains *to maintain the differential*" (italics added). But are coupled nonlinear systems so simply behaved?

Bakker (1983) expressed an even stronger expectation of "unreversed, lock-step" evolution between coupled cursorial predators and their cursorial prey. The lack of reversal speaks to the expectation of unchanging directions of evolutionary change. 'Lock-step' refers to the auxiliary expectation of constancy of rates of evolutionary change so that the advantage differential between predator and prey remains constant. Such expectations led Futuyma and Slatkin (1983) to conclude in their overview of fossil evidence relevant to reciprocal evolutionary interaction that the "ideal paleontological evidence would be a continuous deposit of strata in which each of two species shows *gradual change* in characters that reflect their interaction" (italics added).

Instead, we can show by implementing the theory with coupled, simultaneous equations of predator-induced selection on the prey and prey-induced selection on the predator that there is more than one class of behavior. There are various modes of change, sometimes smooth and continuous (the expectation of gradual change), sometimes sudden and discontinous (unexpected abrupt change and the potential for divergence), and also at times static (unexpected stasis). Yet throughout the simulations there is no change in either the direction of selection (predators 'covet' prey; prey 'covet' escape from predation) nor any change in the rules of the interaction. Novelties and innovations (in mode of predation or mode of escaping predation) are not admitted. Yet the results clearly demonstrate that playing the game changes the rules in ways "like walking through a maze whose walls rearrange themselves with each step you take" (Gleick 1987).

Detailed development of the model is presented elsewhere (DeAngelis, Kitchell, and Post 1985 and in press). The model was developed with three complementary goals in mind: its empirical application to the fossil record, its empirical application to the extant interaction, and its theoretical application to understanding coupled, nonlinear systems. This paper focuses on the latter goal. Graphic presentation of the dynamics of the system (which are many) specifically focus on the selection regimes generated and experienced by the coupled organism and 'environment'.

The principal outcome will be to show that such apparently linear models of escalation have unexpectedly complex behaviors. The selection surface remains unidirectional and smooth only in the absence of coupled organism (prey) and 'environment' (predator). Once the coupling is in place, the selection surface wrinkles and heaves; potential evolutionary trajectories correspond to divergences, abrupt shifts, stasis, *and* continuous change—all in response to a gradual change in even one parameter of the 'environment'.

By systematic exploration using simulation as the experimental tool, we begin to get the 'feel' of such contingent systems in which there is no particular analytical solution. Rather the general behavior of such historical processes corresponds well to the phrase 'It depends'. As argued in a review of the qualitative changes resulting from the use of computer techniques in paleontology, "What is being explored in simulation modelling are the consequences of a given set of assumptions. One creates the outcome as a logical consequence of the theorized process, to discover the way things would be, if the theory of process were operative. . . . Such an approach is necessary to augment and even to develop our limited intuition . . . and becomes indispensable when it expands the limits of our understanding beyond both intuition and the exploration of what did happen to what could happen" (Kitchell, 1990).

The dynamics of the system will be generated by the attempted (but never realized) simultaneous maximization of energy gain by both predator and prey. Energy represents a common currency to define both the intersection of species and 'environment' and the ecological hierarchy (Eldredge and Salthe 1984; Salthe 1985). In general terms, the dynamics of the system are governed by the tendency of each species to solve one of the following equations:

$$\max W_x (X, Y; N_x, N_y) \text{ with respect to } X \quad (1)$$
$$\max W_y (X, Y; N_x, N_y) \text{ with respect to } Y \quad (2)$$

where X refers to the vector of n characters of the prey species that are susceptible to evolutionary change, Y refers to the vector of n characters of the predator species that are susceptible to evolutionary change, and N_x and N_y refer to the population structure of the prey and predator species, respectively. In the analysis to be developed, the two prey ('organism') character traits represent evolution by heterochrony (the X_r axis) and evolution by allometry (the X_s axis). The former is controlled by changes in timing and the latter by changes in relative allocation, both related in turn to changes in development. These two resultant phenotypic characters of the prey are also known to have primary importance in the extant interaction. They will be modeled as evolutionary variables, namely X_r, the reproductive allocation strategy made variable by the age of switching from allocation into growth to allocation principally into reproduction, and X_s, the relative allocation of energy to overall growth in size or to a shift in the allometric relationship of shell thickness with size. Two characters of the predator (as 'environment') will be allowed to vary as well, namely Y_i, the intensity of selection as a consequence of the intensity of predation, and Y_s, the maximum upper size of the predator. The predator's strategy for selection of prey will remain fixed (in keeping with the empirical evidence of this interaction from the fossil record; Kitchell 1986); similarly, no prey innovations (beyond X_r and X_s) will be admitted during an evolutionary run.

We will now sequentially examine the dynamical behaviors, introduced (1) first with the prey in isolation from its 'environment', (2) then with the environment as only one-way (top-down) interaction, and (3) finally with reciprocal or two-way interaction of organism and environment. The consequences are simple to predict in the case of a (prey) species evolving in isolation, with no externally imposed context of a selective regime, either biotic or abiotic. Figure 6.1 shows but a single trajectory from an initial starting position within the field of all possible

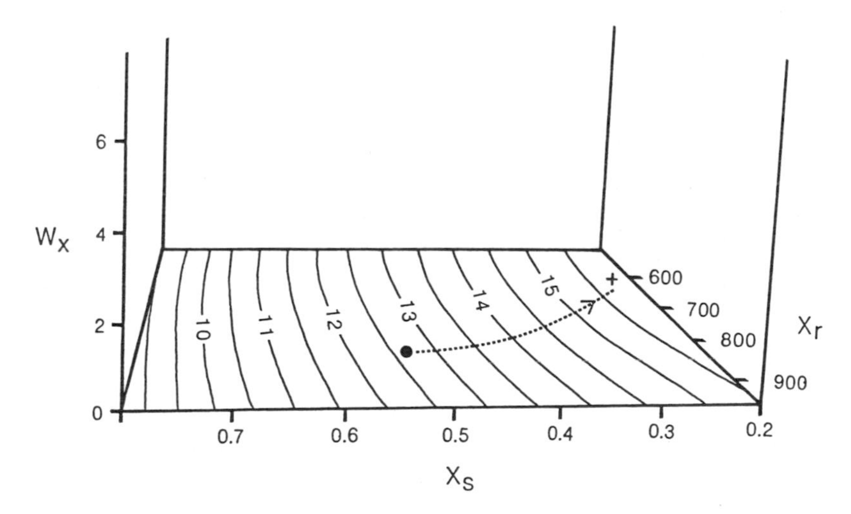

Figure 6.1. The case of no external 'environment': contour diagram of selective surface (W_x of equation 1) of prey for possible combinations of reproductive timing allocation (X_r) and allometric growth allocation (X_s) in the context of no predation. The dotted trajectory shows a possible initial condition and subsequent path to the region of maximum W_x.

combinations of prey traits X_r and X_s. The projection of this trajectory along the two-dimensional field portrayed by the two trait axes maps the changing values of these traits that maximize energy gain and reproductive output for the individual (of the prey species) over its lifetime. The third dimension is portrayed by the contours that express in quantitative terms the relative and absolute amount of reproductive output for all possible prey trait values given the context of no selective regime. The trajectory outlined in figure 6.1 maps the path of steepest ascent to the fixed point of character intersection that provides the maximum reproductive return in a simple world of no external environment.

The situation for one-way interaction (of a selective environment on the prey species) is much more complex and interesting. We then couple the two biotic elements, making one 'driver' and the other 'driven'. What is unexpected is the result that even with a smooth and gradual unidirectional 'driver', the 'driven' response is not always smooth, gradual, nor unidirectional.

We first consider the case in which the selective environment of the prey species changes by systematic increases in predation intensity. The consequences for the prey species at all values of character states are given in figure 6.2 whose contoured surface most noticeably differs from figure 6.1 by its roughness and wrinkling. A smooth change in the environment ('driver') has a nonsmooth consequence (to the 'driven'). Although the display of the full suite of dynamics would best be served by a movie, the panels of figure 6.3 show a series of isolated snapshots of the overall behavior. The differences between figure 6.3A and 6.3B, as well as those between figure 6.3C and 6.3D, are due simply to an increase in predation intensity (Y_i), of the same magnitude. Because the 'environment' of the predator does not represent random selection across all prey traits, the surface is not simply lowered by increasing intensities of selection but is directionally 'driven'. The region of max W_x in figure 6.3A and 6.3C becomes the region of min W_x in figure 6.3B and 6.3D. In addition to changes in overall height there are also shifts in X_r and X_s. As is evident, increased selection by the 'environment' results in a qualitative change in the surface. There need be no single global maximum for a given set of character values. Instead several local maxima appear. Their appearance offers an alternative mode or modes of prey reproductive and energy allocation behavior in the face of the *exact* same 'environment'.

The 'driver' holds to a deterministically steady course. A stochastically varying selection regime may better describe the natural 'driver,' whose populations, linked by planktotrophic larval dispersal, experi-

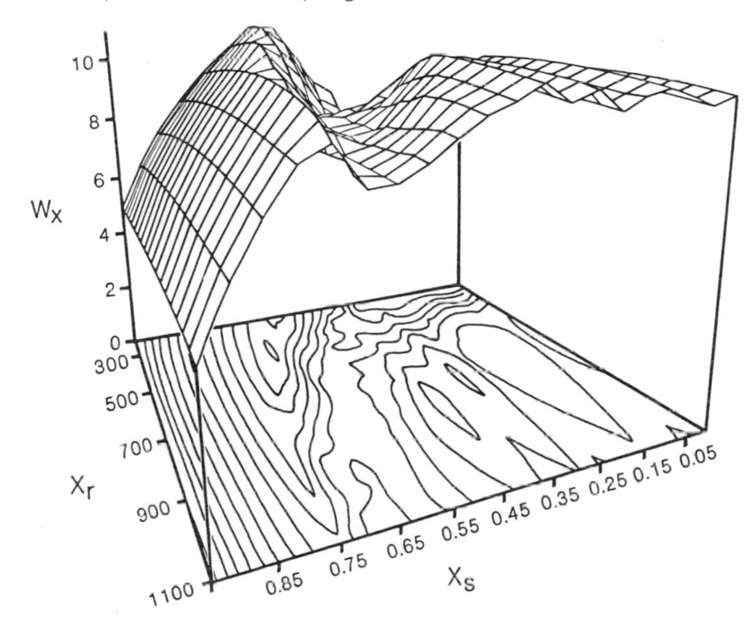

Figure 6.2. The conditions are the same as those of figure 6.1, except that the prey species now experiences the 'environment' of predation. The consequences, expressed as the prey selective surface of W_x and the underlying contour diagram, are quite different from those of figure 6.1. The multiply 'humped' surface indicates that more than one combination of prey traits X_r and X_s are equivalent (or nearly so) within a fixed 'environment' (Y_i = .005479; Y_s = 25 mm).

ence variable recruitment success and consequently variable predation intensity. A static slice through the dynamic motion of the prey surface showing the consequences of low (upper surface), medium, and high (lower surface) intensities of selective predation is given in figure 6.4. These three surfaces, resembling the motion of a wing in flight, illustrate the supervenience (Sober 1984) of character or trait combinations. There is no fixed intrinsic 'value' to any character set and no one character can be identified with any unique physical property (Sober 1984). Instead, each combination, representing different potential units of selection, has a different consequence in a different context. In fact, reproductive and energy allocation behaviors that are best in the context of low predation intensity are worst in the context of high predation intensity, despite no change in the mode of predation nor the rules of prey selection. Instead, those reproductive and energy allocation behaviors that are locally maximal emerge, shift, and disappear within the context of dealing with the same 'driver' but not the same selective regime.

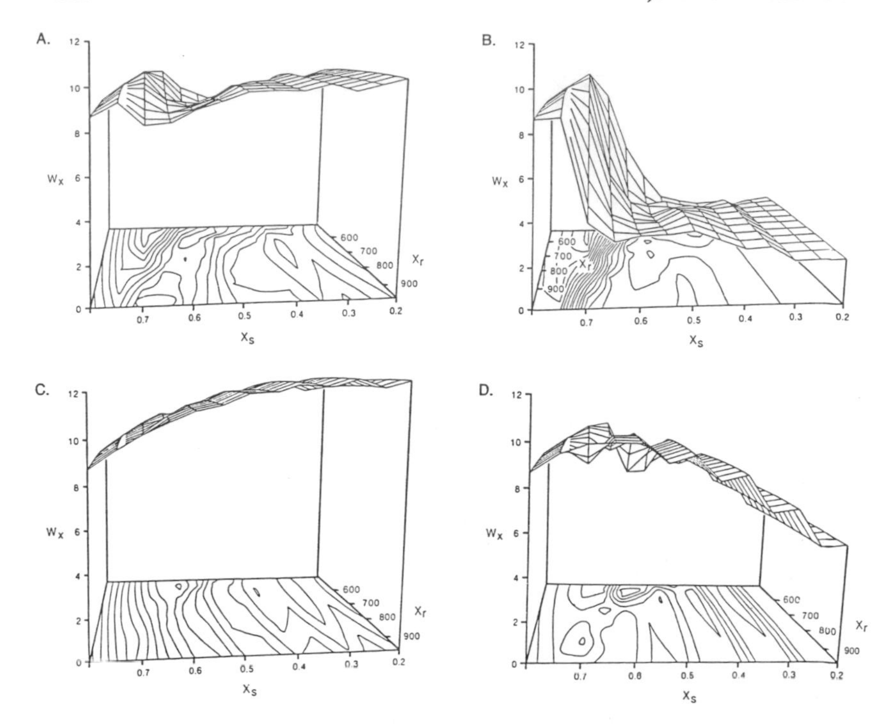

Figure 6.3. The consequences of increasing a single parameter value of the 'environment', predation intensity (Y_i), on the prey selective surface W_x. In **A** and **B**, Y_s remains at 24mm, but Y_i increases from .005479 to .027397. In C and D, Y_s remains at 20 mm, while Y_i similarly increases from .005479 to .027397.

The same holds true if we hold predation intensity constant but introduce a unidirectional change in a character of the 'environment'. A commonly documented evolutionary trend in a character or trait within and between species is increase in maximum size (Cope's rule). Such an empirical trend is also common (although predictably not universal) within fossil lineages of the predator (Kitchell 1986). What are the consequences of assuming a smooth, continuous increase in the predator trait, Y_s (maximum size)? Static snapshots of the induced dynamics as consequences for prey character combinations are given in figure 6.5. The 'driver' is now the evolutionary change of the driver.

In response to a unidirectional increase in a single parameter of the 'environment', the trajectory of prey traits traverses a wide range of character values, bifurcating, diverging, reversing itself, and even, as discussed later, getting trapped. The increase from Y_s of 15 mm to Y_s of 20 mm, for example, induces a shift in both X_r (toward a delay) and

X_ε (toward an increased allometry; figure 6.6). But this region of max W_x bifurcates as Y_s increases further to 25 mm and continues to diverge. Simultaneously, a new local region of max W_x appears. With further increases in Y_s, however, the two diverging 'peaks' both disappear (figure 6.6) At Y_s equal to 35 mm there is the appearance of a new local maximum associated with a reversal in the previous trend of the prey character X_s. Subsequent increases in the 'environment' trait (Y_s), from 50 to at least 90 mm, result in a reversal of the 'organism' trait X_r, from accelerating the onset of reproduction (with age) to delaying it. The dashed lines (figure 6.6) represent potential 'jumps' that might occur (by rapid change over evolutionary time) or not (extinction).

In summary, smooth, continuous change in the 'environment' does not result in like change in the object of that environment. A unidirectional trend in the character that is allowed to vary within the environ-

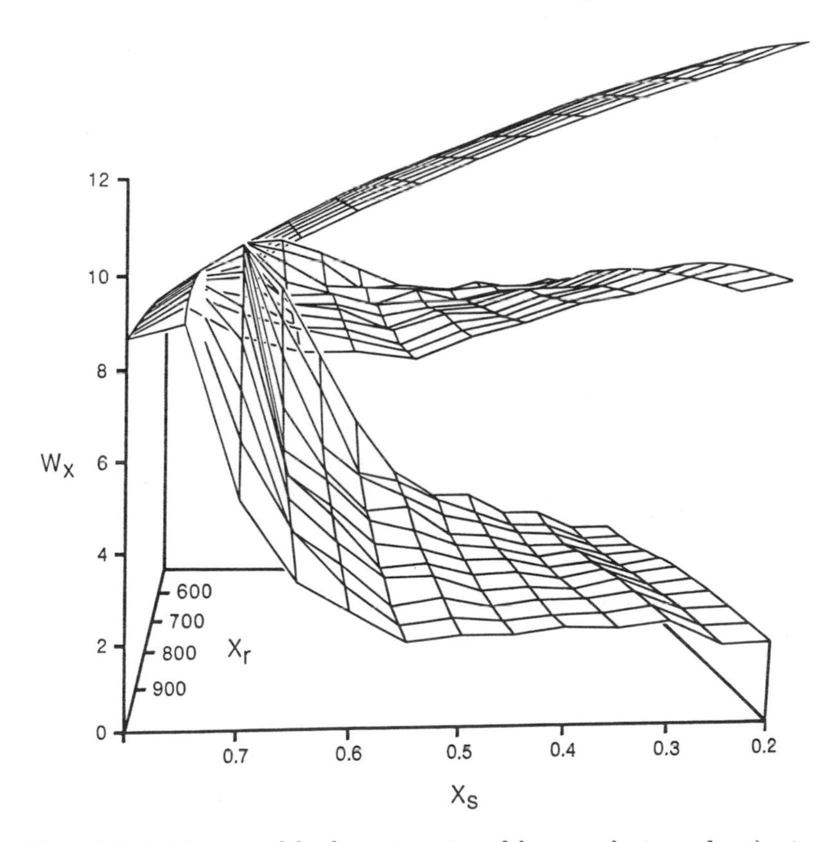

Figure 6.4. A static view of the dynamic motion of the prey selective surface showing the consequences of variable predation intensity (Y_i), from low (upper surface), to medium, to high (lower surface).

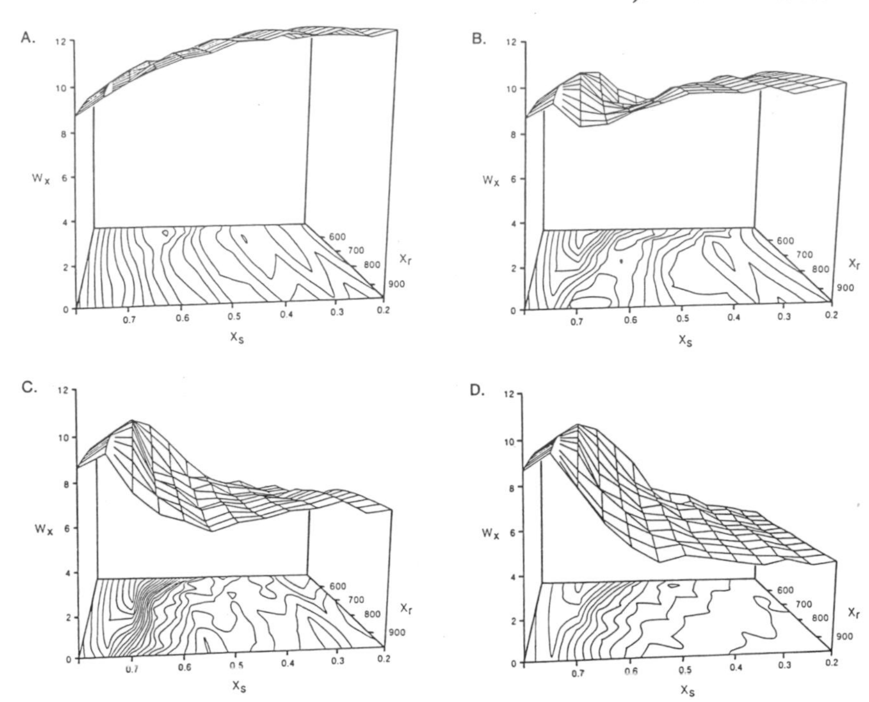

Figure 6.5. The consequences on the prey selective surface of 'driving' the environment with a smooth, continuous increase in a single parameter of the prey environment, namely Y_s, which increases from 20 mm (**A**) to 24 mm (**B**) to 30 mm (**C**) to 36 mm (**D**); Y_i is constant (.005479).

ment does not induce a unidirectional trend in the characters of the species subjected to selection by that environment. Any intuitive expectation of unidirectional escalation is erroneous.

TWO-WAY INTERACTION: WHERE PLAYING THE GAME CHANGES THE RULES

We next consider the situation of two-way interaction between organism and 'environment'. The 'driver' in the preceding analyses has been the predator as environment. It is also possible to couple organism (prey) and environment (predator) so that both simultaneously act as environment for the other. The 'driver' is itself driven and, in being 'driven', drives. The central questions remain: Given a unidirectional change to set the dynamics in motion, will the dynamics be limited to unidirectional escalation, or will more complex behaviors evolve? Also, is there any possibility for stasis in a constantly dynamic interaction in which the selection regimes are both directional?

To change only one feature of the model at a time, namely the introduction of coupling between organism (prey) and 'environment' (predator), we will repeat the conditions of figure 6.6. In figure 6.7 we have an overlay of the results of figure 6.6 (representing the consequences as a function of the prey traits) and the new information of the reciprocal effects of these traits on the predator trait (Y_s). The situation is as follows: these two surfaces are both in motion and are simultaneously responding to and driving the other surface. To 'see' the dynamics would require a movie of the motion of these two coupled surfaces and cannot be effectively captured in two dimensions on the page. However, by inserting ourselves at fixed points between these two surfaces, we can both predict and understand the essential elements of the dynamics. We can begin with some small initial value of Y_s (e.g., 15 mm) to locate the region of max W_x that 'evolves' in response to the 'drive' from the environmental selection field. At this point on figure 6.7, labeled 15, all combinations of possible prey traits have been tried and the one most effective for the prey identified. It is apparent, however, from the diagram that this position of the prey, selected for as a consequence of the predator trait value, is not the most effective value for the predator trait. To rephrase the situation, the predator as 'envi-

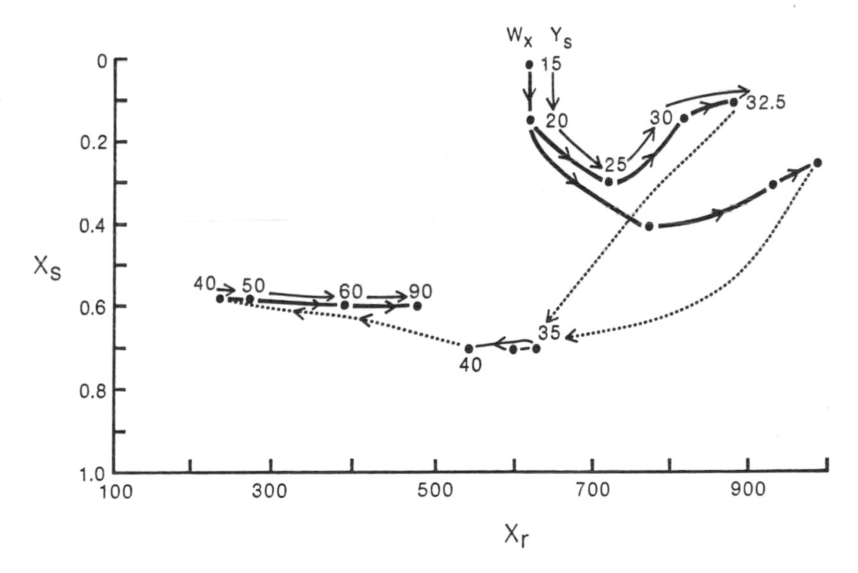

Figure 6.6. A plot of the trajectory of the 'driver' of figure 6.5 (Y_s, solid line) and a corresponding plot of the trajectory of max W_x (bold line) as 'driven'. The dashed lines connect points of sudden disappearance and reappearance of max W_x and indicate the potential for abrupt change (modified from DeAngelis et al., in press).

ronment' has driven the prey traits to values that work against the predator. Instead of an equilibrium or truce mediated by their mutual coupling, the two surfaces are in nonequilibrium. Is this condition local or global for the traits being considered? Because the initiating value of Y_s is 15 mm, whereas the max Y_s at that position on the prey trait surface is 40 mm, the selection on the predator is directional—toward an increase in size. As Y_s increases, however, the action of so doing causes the prey surface to shift. The bold lines follow the earlier diverging paths of figure 6.6, but movement along either of these prey trajectories feeds back on the consequences of the predator's trajectory. As the predatory value of Y_s reaches 32.5 mm, the 'peak' on the prey surface itself disappears, meaning that the gradient of the selection surface abruptly reverses. If the predator responds to this selection gradient by reversing its trend, now toward decreasing values of Y_s, an interesting situation can develop. As the Y_s values pass through 32.5 mm, the selection gradient reverses again, with a reappearance of the previous 'peak'. If the prey traits move back up that selection gradient in response to the moving predator surface, there is now selection on the predator traits to increase, which in turn can cause a repetition of the situation just discussed. As shown in figure 6.7 by the dashed regions, a dynamic stasis could ensue as a consequence of the trapping effect of coupled, reversing selection gradients.

The empirical data on this particular predator-prey interaction better fit the predictions described previously than those of the naive model of perpetual, unidirectional escalation. Data on the behavioral selection of bivalve prey by naticid gastropod predators both within and

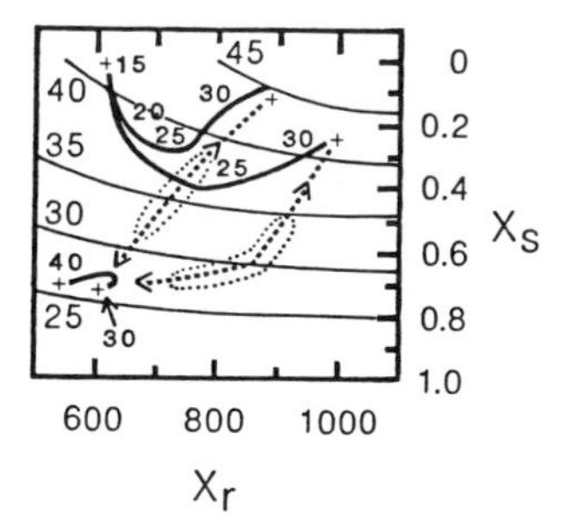

Figure 6.7. The consequences of *coupling* organism (prey) and 'environment' (predator) in two-way interaction. The bold lines outline the trajectory of W_x of the prey (detailed in fig. 6.6). The contours represent isoclines of maximal values of the predator trait Y_s. See text for discussion of the interplay of these two surfaces in continual nonequilibrium. Dashed regions indicate potential stasis by 'trapping' between the two surfaces. (Modified from DeAngelis et al., in press).

among species show that the rules used in the model to drive the interaction pertain to patterns spanning more than 100 myr, despite considerable taxonomic turnover (Kitchell 1986). A naive expectation might be that the prey, over evolutionary time, would through selection have increasingly thick shells, but that the predators might also be able to drill deeper and deeper bore holes. Exceptions would go extinct. To use a familiar line, 'The fossils say no!' Instead, the empirical data exhibit characteristics predicted by the mode: examples of convergence, examples of coexistence of very different reproductive timing-phenotypic allometries, and stasis.

Frequently, when the fossil record fails to fit a general theory, the explanation is that the data are poor. In this case, the theory may have been wanting. Modeling, as a means of exploring the consequences of theory, has shown a complex suite of behaviors not predicted by the simple arms race metaphor. Previously, different rates of change in predator and prey parameters relevant to the interaction, or stasis in these parameters, or multiple means of dealing with the same predator have been considered evidence that the interaction is *not* driving evolutionary change (e.g., Bakker 1983; Stanley, Van Valkenburgh, and Steneck 1983). We have shown, however, that such results are instead characteristic features of a coupled interaction between organism and (biotic) environment as a consequence of nonlinear feedback.

Previous expectations would have made the simplistic prediction that if prey shell thickness is a deterrent to selective predation, then increased predation should elicit an evolutionary response toward increasingly thick shells among prey species (the 1 + 1 = 2 expectation). The modeling effort, however, shows that even though prey shell thickness is a deterrent to selective predation, the counterintuitive result of no change in shell thickness or worse, by the arms race standard, a decrease in shell thickness over evolutionary time, may predictably occur. The modeling predicts that some prey will follow the opposite defense, where rapid growth rates results in increasingly thinned shells. Such predictions match the empirical world, where thin-shelled and thick-shelled bivalve prey, experiencing drilling predation, coexist. There are tradeoffs that result in multiple and dynamic 'solutions'.

Conclusions

If we now reconsider Futuyma and Slatkin's (1983) conclusion that the "ideal paleontological evidence would be a continuous deposit of strata in which each of two species shows gradual change in characters that reflect their interaction," we see that this is not the necessary nor even the expected case. Nor is it the case that the empirical phenomenon of stasis argues against a relevant evolutionary role for biotic interactions, as has been suggested. Instead, there is the potential for both stasis and change within coupled biotic interactions, and the change may be predictably sudden and discontinuous as well as gradual and continuous—even in the absence of a changing abiotic environment.

The purpose of this effort has not been to deny the role or importance of abiotic effects on evolutionary processes. That role is well established. Rather, the purpose has been to begin to develop a new intuition to replace the old intuition born of experience with one-way interactions and linear systems.

Nonlinearity, scaling effects, and hierarchical control are more the rule than the exception in biological systems. The richness of these behaviors is only beginning to be understood. Their development, however, is changing the nature of the evidence needed to advance or reject theories of change. This is why neither part of the question posed at the beginning of this paper is wrong. The hope is that we collectively clear our minds and work in earnest on furthering our understanding of the unexpectedly complex questions we have posed.

Acknowledgments

This research was supported by National Science Foundation Grant BSR-86055310 in collaboration with W. M. Post and D. L. DeAngelis. The graphic output was prepared by W. M. Post and D. L. DeAngelis.

Literature Cited

Bakker, R. T. 1983. The deer flees, the wolf pursues: Incongruencies in predator-prey coevolution. In *Coevolution*, ed. D. J. Futuyama and M. Slatkin, 350–82, Sunderland, Mass.: Sinauer Associates.

Campbell, J. H. 1982. Autonomy in evolution. *Perspectives on evolution*, ed. R. Milkman, 190–201. Sunderland, Mass.: Sinauer Associates.

Campbell, D., J. P. Crutchfield, J. D. Farmer, and E. Jen. 1985. Experimental mathematics: the role of computation in nonlinear science. *Communications of the Assoc. Computing Machinery* 28:374–84.

Carr, T. R., and J. A. Kitchell. 1980. Dynamics of taxonomic diversity. *Paleobiology* 6:427–43.

DeAngelis, D. L., J. A. Kitchell, and W. M. Post. 1985. The influence of naticid predation on evolutionary strategies of bivalve prey: Conclusions from a model. *Am. Nat.* 126:817–42.

DeAngelis, D. L., J. A. Kitchell, and W. M. Post. In press. The nature of stasis and change: Potential coevolutionary dynamics. *Coevolution in ecosystems*, ed. N. Stenseth. Cambridge: Cambridge University Press.

Eldredge, N., and S. N. Salthe. 1984. Hierarchy and evolution. In *Oxford surveys in evolutionary biology*, vol. 1, ed. R. Dawkins and M. Ridley, 184–208. Oxford: Oxford University Press.

Futuyma, D.J., and M. Slatkin. 1983. *Coevolution*. Sunderland, Mass. Sinauer Associates.

Gleick, J. 1987. *Chaos—making a new science*. New York: Viking Penguin.

Gould, S. J. 1977. *Ever since Darwin*. New York: W. W. Norton and Co.

Jerison, H. J. 1973. *Evolution of the brain and intelligence*. New York: Academic Press.

Kitchell, J. A. 1986. The evolution of predator-prey behavior: Naticid gastropods and their molluscan prey. In *The evolution of animal behavior: Paleontological and field approaches*, ed. M. Nitecki and J. A. Kitchell, 88–110. New York: Oxford University Press.

Kitchell, J. A. 1990. Computer applications in palaeontology. In *Palaeobiology. A synthesis*, ed. D. E. G. Briggs and P. R. Crowther, 493–9. Cambridge, Mass.: Blackwell Scientific Publications.

Kitchell, J. A., and T. R. Carr. 1985. Nonequilibrium model of diversification: Faunal turnover dynamics. In *Phanerozoic diversity patterns: Profiles in macroevolution*, ed. J. W. Valentine, 277–309. Princeton, N.J.: Princeton University Press.

Kitchell, J. A., C. H. Boggs, J. F. Kitchell, and J. A. Rice. 1981. Prey selection by naticid gastropods: Experimental tests and application to the fossil record. *Paleobiology* 7:533–52.

Kitchell, J. A., D. L. Clark, and A. Gombos. 1986. Biological selectivity of extinction: A link between background and mass extinction. *Palaios* 1:504–11.

Kohn, A. J. 1987. Progressive adaptation. *Science* 237:1235–6.

Levins, R., and R. Lewontin. 1985. *The dialectical biologist*, Cambridge, Mass.: Harvard University Press.

Lorenz, E. N. 1964. The problem of deducing the climate from the governing equations. *Tellus* 16:1–11.

May, R. M. 1976. Simple mathematical models with very complicated dynamics. *Nature* 261:459–67.

Owen-Smith, N. 1987. Pleistocene extinctions: the pivotal role of megaherbivores. *Paleobiology* 13:351–62.

Salthe, S. N. 1985. *Evolving hierarchical systems*. New York: Columbia University Press.

Sober, E. 1984. *The nature of selection*. Cambridge, Mass.: MIT Press.

Stanley, S. M., B. Van Valkenburgh, and R. S. Steneck. 1983. Coevolution and the fossil record. In *Coevolution*, ed. D. J. Futuyama and M. Slatkin, 328–49. Sunderland, Mass.: Sinauer Associates.

Vermeij, G. J. 1987. *Evolution and escalation*, Princeton, N.J.: Princeton University Press.

Williams, G. C. 1966. *Adaptation and natural selection*. Princeton, N.J.: Princeton University Press.

Part 2

PATTERNS WITHIN INDIVIDUAL TAXA

7

Ecological Processes and Progressive Macroevolution of Marine Clonal Benthos

Jeremy B. C. Jackson and Frank K. McKinney

A trend is a nonrandom, directional change in any quantity over time that can be described statistically as for example, by regression analysis. Progress in an evolutionary sense is merely a trend in adaptation or exaptation (we use adaptation for both concepts because they are usually indistinguishable in practice) driven by natural selection upon individuals or by sorting (selection) of species. Description of a macroevolutionary trend as progressive does not imply that geologically more recent organisms are any better adapted now than their predecessors were before, but that adaptation has proceeded over millions of years in a directional manner.

The Phanerozoic history of life recorded by fossils reveals many apparent patterns of innovation and replacement by major clades extending over periods of 100 my or more (Vermeij 1987). Interpretation of these trends as due to competitive displacement of inferior by superior taxa comprises much of the dogma of historical geology and paleontology. Recently, however, the existence of such trends and their interpretation have been questioned on methodological, theoretical, and philosophical grounds, and for the almost anecdotal nature of much of the supposedly supporting data (Raup 1977; Gould and Lewontin 1979; Gould and Calloway 1980; Gould 1985, 1988; Benton 1987; McKinney 1987).

Much of the controversy about macroevolutionary trends stems from the different kinds of evidence employed and confusion regarding their relative utility. Virtually all patterns are defined in terms of the comparative divesity (numbers) of lower taxa (species, genera, or families) of different major taxa (e.g., genera of bivalves versus genera of brachiopods, species of erect versus species of encrusting bryozoans). Relative abundance is usually ignored or merely estimated because data are typically absent or poor. Thus, for the present, we must rely on individual paleontologists' qualitative field knowledge of the abundances of different taxa, always an arguable quantity. This makes all patterns suspect because diversity and abundance are not always positively correlated; regular sea urchins are among the most abundant and ecologically important grazing animals in the Caribbean, yet there are only nine species in shelf depths (Mayr 1954) compared with hundreds of grazing snails or fishes (Warmke and Abbott 1961; Randall 1967, 1968). Similarly, the number of species of scleractinian corals in shelf depths in the Caribbean is less than twice as great as in the eastern Pacific, yet the development of coral populations and reefs is vastly greater in the Caribbean (Glynn and Wellington 1983; Coates and Jackson 1987). This is not the kind of information that jumps out from most taxonomic monographs.

Another major problem lies in the types of parameters measured and their statistical dependence upon diversity, sample size, and the characteristics of founding members of a clade. Thus clams may display more life habitats now than in the Paleozoic (Stanley 1968, 1975, 1977) simply because there are more clam species, and the same applies to any other such trends (Bambach 1983, 1985). Such bias can be eliminated by drawing from similarly sized populations over time or using other sampling techniques. More fundamentally, however, bivalves are almost certainly living a more diverse and widespread existence because of basic biomechanical innovations such as mantle fusion and siphons that provide a clear biological explanation for their increase (Stanley 1968). This is true regardless of what statistical pitfalls arise.

However counted, there are at least six kinds of fossil evidence available for the definition and interpretation of macroevolutionary pattern. These are the relative diversity or abundance of different (1) higher taxa (e.g., bivalves versus brachiopods); (2) body sizes; (3) functional morphologies (e.g., passive versus active suspension feeders); (4) modes of growth or development (e.g., intrazooidal versus zooidal budding; encrusting versus erect); (5) life habits (e.g., infaunal versus epifaunal); or (6) environmental distributions (e.g., shallow versus deep). All but the first kind of comparison can be made across taxonomic boundaries. Polyphyletic patterns are particularly valuable for helping to rule out the importance of phyletic design constraints or chance.

Once a pattern has been demonstrated, its interpretation should depend upon the strength of the suggested causal evidence (ecological interactions or abiotic environmental changes) and the precision of stratigraphic definition of the relative timing of presumed cause and effect. This does not mean that all factors should necessarily vary in phase. Time lags and thresholds are a notorious difficulty in population and community dynamics (May 1977, 1981). Nevertheless, it should be kept in mind that the published resolution for events within the

"Mesozoic Marine Revolution" (Vermeij 1977) is typically no better than 25 my.

Here we examine numerous macroevolutionary trends in the history of marine clonal benthos for which we believe the biological and paleontological data are adequate to begin to assess the relative importance of ecological interactions versus developmental constraints, extrinsic factors, or chance. Most of our examples are from the Mesozoic to Recent, because less is known paleobiologically about Paleozoic organisms and their environments. We emphasize bryozoans because they are the most throughly studied clonal phylum; remarkably little is known about the paleobiology of corals and coralline algae other than their variable contribution to formation of reefs.

Methods and Sources

Data for patterns presented were drawn primarily from surveys of published studies of fossil faunas and floras, or by examination of museum collections, and are available in the references cited. In addition, we recompiled a more extensive data set for 35 Jurassic to Recent bryozoan faunas based on the sources listed in appendix 7.1. For the Recent faunas, we included only well-calcified species likely to be preserved as fossils. Most of the data are presented as percentages of species bearing the trait of interest to minimize effects of sample size. Regardless, percentages for any fauna are based on at least 22 species, and all but 5 include more than 53 species.

Variations in traits with time were examined statistically by regression analysis, using the midpoints of the intervals (period, epoch, age) for which data were available (DNAG time scale, A. R. Palmer 1983). These analyses do not violate assumptions of independence of observations (Felsenstein 1985) because none of the bryozoan or coral trends are phylogenetically constrained. In all cases except for mobile bryozoans and crustose algae, data are drawn from many evolving clades (families or orders) that persist throughout much or all of the time examined, as well as many shorter-lived clades scattered throughout.

Bryozoans

Bryozoans grow as colonies of minute, asexually produced modular units termed zooids that typically consist of a polypide capable of independent function (feeding, excreting, reproducing) contained within a calcareous skeletal chamber (Ryland 1970; Boardman and Cheetham 1987; McKinney and Jackson 1988). The pattern of zooidal budding

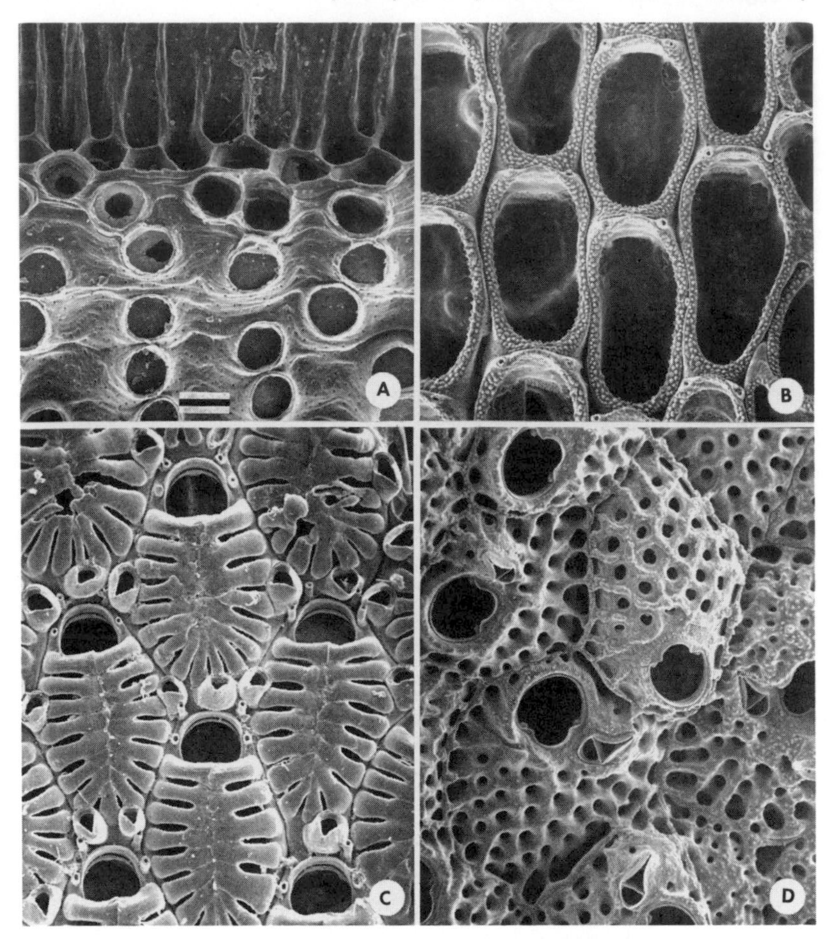

Figure 7.1. Basic zooidal types of calcified bryozoans. **A,** *Plagioecia patina* (Lamarck), a stenolaemate with tubular zooids. **B,** *Conopeum* sp., an anascan cheilostome with noncalcified frontal wall. **C,** *Membraniporella nitida* (Johnston), a cribrimorph cheilostome with fused spines arched above the frontal wall. **D,** *Schizoporella* sp., an ascophoran cheilostome with well-developed frontal wall and frontal budding. Scale bar 100 um; all at same magnification. SEM photographs produced in Applachian State University Electron Microscopy Laboratory.

determines the growth form and, to a first approximation, the life history of the colony (Jackson 1979a; Coates and Jackson 1985). There are two major classes (fig. 7.1).

Stenolaemates have tubular zooids, which during the Paleozoic were unconnected physiologically except at the surface of the colony. In contrast, pores between zooids are common after the Paleozoic. Stenolaemates lack outer (frontal) walls or opercula to close the entrance to zooids, and zooidal polymorphism is absent or rudimentary. Stenolaemates were the only calcified bryozoans during the Paleozoic. A few representatives of several orders, lumped together as Cyclostomata (Boardman 1984), survived the End-Permian extinction and diversified extensively in the Jurassic.

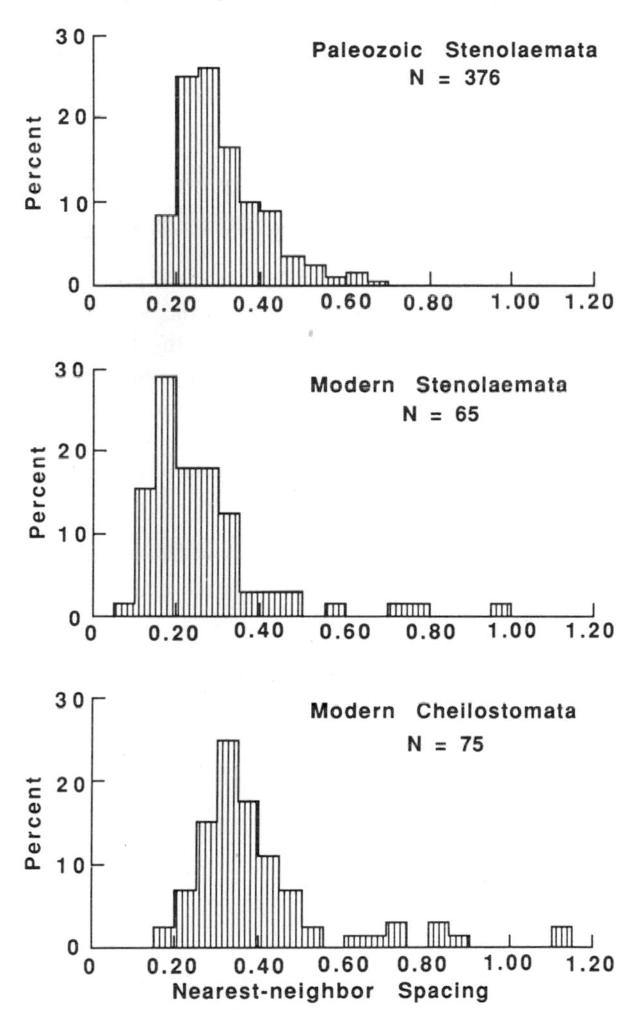

Figure 7.2. Nearest-neighbor spacing of zooidal apertures (stenolaemates) and orifices (cheilostomes) for Paleozoic and Recent bryozoans (R. S. Boardman and F. K. McKinney, unpublished). Measurements for Paleozoic stenolaemates made from illustrations in 15 monographs of Ordovician through Permian bryozoan faunas; those for Recent stenolaemates are from 65 species worldwide. Measurements for Recent cheilostomes are from three monographs of Atlantic bryozoans plus direct measurements of specimens from New Zealand and California.

The major group of gymnolaemates is the Cheilostomata, which first appeared in the Jurassic and underwent an explosive radiation from the Cretaceous to Eocene. Cheilostome zooids are boxlike, with a well-developed system for interzooidal communication through complex systems of pores. They typically form variously elaborate outer (frontal) walls, possess opercula, and have extensive zooidal polymorphism. The design and extent of frontal wall development is the primary basis for recognition of three almost certainly polyphyletic orders: anascans, cribrimorphs, and ascophorans.

Variations in the design of zooids and colonies are extensive and are the major ingredients of the bryozoan evolutionary trends to be discussed (McKinney and Jackson 1988). In contrast, the dimensions of modular units are remarkably constant throughout the Phanerozoic for both stenolaemates and gymnolaemates and are probably physiologically constrained. The modal size of zooids has varied less than two-fold, and the range is similar today to that in the Paleozoic (fig. 7.2). Moreover, the relation of the size and number of tentacles of the lophophore to zooid size is fixed strongly and apparently has been so since the Paleozoic (fig. 7.3; McKinney and Boardman 1985). Similar approximate constancy is evident in the diameter of branches of different

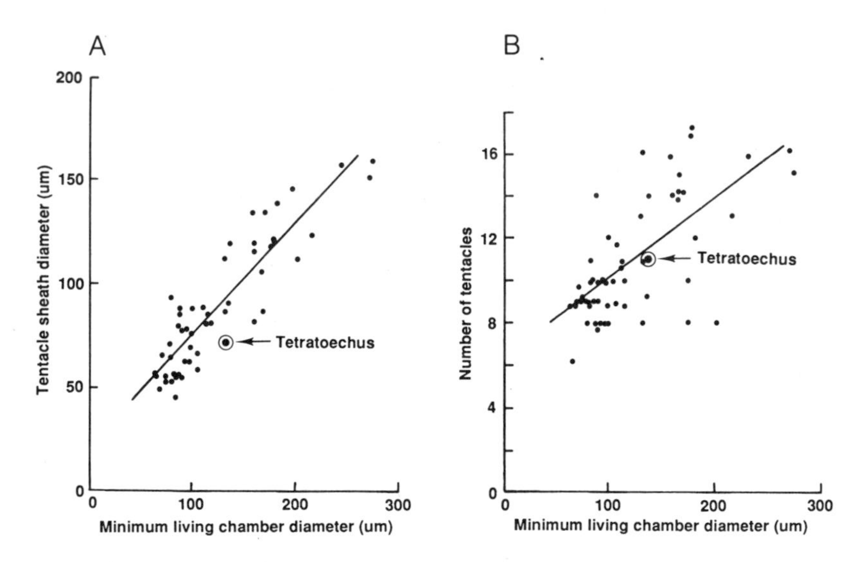

Figure 7.3. Zooidal biometry of living Stenolaemata showing constant dimensional relations of the (A) diameter of the tentacle sheath and (B) number of tentacles against the minimum diameter of the living chamber. Regression line for (A) is Y = 21.3 μm + 0.53X, r = 0.903; regression line for (B) is Y = 6.36 + 0.037X, r = 0.658. Note consistent position of Ordovician *Tetratoechus*, which was not included in computing regressions (After McKinney and Boardman 1985).

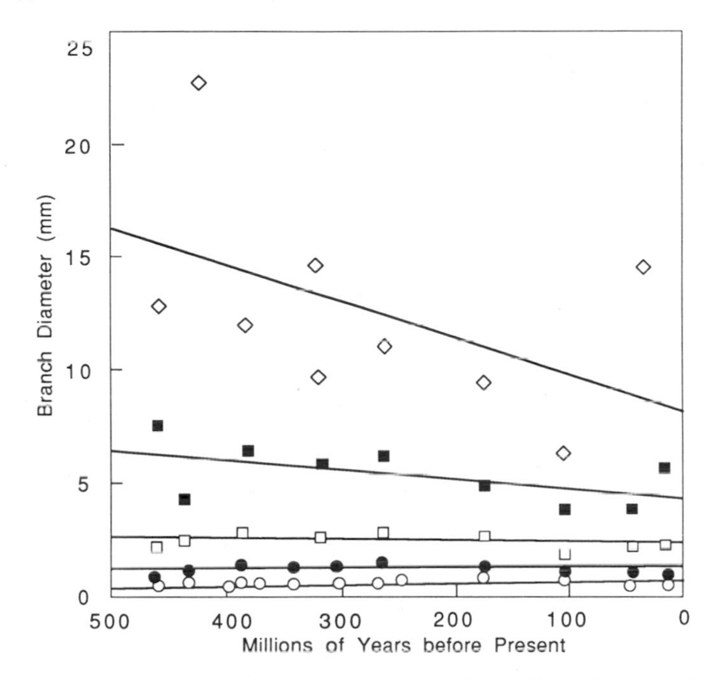

Figure 7.4. Branch diameters of the five major growth forms of erect bryozoans through the Phanerozoic. Data from references in McKinney 1986b and Appendix 1. None of the trends are significant except the trivial "increase" in unilaminates. 0 = unilaminate: diameter = .422 + 0.0004 (time), r = .51, P = .037; • = radial nonmaculate: diameter = 1.21 − .0003 (time), r = −.18, P = .312; □ = bilaminate nonmaculate: diameter = 2.59 − .0005X, r = −.23, P = .279; ■ = radial maculate: diameter = 6.42 − .004 (time), r = .50, P = .083; ◇ = bilaminate maculate: diameter = 16.1 − .0131 (time), r = .43, P = .147. Origin for time = 500 mybp.

erect colonial growth forms throughout the Phanerozoic, regardless of taxonomic affinity (fig. 7.4; McKinney 1986b). Indeed, modular size trends are not a major feature of the macroevolution of clonal organisms (Coates and Jackson 1985).

MACROEVOLUTIONARY TRENDS

Zooidal design and integration. The four "orders" of post-Paleozoic bryozoans form a clear series of increasing skeletal defense and interzooidal integration that parallels their sequence of appearance in the fossil record (table 7.1). Because these are grades of evolution rather than true taxa, the fossil record of diversity of the four groups constitutes a measure of the comparative success of differing levels of bryozoan armament and integration over the last 200 my.

There is a striking, highly significant proportional increase in chei-

Table 7.1. Armament and integration of post-Paleozoic bryozoans

Taxon	Frontal Wall Development	Position of Embryos	Extent of Polymorphism	Communication between Zooids
Cyclostomata	None	Surface	Little	Limited
Anasca	Little	Surface	Moderate	Extensive
Cribrimorpha	Intermediate	Surface	Moderate	Extensive
Ascophora	Extensive	Buried	Extensive	Extensive

lostomes relative to cyclostomes since the Early Cretaceous (fig. 7.5A; Taylor and Larwood 1988). Moreover, since the Turonian, about 89 mya, the relative diversity of cyclostomes, anascans, and cribrimorphs have all decreased and that of ascophorans has increased (fig. 7.5A–D).

In a taxonomically more limited but more detailed morphological survey, Boardman and Cheetham (1973) demonstrated that integration of cheilostome zooids and colonies has increased dramatically throughout their history, but that integration of stenolaemates did not.

Budding patterns of encrusting cheilostomes. Species of encrusting cheilostomes differ in the process of budding by which they grow across the substratum and in their potential for budding vertically away from the substratum (fig. 7.6; Lidgard 1985a, 1985b; McKinney and Jackson 1988; Lidgard and Jackson 1989). Horizontal extension may occur episodically, one zooid at a time. In this intrazooidal budding, the horizontal extension of a colony depends upon the sequential development of individual zooids. Alternatively, extension of a colony may proceed more or less continuously, with a series of developing zooids in various stages of maturity. In this zooidal (or multizooidal) budding, buds may exceed the length of single zooids, and extension of the colony is not dependent upon the sequential maturity of each zooid. Species with zooidal budding exhibit more rapid horizontal growth, greater flexibility of their colony margins, and increased regenerative capacity compared to species with intrazooidal budding (Jackson 1983; McKinney and Jackson 1988). Thus species with zooidal budding are better able to defend themselves against overgrowth, to overgrow other species, and to survive injury than are intrazooidally budding forms.

Species also differ in their capacity to form multilayered colonies by frontal budding of new zooids one on top of another. Species capable of frontal budding are better able to defend against overgrowth by forming thicker colonies or by erecting specialized barriers to overgrowth along their margins than are species that can only grow horizontally (Jackson and Buss 1975; Jackson 1983). Moreover, frontally budding species are better able to defend themselves against grazing predators or to repair injuries from grazers.

Lidgard (1985a, 1985b, 1986) and Lidgard and Jackson (1989) analyzed changes through time in the relative numbers of species with these different modes of growth within the Late Cretaceous to Recent cheilostome faunas of North America (fig. 7.7). The proportions of species with zooidal or frontal budding increased throughout these 100

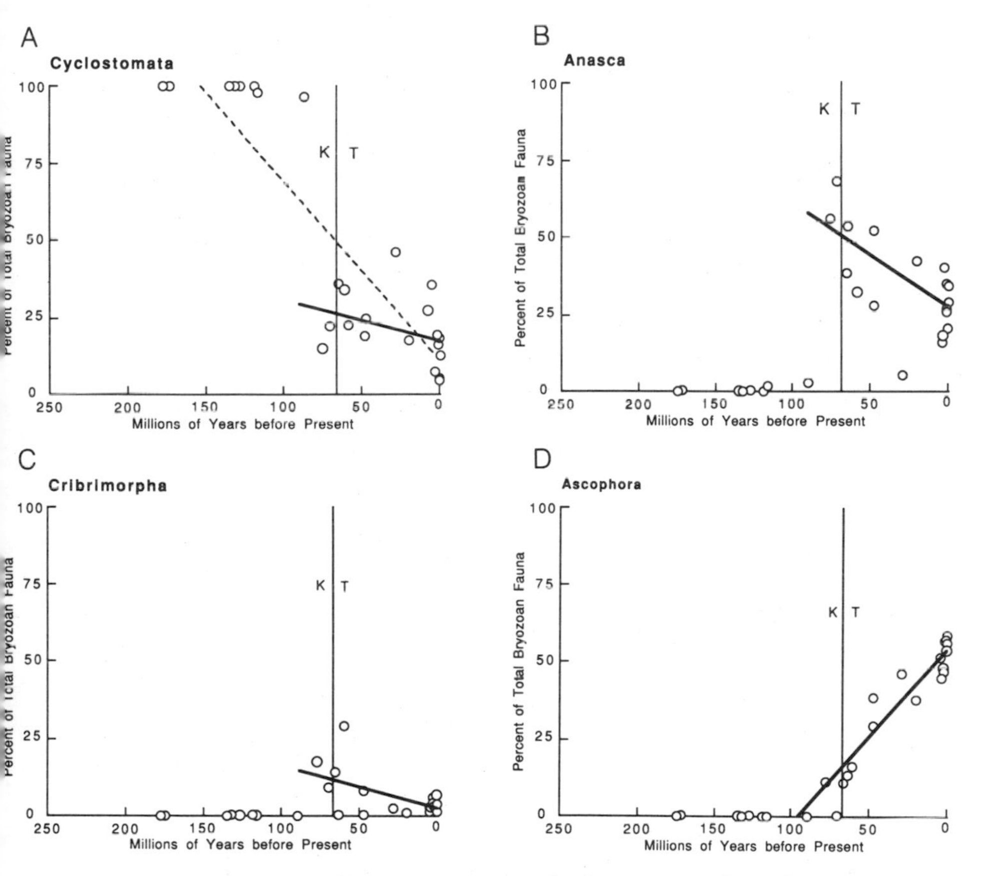

Figure 7.5. Percentage of bryozoan species in the four major "orders" of Jurassic to Recent calcified marine Bryozoa based on data from references in appendix 7.1. The percentage of cheilostomes is the inverse of cyclostomes; total cheilostomes = anascans + cribrimorphs + ascophorans. Regression analyses include only post-Turonian data, except for an additional analysis of cyclostomes that includes all faunas. Origin for X = beginning of Jurassic 208 mybp. **A,** Cyclostomes: *dashed line* Jurassic to Recent, % = 13.1 − .576 (time), r = −.88, P = .0001; *solid line* post-Turonian, % = 44.5 − .128 (time), r = −.35, P = .077. **B,** Anascans: % = 98.6 − .344 (time), r = −.73, P = .0006. **C,** Cribirimorphs: % = 36.7 − .134 (time), r = −.55, P = .0092. **D,** Ascophorans: % = −63.5 + .564 (time), r = .95, P = .0001.

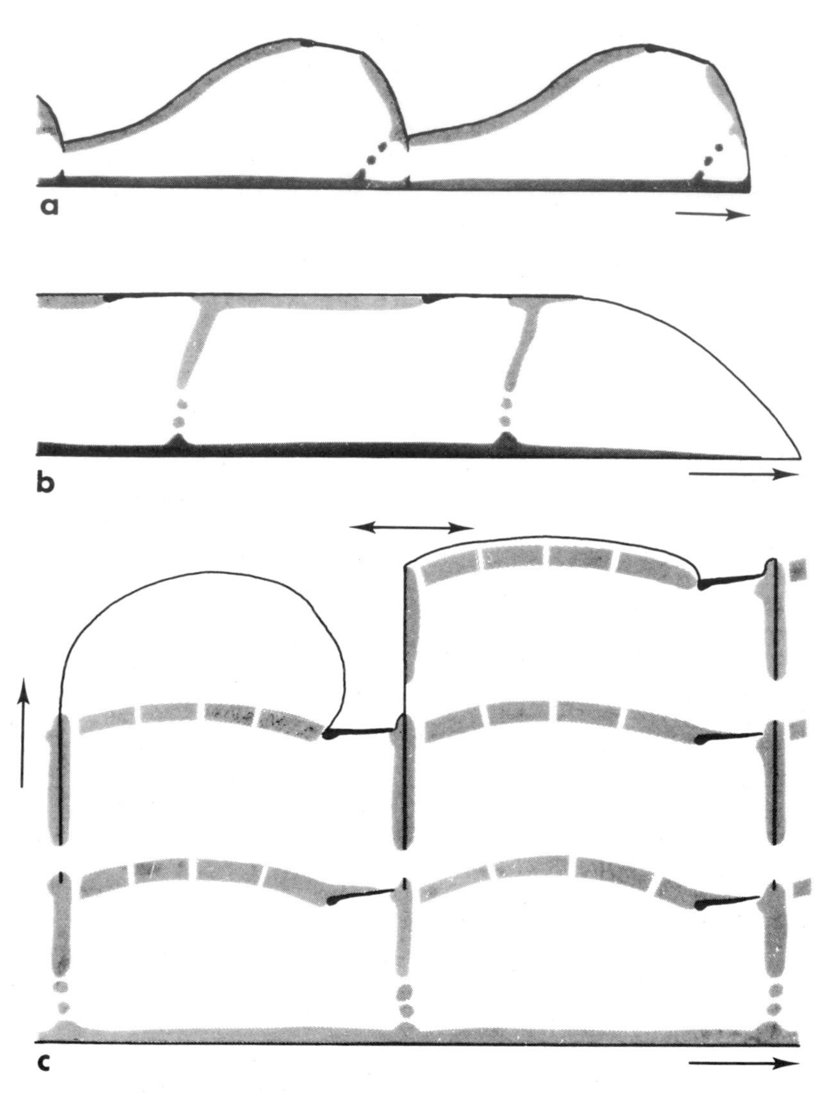

Figure 7.6. Sagittal sections through idealized zooids to show the main types of budding in encrusting cheilostomes. Black lines are cuticle that presently or formerly served as the outer surface of the zooid and colony. Calcified walls are indicted by stippling; cuticle is absent from walls that originally subdivided interior colonial space but is present in walls that originally grew exposed to the exterior environment. **A,** Intrazooidal budding in which lateral, partially isolated chambers serve as points of origin of new zooids. **B,** Zooidal budding in which walls originate interiorly to subdivide continually expanding space into new zooids; during rapid growth the expanding space may extend well beyond the dimensions of a single zooid. **C,** Frontal budding, which increases thickness and height of the colony above the substratum (e.g., fig. 7.1D). Arrows indicate potential growth directions. (After Lidgard 1985a.)

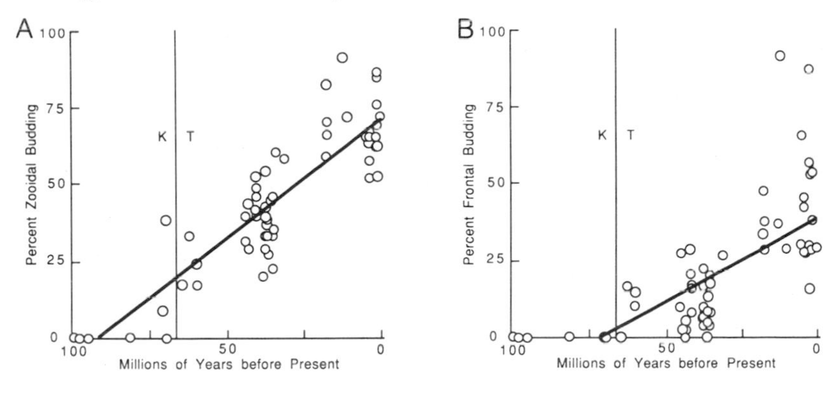

Figure 7.7. Percentage of encrusting cheilostome bryozoan species with different budding types (redrawn from Lidgard 1984, 1986). **A**, Zooidal or multizooidal (not intrazooidal) budding, regression: % = −6.95 + .780 (time), r = .88, P = .0000; **B**, Frontal budding, regression: % = −15.3 + .540 (time), r = .73, P = .0000.

my. Note, however, that variance among contemporaneous faunas is high throughout the entire interval. The faunas sampled include most of the major lineages of cheilostome evolution since the Cretaceous and are therfore highly polyphyletic. Similar patterns almost certainly exist among erect growth forms of cheilostomes but have not yet been analyzed (S. Lidgard, pers. comm. 1989).

Design of bilaminate erect cheilostomes. Erect growth enables bryozoans to rise above the substratum, and thereby increase their surface area, feeding, and reproductive capacity relative to encrusting species (Jackson 1979a; Coates and Jackson 1985; McKinney and Jackson 1988). Erect colonies can also feed above the boundary layer and experience fewer competitive interactions compared with encrusting growth forms. The cost of these advantages is the requirement for a rigid developmental pattern and greater vulnerability to colony destruction by forces of water flow, impacts with solid objects, and predators.

Rigidly erect cheilostomes with narrow bilaminate branches grow in an highly regular treelike pattern (figs. 7.8 and 7.10A; Cheetham, Hayek, and Thomsen 1981; Cheetham and Thomsen 1981; Cheetham and Hayek 1983; Cheetham 1986a). Arborescent bilaminate growth is common and has arisen many times throughout the last 100 my. Many deposits are dominated volumetrically by the remains of these species (Cheetham 1971; Thomsen 1976; McKinney and Jackson 1988). Two features of particular architectural interest are the rate of basal branch thickening and the angles of bifurcation among branches. The rate of branch thickening, rather than variations in mechanical strength of

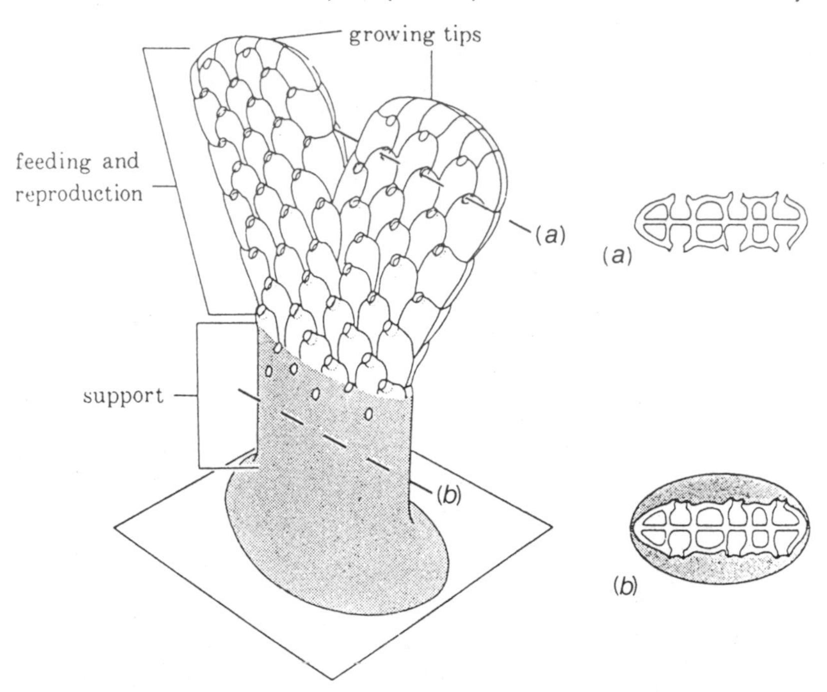

Figure 7.8. Idealized young colony of erect bryozoan with narrow bilaminate branches. Note distal region where zooids can feed and reproduce and proximal (*shaded*) region where zooids are occluded by secondary calcification that provides additional support. **A,** Section through distal region of branch showing zooids open to environment. **B,** Section through proximal region showing occlusion and thickening of branch. See Cheetham and Thomsen (1981) for photographs illustrating these features and mature colonies of erect bilaminate growth. (From Cheetham 1986a.)

skeletal material, is the principal design factor in the resistance of bilaminates to structural failure in response to stress. Greater branch thickening confers greater resistance of colonies to breakage, but at a potential cost of extensive additional calcification and loss of feeding and reproduction by zooids toward the bases of branches. The structural advantages of branch thickening are apparently so great as to confer greater resistance to breakage, even though species that thicken their branches are constructed of weaker materials than species that do not thicken branches. If a colony is not broken, the bifurcation angle between its branches determines how large it can grow before its branches begin to interfere with one another. The optimal angle to avoid such crowding is 90°.

Cheetham (1986a) determined the rate of branch thickening and bifurcation angle for 70 species of arborescent bilaminates of late Cretaceous to Recent age (fig. 7.9). Overall branch thickening increased during this interval; however, the range in values of branch thickening also increased. Bifurcation angles have also increased since the late Cretaceous, whereas the range in angles among contemporaneous species decreases over time (fig. 7.9). Both of these trends are polyphyletic, including numerous major clades of anascans and ascophorans.

Comparative diversity of different erect growth forms. There are five general growth forms of erect bryozoans that have been common throughout the Phanerozoic (McKinney 1986a, 1986b). These are de-

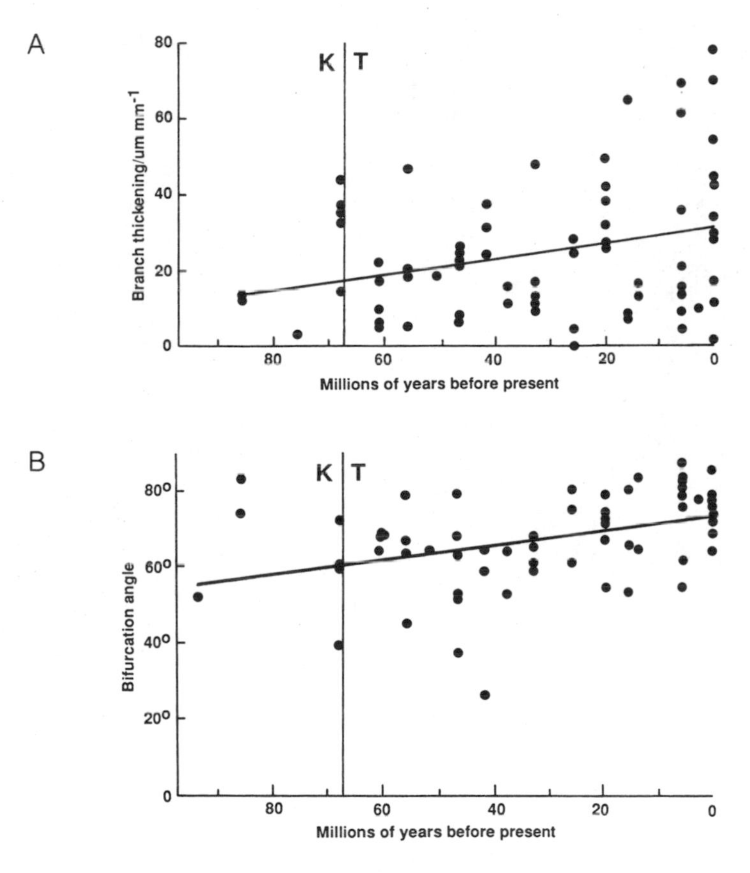

Figure 7.9. Architectural feature of 70 abundant species of Late Cretaceous to Recent rigidly erect bryozoans with narrow bilaminate branches. **A**, Rates of branch thickening, regression: thickening = 10.9 + .21 (time), $r = .29$, $P < .02$. **B**, Bifurcation angles between branches, regression: angle = 54.8 + 0.187 (time), $r = .39$, $P < .002$. (From Cheetham 1986a.)

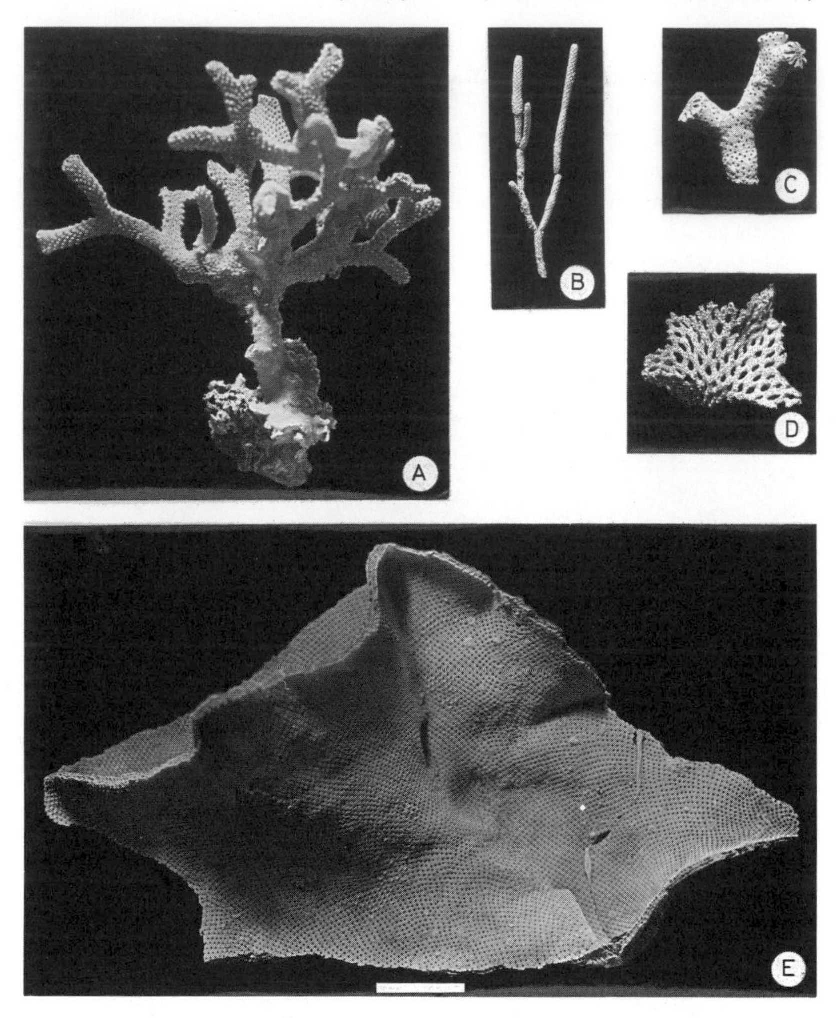

Figure 7.10. Examples of the five principal erect growth forms of calcified erect bryozoans. **A,** Narrow bilaminate branches without maculae. **B,** Narrow radial branches without maculae. **C,** Broad radial branches with maculae. **D,** Narrow unilaminate branches (fenestrate) without maculae. **E,** Broad bilaminate branches (fronds) with maculae. Scale bar 1 cm; all at same magnification.

fined on the basis of branch morphology and dimensions and the presence or absence of regular, elevated surface disruptions termed maculae (fig. 7.10). Among living bryozoans, maculae mark the positions on large branches or sheetlike surfaces where filtered water is expelled at excurrent "chimneys" (Cook 1977; Winston 1978, 1979). The growth forms are (1) narrow radial branches (diameter ≤2 mm) without maculae; (2) broad radial branches (diameter >2 mm), almost all of which bear maculae; (3) narrow bilaminate branches without maculae; (4) broad bilaminate sheets, almost all of which bear maculae, and (5) narrow unilaminate branches that lack maculae. Among the well-calcified species, the three narrowly branched forms can be rigid or articulated, whereas the two broadly branched forms are always rigid. Narrow branches should be more fragile than the other forms (although studies comparable to those for narrow bilaminate colonies have not been done). In contrast, the energetic cost of expelling filtered water is least for colonies with narrow unilaminate branches because of myriad factors related to their porous design.

Proportions of these different erect growth forms since the Jurassic were determined for the faunas in appendix 7.1. Four major, although not independent, trends are evident. Growth forms with narrow branches have increased over the last 200 my (fig. 7.11), as has the proportion of articulated species (fig. 7.12). These two trends demonstrate that erect bryozoans have become, on average, more flexible and probably more fragile over time, notwithstanding the opposite trend among rigid narrow bilaminates in figure 7.9. The growth forms with

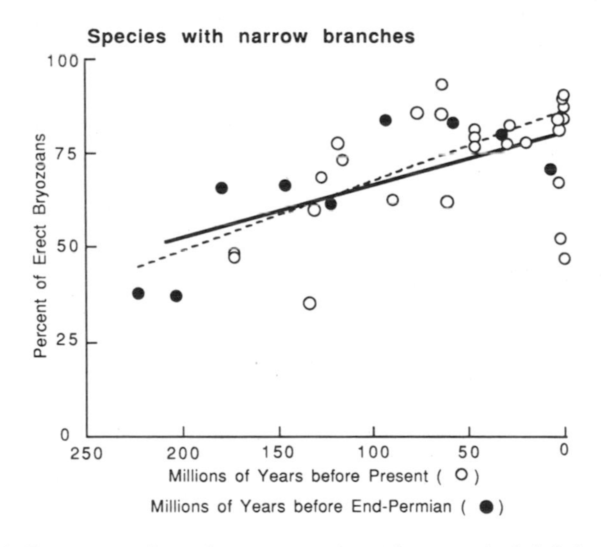

Figure 7.11. Percentage of erect bryozoan species with narrow (radial, bilaminate, and unilaminate) branches throughout the Phanerozoic. Jurassic to Recent based on faunas in Appendix 1, regression (*solid line*, ○): % = 51.5 + .138 (time), r = .51, P = .0034; Paleozoic based on sources in McKinney 1986a, 1986b; regression (*dashed line*, ●): % = 39.2 + .181 (time), r = .81, P = .004.

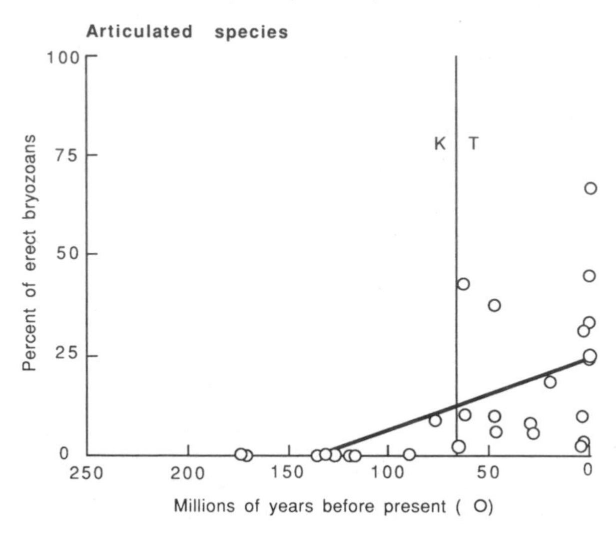

Figure 7.12. Percentage of Jurassic to Recent erect bryozoans with articulated branches (data from faunas in Appendix 1). Regression: % = −13.2 + .179 (time), r = .60, P = .0006.

narrow branches are nonmaculate, so their increase results in greater proportions of nonmaculate to maculate species. In additon, narrowly branched, unilaminate species have increased substantially compared to all others (fig. 7.13). This suggests that feeding efficiency, in terms of decreased cost of expelling filtered water, has prevailed as the overriding factor in the evolution of erect forms relative to their ability to resist breakage due to mechanical stress.

These patterns are paralleled by an offshore shift in the diversity of unjointed, rigid, and cemented versus jointed, flexible and rooted erect bryozoans since the Cretaceous (McKinney and Jackson 1988). Today rigid erect species are most diverse at about 150-m depth, whereas nonrigid species predominate in shallower and very deep waters (fig. 7.14).

Three of the same four trends also occurred during the Paleozoic among entirely different bryozoan clades (McKinney 1986a); the exception is for articulated species, which were uncommon throughout the Paleozoic. The Paleozoic and post-Paleozoic data are plotted together for the proportions of narrow and unilaminate branched species in figures 7.11 and 7.13; the correspondence is extraordinary.

Comparative diversity of encrusting and erect species. The proportion of encrusting species versus erect species decreased throughout the Paleozoic but increased after the Turonian (fig. 7.15). (There are too few

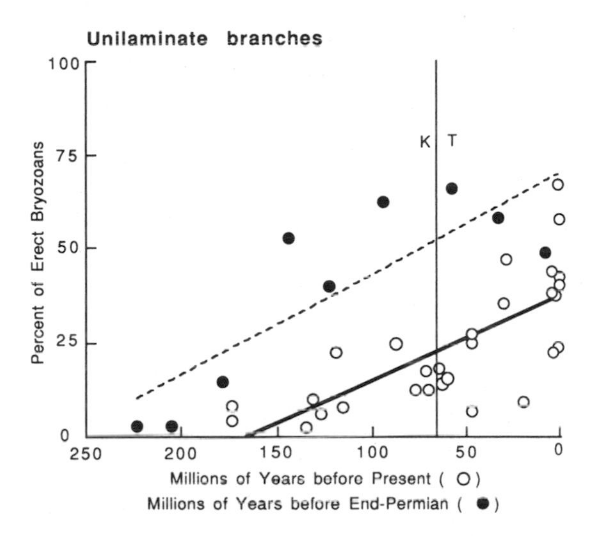

Figure 7.13. Percentage of erect bryozoan species with unilaminate branches throughout the Phanerozoic (data sources same as fig. 7.11). Jurassic to Recent, regression (*solid line*, ○): % = −9.53 + .225 (time), r = .73, P = .0001; Paleozoic, regression (*dashed line*, ●): % = .547 + .267 (time), r = .82, P = .003.

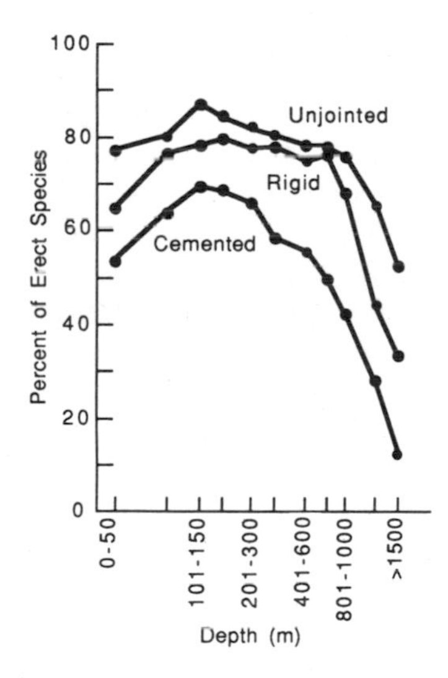

Figure 7.14. Percentages of erect bryozoans that are cemented, rigid, or unjointed as a function of depth based on data contained in 47 surveys of Atlantic and Mediterranean Bryozoa. Percentages of rooted, flexible, or jointed species ae the mirror image of values shown. (From McKinney and Jackson 1988.)

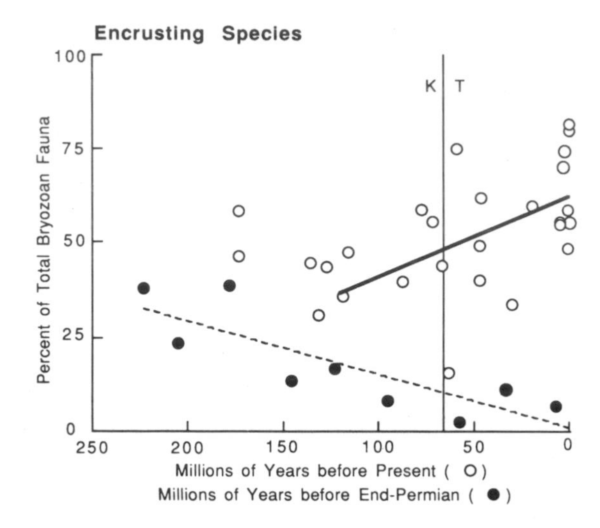

Figure 7.15. Percentage of bryozoan species with encrusting growth form throughout the Phanerozoic (data sources same as fig. 7.11). Jurassic to Recent, regression for post-Turonian (*solid line*, ○): % = 16.8 + .219 (time), r = .42, P = .029; Paleozoic, regression (*dashed line*, ●): % = 37.0 − .141 (time), r = −.83, P = .0031.

data for the pre-Turonian Mesozoic.) These patterns correspond with changes in distribution of growth forms with depth (McKinney and Jackson 1988). Erect bryozoans were diverse and abundant in shallow water throughout the Paleozoic and Mesozoic but became increasingly restricted to midshelf and deeper environments after the Cretaceous. As late as the Maastrichtian, and perhaps until the Neogene, stems of shallow-water seagrasses supported dense populations of rigidly erect bryozoans, whereas Recent seagrass stems are dominated by encrusting, commonly massively multilaminate species and rigidly erect species are rare (Hoffmeister, Stockman and Multer 1967; Voigt 1973, 1979, 1981; Harmelin 1976; A. H. Cheetham, per. comm. 1989). Similarly, erect bryozoans were common on upward-facing surfaces of hardgrounds and reefs but became restricted to cavities, overhangs, vertical walls, and rubble during the late Mesozoic (Ross 1970; Walker and Ferrigno 1973; Kapp 1975; Wilson 1975; Cuffey 1977; Palmer 1982). For Recent seas in general, the depth segregation of encrusting and erect bryozoans is striking (fig. 7.16).

Comparative diversity of free-living species. Free-living bryozoans during the Paleozoic were immobile forms unable to regulate their position on the sediment. These included large laminar colonies on fine sediments in quiet water habitats and circumrotatory colonies on coarse sediments exposed to strong currents (McKinney and Jackson 1988).

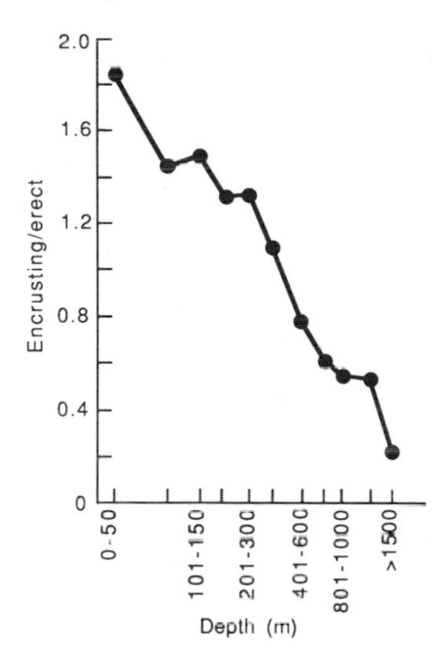

Figure 7.16. Variation with depth in the ratio of encrusting to erect bryozoan species in the Atlantic and Mediterranean (same data base as fig. 7.14.) (From McKinney and Jackson 1988.)

Neither of these growth forms displays any morphological specialization for life on sediments. The diversity of these free-living colonies decreased throughout the Paleozoic. In contrast, free-living bryozoans from the Late Cretaceous onward include mobile colonies highly specialized for life on sediments as well as immobile forms. Moreover, the proportion of mobile to immobile species increased suddenly during the Late Cretaceous and has been high ever since (fig. 7.17). In addition to their diversity, mobile bryozoans are so abundant as to dominate volumetrically many Tertiary to Recent deposits (Lagaaij 1963; Cook and Chimonides 1983; McKinney and Jackson 1988).

Scleractinian Corals

Most scleractinians are clonal and colonial modular animals like bryozoans, but there are also solitary, aclonal species (Coates and Jackson 1985, 1987; Oliver and Coates 1987). Scleractinian modules and colonies are about one order of magnitude larger than bryozoans. Nevertheless, like bryozoans, the modular dimensions of colonial scleractinians have been quite constant throughout their history, but most colony-level features have not (Coates and Jackson 1985).

Integration of coral modules and proportions of growth forms have been studied sufficiently for comparison with bryozoans. Integration of corals is defined on the basis of the skeletal morphology of the corallites

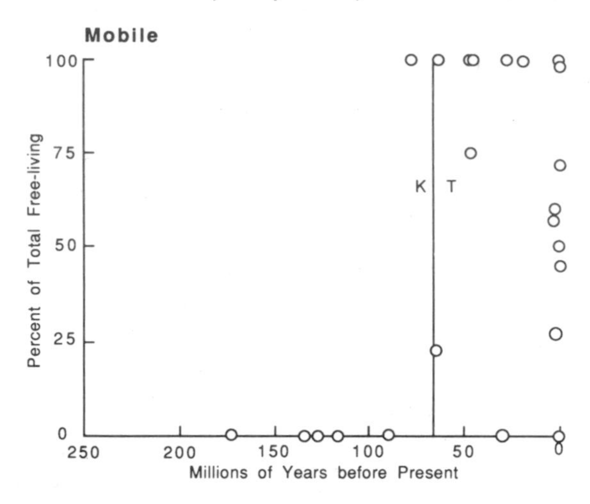

Figure 7.17. Percentage of Jurassic to Recent free-living bryozoans capable of colony movement (data from appendix 7.1).

that house the coral polyps (Coates and Oliver 1973; Coates and Jackson 1985). Highly integrated corallites have porous walls between corallites, lack them entirely, or are meandroid. Poorly integrated corallites are separated by well-defined wall structures. High integration appears to be associated with flexibility of growth, especially in large erect forms like acroporids, but this has not been analyzed directly. Growth forms are defined as for the most general categories of bryozoans, including erect, encrusting, and free-living colonies.

Data for analysis of temporal variation in integration and growth forms are the same as those used by Coates and Jackson (1985). The proportion of species with high integration has increased significantly since the Triassic (fig. 7.18). This pattern parallels the increase in cheilostomes versus cyclostomes, suggesting a general premium on improving physiological interaction among modules regardless of phylum. In contrast, the percentage of erect species decreased until the Turonian and then increased (although the post-Turonian regression is not significant), which is opposite to the pattern for bryozoans.

Coralline Algae

Crustose coralline algae are nonarticulated calcareous red algae that are probably the most abundant inhabitants upon hard substrata within the photic zone of Recent seas (Steneck 1983,1986). Two principal groups have existed throughout the Phanerozoic. The Solenoporaceae lived from the Cambrian to Miocene and were most abundant and

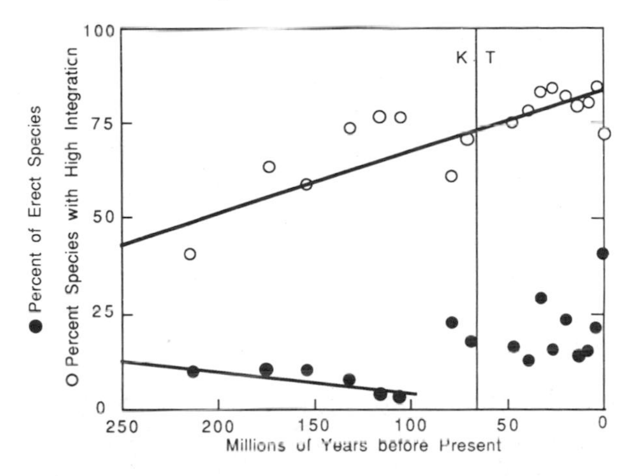

Figure 7.18. Morphological features of Triassic to Recent scleractinian corals (data from sources in Coates and Jackson 1985). Integration of corallites, regression (○); % = 43.7 + .162 (time), r = .79, P = .0002. Percent erect species (●), Triassic through Turonian regression: % = 12.6 −.058 (time), r = −.80, P = .029; post-Turonian regression: % = 4.87 + .071 (time), r = .25, P = .231. Note, however, that all post-Turonian values are higher than for any previous fauna.

diverse before the Cenozoic. They were characterized by having their meristem and reproductive structures (conceptacles) at the surface of the thallus (perithallus) and lack of a specialized system for translocation of photosynthates through the thallus. The Corallinaceae extend from the Lower Carboniferous to Recent. Their meristem and conceptacles are located beneath the surface of the plant in the hypothallus, and they also possess fusion cells that aid in translocation of photosynthates. Thus solenopores are roughly analagous to cyclostomes and corallines to cheilostomes with regard to their differences in the topology of growth, reproduction, and physiological integration. The two algal groups also turn over in relative diversity at the same time as the bryozoans (Late Cretaceous), and there was a similar decrease in branching and leafy crusts (approximately equivalent to foliose bryozoans) relative to thick encrusting forms throughout the Cenozoic.

Discussion

Four general conclusions can be drawn from these data that generally parallel those for aclonal benthic organisms such as snails, bivalves, and barnacles (Vermeij 1987; Jackson 1988) and for terrestrial plants (Lidgard and Crane 1988a, 1988b).

PROGRESSIVE TRENDS EXIST

Many long-term trends are apparent in the history of the principal fossilizable clonal benthos (see also Key 1988). These include major changes in the percentages and distributions of species with different zooidal morphologies and levels of integration, growth forms, modes of growth, and habitats—not just the addition of "a rib, a bump, or a millimeter over millions of years" (Gould 1988). Although not all intervals are equally represented by the data, the trends are statistically significant and are too numerous to be ascribed to chance. Most data on abundance of fossils are qualitative and too incomplete to compare with species diversity for any of the trends (references in McKinney 1986a, 1986b; appendix 7.1). In general, the patterns match our impressions of the relative abundances of different kinds of bryozoans, and there are no obvious contradictions (McKinney and Jackson 1988).

Unlike aclonal animals such as bivalves, the diversity of modes of life of clonal benthos has not dramatically increased. Mobile sediment-dwelling bryozoans are new, but they more or less replaced immobile sediment dwellers, and the same was true for corals (Gill and Coates 1977; Jackson 1985; McKinney and Jackson 1988). Except for other minute forms of sediment-dwelling bryozoans (reviewed in McKinney and Jackson 1988), there have been no other major innovations in growth forms since the Paleozoic, nor increase in tiering, nor infaunalization as in aclonal animals (Stanley 1968, 1975; Thayer 1983; Bottjer and Ausich 1986; Vermeij 1987). Nevertheless, long-term persistence of clades in shallow water has been accompanied by major shifts in morphology and design, such as the development of massive encrusting cheilostomes and coralline algae with subsurface sexual reproduction and plumbing between modules, or the improvements in design of rigidly erect bilaminate cheilostomes. Clades whose morphology did not change substantially became restricted to offshore or cryptic habitats after the Jurassic, just as did brachiopods and sessile echinoderms during the Paleozoic (Jackson, Goreau, and Hartman 1971; Jablonski et al. 1983; Sepkoski and Miller 1985; Vermeij 1987; Jablonski and Bottjer, this volume).

Thus the shallow water biological landscape has become progressively more montonous, at least since the Cretaceous. Consider, for example, Caribbean coastal environments (Kinzie 1973; Kaplan 1982). Corals, gorgonians, and sponges are the ony common erect forms shallower than 10 to 30 m, except on vertical walls or in caves, which is why the latter are so popular with sports divers. The great majority of noncryptic hard substrata are covered by pavements of crustose coralline algae that appear so similar to cement as to be ignored by zoologists (Kinzie's "barren zone"), and even most encrusting as well as mobile animals are restricted to cryptic habitats. Sediment surfaces are also largely barren of animals except for the mounds that signal the presence of deep burrowers below. A snorkler in the Silurian or Devonian tropics would have seen a much more varied biota living out in the open than can be seen today (cartoons in McKerrow 1978).

PROGRESSIVE TRENDS CAN BE EXPLAINED ECOLOGICALLY

All of the trends described can be readily explained as the consequence of escalation in the defenses employed in biological interactions (sensu Dawkins and Krebs 1979; Fox 1981; Vermeij 1987) or by progressive adaptation to *routine* physical processes such as wave action or currents. The importance of biological interactions is supported by the following observations.

1. Interactions among Recent clonal benthos strongly affect their distribution and abundance (Jackson 1977, 1983; Sutherland and Karlson 1977; Steneck 1983, 1986; Buss 1986; Sebens 1986; Jackson and Kaufmann 1987).

2. Variations in functional morphology affect the outcome of biological interactions among clonal benthos (Jackson 1977, 1983; Steneck 1983, 1986). This is most obvious in direct interactions when colonies come into contact. Examples include greater overgrowth potential and defense by encrusting bryozoans that can raise their growing margins (as for cheilostomes with zooidal or frontal budding) and greater regenerative capacity of bryozoans and crustose algae with extensive physiological integration. However, the same arguments also apply to many indirect interactions. Clades that have evolved greater armament against predators like ascophoran cheilostomes can survive better in areas exposed to predators than less armored forms such as cyclostomes or anascans (Harmelin 1985; Jackson and Hughes 1985). In effect, they aquire greater "enemy-free space" (Jeffries and Lawton 1984) and thus potentially greater access to other resources such as food (Vermeij 1987). Moreover, the better the armor of any group, the more predators are likely to shift their attention to less armored forms, thereby actually increasing predation pressures upon them. Similar arguments apply to innovative modes of resource acquisition (suspension feeding by unilaminate bryozoans) or improvements in structural design (rigid branched bilaminates), but these have not been studied in the field.

3. There is abundant direct evidence for past interactions involving clonal benthos, including mutual (contemporaneous) overgrowths and

scars of grazing predators (Vermeij 1977, 1987; Jackson 1983; Steneck 1983, 1986).

4. Both the diversity and abundance of animals able to scrape, gouge, drill, or crush calcareous substrata (e.g., coral, bryozoan, or coralline algal skeletons, echinoderm tests, molluscan shell); the depth of their excavations preserved as trace fossils; and their inferred rate of excavation have all increased greatly since the Jurassic (Vermeij 1977, 1987; Steneck 1983, 1986) and to a lesser degree during the mid-Paleozoic (Signor and Brett 1984). Armament that had provided ample protection throughout the Paleozoic became easily penetrable by the Late Cretaceous. Similarly, the diversity and abundance of animals able to move through and rework sediments, and the depth and rate of reworking that they are inferred to have effected, have increased throughout the Phanerozoic, perhaps at an even greater rate after the Permian (Thayer 1983). Sediments have become progressively more unstable as a platform for immobile epifauna and have provided less protection for shallow infauna, particularly since the Triassic.

Unlike these sources of disturbance, however, and despite important advances in paleoceanography (e.g., Berner and Raiswell 1983), it is still not possible to document changes in the resource base (forms, abundance, and availability of organic materials, nutrients, and energy) over Phanerozoic time.

5. Morphological and distributional trends in clonal benthos interpreted as adaptations against increased armament in biological interactions are more or less contemporaneous and coincide approximately (within the limited stratigraphic resolution of the available data) with independent evidence for changes in armament. *Slowly* since the Cretaceous all these events described in this paper occurred: cheilostomes largely replaced cyclostomes, and ascophorans increased greatly in diversity relative to other cheilostomes; narrowly branched or articulated erect bryozoans became more diverse and abundant than broadly branched or rigidly erect species, and the latter became more restricted to deeper water or cryptic habitats where predation and bioturbation are less intense; encrusting species of bryozoans and coralline algae became more diverse and abundant than erect forms; mobile free-living bryozoans and corals became more successful than immobile free-living species, and most of the increases in predatory armament and and bioturbation documented by Vermeij (1987) and Thayer (1983) occurred. In some cases, newly successful forms even became dependent upon the organisms that graze upon them, as with some crustose corallines that become smothered by filamentous algae if they are not grazed by limpets (Steneck 1983, 1986).

Sometimes opposing trends in different taxa strengthen the ecological explanation for trends, as with the decrease in proportion of erect bryozoans since the Late Cretaceous while erect corals increased in diversity and abundance (figs. 7.15 and 7.18). Erect corals are not only taller than erect bryozoans but can also grow much faster. Some acroporids, for example, reach 1 to 2 m, grow as fast as 20 cm/y, and thrive in areas of intense physical and biological disturbance (Adey 1978; Tunnicliffe 1981; Stimson 1985). In this way they commonly outgrow the extent of injuries due to abundant predators such as coralivorous snails, worms, sea urchins, and fish, unless the coral populations are otherwise greatly reduced so that effects of predators are concentrated upon the few survivors (Kaufman 1981; Knowlton et al. 1981; Knowlton, Lang, and Keller 1990). In contrast, erect bryozoans are so small that they are readily eliminated by these predators (Vance 1979).

In general, erect growth was more successful than encrusting habits until the Late Cretaceous, at least as measured by relative abundance of species. Afterward, diverse assemblages of large, erect, rapidly growing stony corals, gorgonians, and sponges have thrived in shallow, noncryptic habitats, whereas smaller erect bryozoans, hydroids, colonial ascidians, and coralline algae have not. Abundant small erect animals on artificially isolated habitats like piers, rafts, buoys, and fouling panels are exceptions that support the rule of restriction of such animals to cryptic or deep substrata in more natural situations (McKinney and Jackson 1988).

OTHER FACTORS DO NOT EXPLAIN THE TRENDS

No evidence indicates that changes in seawater or atmospheric composition, climate, sea level, vulcanism, continental distributions, or other abiological processes have varied in a similar pattern to the major trends we have described (Berner and Raiswell 1983; Crowley 1983; Fischer 1984; Hallam 1984; Holland and Trendall 1984; Wilde and Berry 1984). Moreover, the only "episodic event" that clearly substantially affected the biological trends is whatever caused the great extinctions at the end of the Permian (McKinney 1986a), and the same is true of the numbers of bryozoan families throughout the Phanerozoic (Taylor and Larwood 1988). Otherwise, all of the trends are remarkably independent of mass extinctions, even at the end of the Cretaceous. These observations say nothing about the importance of abiological events for mass extinctions, only that mass extinctions seem to be of little importance to the trends (see also Jablonski 1986:323).

Most of the trends reviewed here include more than simple changes in size and shape, and all involve different families, orders, or classes. For example, trends in cheilostome budding patterns include both anascans and ascophorans, and trends in proportions of erect growth forms, their branch dimensions, and articulation involve practically all major bryozoan taxa. Some of the trends even involve both plants and animals. Thus it is exceedingly unlikely that any of these patterns are due merely to radiations in the evolution of particular body plans.

Many of the morphological or environmental shifts that comprise these trends must involve considerable energetic costs, although these have not been investigated for clonal benthos. Examples include increased restriction to deep water or cryptic habitats where plankton and other suspended food are less abundant than out in the open in shallow water (Parsons, Takahashi, and Hargrave 1984; Buss and Jackson 1981) and deposition of relatively enormous volumes of calcareous cement for defensive armament (A. R. Palmer Jr. 1981, 1983). The origin, progressive innovation, and survival of such costly changes are inexplicable in evolutionary terms if they did not contribute to survival and reproduction (Fox 1981; Strong, Lawton, and Southwood 1984).

PROGRESSIVE TRENDS MAY BE EXPLAINED BY SPECIES SELECTION

To the best of our knowledge, no one has documented any such adaptive trends in the evolution of a single monophylectic clade of marine organism. Moreover, the only rigorous phenetic study of the tempo of evolution among marine bryozoans showed remarkable stasis of all morphological characters within morphospecies and punctuational change across morphospecies boundaries (Cheetham 1986b). Thus we have no idea whether changes in design or distribution occur in association with speciation or only by differential survival of species that share traits which confer competitive superiority over other species (Stanley 1979; Levinton 1988).

Regardless of whether speciation plays an active role, the progressive trends documented here are all compatible with the concept of species selection as a competitive process between species (sensu Stanley and Newman 1980; Stanley, Van Valkenberg, and Steneck 1983). This is evident in the high variance maintained throughout many trends by the continued appearance of new species (often from distantly related clades) with more "primitive" characteristics and the striking species specificity of most of the morphological traits involved, including modes of budding, frontal wall development of zooids, rates of thickening and bifurcation angles of narrow bilaminate branches, and even branch dimensions and growth form (Cheetham 1986a; Lidgard 1986).

The only way trends could be maintained against such a chaotic input of new species is by sustained, differentially high extinction of the more primitive species (Stanley and Newman 1980; Stanley, Van Valkenberg, and Steneck 1983).

Gould (1988) has suggested that trends with high variance, particularly increasing variance, derive simply from differential expansion of clades into new morphological dimensions or habitats as numbers of species increase. This statistical reasoning probably suffices for most trends of size increase (Stanley 1973) but does not explain the trends reported here because (1) the numbers of species per fauna in appendix 7.1 and other sources of data do not increase systematically over time; (2) variance does not increase over time for most trends, and there are striking shifts in extreme values as well as means; and (3) most trends involve shifts in percentages of species with a given trait rather than a dimensional character such as size or depth distribution. Moreover, many dimensional characters such as branch thickness are tightly associated with developmental characters such as mode of budding or zooidal form that are not size dependent per se.

The most striking feature of the trends reported here and elsewhere (Thayer 1983; Vermeij 1987) is their great duration. All the trends extend over 80 my or more, and some for hundreds of million years. It is inconceivable that such patterns could result from tightly coupled ecological interactions (natural selection) between individual clades over such extended periods. Nevertheless, ecological interactions constitute the only plausible mechanism (Jackson 1988).

The great majority of species interact with large numbers of other species so that coevolutionary responses to one or a few enemies or hosts must be the exception rather than the rule (Fox 1981; Futuyma and Slatkin 1983; Strong, Lawton, and Southwood 1984). Conspecific colonies of bryozoans may encounter 100 or more other species on a single reef (Buss and Jackson 1979; Jackson 1979b, 1984), and similar diversity of interaction is often characteristic of other sessile organisms (Ohlhorst 1980; Sebens 1986). The situation is even more complex because most species occur in a variety of depths, habitats, and geographical regions (Goreau and Wells 1967; Jackson, Winston and Coates 1985), where they encounter additional predators and competitors. The result of such variability is greatly prolonged coexistence beyond the ability of ecologists to measure during their lifetimes (Hutchinson 1961; Chesson and Warner 1981; Warner and Chesson 1985). Such diffuse interaction is compatible with the theory of ecologically based species selection acting against the continued production of morphologically distinct species as the driving force of macroevolutionary trends (Stanley and Newman 1980; Stanley, Van Valkenberg, and Steneck 1983).

The trends discussed here are based on extensive study of the functional ecology and paleoecology of the groups involved, coupled with detailed compilations of species from primary sources. All the trends need further investigation. Crucial is the analysis of many more Early Cretaceous to Paleocene biotas to better resolve the origin of trends and details of their passage through the end-Cretaceous extinction and the inclusion of data on relative abundance. Regardless of how our interpretations may change with additional data, however, it seems clear that biological interactions can drive adaptive trends that define macroevolutionary progress over very long periods of time. This contradicts the view that global catastrophes and mass extinctions repeatedly cancel out the effects of Darwinian evolution (Gould 1985).

Acknowledgments

Discussions with Alan Cheetham provided much critical focus. He, Tony Coates, Marcus Key, Nancy Knowlton, and Mary Jane West-Eberhard reviewed the manuscript. Karl Kaufmann helped with statistics, and Dennis Gordon and John and Dorothy Soule provided some of the specimens for figure 7.2.

Literature Cited

Adey, W. H. 1978. Coral reef morphogenesis: a multidimensional model. *Science* 177:1000–1002.

Bambach, R. K. 1983. Ecospace utilization and guilds in marine communities through the Phanerozoic. In *Biotic interactions in recent and fossil benthic communities*, ed. M. J. S. Tevesz and P. L. McCall, 719–46. New York: Plenum.

———. 1985. Phanerozoic marine communities. In *Patterns and processes in the history of life*, ed. D. M. Raup and D. Jablonski, 407–28. Berlin: Springer-Verlag.

Benton, M. J. 1987. Progress and competition in macroevolution. *Biol. Rev.* 62:305–38.

Berner, R. A., and R. Raiswell. 1983. Burial of organic carbon and pyrite sulfur in sediments over Phanerozoic time: a new theory. *Geochim. Cosmochim. Acta* 47:855–62.

Boardman, R. S. 1984. Origin of the post-Triassic Stenolaemata (Bryozoa): a taxonomic oversight. *J. Paleontol.* 58:19–39.

Boardman, R. S., and A. H. Cheetham. 1973. Degrees of colony dominance in stenolaemate and gymnolaemate Bryozoa. In *Animal colonies*, ed. R. S.

Boardman, A. H. Cheetham, and W. A. Oliver, Jr., 121–220. Stroudsburg, Pa.: Dowden, Hutchinson and Ross.

Boardman, R. S., and A. H. Cheetham. 1987. Phylum Bryozoa. In *Fossil invertebrates*, ed. R. S. Boardman, A. H. Cheetham, and A. J. Rowell, 497–549. Oxford: Blackwell Scientific Publications.

Bottjer, D. J., and W. I. Ausich. 1986. Phanerozoic development of tiering in soft substrata suspension-feeding communities. *Paleobiology* 12:400–420.

Buss, L. W. 1986. Competition and community organization on hard surfaces in the sea. In *Community ecology*, ed. J. Diamond and T. J. Case, 517–36. New York: Harper and Row.

Buss, L. W., and J. B. C. Jackson. 1979. Competitive networks: Nontransitive competitive relationships in cryptic coral reef environments. *Am. Nat.* 113:223–34.

———. 1981. Planktonic food availabilty and suspension-feeder abundance: Evidence of in situ depletion. *J. Exper. Mar. Biol. Ecol.* 49:151–61.

Cheetham, A. H. 1971. Functional morphology and biofacies distribution of cheilostome Bryozoa in the Danian Stage (Paleocene) of southern Scandanavia. *Smithsonian Contrib. Paleobiol.* 6:1–87.

———. 1986a. Branching, biomechanics, and bryozoan evolution. *Proc. R. Soc. London B* 228:151–71.

———. 1986b. Tempo of evolution in a neogene bryozoan: Rates of morphologic change within and across species boundaries. *Paleobiology* 12:190–202.

Cheetham, A. H., and L. -A. C. Hayek. 1983. Geometric consequences of branching growth in adeoniform Bryozoa. *Paleobiology* 9:240–60.

Cheetham, A. H., L. -A. C. Hayek, and E. Thomsen. 1981. Growth models in fossil arborescent cheilostome bryozoans. *Paleobiology* 7:68–86.

Cheetham, A. H., and E. Thomsen. 1981. Functional morphology of arborescent animals: Strength and design of cheilostome bryozoan skeletons. *Paleobiology* 7:355–83.

Chesson, P. L., and R. R. Warner. 1981. Environmental variability promotes coexistence in lottery competitive systems. *Am. Nat.* 117:923–43.

Coates, A. G., and J. B. C. Jackson. 1985. Morphological themes in the evolution of clonal and aclonal marine invertebrates. In *Population biology and evolution of clonal organisms*, ed. J. B. C. Jackson, L. W. Buss, and R. E. Cook, 67–106. New Haven, Conn.: Yale University Press.

———. 1987. Clonal growth, algal symbiosis, and reef formation by corals. *Paleobiology* 13:363–78.

Coates, A. G., and W. A. Oliver, Jr. 1973. Coloniality in zoantharian corals. In *Animal colonies*, ed. R. S. Boardman, A. H. Cheetham, and W. A. Oliver, Jr. 3–27. Stroudsburg, Pa.: Dowden, Hutchinson, and Ross.

Cook, P. L. 1977. Colony-wide water currents in living Bryozoa. *Cah. Biol. Mar.* 18:31–47.

Cook, P. L., and P. J. Chimonides. 1983. A short history of the lunulite Bryozoa. *Bull. Mar. Sci.* 33:566–81.

Crowley, T. J. 1983. The geologic record of climatic change. *Rev. Geophys. Space Phys.* 21:828–77.

Cuffey, R. J. 1977. Bryozoan contributions to reefs and bioherms through time. *Studies Geol.* 4:181–94.

Dawkins, R., and J. R. Krebs. 1979. Arms races between and within species. *Proc. R. Soc. London B* 205:489–511.

Felsenstein, J. 1985. Phylogenies and the comparative method. *Am. Nat.* 125:1–15.

Fischer, A. G. 1984. The two Phanerozoic supercycles. In *Catastrophes and earth history,* ed. W. A. Berggren and J. A. van Couvering, 129–50. Princeton, N.J.: Princeton University Press.

Fox, L. R. 1981. Defense and dynamics in plant-herbivore systems. *Am. Zool.* 21:853–64.

Futuyma, D. J., and M. Slatkin, eds. 1983. *Coevolution.* Sunderland, Mass.: Sinauer Associates.

Gill, G. A., and A. G. Coates. 1977. Mobility, growth patterns and substrate in some fossil and Recent corals. *Lethaia* 10:119–34.

Glynn, P. W., and G. M. Wellington. 1983. *Corals and coral reefs of the Galapagos Islands.* Berkeley and Los Angeles: University California Press.

Goreau, T. F., and J. W. Wells. 1967. The shallow-water Scleractinia of Jamaica: Revised list of species and their vertical distribution range. *Bull. Mar. Sci.* 17:442–53.

Gould, S. J. 1985. The paradox of the first tier: an agenda for paleobiology. *Paleobiology* 11:2–12.

———. 1988. Trends as changes in variance: a new slant on progress and directionality in evolution. *J. Paleontol.* 62:319–29.

Gould, S. J., and C. B. Calloway. 1980. Clams and brachiopods—Ships that pass in the night. Paleobiology 6:383–96.

Gould, S. J., and R. C. Lewontin. 1979. The spandrels of San Marco and the Panglossian paradigm: a critique of the adaptationist program. *Proc. R. Soc. London B* 205:581–98.

Hallam, A. 1984. Pre-Quarternary sea-level changes. *Ann. Rev. Earth Planetary Sci.* 12:205–43.

Harmelin, J. -G. 1976. Le sous-ordre des Tubuliporina (Bryozoaires, Cyclostomes) en Mediterranee: Ecologie et systematique. *Mem. Inst. Oceanogr.* Monaco 10:1–326.

———. 1985. Bryozoan dominated assemblages in Mediterranean cryptic environments. In *Bryozoa: Ordovician to Recent,* ed. C. Nielsen and G. P. Larwood, 135–43. Fredensborg, Denmark: Olsen and Olsen.

Hoffmeister, J. E., K. W. Stockman, and H. G. Multer. 1967. Miami limestone of Florida and its Recent Bahamian counterpart. *Geol. Soc. Am. Bull.* 78:175–90.

Holland, H. D., and A. F. Trendall, eds. 1984. *Patterns of change in earth evolution.* Berlin: Springer-Verlag.

Hutchinson, G. E. 1961. The paradox of the plankton. *Am. Nat.* 95:137–45.

Jablonski, D. 1986. Evolutionary consequences of mass extinctions. In *Patterns and processes in the history of life*, ed. D. M. Raup and D. Jablonski, 313–29. Berlin: Springer-Verlag.

Jablonski, D., J. Sepkoski, Jr., D. J. Bottjer, and P. M. Sheehan. 1983. Onshore-offshore patterns in the evolution of Phanerozoic shelf communities. *Science* 222:1123–5.

Jackson, J. B. C. 1977. Competition on marine hard substrata: the adaptive significance of solitary and colonial strategies. *Am. Nat.* 111:743–67.

———. 1979a. Morphological strategies of sessile animals. In *Biology and systematics of colonial organisms*, ed. C. Larwood and B. R. Rosen, 499–555. London: Academic Press.

———. 1979b. Overgrowth competition between encrusting cheilostome ectoprocts in a Jamaican cryptic reef environment. *J. An. Ecol.* 48:805–23.

———. 1983. Biological determinants of present and past sessile animal distributions. In *Biotic interactions in Recent and fossil benthic communities*, ed. M. J. S. Tevesz and P. L. McCall, 39–120. New York: Plenum.

———. 1984. Ecology of cryptic coral reef communities. III. Abundance and aggregation of encrusting organisms with particular reference to cheilostome Bryozoa. *J. Exper. Mar. Biol. Ecol.* 75:37–57.

———. 1985. Distribution and ecology of clonal and aclonal benthic invertebrates. In *Population biology and evolution of clonal organisms*, ed. J. B. C. Jackson, L. W. Buss, and R. E. Cook, 297–355. New Haven, Conn.: Yale University Press.

———. 1988. Does ecology matter? *Paleobiology* 14:307–12.

Jackson, J. B. C., and L. W. Buss. 1975. Allelopathy and spatial competition among coral reef invertebrates. *Proc. Nat. Acad. Sci. U.S.A.* 72:5160–3.

Jackson, J. B. C., and T. P. Hughes. 1985. Adaptive strategies of coral-reef invertebrates. *Am. Sc.* 73:265–74.

Jackson, J. B. C., and K. W. Kaufmann. 1987. *Diadema antillarum* was not a keystone predator in cryptic reef environments. *Science* 235:687–9.

Jackson, J. B. C., T. F. Goreau, and W. D. Hartman. 1971. Recent brachiopod-sclerosponge communities and their paleoecological significance. *Science* 173:623–5.

Jackson, J. B. C., J. E. Winston, and A. G. Coates. 1985. Niche breadth, geographic range, and extinction of Caribbean reef-associated cheilostome Bryozoa and Scleractinia. *Proc. 5th Int. Coral Reef Cong.* 4:151–8.

Jeffries, M. J., and J. H. Lawton. 1984. Enemy free space and the structure of ecological communities. *Biol. J. Linn. Soc.* 23:269–86.

Kaplan, E. 1982. *A field guide to coral reefs of the Caribbean and Florida*. Boston: Houghton Mifflin.

Kapp, U. S. 1975. Paleoecology of Middle Ordovician stromatoporoid mounds in Vermont. *Lethaia* 8:195–206.

Kaufman, L. 1981. There was biological disturbance on Pleistocene coral reefs. *Paleobiology* 7:527–32.

Key, M. M. 1988. Progressive macroevolutionary patterns in colonial animals. *Geol. Soc. Am. Abstracts with Programs* 20:A201.

Kinzie, R. A., III. 1973. The zonation of West Indian gorgonians. *Bull. Mar. Sci.* 23:93–155.

Knowlton, N., J. C. Lang, M. C. Rooney, and P. Clifford. 1981. Evidence for delayed mortality in hurricane-damaged Jamaican staghorn corals. *Nature* 294:251–2.

Knowlton, N., J. C. Lang, and B. D. Keller. 1990. Case study of natural population collapse: Post-hurricane predation on Jamaican staghorn corals. *Smithsonian Contrib. Mar. Sci.* 31:1–25.

Lagaaij, R. 1963. *Cupuladria canariensis* (Busk)—portrait of a bryozoan. *Palaeontology* 6:172–217.

Levinton, J. 1988. *Genetics, paleontology, and macroevolution*. Cambridge, Mass.: Cambridge University Press.

Lidgard, S. 1985a. Budding processes and geometry in encrusting cheilostome bryozoans. In *Bryozoans: Ordovician to Recent*, ed. C. Nielsen and G. P. Larwood, 175–82. Fredensborg, Denmark: Olsen and Olsen.

———. 1985b. Zooid and colony growth in encrusting cheilostome bryozoans. *Palaeontology* 28:255–91.

———. 1986. Ontogeny in animal colonies: a persistent trend in the bryozoan fossil record. *Science* 232:230–2.

Lidgard, S., and P. R. Crane. 1988a. Quantitative analyses of the early angiosperm radiation. *Nature* 331:344–6.

———. 1988b. What was the pattern of the angiosperm diversification? A comparative test using palynofloras and leaf macrofloras. *Geol. Soc. Am. Abstracts with Programs* 20:A257.

Lidgard, S., and J. B. C. Jackson. 1989. Growth in encrusting cheilostome bryozoans: I. Evolutionary trends. *Paleobiology* 15:255–82.

May, R. M. 1977. Thresholds and breakpoints in ecosystems with a multiplicity of stable states. *Nature* 269:471–7.

———. ed. 1981. *Theoretical ecology*. Sunderland, Mass.: Sinauer Associates.

Mayr, E. 1954. Geographic speciation in tropical echinoids. *Evolution* 8:1–18.

McKerrow, W. S., ed. 1978. *The ecology of fossils*. Cambridge, Mass.: MIT Press.

McKinney, F. K. 1986a. Evolution of erect marine bryozoan faunas: Repeated success of unilaminate species. *Am. Nat.* 128:795–809.

———. 1986b. Historical record of erect bryozoan growth. *Proc. R. Soc. London B* 228:133–48.

———. 1987. "Progress" in evolution. *Science* 237:575.

McKinney, F. K., and R. S. Boardman. 1985. Zooidal biometry of Stenolaemata. In *Bryozoans: Ordovician to Recent*, ed. C. Nielsen and G. P. Larwood, 193–203. Fredensborg, Denmark: Olsen and Olsen.

McKinney, F. K., and J. B. C. Jackson. 1988. *Bryozoan evolution*. Boston: Unwin Hyman.

Ohlhorst, S. L. 1980. Jamaican coral reefs: Important biological and physical parameters. Ph.D. diss., Yale University, New Haven, Conn.

Oliver, W. A., Jr. and A. G. Coates. 1987. Phylum Cnidaria. In *Fossil inver-*

tebrates, ed. R. S. Boardman, A. H. Cheetham, and A. J. Rowell, 140–93. Oxford: Blackwell Scientific Publications.

Palmer, A. R. 1983. *The decade of North American geology time scale*. Boulder, Colo.: Geological Society of America.

Palmer, A. R., Jr. 1981. Do carbonate skeletons limit the rate of body growth? *Nature* 292:150–2.

———. 1983. Relative cost of producing skeletal organic matrix versus calcification: evidence from marine gastropods. *Mar. Biol.* 75:287–92.

Palmer, T. J. 1982. Cambrian to Cretaceous changes in hardground communities. *Lethaia* 15:309–23.

Parsons, T. R., M. Takahashi, and B. Hargrave. 1984. *Biological oceanographic processes*, 3d ed. Oxford: Pergamon Press.

Randall, J. E. 1967. Food habits of reef fishes in the West Indies. *Studies Trop. Oceanogr.* 5:655–847.

———. 1968. *Caribbean reef fishes*. Neptune City, N.J.: T. F. H. Publications.

Raup, D. M. 1977. Stochastic models in evolutionary paleontology. In *Patterns of evolution as illustrated by the fossil record*, ed. A. Hallam, 59–78. Amsterdam: Elsevier.

Ross, J. P. 1970. Distribution, paleoecology, and correlation of Champlanian Ectoprocta (Bryozoa), New York State. III. *J. Paleontol.* 44:346–82.

Ryland, J. S. 1970. *Bryozoans*. London: Hutchinson University Library.

Sebens, K. P. 1986. Spatial relationships among encrusting marine organisms in the New England subtidal zone. *Ecol. Monogr.* 56:73–96.

Sepkoski, J. J., Jr, and A. I. Miller. 1985. Evolutionary faunas and the distribution of Paleozoic benthic communities in space and time. In *Phanerozoic diversity patterns*, ed. J. W. Valentine, 153–90. Princeton, N.J.: Princeton University Press.

Signor, P. W., III, and C. E. Brett. 1984. The mid-Paleozoic precursor to the Mesozoic marine revolution. *Paleobiology* 10:229–45.

Stanley, S. M. 1968. Post-Paleozoic adaptive radiation of infaunal bivalve molluscs—a consequence of mantle fusion and siphon formation. *J. Paleontol.* 42:214–29.

———. 1973. An explanation for Cope's rule. *Evolution* 27:1–26.

———. 1975. Adaptive themes in the evolution of the Bivalvia (Mollusca). *Ann. Rev. Earth Planetary Sci.* 3:361–85.

———. 1977. Trends, rates, and patterns of evolution in the Bivalvia. In *Patterns of evolution as illustrated by the fossil record*, ed. A. Hallam, 209–50. Amsterdam: Elsevier.

———. 1979. *Macroevolution: Pattern and process*. San Francisco: W. H. Freeman.

Stanley, S. M., and W. A. Newman. 1980. Competitive exclusion in evolutionary time: the case of the acorn barnacles. *Paleobiology* 6:173–83.

Stanley, S. M., B. Van Valkenberg, and R. S. Steneck. 1983. Coevolution and the fossil record. In *Coevolution*, ed. D. J. Futuyama and M. Slatkin, 328–49. Sunderland, Mass.: Sinauer Associates.

Steneck, R. S. 1983. Escalating herbivory and resulting adaptive trends in calcareous algae. *Paleobiology* 9:44–61.

———. 1986. The ecology of coralline algal crusts: Convergent patterns and adaptive strategies. *Ann. Rev. Ecol. Systematics* 17:273–303.

Stimson, J. 1985. The effect of shading by the table coral *Acropora hyacinthus* on understory corals. *Ecology* 66:40–53.

Strong, D. R., J. H. Lawton, and R. Southwood. 1984. *Insects on plants.* Cambridge, Mass.: Harvard University Press.

Sutherland, J. P., and R. H. Karlson. 1977. Development and stability of the fouling community at Beaufort, North Carolina. *Ecol. Monogr.* 47:425–46.

Taylor, P. D., and G. P. Larwood. 1988. Mass extinctions and the pattern of bryozoan evolution. In *Extinctions and survival in the fossil record,* ed. G. P. Larwood, 99–119. Oxford: Clarendon Press.

Thayer, C. W. 1983. Sediment-mediated biological disturbance and the evolution of marine benthos. In *Biotic interactions in Recent and fossil benthic communities,* ed. M. J. S. Tevesz and P. L. McCall, 479–685. New York: Plenum.

Thomsen, E. 1976. Depositional environment and development of Danian bryozoan biomicrite mounds (Karlby Klint, Denmark). *Sedimentology* 23:485–509.

Tunnicliffe, V. 1981. Breakage and propagation of the stony coral *Acropora cervicornis. Proc. Nat. Acad. Sci. U.S.A.* 78:2427–31.

Vance, R. R. 1979. Effects of grazing by the sea urchin, *Centrostephanus coronatus,* on prey community composition. *Ecology* 60:537–46.

Vermeij, G. J. 1977. The Mesozoic marine revolution: Evidence from snails, predators, and grazers. *Paleobiology* 3:245–58.

———. 1987. *Evolution and escalation.* Princeton, N.J.: Princeton University Press.

Voigt, E. 1973. Environmental conditions of bryozoan ecology of the hardground biotope of the Maastrichtian Tuff Chalk near Maastricht (Netherlands). In *Living and fossil Bryozoa,* ed. G. P. Larwood, 185–97. London: Academic Press.

———. 1979. The preservation of slightly or non-calcified Bryozoa (Ctenostomata and Cheilostomata) by bioimmuration. In *Advances in bryozoology,* ed. G. P. Larwood and M. B. Abbott, 541–64. London: Academic Press.

———. 1981. Upper Cretaceous bryozoan-seagrass association in the Maastrichtian of the Netherlands. In *Recent and fossil bryozoa,* ed. G. P. Larwood and C. Nielsen, 281–98. Fredensborg, Denmark: Olsen and Olsen.

Walker, K. R., and K. F. Ferrigno. 1973. Major Middle Ordovician reef tract in east Tennessee. *Am. J. Sci.* 273-A:294–325.

Warmke, G. L., and R. T. Abbott. 1961. *Caribbean seashells.* Narberth, Pa.: Livingston.

Warner, R. R., and P. L. Chesson. 1985. Coexistence mediated by recruitment fluctuations: a field guide to the storage effect. *Am. Nat.* 125:769–87.

Wilde, P., and W. B. N. Berry. 1984. Destabilization of the oceanic density structure and its significance to marine "extinction" events. *Palaeogeogr. Palaeoclimatol. Palaeoecol.* 48:143–62.

Wilson, J. L. 1975. *Carbonate facies in geologic history.* Berlin: Springer-Verlag.

Winston, J. E. 1978. Polypide morphology and feeding behavior in marine ectoprocts. *Bull. Mar. Sci.* 28:1–31.

———. 1979. Current-related morphology and behaviour in some Pacific coast bryozoans. In *Advances in bryozoology,* ed. G. P. Larwood and M. B. Abbott, 247–68. London: Academic Press.

Appendix

A7.1. Jurassic to Recent bryozoan faunas and sources of data used for analyses of trends

Age	Number of Species	Sources
Bajocian-Callovian	54	17
Bajocian-Callovian	69	29
Valanginian	75	12
Neocomian	28	30
Hauterivian	28	21
Barremian-Aptian	28	31, 32, 33
Aptian	53	12, 23
Turonian	28	24
Coniacian-Maastrichtian	544	3, 27
Maastrichtian	22	11
Danian	181	2, 3
Danian-Montian	58	28
Danian-Selandian	65	9
Thanetian	80	15
Eocene	60	13
Eocene	53	35
Eocene	466	9
Rupelian	39	26
Oligocene	134	9
Aquitanian-Burdigalian	79	5, 6, 7, 8
Pliocene	124	22
Pliocene	122	1, 4
Pliocene-Pleistocene	129	10
Pleistocene	71	10
Recent	361	14
Recent	114	16

A7.1. *(continued)*

Age	Number of Species	Sources
Recent	125	18
Recent	197	19, 20, 25
Recent	56	34

1. Balson, P. S., and P. D. Taylor. 1982. Paleobiology and systematics of large cyclostome bryozoans from the Pliocene Crag of Suffolk. *Palaeontology* 25:529–54.
2. Berthelsen, O. 1962. Cheilostome Bryozoa in the Danian deposits of east Denmark. *Danmarks Geologiske Undersolgelse* (ser. 2) 83:1–290.
3. Brood, K. 1972. Cyclostomatous Bryozoa from the Upper Cretaceous and Danian in Scandinavia. *Stockholm Contrib. Geol.* 26:1–464.
4. Busk, G. 1859. A *monograph of the fossil Polyzoa of the Crag*. London: Palaeontographical Society.
5. Canu, F. 1907. Les bryozoaires fossiles des terrains du Sud-Ouest de la France. I. Aquitanien. *Bull. Soc. Geol. France* (serv. iv) 6:510–8.
6. ———. 1909. Les bryozoaires fossiles des terrains du Sud-Ouest de la France. III. Burdigalien. *Bull. Soc. Geol. France* (ser. iv) 9:442–9.
7. ———. 1915. Les bryozoaires fossiles des terrains du Sud-Ouest de la France. IX. Aquitanien. *Bull. Soc. Geol. France* (ser. iv) 15:320–34.
8. ———. 1917. Les bryozoaires fossiles des terrains du Sud-Ouest de la France. X. Burdigalien. *Bull. Soc. Geol. de France* (ser. iv) 16:127–52.
9. Canu, F., and R. S. Bassler. 1920. North American early Tertiary Bryozoa. *Bull. U.S. Nat. Mus.* 106:1–302.
10. ———. 1923. North American later Tertiary and Quarternary Bryozoa. *Bull. U.S. Nat. Mus.* 125:1–302.
11. ———. 1926. Class Bryozoa. *U.S. Geol. Surv. Prof. Pap.* 137:32–39.
12. ———. 1928. Studies on the cyclostomatous Bryozoa. 2. Lower Cretaceous cyclostomatous Bryozoa. *Proc. U.S. Nat. Mus.* 67:1–124.
13. ———. 1929. Bryozoaires eocenes de la Belgique. *Mem. Mus. R. Hist. Nat. Belg.* 39:1–69.
14. ———. 1929. Bryozoa of the Philippine region. *Bull. U.S. Nat. Mus.* 100:1–685.
15. ———. 1933. The bryozoan fauna of the Vincentown Limesand. *Bull. U.S. Nat. Mus.* 165:1–108.
16. Cook, P. L. 1986. Bryozoa from Ghana, a preliminary survey. *Kon. Mus. Mid. Afrika (Tevuren, Belgie), Zoolog. Wetenschap. Ann.* 238:1–315.
17. Gregory, J. W. 1896. *Catalogue of the fossil Bryozoa in the Department of Geology, British Museum (Natural History). The Jurassic Bryozoa*. London: British Museum (Natural History).
18. Hayward, P. J., and P. L. Cook. 1983. The South African Museum's Meiring Naude Cruises. Part 13. Bryozoa II. *Ann. S. African Mus.* 91:1–161.

19. Hayward, P. J., and J. S. Ryland. 1979. *Br. ascoph. bryoz.* London: Academic Press.
20. Hayward, P. J., and J. S. Ryland. 1985. *Cyclost. bryoz.* London: E. J. Brill/Dr. W. Backhuys.
21. Hillmer, G. 1971. Bryozoen (Cyclostomata) aus dem Unter-Hauterive von Nordwestdeutschland. *Mitteil. Geolog.-Palaontolog. Inst. Univ. Hamburg* 40:5–106.
22. Lagaaij, R. 1952. The Pliocene Bryozoa of the low countries. *Meded. Geolog. Stich.* C5:1–233.
23. Pitt, L. J. 1976. A new cheilostome bryozoan from the British Aptian. *Proc. Geol. Assoc.* 87:65–68.
24. Prantl, F. 1938. Lower Turonian Bryozoa from Predboj (Bohemia). *Rozpravy Stat. Geolog. Ust. Ceskoslov. Repub.* 8:1–71.
25. Ryland, J. S., and P. J. Hayward. 1977. *British anascan bryozoans.* London: Academic Press.
26. Vavra, N. 1983. Bryozoen aus dem Unteren Meeressand (Mitteloligozan) von Eckelsheim (Mainzer Becken, Bundersrepublik Deutschland). *Mainzer Naturwissenschaft. Arch.* 21:67–123.
27. Voigt, E. 1930. Morphologosche und stratigraphische untersuchungen uber die bryozoenfauna der oberen Kreide. 1. Teil. *Leopoldina* 6:379–579.
28. ———. 1987. Die bryozoen des klassischen Dano Montiens von Mons (Belgien). *Toelicht. Verhand. Geolog. kaart en Mijnkaart Belg.* 17:1–161.
29. Walter, B. 1969. Les bryozoaires jurassiques en France. *Doc. Lab. Geolog. Fac. Sci. Lyon* 35:1–328.
30. ———. 1972. Les bryozoaires neocomiens du Jura Suisse et Francais. *Geobios* 5:277–354.
31. ———. 1977. Un gisement de bryozoaires aptiens dans le Gard. *Geobios* 10:325–36.
32. Walter, B., and B. Clavel. 1979. Nouveaux apports à la connaissance de la faune aptienne bryozoaires du sud-est de France. *Geobios* 12:819–37.
33. Walter, B., A. Arnaud-Vanneau, H. Arnaud, R. Busnaro, and S. Ferry. 1975. Les bryozoaires Barremo-Aptiens du sud-est de la France. Gisements et paleoecologie, biostratigraphie. *Geobios* 8:83–117.
34. Winston, J. E. 1982. Marine bryozoans (Ectoprocta) of the Indian River area (Florida). *Bull. Am. Mus. Nat. Hist.* 173:99–176.
35. Ziko, A. 1985. Eocene Bryozoa from Egypt, a paleontological and paleoecological study. *Tubinger Mikropalao. Mitteil.* 4:1–183.

8

Tectonic Events and Climatic Change: Opportunities for Speciation in Cenozoic Marine Ostracoda

Thomas M. Cronin and Noriyuki Ikeya

In his paper "The adaptation of populations to varying environments," Lewontin (1957) pointed out the simplifying assumption of constant environment in classic theories of evolutionary dynamics of Lotka-Volterra and Fisher. He stated "The population geneticist today, however, is faced with the problem of reintroducing complexities of nature into his study of evolution" (Lewontin 1957:395). More than 30 years later, a similar situation exists for paleontologists, who usually work at higher taxonomic levels. Paleontologists are trying to understand the roles of biotic and abiotic environmental factors in long-term patterns of diversification and extinction, and in the processes of selection and "sorting" (Vrba 1984a; Vrba and Eldredge 1984), without paying sufficient attention to environmental change.

If we are to test the environment's role in evolution, we must give more consideration to characterizing temporal environmental change. The geological record, justifiably valued for its time dimension, contains an equally valuable record of paleoenvironment and should be cultivated as a source of "natural experiments." Climatic and tectonic events are classic examples of natural experiments. They influence the Earth's oceans, geography, cryosphere, and biosphere at virtually all time scales and furnish the physical environmental milieu in which biotic evolution occurs. Climatic and tectonic events obviously influence the zoogeography of species through changes in physical and chemical parameters and the formation and destruction of environmental barriers—they create *opportunities* for speciation through the isolation of populations. Not surprisingly, empirical patterns of radiation and extinction in the paleontological record have often been causally linked with climatic and tectonic events. Yet with few exceptions (Vrba 1984b; Cronin 1985; Hoffman and Kitchell 1985; Stehli and Webb 1985), the association of environmental change with evolutionary patterns has come more through casual observation than hypothesis and experiment. For example, reference to "environmental deterioration" (Stanley 1986:104) as a factor influencing rates of speciation in bivalves seems ambiguous given current data on Neogene climatic change. What does deterioration mean to a bivalve species? If unfavorable environmental conditions are inferred from an observed high extinction rate, the argument for causality becomes circular.

In this paper, our hypothesis is that environmental events affect ostracode taxa differentially; our experiments are found in the geological record. We attempt to use the geological record to calculate a frequency of speciation index (FSI) derived from the *number of opportunities* for new species to evolve. The FSI for any particular clade equals the number of speciation events that actually occurred divided by the total number of opportunities for speciation in that clade. Opportunities are defined by specific tectonic and climatic events to which species were subjected, most often by vicariance. No two intervals of time have had the same sequence of climatic and tectonic events and no two regions have had the same environmental changes for any interval of time. Thus measuring frequencies of speciation against carefully defined opportunities is an alternative to comparing evolutionary rates as a function of absolute or relative time. The calculation of the FSI is discussed in the Methods section.

The rationale behind this approach stems from well-known models of allopatric and parapatric speciation that hold (1) that speciation is a geographical process in which complete or partial isolation by distance, barriers, or both are key factors in the origin of new species and (2) that extrinsic events usually control isolating barriers, especially in vicariance models. Allopatric speciation (advocated by Mayr 1942, 1982) has also been applied to the punctuated equilibrium model of Eldredge and Gould (1972) to explain patterns in the fossil record. Mayr (1947) long ago addressed the pivotal role of barriers and environmental change in a lucid discussion of "ecological speciation." The timing of the formation of barriers between populations is still a focal point in debates between proponents of various models of speciation (White 1978; Endler 1977; Barigozzi 1982; Mayr 1982). Advances in paleoclimatology and tectonic history now allow us to be more specific, and sometimes quantitative, about environmental change, its effect on species' zoogeography, and the barriers between populations.

We have four goals in this paper:

1. To define the concept of *opportunities*, using well-known environmental barriers that formed as a result of specific tectonic and cli-

matic events—the formation of the Isthmus of Panama, a land barrier to marine organisms, and Milankovitch climatic cycles of the past 2.5 my and their related thermal barriers to temperature-sensitive species.

2. To compare FSI values for the isthmus with those for climatic cycles in ostracode clades subjected to these two distinct types of environmental events.

3. To compare FSI values *among clades* having different biological characteristics or ecological requirements that were subjected to the same climatic changes, that is, clades having the same opportunities to speciate. According to Vrba and Eldredge (1984), heritable emergent characters of species that might affect differential rates of speciation include population size, spatial segregation, and the nature of the species' periphery. Jablonski (1987) claimed that geographical range was an important emergent species-level character in the evolution of molluscs and distinguished it from individual-level characters of enzyme kinetic properties that may also influence geographical range. Vermeij (1987) contrasted speciation and extinction rates among tropical Pacific mollusc groups having different dispersal capabilities. Here we compare FSI values between arctic and temperate clades, between eurytopic and stenotopic clades, and between numerically rare and common taxa.

4. To compare FSI values *within clades* of circumpolar arctic taxa that evolved independently in high-latitude arctic and subarctic seas in, and adjacent to, the North Atlantic and the North Pacific oceans. These areas were subjected to similar global climatic changes during the last few million years and we wish to learn if circumpolar genera are more diverse off eastern Asia than off northeastern North America. If so, what factors account for the observed patterns?

Material: Neogene Ostracoda

Neogene and Quaternary marine ostracode genera form the taxonomic basis for our study. Ostracoda, small, bivalved Crustacea, are ideal for our studies for the following reasons:

1. Species level taxonomy is possible in many genera using the fossilized carapace (Cronin 1987a); there is often a high correlation between carapace and soft part (including copulatory apparatus) morphology within a species (Tsukagoshi and Ikeya 1987; Tsukagoshi 1988). Also, males and females of individual species can often be distinguished on the basis of the adult carapace, as in the cryophilic genus *Finmarchinella*, shown in figure 8.1.

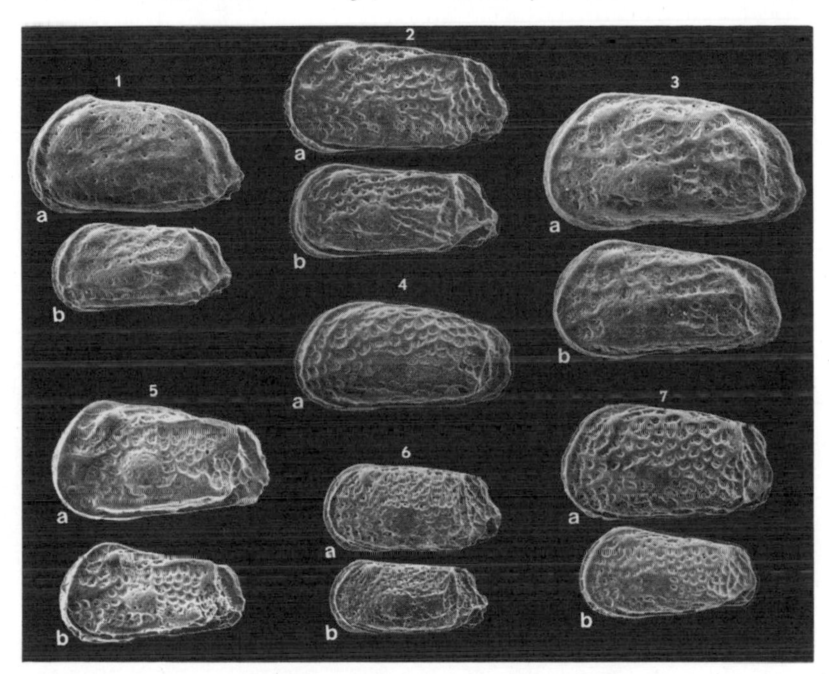

Figure 8.1. Interspecific variation and sexual dimorphism in circumpolar genus *Finmarchinella* Swain 1963. All photos are lateral views of adult, left valves, (X50). Note sexual dimorphism in carapace size and shape but similar ornament pattern in male and female of same species. **1** *F. hanaii* Okada, 1979, Sawane Fm., Pliocene, Japan, a = female, b = male. **2** *F. daishakaensis* Tabuki 1986, Setana Fm., Pliocene, Japan, a = female, b = male. **3** *F. logani* Brady and Crosskey, 1871, Champlain Sea, late Pleistocene, Quebec, a = female, b = male. **4** *F. uranipponica* Ishizaki, 1969, Tomikawa Fm., Pleistocene, Japan, a = female. **5** *F. angulata* (Sars 1865), Tomikawa Fm., Pleistocene, Japan, a = female, b = male; **6** *F. rectangulata* Tabuki, 1986, Setana Fm., Pliocene, Japan, a = female, b = male; 7 *F. nealei* Okada, 1979, Tomikawa Fm., Pleistocene, Japan, a = female, b = male.

2. Ostracodes are commonly fossilized in large numbers in marine sedimentary deposits allowing study of large populations.

3. Ostracodes grow by molting, which gives an understanding of species ontogeny through the preservation of preadult valves; this assures us that adults are being compared among species.

4. Many Neogene fossilized species are extant, so we commonly have their modern ecological and zoogeographical data.

5. Many ostracodes species have limited temperature ranges and are sensitive to climatic change.

Recently, strong skepticism has been expressed about the value of paleontology for speciation studies. For example, Levinton et al. (1987:176), citing mainly work by Skevington (1967) on graptolites for support, do not believe morphological differences equate with reproductive isolation and concluded "we do not believe it is possible to study speciation in the fossil record." Although this may or may not be the case for graptolites, we believe their conclusions are overly pessimistic for ostracodes and probably many other extant groups that have a fossil record. For example, the integrated morphological, ecological, and ethological studies of *Loxoconcha* by Kamiya (Kamiya 1988a, 1988b, 1988c) and the studies of copulatory apparatus, carapace, and zoogeographical variation in *Cythere* by Tsukagoshi and Ikeya (Tsukagoshi and Ikeya 1987; Ikeya and Tsukagoshi 1988; Tsukagoshi 1988) show carapace features distinguish reproductively and ecologically distinct species. The ostracode species discussed in this paper, many of which are extant, are considered valid biological species based on our experience with their taxonomy, morphology, ecology, zoogeography, and biostratigraphy. Additional studies on the evolutionary biology of ostracodes can be found in Hanai, Ikeya, and Ishizaki (1988).

We used collections in Shizuoka University Institute of Geosciences, Tokyo University Museum, the U. S. Geological Survey, and the Smithsonian Institution for our study. Primary taxonomic papers for each genus are listed as references in the appendices; complete species lists are given in appendixes 8.1–8.6.

We carefully selected ostracode clades—genera and groups of closely related species—for which we or our colleagues recently had performed taxonomic, paleontological, and zoogeographical studies. For many genera, all known species were studied from their entire stratigraphic and geographical ranges. For other clades, we examined all members from a particular region that had evolved from a common ancestor. Despite a few poorly understood species in some clades, each clade met the condition of monophyly, which assured us a closed phylogenetic group.

The first stratigraphic appearance datum (FAD) of a species in a single section or in a limited geographical region often represents migration from another area and not a true evolutionary origination. By examining the entire geographical and stratigraphic ranges for each ostracode clade, we minimize this problem and can often distinguish true evolutionary originations from migration events. We studied only clades for which this could be accomplished.

Our data limitations fall into two categories.

1. The discontinuous stratigraphic record of shallow-water marine deposits limits the accuracy in dating first-appearance datums. Thus

estimates of the number of opportunities to speciate caused by climatic cycles may be slightly over- or underestimated for a species.

2. The advantages of using extant clades—such as knowledge of the modern ecology and zoogeography of extant members and the ability to compare fossilized and nonfossilized aspects of the phenotype (soft parts and hard parts) (see also Vrba 1984)—are offset by the fact that it is impossible to examine the entire shape of the clade because it is still evolving. We do not know what the next 10 million years will bring for each clade, and thus comparison of clade shape to extinct clades or to stimulated clades (sensu Gould, Gilinsky, and German 1987) is impossible.

Methods: Opportunities for Speciation—Tectonic and Climatic Events

The last 3 million years represent an excellent interval of time to identify environmental changes that influenced organisms in general and ostracodes in particular. Table 8.1 summarizes the characteristics of the isthmus and climatic events as they relate to isolation of populations.

The frequency of speciation index was computed as follows. For a species in any clade, O_{s1i} is the number of opportunities for that same species, *i*, to split into a new species during the portion of the species stratigraphic range that overlaps the total interval of time under study (2.5 my for the following climatic examples). The total number of opportunities for the entire clade is expressed as

$$O_{TOT} = \sum_{i=1}^{n} O_i$$

where n is the total number of species in the clade.

S_{NEW} represents the total number of new species evolving during the total interval of time under study. The FSI for a clade is calculated as

$$FSI = S_{NEW} / O_{TOT}$$

Table 8.1. Major characteristics of tectonic and climatic events

Character	Isthmus	Climatic
Incidence	One time	Periodic
Degree of isolation	Complete	Partial, variable
Type of barrier	Land	Thermal, distance
Populations	Large	Small

The following paragraphs discuss in more detail how the Isthmus and climatic cycles are used to calculate opportunities for speciation.

TECTONIC EVENTS

The formation of the Isthmus of Panama is the prototype of a natural experiment in which a land barrier isolates previously contiguous populations of tropical marine species (fig. 8.2). It is generally agreed that

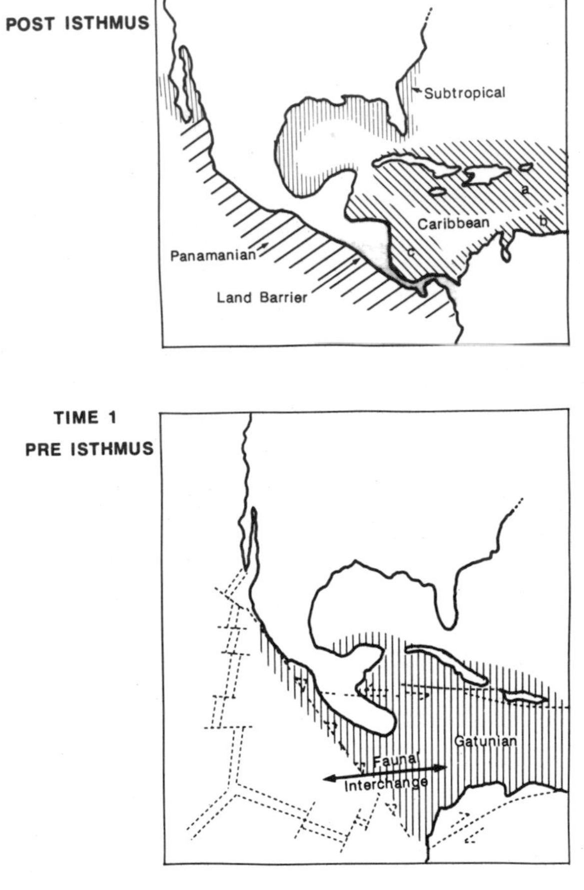

Figure 8.2. The formation of the Isthmus of Panama. Time 1 shows paleogeography and structural features of the pre-Isthmus interval, about 8 mya. Time 2 is modern geography showing isthmus as a land barrier and modern ostracode tropical and subtropical provinces (From Cronin 1987a.) (Time 1 Gatunian molluscan province from E. Petuch, *Palaeogeog., Palaeoclim., Palaeoecol.* 37 (1982): 277–312; time 2 based on Cronin 1987a).

the isthmus emerged gradually between 3.5 and 3.0 million years ago (mya) Keigwin 1978; Jones and Hasson 1985). Keller, Zenker, and Stone (1989) recently suggested on the basis of divergence of faunal provinces on either side that cessation of sustained surface flow occurred as late as 2.4 mya, with intermittent littoral and neritic exchange as late as 1.8 mya. However, it is not clear whether a delay would occur between closure of marine exchange and faunal divergence, and we prefer to use the date of 3.0 mya for the isthmus pending direct stratigraphic studies of the important Central American record of shallow-water deposits.

Since it formed, the isthmus has isolated marine organisms in the Caribbean from those in the eastern Pacific. This event had major implications for global oceanography and climate (Keigwin 1978; Cronin 1988a), as well as for the termination of the Atlantic-Pacific tropical marine zoogeographical connection and the interchange of North American–South American terrestrial organisms (Stehli and Webb 1985; Cronin 1988b).

We estimated the number of opportunities for new marine ostracode species by first determining the number of pre-Isthmus species in the selected clades, that is, how many species were living before about 3.0 mya, and thus had contiguous ranges across the tropics of the eastern Pacific and the Caribbean. Fossil evidence from many well-studied late Miocene and early Pliocene formations in the Caribbean, Central American, and the Gulf of California region provided the data for this analysis. For each pre-isthmus species occurring on both sides, the formation of the land barrier would in theory create two opportunities. Populations from the Caribbean, the eastern Pacific, or both could evolve into a new species. Then, by determining the number of new species in these genera that appeared since 3.0 mya, we obtain an estimate of the percentage of species that split to form new species in each clade (a maximum of two times the original number of species). Figure 8.2 schematically shows how the isthmus led to increased provinciality in marine invertebrates.

CLIMATIC OSCILLATIONS

We know from the deep-sea isotopic and paleontological record that astronomical cycles have dominated climatic and oceanographic history since about 2.4 mya, a time when high-amplitude, high-frequency glacial-interglacial cycles began in the Atlantic (Shackleton et al. 1984) and Pacific (Rea and Schrader 1985). These cycles were caused by variation in the Earth's eccentricity, tilt, and precession and were originally postulated as a major influence on Earth climate by Milankovitch (1941; see Hays, Imbrie, and Shackleton 1976; Berger et al. 1984). Figure 8.3 schematically depicts an interglacial-glacial cycle and how it constitutes an opportunity for speciation in temperature-sensitive arctic ostracode species. Each odd-numbered oxygen isotope stage represents a period of warm climate, low ice volume, and high sea level, such as today's interglacial period. Each even-numbered iso-

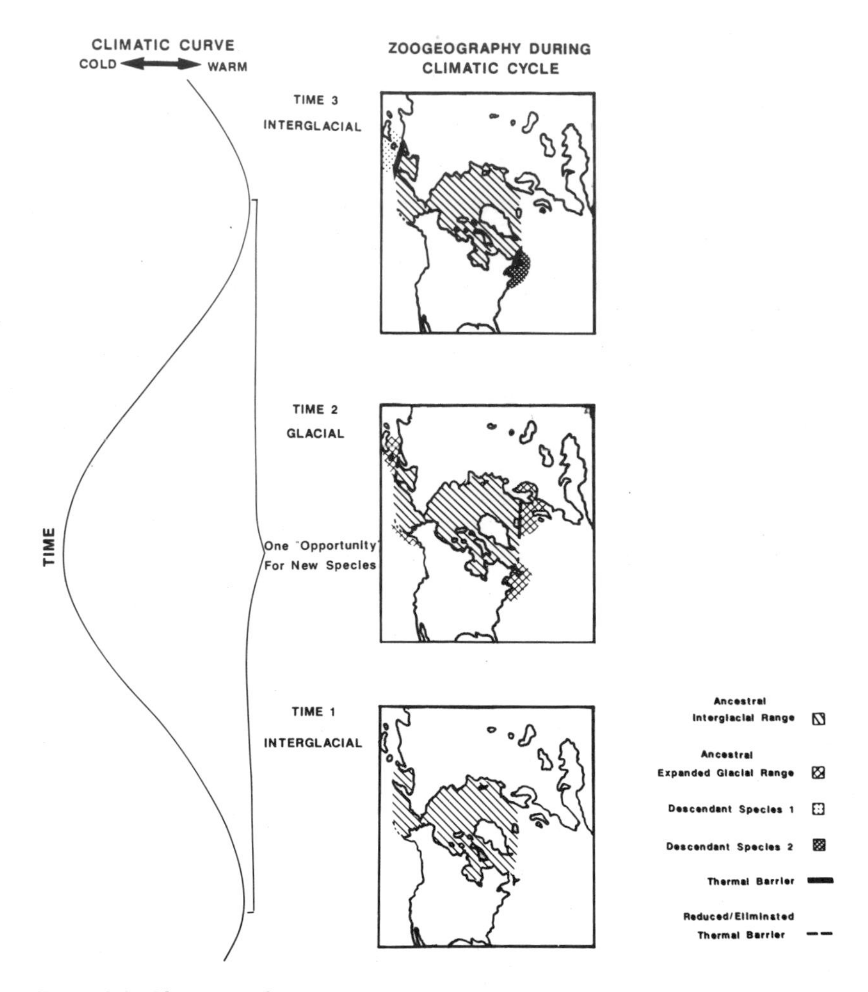

Figure 8.3. Climatic cycles are opportunities for speciation. At left, sine wave schematically depicts a Milankovitch climatic cycle. One cycle is defined as interval between successive interglacial intervals (time 1 and time 3). Time 2 is the intervening glacial period. At right, paleozoogeography of circumpolar arctic species shows southward range expansion during time 2, leading to splitting off of descendant species during climatic warming of next interglacial. See text for discussion.

Table 8.2. Stratigraphic Intervals and Speciation Opportunities

Stratigraphic Range	Number Of Opportunities
Early Pilocene-Recent	46
Early Pliocene–Middle Pleistocene	43
Early Pliocene–Early Pleistocene	37
Early Pleistocene–Recent	32
Early Pleistocene–Late Pleistocene	31
Early Pleistocene–Middle Pleistocene	29
Early Pleistocene	23
Early Pliocene–Late Pliocene	14
Late Pliocene	14
Middle Pleistocene–Recent	9
Middle Pleistocene	6
Late Pleistocene–Recent	3
Recent (= Holocene)	1

Note: The early-late Pliocene boundary is informally placed at 2.5 mya to coincide with the beginning of high-frequency, high-amplitude climatic cycles. The ages of stratigraphic intervals are as follows: early Pliocene, 5.3–2.5 mya; late Pliocene, 2.5–1.7 mya; early Pleistocene 1.7–10.7 mya; middle Pleistocene, 0.7–0.014 mya; late Pleistocene, 0.014–0.001 mya; Recent (= Holocene), 10,000–present.

tope stage is a period of cold climate, greater ice volume, and low sea level. Together, one even stage and one odd stage comprise a complete cycle. Sixty-three oxygen isotope stages have been formally numbered for the interval between 1.7 mya and the present (Shackleton and Opdyke 1973; Ruddiman, Raymo, and McIntyre 1986). Thus a total of about 32 complete glacial-interglacial cycles has occurred since 1.7 mya. Ruddiman, Raymo, and McIntyre (1986) show about 14 additional major warm intervals between 2.5 and 1.7 mya; thus, about 46 complete cycles have occurred since 2.4 mya. The 23 cycles between 1.7 and 0.7 mya have a 41,000-year periodicity, and those since about 700,000 years ago have a predominant 100,000-year periodicity.

Table 8.2 shows the number of climatic cycles (= opportunities) that arctic and temperate species were subjected to for particular stratigraphic intervals. By dividing the number of speciation events (FADs) for each clade since 2.5 mya by the number of opportunities, we can estimate FSI values for that group.

Plio-Pleistocene climatic cycles caused significant shifts in geographical distributions of shallow-water marine organisms, including ostracodes. In the western North Pacific abundant evidence in the papers of Hanai (1959), Okada (1979), Tabuki (1986), and Cronin and Ikeya (1987) demonstrate climatically related zoogeographical shifts in temperature-sensitive ostracode species. Ostracode faunas in the North Atlantic Ocean similarly responded to climatic change (Hazel 1968; Valentine 1971; Cronin 1981, 1989). For example, during the glacial phase of each cycle, cryophilic species inhabiting high latitudes migrate 5° to 6° latitude to the south (fig. 8.3, time 2) (Hazel 1968; Cronin 1988a) and at the same time inhabit a greatly reduced continental shelf area due to sea level lowering of 75 to 125 m (Cronin 1983). Upon completion of the cycle, warmer climates force cryophilic species to migrate north to their primary range and to expand up the continental shelf.

Conversely, during warm interglacials, thermophilic species normally having tropical distributions migrate north to midlatitudes (Valentine 1971; Cronin 1988a). Each climatic cycle thus involves range expansion and contraction. During this process, the distalmost populations of species may become partially isolated by distance from the main range of the species during range expansion when thermal barriers break down during climatic change. Subsequent range contraction and the reestablishment of thermal barriers therefore constitute a chance for daughter species to split off in populations that migrated to peripheral regions. This process probably describes the way many species evolved in benthic groups that inhabit north-south oriented continental shelves (Valentine 1984; Cronin and Ikeya 1987). Figure 8.3 shows two resulting descendant species, one endemic to the Pacific and one to the Atlantic.

Based in part on zoogeographical evidence of ostracode migrations in response to such climatic change, we have treated each complete climatic cycle as a single opportunity to speciate. Using species' biostratigraphic ranges and the climatic history for that stratigraphic interval, we then computed the total opportunities for each species and for each clade.

In summary, climatic cycles and the Isthmus of Panama offer strongly contrasting types of perturbations to organisms' environments, that is, distinct types of opportunities. The formation of the isthmus was a one-time event, after which evolutionary divergence of populations on both sides had about 3 million years, or slightly less, in which to occur. Climatic cycles are periodic and relatively short term. The land barrier created virtually complete isolation for shallow-water marine ostracode species, which could not disperse around South America due to their adaptation to warm tropical climates. There is no evidence for dispersal across Central America by birds, a process known to occur in some fresh-water and brackish water ostracodes. Conversely, climatically induced formation and breakdown of thermal barriers are rapid, transient events. In probabilistic terms, the land barrier creates a very low chance of interbreeding, whereas thermal barriers are much less likely to completely prevent genetic interchange. As stated by Mayr (1947;268): "The interruption of gene flow even between two 'isolated' populations is thus always relative and incomplete."

Finally, if we equate population size with areas, populations isolated by climatic events were probably smaller than those split by the isthmus because the former inhabited small segments of a much reduced continental shelf. Caribbean–Gulf of Mexico and eastern Pacific populations of tropical species that were isolated from one another by the isthmus had extensive areas of shallow shelf and lagoons to inhabit. We assume that large populations existed in these environments. Populations isolated by climatic events are considered smaller than those isolated by the isthmus but larger than "founder" populations.

This approach necessarily simplifies complex environmental changes for any particular region. For example, it neglects the interaction of organisms and short-term, environmental changes such as short-term climatic cycles of 10^2 to 10^4 years. Nonetheless, Milankovitch cycles were predominant in their effect on temperature-sensitive marine organisms over the past few million years, and their periodicities of 20,000, 40,000, and 100,000 years constitute a fitting time scale within which to examine patterns of speciation.

In the case of tropical clades the isolation by land was certainly not the only factor influencing the evolution of tropical genera in this region. In some cases, we could identify other factors that account for a species origination. For example, isolation by deep water led to speciation in some clades on the Hawaiian, Clipperton, and Galapagos Islands. In future studies, it would be useful to analyze only those species living in the immediate area of the isthmus.

We also recognize that biotic factors—particularly interactions between species and intrinsic genomic traits—are considered by many to be, in Schopf's words, "half the story" (1984) and will always be difficult to separate from those due to environmental changes. Indeed, climatic cooling and associated glacioeustatic sea level lowering may intensity biotic interactions among shelf-dwelling species competing for resources, perhaps leading to increased selective pressure. Notwithstanding biotic factors, we believe Milankovitch climatic cycles and the Isthmus of Panama were primary events that were the catalysts for zoogeographical shifts and population isolation, which in turn set the speciation process in motion.

Results

ISTHMUS OF PANAMA

We studied eight tropical clades for the effects of the isthmus on their evolution. Appendix 8.1 lists for each clade the number of pre-isthmus species, the number of opportunities for new species (twice the number of pre-isthmus species), and the number of new post-isthmus species that evolved in the Atlantic and Pacific. Post-isthmus species could have evolved anytime since about 3.0 mya. Other species that evolved in tropical regions adjacent to the Central American–Caribbean area, as in the Galapagos or Clipperton Islands, are not considered to have evolved due to isolation by the land barrier.

A total of 24 species evolved in the eight clades since the isthmus formed, 11 species in the Atlantic-Caribbean and 13 species in the eastern Pacific. These values give a speciation frequency index of .245 (one new species per 4.1 opportunities) for the total, .224 (one species for 4.5 opportunities) for the Atlantic, and .265 (one species per 3.8 opportunities) for the Pacific. This is equivalent to 3.7 new species per million years since closure on the Atlantic side and 4.3 per million years on the Pacific side.

Tropical ostracode genera vary greatly in their absolute abundances in samples. For example, *Caudites* and *Radimella* often comprise 20% to 30% of a sample, *Neocaudites* less than 5%. If one compares totals for clades of relatively common species with rare species, there is a small difference in the FSI values: .272 for the rare species (one new species per 3.7 opportunities) and .222 (one species per 4.5 opportunities).

ARCTIC AND TEMPERATE CLADES: CLIMATIC CHANGE

We analyzed 12 arctic (high-latitude, northern hemisphere) clades having circumpolar zoogeographical distributions and 6 temperate clades endemic to the U. S. Atlantic Coast for their response to climatic change (appendixes 8.2 and 8.3). Table 8.3 and Figure 8.4 show the resulting FSI values for various inter- and intraclade comparisons. Data

Table 8.3 Speciation Frequency for Marine Ostracodes

Group	S_{TOT}	O_{TOT}	S_{NEW}	FSI	Opp/Spp	Nsp/my
1. Total Atlantic	83	2,067	35	.0169	59.1	14
2. Arctic Atlantic	22	416	11	.0264	37.8	4.4
3. Temperate Atlantic	61	1,651	24	.0145	68.8	9.6
4. Arctic total	75	2,240	53	.0237	42.3	21.2
5. Arctic eurytopic	10	267	4	.0150	66.8	1.6
6. Arctic stenotopic	66	1,973	49	.0248	40.3	19.6
7. Arctic Pacific	53	1,825	42	.0230	43.4	16.8
8. Arctic/temperate rare	27	819	25	.0305	32.8	10
9. Arctic/temperate common	109	3,072	65	.0212	47.3	26

Note: nsp/my = new species per million years since 2.5 mya; S_{TOT} is total species studied for the group; FSI is frequency of speculation index.

for circumpolar clades are separated into Atlantic and Pacific groups to permit intra- and interclade comparison.

INTERCLADE: ARCTIC VERSUS TEMPERATE

The control of shallow marine ostracode zoogeography by water temperature makes zoogeographical distribution a useful character for comparing clades. Our results show that arctic clades (group 4, table 8.3) have a higher FSI (.0237) than do temperate clades from the Atlantic (group 3, FSI = .0145). One new species evolved in arctic clades every 42.3 opportunities but only one per 68.6 opportunities for temperate clades during the same climatic events.

INTERCLADE: EURYTOPIC VERSUS STENOTOPIC

Population ecology has given us the familiar division of *r*- and K-selective organisms whose biotic characteristics are related to their particular selective strategy. K-selective species require a constant, predictable environment, their mortality is density dependent, and they have constant population size and are slow developers with large body size and few progeny; *r*-selective species thrive in variable, unpredictable environments, have density-independent mortality, variable unsaturated population size, early reproduction, many offspring, and short life spans (Pianka 1983). Some ostracodes are well known for their tolerance of a wide range of environmental conditions, especially salinity and temperature. Other taxa inhabit constant environments, occur in low to moderate numbers, and appear to be more specialized in their environmental preference. These two groups are considered *r*-selective and K-selective, respectively, and serve as a basis for comparing frequency of speciation, particularly because climatic change causes periodic catastrophic mortality of density independent taxa (Schopf 1984).

Three circumpolar genera—*Palmenella*, *Heterocyprideis*, and *Sarsicytheridea*—tolerate reduced, variable salinities and relatively broad depth ranges. The frequency of speciation for this group (group 5, table 8.3) is .0150 (one new species per 66.8 opportunities). The remaining nine circumpolar arctic clades are more stenotopic and have values of .0248 (one new species per 40.3 opportunities). These values quantify how much stenotopic, K-selective clades are more likely to speciate during cyclic climatic changes than are eurytopic, *r*-selective clades.

A related characteristic of ostracodes is their relative abundance in assemblages. Qualitative observations on the relative abundance let us subdivide the arctic and temperate groups into rare and common clades. If we assume that abundance reflects an underlying biological feature of the organism related to population structure, the data show FSI values of .0305 (one new species per 32.8 opportunities) for rare taxa and .0212 (one new species per 47.3 opportunities) for common taxa. Common taxa subjected to the same climatic changes speciate less frequently than do rare taxa.

INTRACLADE: ARCTIC PACIFIC VERSUS ARCTIC ATLANTIC

The occurrence of populations of circumpolar arctic species in high latitudes of the North Atlantic (group 2, table 8.3) and North Pacific (group 7, table 8.3) regions allows us to compare the independent evolution of the same clade—in a few cases even the same species that were subjected to the same global climatic changes in two areas. FSI values were .0264 (one new species per 37.8 opportunities) for the Atlantic and .0230 (one new species per 43.4 opportunities) for the Pacific. The similarity in values is interesting in light of the higher species diversity in circumpolar clades in the Pacific, where 16.8 new species evolve per million years and the comparable value for the Atlantic is 4.4 per million years. This apparent higher speciation rate is actually due to a higher pre-3.0 million year diversity in circumpolar clades (many probably originated in the North Pacific), despite the same number of climatic cycles. Higher diversity led to a much greater total number of opportunities in the Pacific. As measured by FSI values, rates of speciation in the northern Pacific were actually slightly lower than in the North Atlantic.

Discussion

For the marine ostracodes, our results suggest that (1) the Isthmus of Panama led to a speciation rate about 10 times that caused by climatic change; (2) the lowest frequencies of speciation among groups subjected to climatic cycles were temperate Atlantic clades and eurytopic arctic clades; (3) the highest frequencies of speciation were found among rare taxa, arctic stenotopes, and North Atlantic circumpolar taxa; and (4) within clades evolving independently in the North Pacific and Atlantic Oceans, frequencies of speciation were roughly equivalent in both regions.

Lower FSI values for clades subjected to climatic change than for those separated by the isthmus raise the issue of within-species stasis during environmental change. Only one species arising per 60 or 70 climatic cycles indicates speciation is a relatively infrequent event given the magnitude of the latitudinal zoogeographical shift and the drop in sea level. One of us (Cronin 1985) had postulated that high-frequency Neogene climatic cycles may modulate evolutionary events in marine ostracodes and other groups by fostering stability, in essence lowering the rate of new species origination. Elsewhere, it is argued that within-species stasis during climatic cycles is the very reason ostracodes, foraminifers, and other commonly fossilized groups are so useful in paleoclimatology (Cronin 1987b). Highly integrated genetic units comprising ecologically specialized species are excellent tools to track environmental change over many climatic cycles occurring over millions of years. The results presented here provide a means to quantify this stasis using a measure of within-species stability in terms other than absolute time. Our initial results illustrate different degrees of stasis within and among different clades and provide a basis for future comparisons with other groups.

As a fundamentally different type of isolating event, the Isthmus of Panama yielded high FSI values, but the duration of isolation, 3.0 my, and the more effective isolating land barrier were more conducive to speciation. It should also be pointed out that several ostracode genera subjected to isolation by the isthmus actually had more new species evolve as the result of isolation on islands, such as the case of the genus *Caudites* in the Galapagos (Schneider and Cronin, in prep.) and *Puriana* on Caribbean islands (Cronin 1987a). This appears to result from speciation from small founder populations; however, a method to estimate the number of founder opportunities has not yet been developed, and FSI values are not yet determined for this type of event. The frequency of speciation that resulted from the isthmus may eventually be found to be relatively low compared to that caused by founder events.

One general conclusion to emerge from this study is that the hypothesis that environmental change differentially affects different taxa can be tested quantitatively and confirmed. One should not therefore ask whether abiotic factors influence evolution, for geological and paleozoogeographical evidence for the disruption of organisms' physical and chemical habitat is pervasive. Rather, by selecting known environmental events, we can ask how certain types of events differentially influence clades having particular biotic characters. Certain biotic traits in ostracodes related to zoogeography, temperature tolerance, and population structure convey upon some species an ability to speciate more or less frequently than other species. We find that fundamentally different types of environmental events impact organisms differentially, a fact that mitigates against the measurement of evolutionary change simply as a measure of time.

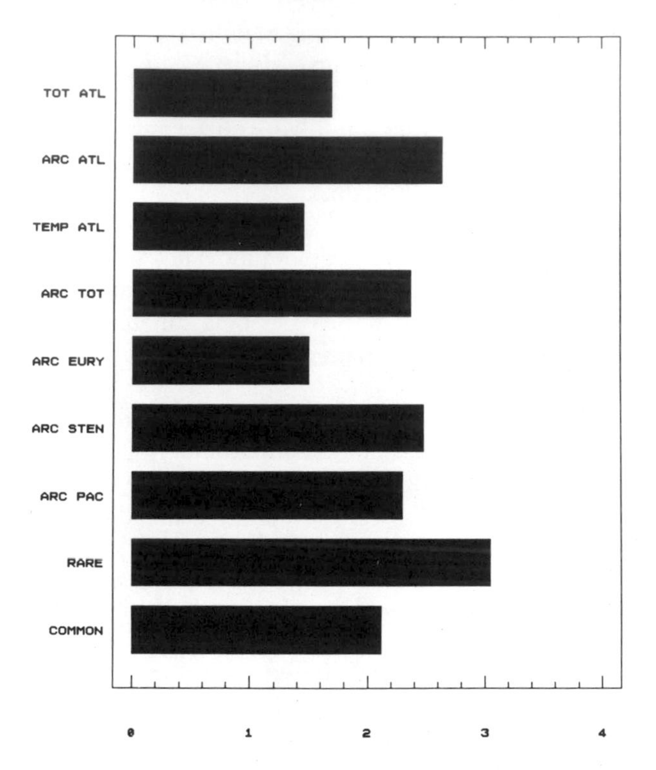

Figure 8.4. Histogram showing FSI values for various groups of ostracode taxa. The higher the FSI value, the greater number of new species that evolved per opportunity. See text for comparison among groups.

In conclusion, we propose that studies of FSI values of other organisms subjected to events such as Milankovitch cycles and the Isthmus of Panama would provide interesting comparisons with the ostracode data. Although we reiterate our caution about the inherent difficulties and the need for simplification in quantifying opportunities, earth history is replete with other types of events that can be viewed as opportunities for speciation and for which FSI values can be obtained. One candidate is the tectonic rifting that opened the Gulf of California about 8 mya, providing opportunities for colonization in a semiisolated tropical environment. The late Miocene–early Pliocene fauna of the Gulf of California region provides an ideal record of the Gulf's earliest inhabitants and its modern fauna, their descendants. Another example is isolation of tropical shallow-water invertebrates in atolls and islands via passive dispersal, which isolates "founder" type populations, as in the cases alluded to previously. No matter which events one chooses, the complexities of environmental change are recorded more fully in the geological record than most paleontologists realize—fully enough to provide the experiments to test theories of speciation in nature.

Acknowledgments

Many colleagues freely assisted our taxonomic and zoogeographic studies of ostracodes, including R. H. Benson, G. P. Coles, T. Hanai, J. E. Hazel, K. Hayashi, D. Horne, A. R. Lord, J. W. Neale, R. Tabuki, A. Tsukagoshi, and W. van den Bold. We are grateful to M. A. Buzas and R. H. Benson (Smithsonian Institution), G. Boxshall (British Museum), and I. Hayami (Tokyo University Museum) for access to type collections. H. Dowsett provided helpful comments on climatic change and opportunities. John Pojeta, Jr., C. E. Schneider, R. M. Ross, G. Vermeij, and an anonymous reviewer provided helpful comments on the manuscript. Funding for field work in Japan was provided by National Geographic Society Grant 2846-84 to T. M. Cronin. N. Ikeya was supported in part by a Grant-in-Aid for Scientific Research (Project 59440006) from the Ministry of Education, Science and Culture, Government of Japan.

Literature Cited

Barigozzi, C., ed. 1982. *Mechanisms of speciation*. New York: Alan R. Liss.

Berger, A. L., J. Imbrie, J. Hays, G. Kukla, and B. Saltzman, eds. 1984, *Milankovitch and climate*, Part I. Boston: D. Reidel Publishing

Cronin, T. M. 1981. Paleoclimatic implications of late Pleistocene ostracodes from the St. Lawrence Lowlands. *Micropaleontology* 27:384–418.

Cronin, T. M. 1983. Rapid sea level and climate change: Evidence from continental and island margins, *Quat. Sci. Rev.*, 1(3): 177–214.

Cronin, T. M. 1985. Speciation and stasis in marine Ostracoda: Climatic modulation of evolution. *Science* 227: 60–63.

Cronin, T. M. 1987a. Evolution, biogeography and systematics of *Puriana*: Evolution and speciation in Ostracoda. *J. Paleontol.* Memoir no. 21, p. 71.

Cronin, T. M. 1987b. Speciation and cyclic climatic change. In *Climate: History, periodicity and predictability*, ed. M. R. Rampino, J. E. Sanders,

W. S. Newman, and L. K. Konigsson, 333–42. New York: Van Nostrand Reinhold.

Cronin, T. M. 1988a. Evolution of marine climates of the U. S. Atlantic coast during the past four million years. *Phil. Trans. R. Soc. Lond. Series B*, no. 318:661–78.

Cronin, T. M. 1988b. Geographic isolation in marine species: Evolution and speciation in Ostracoda I. In *Proceedings of the 9th International Symposium on Ostracoda*, ed. T. Hanai, N. Ikeya, and K. Ishizaki, 871–89. Tokyo: Kodansha Ltd. Press; Amsterdam: Elsevier.

Cronin, T. M. 1989. Paleozoogeography of post-glacial Ostracoda from northeastern North America. In "*The Late Quaternary Development of the Champlain Sea Basin*," ed. N. R. Gadd. *Geological Assoc. Can. Spec. Paper.* 35:125–44.

Cronin, T. M., and N. Ikeya. 1987. The Omma-Manganji ostracode fauna (Plio-Pleistocene) of Japan and the zoogeography of circumpolar species. *J. Micropalaeontol.* 6 (2):65–88.

Eldredge, N., and S. J. Gould. 1972. Punctuated equilibria: an alternative to phyletic gradualism. In *Models in paleobiology*, ed. T. J. M. Schopf, 82–115. San Francisco: Freeman, Cooper and Co.

Endler, J. A. 1977. Geographic variation, speciation and clines. *Monogr. Pop. Biol.*, no. 10. Princeton, N. J.: Princeton University Press.

Gingerich, P. D. 1985. Species in the fossil record: Concepts, trends, and transitions. *Paleobiology* 11(1): 27–41.

Gould, S. J., N. L. Gilinsky, and R. Z. German. 1987. Asymmetry of lineages and the direction of evolutionary time. *Science* 236: 1437–41.

Hanai, T. 1959. Studies on the Ostracoda from Japan. *J. Faculty Sci. Univ. Tokyo* 11(4): 409–18.

Hanai, T., N. Ikeya, and K. Ishizaki, eds. 1988. *Evolutionary biology of Ostracoda, its fundamentals and applications*. Tokyo: Kodansha Ltd. Press; Amsterdam: Elsevier.

Hazel, J. E. 1968. Pleistocene ostracode zoogeography in Atlantic coast submarine canyons. *J. Paleontol.* 42: 1264–71.

Hazel, J. E. 1983. Age and correlation of the Yorktown (Pliocene) and Croatan (Pliocene and Pleistocene) formations at the Lee Creek Mine. In *Geology and paleontology of the Lee Creek Mine, North Carolina*, vol. 1, ed. C. E. Ray. *Smithsonian Contrib. Paleobiol.* 53: 81–199.

Hays, J. D., J. Imbrie, and N. J. Shackleton. 1976. Variations in the Earth's orbit: Pacemaker of the ice ages. *Science* 194:1121–32.

Hoffman, A., and J. A. Kitchell. 1985. Evolution in a pelagic system: a paleobiologic test of models of multispecies evolution. *Paleobiology* 10(1): 9–33.

Ikeya, N., and A. Tsukagoshi. 1988. The interspecific relations between three close species of the genus *Cythere* G. F. Muller, 1785. In *Proceedings of the 9th International Symposium on Ostracoda*, ed. T. Hanai, N. Ikeya, and K. Ishizaki, 891–917. Tokyo: Kodansha Ltd. Press; Amsterdam: Elsevier.

Jablonski, D. 1987. Heretability at the species level: Analysis of geographic ranges of Cretaceous mollusks. *Science* 238:360–3.

Jones, D. S., and P. F. Hasson. 1985. History and development of the marine invertebrate faunas separated by the Central American Isthmus. In *The great American biotic interchange*, ed. F. G. Stehli and S. D. Webb, 325–55. New York. Plenum.

Kamiya, T. 1988a. Morphological and ethological adaptations of Ostracoda to microhabitats in *Zostrea* beds. In *Proceedings of the 9th International Symposium on Ostracoda*, ed. T. Hanai, N. Ikeya, and K. Ishizaki, 303–18. Tokyo: Kodansha Ltd. Press; Amsterdam: Elsevier.

Kamiya, T. 1988b. Different sex-ratios in two Recent species of *Loxoconcha* (Ostracoda). *Senckenbergiana Lethaea* 68(5/6): 377–45.

Kamiya, T. 1988c. Contrasting population ecology of two species of *Loxoconcha* (Ostracoda, Crustacea) in Recent *Zostrea* (eelgrass) beds: Adaptive differences between phytal and bottom-dwelling species. *Micropaleontology* 34(4): 316–31.

Keigwin, L. D. 1978. Pliocene closing of the Isthmus of Panama based on biostratigraphic evidence from nearby Pacific Ocean and Caribbean cores. *Geology* 6(10): 630–4.

Keller, G., C. E. Zenker, S. M. Stone. 1989. Late Neogene history of the Pacific-Caribbean gateway. *J. South Am. Earth Sci.* 2(1): 73–108.

Levinton, J. S., et al. 1986. Organismic evolution: the interaction of microevolutionary and macroevolutionary processes. In *Patterns and processes in the history of life*, ed. D. M. Raup and D. Jablonski, 167–82. Berlin: Springer-Verlag.

Lewontin, R. C. 1957. The adaptations of populations to varying environments. *Cold Spring Harbor Symp. Quant. Biol.* 22:395–408.

Mayr, E. 1942. *Systematics and the origin of species*. New York: Columbia University Press.

Mayr, E. 1947. Ecological factors in evolution. *Evolution* 1(4):263–88.

Mayr, E. 1982. Processes of speciation in animals. In *Mechanisms of speciation*, ed. C. Barigozzi, 1–19. New York: Alan R. Liss.

Milankovitch, M. 1941. Kanon der Erdbestrahlung und seine Anwendung auf des Eiszeitproblem. *Academie Royal Serbe*. Spec. Ed. no. 133.

Okada, Y. 1979. Stratigraphy and Ostracoda from late Cenozoic strata of the Oga Peninsula, Akita Prefecture. *Trans. Proc. Palaeontol. Soc. Japan* N.S. 115: 143–73.

Pianka, E. 1983. *Evolutionary ecology*, 3d ed. New York: Harper and Row.

Rea, D. K., and H. Schrader. 1985. Late Pliocene onset of glaciation-ice rafting and diatom stratigraphy of DSDP North Pacific cores. *Palaeogeogr. Palaeoclimatol. Palaeoecol.* 49:313–25.

Repenning, C. A., E. M. Brouwers, L. D. Carter, L. Marincovich, Jr., and T. A. Ager. 1987. The Beringian ancestry of *Phenacomys* (Rodentia: Cricetidae) and the beginning of the modern arctic ocean borderland biota. *U. S. Geolog. Surv. Bull.* 1687.

Ruddiman, W. F., M. Raymo, and A. McIntyre. 1986. Matuyama 41,000-

year cycles: North Atlantic Ocean and northern hemisphere ice sheets. *Earth Planetary Sci. Letters* 80, 117–29.

Schneider, C. E., and T. M. Cronin. In Preparation. Speciation in the ostracode *Caudites* in the Galapagos.

Schopf, T. J. M. 1984. Climate is only half the story in the evolution of organisms through time. In *Fossils and climate*, ed. P. J. Brenchley, 279–89. Chichester, England: John Wiley and Sons.

Shackleton, N. J., and N. D. Opdyke. 1973. Oxygen isotope and paleomagnetic stratigraphy of equatorial Pacific core V28–238: Oxygen isotope temperatures and ice volumes on a 10^5 and 10^6 year scale. *Quaternary Res.* 3: 39–55.

Shackleton, N. J., et al. 1984. Oxygen isotope calibration of the onset of ice-rafting and history of glaciation in the North Atlantic region. *Nature* 307: 620–3.

Skevington, D. 1967. Probable instance of genetic polymorphism in the graptolites. *Nature* 213: 810–2.

Stanley, S. M. 1986. Population size, extinction, and speciation: the fission effect in Neogene Bivalvia. *Paleobiology* 12(1): 89–110.

Stehli, S. G., and S. David Webb, eds. 1985. *The great American biotic interchange*. New York: Plenum.

Tabuki, R. 1986. Plio-Pleistocene Ostracoda from the Tsugaru Basin North Honshu, Japan. *Bull. College Educ. Univ. Ryukyus* 29(2): 27–160.

Teeter, J. W. 1973. Geographic distribution and dispersal of some shallow-water marine Ostracoda. *Ohio J. Sci.* 73(1): 46–54.

Tsukagoshi, A. 1988. Reproductive character displacement in the ostracod genus *Cythere*. *J. Crustacean Biol.* 8(4): 563–75.

Tsukagoshi, A., and N. Ikeya. 1987. The ostracode genus *Cythere* G. F. Müller, 1785 and its species. *Trans. Proc. Palaeontol. Soc. Japan*, N.S. 148: 197–222.

Valentine, P. C. 1971. Climatic implication of a late Pleistocene ostracode assemblage from southeastern Virginia. *U. S. Geolog. Surv. Prof. Pap.* 683-D.

Valentine, J. W. 1984. Climate and evolution in the shallow sea. In *Fossils and climate*, ed. P. J. Brenchley, 265–77. Chichester, England: John Wiley and Sons.

Vermeij, G. 1987. The dispersal barrier in the tropical Pacific: Implications for molluscan speciation and extinction. *Evolution* 41(5): 1046–58.

Vrba, E. S. 1984a. What is species selection? *Systematic Zool.*, 33(3): 318–28.

Vrba, E. S. 1984b. Evolutionary pattern and process in the sister-group Alcelaphini-Aepycerotini (Mammalia: Bovidae). In *Living fossils*, ed. N. Eldredge and S. M. Stanley, pp. 62–79. New York: Springer-Verlag.

Vrba, E. S., and N. Eldredge. 1984. Individuals, hierarchies and processes: Towards a more complete evolutionary theory. *Paleobiology* 10(2): 146–71.

White, M. J. D. 1978. *Modes of speciation*. San Francisco: W. H. Freeman and Co.

Appendixcs

Taxonomic references for each clade are given at the end of the appendixes. Additional references are in Literature Cited.

A8.1. Tropical Clades: Isolation by the Isthmus of Panama

Genus/Author (References)	Pre-Isthmus Species	O_{TOT}	Post-Isthmus S_{NEW} Atlantic	Pacific	Other
Cativella Coryell and Fields 1937 (Cronin 1988b)	3	6	0	0	0
Caudites Coryell and Fields 1937 (Pokorny 1970; Schneider and Cronin, in prep.)	13	26	2	6	7 Ga
Hermanites Puri 1955 (Cronin 1988b)	1	2	0	1	0
Neocaudites Puri 1960 (Witte, pers. comm.: Cronin, unpub.)	5	10	2	1	2WPac 2EAtl
Orionina Bold, 1963 (Bold 1963; Cronin and Schmidt 1988)	3	6	2	0	3WPac
Puriana Coryell and Fields 1953 (Cronin 1987a)	16	32	3	1	0
Radimella Pokorny 1968 (Pokorny 1968; Bold 1975)	7	14	2	3	1Cl 1Haw, 10Ga
Touroconcha Ishizaki and Gunther 1976 (Cronin, 1988b)	1	2	0	1	0
Totals	49	98	11	13	26

Note: Abbreviations: Pre-Isthmus species = number of species before isthmus formed; O_{TOT} = total opportunities for genus; S_{NEW} = new species; Ga = Galapagos Islands; WPac = western Pacific; EAtl = eastern Atlantic; Cl = Clipperton Islands; Ha = Hawaiian Islands.

A8.2. Circumpular Arctic Clades: Isolation by Climatic Change

Genus/Author (References)	Pacific			Atlantic		
	Total Species	O_{TOT}	S_N	Total species	O_{TOT}	S_N
Acanthocythereis Howe (Cronin and Ikeya 1987)	2	89	2	1	32	0
Baffinicythere Hazel 1967 (Neale and Howe 1975; Cronin and Ikeya 1987)	2	86	2	1	32	0
Cythere Muller 1785 (Ikeya and Tsukagoshi 1988; Tsukagoshi and Ikeya 1987)	12	406	9	2	41	2
Elofsonella Pokorny 1955 (Tabuki 1986)	1	43	1	1	9	0
Finmarchinella Swain 1963 (Neale 1974; Tabuki 1986)	8	333	6	4	53	3
Hemicythere Sars 1922–28 (Schnorikov 1974; Cronin and Ikeya 1987)	10	280	8	3	50	2
Heterocyprideis Elofson 1941 (Cronin 1981; Cronin and Ikeya 1987)	1	32	1	2	18	1
Normanicythere Neale 1959 (Neale 1959; Tabuki 1986)	5	175	5	1	32	0
Palmenella Hirshmann 1909 (Tabuki 1986; Yajima 1987)	1	46	0	1	32	0
Rabilimis Hazel 1967 (Cronin and Ikeya 1987)	2	63	2	2	12	1
Robertsonites Swain 1963 (Tabuki 1986)	7	206	5	1	32	1

A8.2. (*continued*)

Genus/Author (References)	Pacific			Atlantic		
	Total Species	O_{TOT}	S_N	Total species	O_{TOT}	S_N
Sarsicytheridea Athersuch 1982 (Athersuch 1982)	2	66	1	3	73	3
Totals	53	1,825	42	22	416	11

Note: For some circumpolar genera, little pre-late Pleistocene fossil record exists in northeastern North America. We assume migration occurred from Pacific to Atlantic via the Arctic by about the Plio-Pleistocene boundary, 1.7 mya. Evidence for post-3.0 mya migration exists from molluscs (Gladenkov 1981) and ostracodes (Cronin and Ikeya 1987; Repenning et al. 1987). This interpretation is supported by geological studies showing the Bering Strait opened about 3.0 to 2.5 mya, after which cryophilic Pacific species expanded their ranges eastward across the Arctic.

A8.3. Western Atlantic Endemic Temperate/Subtropical Clades: Climatic Change

Genus/Author (Reference)	Tot sp.	O_{TOT}	S_{NEW}
Bensonocythere Hazel 1967 (Hazel 1983; Cronin, 1990)	17	489	7
Climacoidea Puri 1956 (Plusquellec and Sandberg 1969; Hazel and Cronin, 1988)	15	377	7
Malzella Hazel 1983 (Ikeya and Hazel, unpubl.)	10	173	7
Muellerina Bassiouni 1965 (Hazel, 1983; Coles and Cronin 1987)	8	184	2
Neocaudites Puri 1960 (Hazel 1983; Cronin, unpubl.)	6	198	1
Puriana Coryell and Fields 1953 (Cronin 1987a)	5	230	0
Totals	61	1651	24

A8.4. Tropical Clades: Isthmus of Panama

Species/Author	FAD	LAD	Origin	Paleozoogeography
1. +*Cativella navis* Coryell and Fields 1937	LMio	E	Carib	GulfMex, WAtl, EPac, Carib
2. +*C. pulleyi* Teeter 1975	Olig	E	Carib	Carib
3. +*C. semitranslucens* Crouch 1949	Plio	E	EPac	EPac
4. #*Caudites albatrossi* Pokorny 1970	Hol	E	Gal	Gal
5. +*C. angulata* Teeter 1975	Mio	E	Carib	Fla, Carib
6. #*C. anteroides* Pokorny 1970	Hol	E	Gal	Gal
7. #*C. asymmetricus* Pokorny 1970	Pl	E	Gal	Gal
8. #*C. aff. asymmetricus* Pokorny 1970	Hol	E	Gal	Gal
9. /*C. chipolensis* Puri 1953	Mio	Mio	Fla	Fla
10. #*C. croca* Bate, Whittaker, and Mayes 1980	Hol	E	Gal	Gal
11. **C. fragilis* Leroy 1943	EPl	E	Cal	EPac
12. **C. highi* Teeter 1975	Hol	E	Carib	Carib
13. +*C. howei* Puri 1960	EPli	E	Carib	Carib, CA, Fla
14. +*C. medialis* Coryell and Fields 1937	Mio	LPli	Carib	Carib
15. +*C. nipensis* Bold 1946	Mio	E	Carib	Carib, WAtl, EPac
16. **C. paraasymmetricus* Hazel 1983	LPli	LPl	Fla	Fla
17. #*C. paranteroides* Pokorny 1970	Pl	E	Gal	Gal

A8.4. (*continued*)

Species/Author	FAD	LAD	Origin	Paleozoogeography
18. *C. *purii* McKenzie and Swain 1967	Hol	E	EPac	EPac
19. +C. *rectangularis* Brady 1869	Mio	E	Carib	Carib
20. +C. *rosaliensis* Swain and Gilby 1974	Mio	E	GulfCal	EPac, GulfCal
21. /C. *sacer* Bold 1970	Mio	Mio	Carib	Carib
22. #C. *sanctaecrucis* Pokorny 1970	Hol	E	Gal	Gal
23. /C. *sellardsi* (Howe and Neill 1935)	EPli	EPli	GulfMex	GulfMex
24. *C. *serrata* Benson and Kaesler 1963	Hol	E	EPac	EPac
25. +C. *symmetricus* Bate, Whittaker, and Mayes	EPli	E	Carib	EPac
26. +C. *tricostata* Bate, Whittaker, and Mayes	EPli	E	SA	SA
27. +C. *obliquecostata* Bold 1963	Mio	E	Carib	Carib
28. *C. sp. B of Valentine 1976	LPl	E	Cal	Cal
29. +C. sp. D of Valentine 1976	EPli	E	EPac	EPac, GulfCal
30. *C. sp. E of Valentine 1976	Hol	E	GulfCal	EPac
31. /C. n. sp. A	Mio	Mio	Mex	Mex
32. /C. n. sp. B	LMio	LMio	GulfCal	GulfCal
33. /C. n. sp. C	LMio	EPli	Baja	Baja
34. /C. n. sp. D	LMio	EPli	Baja	Baja
35. /C. n. sp. E	EPli	EPli	Carib	Carib
36. +C. n. sp. F	EPli	MPl	CA	CA, WAtl

A8.4. (*continued*)

Species/Author	FAD	LAD	Origin	Paleozoogeography
37. +*C.* n. sp. G	Pli	Pli	Ec	Ec
38. **C.* n. sp. H	Hol	E	Baja	Baja
39. /*C.* n. sp. I	EPli	EPli	Jam	Jam
40. /*C.* n. sp. J	EPli	EPli	Jam	Jam
41. +*C.* n. sp. K	EPli	MPl	GulfMex	WAtl, GulfMex
42. /*C.* n. sp. L	EPli	EPli	Carib	Carib
43. +*Hermanites hornibrooki* Puri 1960	EPli	E	Carib	EPac, Carib
44. **H. tricornis* (Swain 1967)	Hol	E	GulfCal	EPac
45. +*Neocaudites angulatus* Hazel 1983	EPli	LPli	WAtl	WAtl
46. **N. atlantica* Cronin 1979	MPl	E	WAtl	WAtl
47. **N. henryhowei* McKenzie and Swain 1967	Hol	E	EPac	EPac
48. #*N.* sp. A Witte pers. comm.	Hol	E	EAtl	EAtl
49. **N. pulchra* Teeter 1975	Hol	E	Carib	Carib
50. +*N. subimpressus* (Edwards 1944)	EPli	E	WAtl	Circumtrop
51. +*N. triplistriatus* (Edwards 1944)	EPli	E	Carib	WAtl, Carib
52. /*N. aff triplistriatus* (Edwards 1944)	Mio	LMio	Carib	Carib
53. #*N. tubercularis* Omatsola 1972	Hol	E	EAtl	EAtl
54. +*N. variabilus* Hazel 1983	EPli	EPl	WAtl	Watl
55. +*N.* n. sp. 1	EPli	E	WAtl	WAtl
56. /*N.* n. sp. 2	Mio	Mio	Carib	Carib
57. /*Orionina armata* Poag 1972	Oli	Oli	GulfMex	GulfMex
58. **O. boldi* Cronin and Schmidt 1988	MPl	E	Fla	Carib

A8.4. (*continued*)

Species/Author	FAD	LAD	Origin	Paleozoogeography
59. +*O. bradyi* Bold 1963	Mio	E	Carib	Carib, WAtl, GulfMex
60. /*O. brouwersae* Cronin and Schmidt 1988	LMio	E	MarshI	WPac
61. /*O. butlerae* Bold 1965	Oli	Mio?	Carib	Carib, GulfMex
62. **O. ebanksi* Teeter 1975	Hol	E	Carib	Carib
63. /*O. eruga* Bold 1963	MMio	Mio	Carib	Carib
64. #*O. flabellacosta* Holden 1976	Mio	Plio	Haw	WPac
65. +*O. fragilis* Bold 1963	LMio	EPli	Carib	Carib
66. +*O. vaughani* (Ulrich and Bassler 1904)	Mio	E	Carib	Carib, WAtl
67. +*Puriana bajaensis* Cronin 1987	LMio	E	GulfCal	EPac
68. +*P. carolinensis* Hazel 1983	EPli	E	WAtl	WAtl, GulfMex
69. +*P. convoluta* Teeter 1975	EPli	E	WAtl	WAtl, Carib, SA
70. /*P. elongorugata* (Howe 1936)	Oli	Mio	GulfMex	GulfMex
71. +*P. floridana* Puri 1960	EPli	E	WAtl	WAtl, GulfMex, Carib
72. /*P. formosa* Bold 1963	MMio	Mio	Carib	Carib
73. +*P. gatunensis* Coryell and Fields 1953	Mio	Pl	Carib	Carib
74. **P. horrida* Benson and Kaesler 1963	Hol	E	GulfCal	EPac
75. **P. matthewsi* Teeter 1975	Hol	E	Carib	Carib
76. +*P. mesacostalis* (Edwards 1944)	EPli	E	WAtl	WAtl, GulfMex
77. +*P. minuta* Bold 1963	MMio	Plio	Carib	Carib

A8.4. (*continued*)

Species/Author	FAD	LAD	Origin	Paleozoogeography
78. +*P. pacifica* Benson 1959	LMio	E	EPac	EPac
79. +*P. paikensis* Cronin 1987	EPli	E	EPac	EPac
80. +*P. rugipunctata* (Ulrich and Bassler, 1904)	Mio	E	WAtl	WAtl, Carib
81. **P. variabilis*	Hol	E	Brazil	SA
82. +*P.* sp. A of Cronin, (1987)	EPli	EPli	Carib	Carib
83. +*P.* sp. B of Cronin, 1987	Mio	Mio	Carib	Carib
84. +*P.* sp. C of Cronin, 1987	Mio	Mio	Carib	Carib
85. +*P.* sp. D of Cronin, 1987	LMio	LMio	Carib	Carib
86. +*P.* sp. E of Cronin, 1987	EPli	EPli	Carib	Carib
87. +*P.* sp. F of Cronin, 1987	EPli	EPli	Carib	Carib
88. **P.* sp. G of Cronin, 1987	Hol	E	Berm	Berm
89. #*Radimella aequatoris* Pokorny 1969	Hol	E	Gal	Gal
90. #*R. clippertonensis* Allison and Holden 1971	Hol	E	Clip	Clip
91. **R. convergens* (Swain 1967)	Hol	E	GulfCal	EPac
92. +*R. confragosa* (Edwards 1944)	LMio	E	Carib	Carib, WAtl, CA, SA
93. +*R. confragosa* form A of Bold 1975	LMio	E	Carib	Carib, CA, NA
94. #*R. darwini* Pokorny 1969	Hol	E	Gal	Gal
95. #*R. deminuta* Pokorny 1969	Hol	E	Gal	Gal
96. #*R. dictyon* Pokorny 1968	Hol	E	Gal	Gal

A8.4. (*continued*)

Species/Author	FAD	LAD	Origin	Paleozoogeography
97. #*R. difficilis* Pokorny 1969	Hol	E	Gal	Gal
98. #*R. españioliensis* Pokorny 1969	Hol	E	Gal	Gal
99. #*R. helenae* Pokorny 1969	Hol	E	Gal	Gal
100. #*R. kingmaina* Pokorny 1969	Hol	E	Gal	Gal
101. +*R. labrinthus* Carreño 1985	EPli	MPli	EPac	EPac
102. +*R. mariae* Carreño 1985	EPli	EPli	EPac	EPac
103. #*R. oahuensis* Holden 1967	LPl	E?	Haw	Haw
104. **R. palsoensis* (Leroy 1943)	EPl	E?	EPac	EPac
105. #*R. ponderosa* Pokorny 1968	Hol	E	Gal	Gal
106. #*R. pumilio* Pokorny 1969	Hol	E	Gal	Gal
107. **R. wantlandi* Teeter 1975	EPl	E	Carib	Carib
108. +*R.* sp. 1 of Bold 1975	EPli	EPli	SA	SA
109. **R.* sp. 3 of Bold 1975	Hol	E	SA	SA, CA, NA
110. **R.* sp. A of Valentine 1976	Hol	E	EPac	EPac, GulfCal
111. +*R.* n. sp. 1	LMio	EPli	CA	CA
112. +*R.* n. sp. 2	EPli	EPli	EPac	EPac
113. **Touroconcha emaciata* (Swain 1967)	Hol	E	GulfCal	EPac
114. +*T. lapidiscola* (Hartmann 1959)	LMio	E	GulfCal	GulfCal, GulfMex, Carib

+Pre-isthmus species; extant when the isthmus formed.

*Evolved post-isthmus in Caribbean/eastern Pacific off North or Central American continental margins, presumably due to land barrier.

#Evolved post-isthmus off isolated islands; origin is unrelated to formation of Isthmus.

A8.4. (*continued*)

/Extinct in pre-isthmus time (pre-early Pliocene); not included in Appendix 8.1 and calculations.

Abbreviations Regions: Carib = Caribbean; WAtl = western North Atlantic; EPac = eastern Pacific; GulfMex = Gulf of Mexico; GulfCal = Gulf of California; Ec = Ecuador; Jam = Jamaica; Fla = Florida; Haw = Hawaii; Sa = South America; CA = Central America; MarshI = Marshall Islands; Clip = Clipperton Islands; circumtrop = circumtropical; Gal = Galapagos; Berm = Bermuda;

Ages: Oli = Oligocene; Mio = Miocene; EPli = early Pliocene; LPli = late Pliocene; EPl = early Pleistocene; MPl = middle Pleistocene; LPl = late Pleistocene; Hol = Holocene; E = extant; FAD = first appearance datum; LAD = last appearance datum.

A8.5. Circumpolar Arctic Clades: Climatic Change

	North Pacific			*North Atlantic*		
Species/Author	FAD	LAD	Opps.	FAD	LAD	Opps.
1. *Acanthocythereis dunelmensis* (Norman, 1865)	LPli	E	46	EPl*	E	32
2. A. *mutsuensis* Ishizaki, 1966	LPli	MPl	43			
3. *Baffinicythere howei* Hazel, 1967	LPli	MPl	43	EPl*	E	32
4. *B.*? sp. A of Cronin and Ikeya, 1987	LPli	MPl	43			
5. *Cythere alveolivalva* Smith, 1952	LPl	E	3			
6. *C. cronini* Tsukagoshi and Ikeya, 1987				MPl	E	9
7. *C. hanaii* Tsukagoshi and Ikeya, 1987	EPli	E?	46			
8. *C. golikovi* Schornikov, 1974	LPli	E	46			
9. *C. kamikoaniensis* Tsukago-	LMio	LMio	0			

A8.5. (*continued*)

Species/Author	North Pacific FAD	North Pacific LAD	North Pacific Opps.	North Atlantic FAD	North Atlantic LAD	North Atlantic Opps.
shi and Ikeya, 1987						
10. *C. lutea* Müller, 1794				EPl*	E	32
11. *C. nishinipponica* Okubo, 1976	LPl	E	3			
12. *C. nopporensis* Tsukagoshi and Ikeya, 1987	LPli	E?	46			
13. *C. omotenipponica* Hanai, 1959	LMio	E	46			
14. *C. sanrikuensis* Tsukagoshi and Ikeya, 1987	LPli	E	46			
15. *C. schornikovi* Ikeya and Tsukagoshi, 1988	LPli	E	46			
16. *C. urupensis* Schornikov, 1974	LPli	E	46			
17. *C. uranipponica* Hanai, 1959	Mio	E	46			
18. C. sp.	EPl	E	32			
19. *Elofsonella concinna* (Jones, 1856)	LPli	MPl	43	MPl	E	9
20. *Finmarchinella angulata* (Sars, 1865)	Mio	E	46	EPl	E	32
21. *F. aff. angulata* (Sars, 1865)	LPli	E	46			
22. *F. barentzovoensis* (Mandelstam, 1957)				LPl	E	3
23. *F. daishakaensis* Tabuki 1986	LPli	E	46			
24. *F. finmarchica* (Sars 1865)				MPl	E	9

A8.5. (*continued*)

Species/Author	North Pacific FAD	North Pacific LAD	North Pacific Opps.	North Atlantic FAD	North Atlantic LAD	North Atlantic Opps.
25. *F. hanaii* Okada 1979	LPli	E	46			
26. *F. logani* (Brady and Crosskey 1871)				MPl	E	9
27. *F. aff. logani* (Brady and Crosskey 1871)	LPli	LPli	14			
28. *F. nealei* Okada 1979	LPli	E	46			
29. *F. rectangulata* Tabuki 1986	LPli	MPI	43			
30. *F. uranipponica* (Ishizaki 1969)	LPli	E	46			
31. *Hemicythere borealis* (Brady 1868)				EPl*	E	32
32. *H. emarginata* (Sars 1865)	EPli	MPl	43	MPl	E	9
33. *H. aff. emarginata* (Sars 1865)	LPli	LPli	14			
34. *H. gorokuensis* Ishizaki 1966	Mio	Mio	0			
35. *H. gurajnovae* Schornikov 1974	LPli	E	46			
36. *H. kitanipponica* (Tabuki 1986)	LPli	EPl	37			
37. *H. nana* Schornikov 1974	Hol	E	1			
38. *H. ochotensis* Schornikov 1974	LPli	E	46			
39. *H. orientalis* Schornikov 1974	LPli	E	46			

A8.5. (*continued*)

Species/Author	North Pacific FAD	LAD	Opps.	North Atlantic FAD	LAD	Opps.
40. *H. posterovestibulata* Schornikov 1974	Hol	E	1			
41. *H. quadrinodosa* Schornikov 1974	LPli	E	46			
42. *H. villosa* (Sars 1865)				MPl	E	9
43. *Heterocyprideis fascis* (Brady and Norman 1889)	EPl	E	32	MPl	E	9
44. *H. sorbyana* (Jones 1857)				MPl	E	9
45. *Normanicythere concinella* Swain 1963	EPl	E	32			
46. *N. japonica* Tabuki 1986	LPli	MPl	43			
47. *N. leioderma* (Norman 1869)	LPli	MPl	43	EPl*	E	32
48. *N. aff. japonica*	LPli	LPli	14			
49. N. sp. A	LPli	MPl	43			
50. *Palmenella limicola* (Norman 1865)	Mio	E	46	EPl*	E	32
51. *Rabilimis mirabilis* (Brady 1868)				MPl	E	9
52. *R. paramirabilis* (Swain 1963)	EPl	LPl	31			
53. *R. septentrionalis* (Brady 1866)	EPl	E	32	LPl	E	3
54. *Robertsonites hanaii* Tabuki 1986	LPli	MPl	43			
55. *R. logani* of Swain 1963	EPl	E?	32			

A8.5. (*continued*)

Species/Author	North Pacific FAD	North Pacific LAD	North Pacific Opps.	North Atlantic FAD	North Atlantic LAD	North Atlantic Opps.
56. *R. reticuliforma* (Ishizaki 1966)	Mio	MPl	43			
57. *R. tsugaruana* Tabuki 1986	EPli	EPl	37			
58. *R. tuberculatus* (Sars 1865)				EPl*	E	32
59. *R.* sp. 1 of Tabuki 1986	LPli	LPli	14			
60. *R.* n. sp. A	EPl	EPl	23			
61. *R.* n. sp. B	LPli	LPli	14			
62. *Sarsicytheridea bradii* (Norman 1865)	LPli	MPl	43	EPl*	E	32
63. *S. macrolaminata* (Elofson 1939)	EPl	EPl	23	EPl*	E	32
64. *S. punctillata* (Brady 1865)				MPl	E	9
Totals			1825			416

*Assumes migration from Pacific to Atlantic via Arctic occurred about at the Pliocene–Pleistocene boundary, 1.7 mya. FAD = first appearance datum, LAD = last appearance datum.

A8.6. Endemic Western North Atlantic Clades: Climatic Change

Species/Author	FAD	LAD	Opportunities
1. *Bensonocythere americana* Hazel 1967	EPli	E	46
2. *B. aff americana* Hazel 1967	LPl	E	3
3. *B. arenicola* (Cushman 1906)	EPli	E	46
4. *B. blackwelderi* Hazel 1983	EPli	MPl	43
5. *B. aff. blackwelderi* Hazel 1983	EPli	E	46
6. *B. bradyi* Hazel 1983	LPli	LPli	14
7. *B. calverti* (Ulrich and Bassler 1904)	LMio	EPli	0
8. *B. florencensis* Cronin, in press	EPli	LPli	14

A8.6. (continued)

Species/Author	FAD	LAD	Opportunities
9. *B. gouldensis* Hazel 1983	EPli	EPl	37
10. *B. hazeli* Cronin, in press	LPli	E	46
11. *B. holleyensis* Cronin, in press	LPli	LPli	14
12. *B. ricespittensis* Hazel 1983	LPli	EPl	37
13. *B. rugosa* Hazel 1983	EPli	EPl	37
14. *B. sapeloensis* (Hall 1965)	EPl	E	32
15. *B. trapezoidalis* (Swain 1974)	EPli	LPli	14
16. *B. valentinei* Cronin, in press	LPli	E	46
17. *B.* sp. D of Cronin (in press)	EPli	LPli	14
18. *Climacoidea concinnoidea* (Swain 1955)	LPl	E	3
19. *C. edwardsi* (Plusquellec and Sandberg 1969)	LPl	E	3
20. *C. floridana* (Puri 1960)	EPl	E	32
21. *C. foresteri* Hazel and Cronin 1988	Hol	E	1
22. *C. gigantica* (Edwards 1944)	EPli	E	46
23. *C. jamesensis* (Hazel 1983)	EPli	EPl	37
24. *C. mimica* (Plusquellec and Sandberg 1969)	EPli	E	46
25. *C. multicarinata* (Swain 1955)	Hol	E	1
26. *C. multipunctata* (Edwards 1944)	EPli	E	46
27. *C. nelsonensis* (Grossman 1967)	EPli	E	46
28. *C. plueurata* (Puri 1956)	EPli	EPl	37
29. *C. purii* (Keyser 1975)	Hol	E	1
30. *C. redbayensis* (Puri 1956)	Mio	Mio	0
31. *C. reticulata* Hazel and Cronin 1986	EPl	E	32
32. *C. tuberculata* (Puri 1960)	EPli	E	46
33. *Malzella bellegladensis* (Kontrovitz 1978)	LPli	MPl	43
34. *M. chetumalensis* (Teeter 1975)	Mpl	E	9

A8.6. (*continued*)

Species/Author	FAD	LAD	Opportunities
35. *M. conradi* (Howe and McGuirt 1935)	LMio	LPli	14
36. *M. evexa* Hazel 1983	EPli	EPl	37
37. *M. floridana* (Benson and Coleman 1963)	MPl	E	9
38. *M. littorala* (Grossman 1967)	MPl	E	9
39. M. n. sp. A	EPli	EPl	37
40. M. n. sp. B	MPl	E	9
41. M. n. sp. C	LPl	E	3
42. M. n. sp. D	LPl	E	3
43. *Muellerina abyssicola* (Sars 1865)	Mio	E	0*
44. *M. bassiounii* Hazel 1983	LPli	MPl	32
45. *M. blowi* Hazel, 1983	Mio	LPli	14
46. *M. canadensis* (Brady 1870)	MPl	E	9
47. *M. hazeli* Coles and Cronin 1987	EPl	E	32
48. *M. ohmerti* Hazel 1983	Mio	E	46
49. *M. petersburgensis* Hazel 1983	Mio	LPli	14
50. *M. wardi* Hazel 1983	EPli	EPl	37
51. *Neocaudites angulatus* Hazel 1983	EPli	LPli	14
52. *N. atlantica* Cronin 1979	MPl	E	9
53. *N. subimpressus* (Edwards 1944)	EPli	E	46
54. *N. triplistriatus* (Edwards 1944)	EPli	E	46
55. *N. variabilus* Hazel 1983	EPli	EPl	37
56. N. n. sp. 1	EPli	E?	46
57. *Puriana carolinensis* Hazel 1983	EPli	E	46
58. *P. convoluta* Teeter 1975	EPli	E	46
59. *P. floridana* Puri 1960	EPli	E	46
60. *P. mesacostalis* (Edwards 1944)	EPli	E	46
61. *P. rugipunctata* (Ulrich and Bassler 1904)	Mio	E	46
Total			1651

*Not included in totals because it is a deep-water species.

FAD = first appearance datum; LAD = last appearance datum.

Taxonomic references:

Athersuch, John 1982. Some ostracod genera formerly of the family Cytherideidae Sars, In *Fossil and Recent Ostracods*, eds. Bate, R. H., E. Robinson, L. M. Sheppard. Chichester, England: Ellis Horwood, pp. 231–75.

Bold, van den W. A. 1963. The ostracode genus *Orionina* and its species. *J. Paleontol.*: 37: 33–50.

Bold, van den W. A. 1975. Distribution of the *Radimella confragosa* group (Ostracoda Hemicytherinae) in the late Neogene of the Caribbean. *J. Paleontol.*: 49: 692–700.

Coles G. P. and Cronin, T. M. 1987. On *Muellerina hazeli* n. sp. Coles and Cronin. *Stereo-Atlas of Ostracod Shells* 14: pp. 21–24.

Cronin T. M. and Schmidt N. 1988. Evolution and biogeography of *Orionina* in the Atlantic, Pacific and Caribbean: Evolution and speciation in Ostracoda II. *Proc. 9th Intl. Symp. Ostracoda*, eds. T. Hanai, N. Ikeya, K. Ishizaki. Tokyo: Kodansha, pp. 927–38.

Gladenkov, Y. 1981. Marine Plio-Pleistocene of Iceland and problems of its correlation. *Quat. Res.* 15: 18–23.

Hazel, J. E. 1983. Age and correlation of the Yorktown (Pliocene) and Croatan (Pliocene and Pleistocene) at the Lee Creek Mine. *Smithson. Contrib. Paleobiol.* 33: 81–199.

Hazel J. E. and Cronin, T. M. 1988. The North American ostracode genus *Climacoidea* Puri, 1956 and the tribe Campylocytherini Puri, 1960 (Neogene and Quaternary). In *Proc. 9th Intl. Symp. Ostracoda*, eds. T. Hanai, N. Ikeya, and K. Ishizaki. Tokyo: Kodansha, pp. 39–56.

Ikeya N. and Tsukagoshi, A. 1988. The inter-specific relations between three close species of the genus *Cythere* O. F. Muller. In *Proc. 9th Intl. Symp. Ostracoda*, eds. T. Hanai, N. Ikeya, K. Ishizaki. Tokyo: Kodansha, pp. 891–917.

Neale, J. W. 1959. *Normanicythere* gen. nov. (Pleistocene and Recent) and the division of the ostracod family Trachyleberididae. *Paleontol.* 2: 72–93.

Neale, J. W. 1974. The genus *Finmarchinella* Swain, 1963 (Crustacea: Ostracoda) and its species. *Bull. Br. Mus. (Nat. Hist.)* 27: 83–93.

Neale J. W. and Howe, H. V. 1975. The marine Ostracoda of Russian Harbour, Novaya Zemlya and other high latitude faunas *Bull. Amer. Paleontol.*, 65 (282): 381–431.

Plusquellec, P. A. and Sandberg, P. A. 1969. Some genera of the ostracode subfamily Campylocytherinae. *Micropaleontol.* 15: 427–80.

Pokorny, V. 1969. The genus *Radimella* Pokorny, 1969 (Ostracoda, Crustacea) in the Galapagos Islands. *Acta Univ. Carolinae*, Geologica no. 4: 293–334.

Pokorny, V. 1970. The genus *Caudites* (Ostracoda: Crustacea) in the Galapagos Islands. *Acta Univ. Carolinae*, Geologica no. 4:267.

Schornikov, E. I. 1974. On the study of Ostracoda (Crustacea) from the intertidal zone of the Kuril Islands. *Acad. Sci. U.S.S.R., Far Eastern Sci. Center, Inst. Mar. Biol.* no. 1: 137–214 (in Russian).

Tsukagoshi, A., and Ikeya, N. 1987. The ostracod genus *Cythere* O. F.

Muller, 1785 and its species. *Trans. Proc. Palaeontol. Soc. Japan*, 148: 197–222.

Yajima, M. 1988. Preliminary notes on the Japanese Miocene Ostracoda. In *Proc. 9th Intl Symp. on Ostracoda*, eds. T. Hanai, N. Ikeya, K. Ishizaki, pp. 1073–86.

9

Biotic and Abiotic Factors in the Evolution of Early Mesozoic Marine Molluscs

Anthony Hallam

The paleontological approach to the study of the respective roles of biotic and abiotic factors in evolution depends heavily on a suitable choice of subject matter—if any hopes are to be entertained of arriving at valid and significant conclusions. Thus it is essential to restrict investigation, at least initially, to fossil groups that occur in abundance and whose taxonomy, stratigraphic, and geographical distribution, and likely biology are at least reasonably well known. Both biostratigraphic and facies control should be of a high order, so that environmental changes in space and time can be well monitored.

These requirements are satisfactorily met for ammonites and bivalves in the Lower and Middle Jurassic rocks of western Europe, whose stratigraphy and facies have been intensively studied for many years (Hallam 1975a). These two groups are by far the most abundant and diverse of all the macroinvertebrates present. Other molluscan groups, such as gastropods, belemnites, nautiloids, and scaphopods, are unsuitable for study for reasons of low diversity, sporadic occurrence, and/or inadequate taxonomic investigation. This account is therefore restricted essentially to Jurassic ammonites and bivalves, but brief reference is made also to Triassic ammonites, bivalves, and arachaeogastropods in evaluating the significance of the end-Triassic extinction event. The stratigraphic interval under consideration embraces two mass extinction events with longer intervals before, after, and in between characterized by more modest turnover, which often involves extinction in the case of ammonites but more usually marks benthic community replacement (in the sense of Miller [1986]) for other groups.

This study expands on ammonite data presented in Hallam (1987) in the context of the gross pattern of faunal turnover and presents new data from a bivalve diversity study that is thought to have a bearing on the subject of competition. A discussion of the possible interaction of biotic and abiotic factors takes account of earlier published work on phyletic size increase in ammonites and bivalves and the relationship of faunal turnover to facies changes up the stratigraphic succession.

The Role of Competition

The extent to which competition plays a role in communities of living organisms is a matter of considerable controversy among ecologists, and it is proving rather hard to establish with a high measure of confidence in many instances. Consequently, to assess the role of competition in fossil communities might seem on the face of it a dauntingly difficult task. Nevertheless, the situation is not necessarily as hopeless as might initially appear and requires further exploration. I shall attempt to do this by reference to a survey I made some years ago of the stratigraphic distribution and ecology of Jurassic bivalves in western Europe (Hallam 1976).

For the purposes of the survey, it was necessary to establish what appeared to be good biospecies, assuming, as paleontologists are obliged to, that there is a good correlation between shell morphology and other biological attributes, including reproductive isolation. That this correlation is less than perfect is well appreciated, and sibling species are likely to go undetected in the fossil record. Provided, however, that such species are in a minority, there is no reason why general patterns that reflect genuine biological entities should not emerge from a study of "morphospecies." In the present context it is worth bearing in mind that malacologists classify their species (though not, of course, higher taxa) almost exclusively on shell characters, not soft parts, and that nothing is known of the genetics of the most living molluscs. Patterns of molluscan species distribution are, however, clearly recognizable and can prove of great value in ecological and biogeographical research.

It soon became apparent that most of the thousands of species names that had been published for Jurassic bivalves over the last two centuries were likely to be invalid for a number of reasons and that continuing use of these names would serve only to confuse progress in looking for patterns of potential biological significance. The most important reason was the narrow typological concept of fossil species that held sway until quite recently, with all kinds of trivial morphological attributes being held to have taxonomic significance. A further factor was the failure of taxonomists to leave their museum bench and do their own field collecting. Additionally, an unduly parochial approach to their subject matter led to the same morphological entities being given different names at different stratigraphic horizons and geographical localities. Foreign literature was evidently not read as thoroughly as literature emanating from the home country, and 19th-century German, French, and British workers showed a strong tendency to prefer the use of names proposed by their compatriots.

The strategy I adopted was to use the modern population approach to bivalve taxonomy, which allows fully for a substantial range of morphological and size variation within given species, and to look for discontinuities as revealed, for instance, by multimodal distributions in quantifiable characters measured from specimens collected at given stratigraphic horizons in particular localities. The stratigraphic and geographical range of the morphological entities so recognized was then determined as precisely as possible by field collecting and examination of museum material in several countries. The earliest valid name published in the literature was used for the species in question. As a result of this survey, only a few hundred Jurassic bivalve species were recognized as valid, but this was a sufficiently large number to reveal a variety of interesting patterns that would have remained obscured if all the published names had been used uncritically. Of course such a wide-ranging survey runs the risks of superficiality, but it has the advantage of consistency, having been conducted by only one worker who "develops an eye" for what are likely to be the key units of evolution for a given group of fossils. In fact, more detailed research on particular families and genera (Johnson 1984, 1985; Hallam 1982) has broadly confirmed the substantial stratigraphic and geographical species ranges recognized in the initial survey.

With few exceptions, Jurassic bivalve genera are easy to distinguish, and it is only the distinction of species that is likely to provoke controversy. One of the most interesting results of my European survey, confirmed by later work in other continents, was the recognition of a rule that appears to have quite general validity, namely, that there tends to be only one species of a given genus at a given stratigraphic horizon and within a given region. The horizon may be as much as several ammonite zones and the region may be continent wide. Species of the same genus are normally quite distinct from each other and maintained their morphology for millions of years in a condition of stasis, though often with some phyletic size increase, until a comparatively abrupt change to another related species, either a descendant or an immigrant (Hallam 1978b).

Although differing in matters such as choice of species name, the more intensive research by Johnson (1984, 1985) on Jurassic scallops led to essentially similar results. For the genus *Gryphaea*, for which 20 species were recognized in the Jurassic world wide, the same rule applies. *Gryphaea* species are exceptional among Jurassic bivalves in revealing anagenetic trends apart from size increase, albeit within a framework of punctuated equilibria (Hallam 1982).

Assuming that the species recognized correspond at least for the most part to genuine biological entities, what does the rule imply? Clearly the ecology of fossil species cannot be inferred directly, but the similar ecology of similar species suggests that some kind of competitive exclusion may have operated in the Jurassic. This conclusion is at variance with what Stanley (1973a) has inferred on the basis of his field studies of living suspension-feeding bivalves, which revealed many examples of congeneric forms living sympatrically—namely, that interspecies competition is low. It should, however, be noted, in opposition to Stanley's conclusion, that ecological studies of birds have shown cases in which multiple species of one genus co-occur but are interpreted to have divided the habitat because of strong competition. Bird ecology, however, is likely to differ significantly from that of bivalves, which experience a relatively passive suspension—or deposit—feeding mode of life.

The matter has now been investigated further by means of diversity study of a series of such bivalve assemblages taken from a variety of facies throughout western Europe. The full data are as yet unpublished, but the principal relevant results are outlined here.

None of the 30 assemblages analyzed contains more than a small fraction of the total numbers of species in existence in Europe at the time they were accumulated. Thus, from early mid-Jurassic times onward the species richness for successive stages fluctuated, at a very conservative estimate, at around 100 (Hallam 1976). A large proportion of these are by no means rare, yet the richest assemblage recorded has only 22 species and most, substantially less. The fact that condensation might have caused the mixing of biotopes of different ages in the case of at least some of the richer faunas only serves to underline the contrast. No geographical provincialism within Europe has been recognized, and the contemporary faunal differences in various localities relate solely to facies. Hence, one might reasonably have expected the richer assemblages to contain a higher proportion of the total standing diversity if competition were insignificant.

Though this argument is not conclusive, it is reinforced by a second set of facts. The dominant species (defined as exceeding 5% of the total) of the 25 polytypic assemblages almost never belong to the same family, let alone genus, implying at least moderate competition between closely related taxa. The only exceptions are *Plagiostoma gigantea* and

Pseudolimea pectinoides (Limidae) co-occurring in the Hettangian of Dorset and *Lopha gregarea* and *L. marshi* (Ostreidae) in the Bathonian of Campagnettes, Normandy. In both cases there is a large difference in size, so that the life mode might have been sufficiently different to exclude close competition for resources or space.

The presence in a given assemblage of certain species rather than other closely related forms is more likely to relate to chance colonization and opportunism than to a host of subtle environmental differences excluding particular species. This is because many bivalve taxa appear to have been moderately eurytopic and able to tolerate a variety of substrates and water conditions (Hallam 1976). Once a certain group of species became established in a particular area, they might have been able to exclude for a long time other species equally capable of flourishing there. Thus the common species in the Hettangian of Dorset and the Pliensbachian of Yorkshire were observed to persist as dominants through a considerable thickness of strata.

Predation presently is regarded as by far the most important limiting factor on subtidal, suspension-feeding bivalves (Stanley 1973a) and, in at least some tropical seas, population densities diminish drastically for this reason at depths of only a few meters compared with shallower, more inshore areas (Jackson 1972). The situation is likely to have been substantially different in the Jurassic, because predation pressures apparently increased strikingly in the late Cretaceous and Cenozoic following adaptive radiation of neogastropods, crabs, and teleosts (Vermeij 1977). The defensive response of the bivalves was primarily in the increase of infaunalization, and Jurassic bivalve-faunas have a much higher proportion of epifauna than Recent faunas (compare the data in Stanley [1970] and Hallam [1976]). Furthermore, the order Pterioida, exclusively epifaunal, accounts for an astonishingly high proportion of the total number of individuals counted in the 25 polytypic assemblages—no less than 58%. This figure would be even higher if the five monotypic assemblages (from oyster beds) were also taken into account.

With regard to Jurassic ammonites, greater problems arise because of the considerable amount of taxonomic oversplitting that has been undertaken. In my personal experience of Lower Jurassic ammonite collecting, the only taxa I was convinced were true biological entities were, with few exceptions, genera. Now that considerable morphological variability has been recognized in what are probably true ammonite biospecies (Reeside and Cobban 1960; Howarth 1973; Westermann 1966; Callomon 1985; Bayer and McGhee 1985) it is likely that the same competitive exclusion principle applies as for bivalves. Callomon's (1985) study of cardioceratids is especially illuminating in this regard. Whereas he recognizes up the Middle and Upper Jurassic succession numerous species of stratigraphic and evolutionary significance, at a given horizon within the whole boreal realm there is only one species showing continuous variation in its morphological characters. The burden of proof, involving substantial morphometric analysis, should now rest on those who would argue for a multiplicity of contemporary congeneric species.

In accepting a role for competition among Jurassic molluscs it is important to distinguish between two kinds. What Darwin (1859) had in mind when he discussed the "struggle for existence" was the eventual success of adaptively superior organisms, a concept that has been uncritically accepted by generations of evolutionary biologists and has received a modern formulation by Van Valen (1973) with his familiar Red Queen hypothesis. This type of competition implies that later arrivals on the evolutionary scene can displace the inhabitants of given ecological niches and is hence appropriately termed *displacive competition*. On the other hand, my bivalve work has led me to believe that the most important thing was to be "first in the field." The earlier species occupant of a niche would stay there until some physical disturbance caused its elimination. Only after the niche had been so vacated could another species, which could be a direct evolutionary descendant, come to reoccupy it. This alternative phenomenon can be termed *preemptive competition* and implies that the prime motor of evolutionary change is, contrary to Darwin's belief, the physical not the biotic environment. Indeed, evolution could conceivably grind to a halt in the absence of abiotic change, a thought that has led Stenseth and Maynard Smith (1984) to formulate the so-called stationary model of evolution, as opposed to the Red Queen model characterized by continuous biotic interactions.

The rival Red Queen and stationary hypotheses carry different implications regarding changes of faunal diversity with time and can be formulated as testable models (fig. 9.1). In the preemptive model (fig. 9.1A) two groups successively rise to peaks of diversity and then decline, with most of the decline of the older group (X) taking place before the diversity of the younger group (Y) had begun to rise significantly. The implication is that a niche had to be at least substantially vacated before the younger group became dominant. For the displacive model (fig. 9.1B) the stratigraphic ranges of the two diversity curves overlap considerably. This pattern, unlike that of fig. 9.1A, is consistent with a prediction of the displacive competition hypothesis that the rise to dominance of the younger group is the cause of the decline of the older group. The high diversity and turnover rate of ammonites makes them suitable subject matter for this kind of analysis.

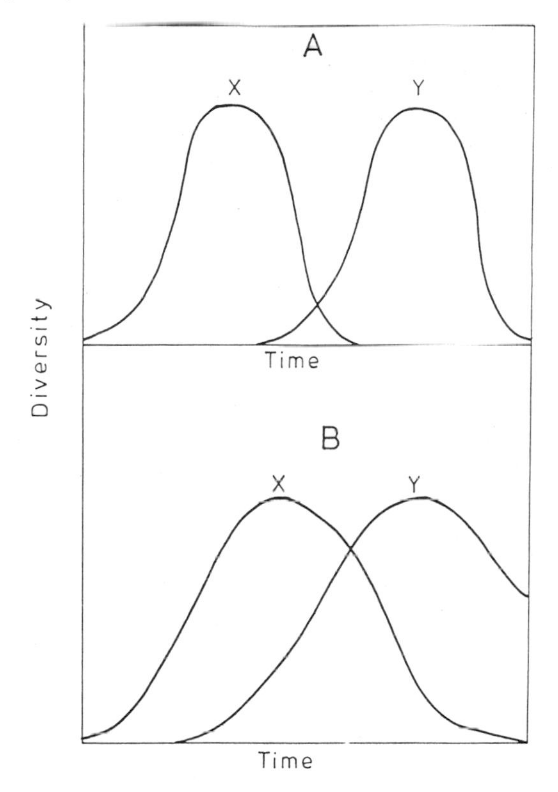

Figure 9.1. Alternative competition models for patterns of diversity change through time of groups X and Y, A Preemptive competition. B Displacive competition.

Ammonite Turnover

Throughout the early Mesozoic a series of ammonite superfamilies appeared in more or less regular succession, with one major interruption at the end of the Triassic, when nearly all ammonites went extinct (fig. 9.2); possibly no more than one phylloceratid genus survived to be the root stock of younger Mesozoic groups (Tozer 1981). The turnover at both superfamily and family level can be examined in more detail for the early Jurassic and early mid-Jurassic respectively (figs. 9.3 and 9.4). The taxa in question are probably coherent clades and exhibit a kind of evolutionary relay similar to that described for Devonian ammonoids by House (1985).

With regard to the alternative models under discussion, the results of figures 9.3 & 9.4 show a fair degree of overlap for successive superfamilies but less for successive families. If genera, species (where they can be reliably distinguished), and numbers of individuals are consid-

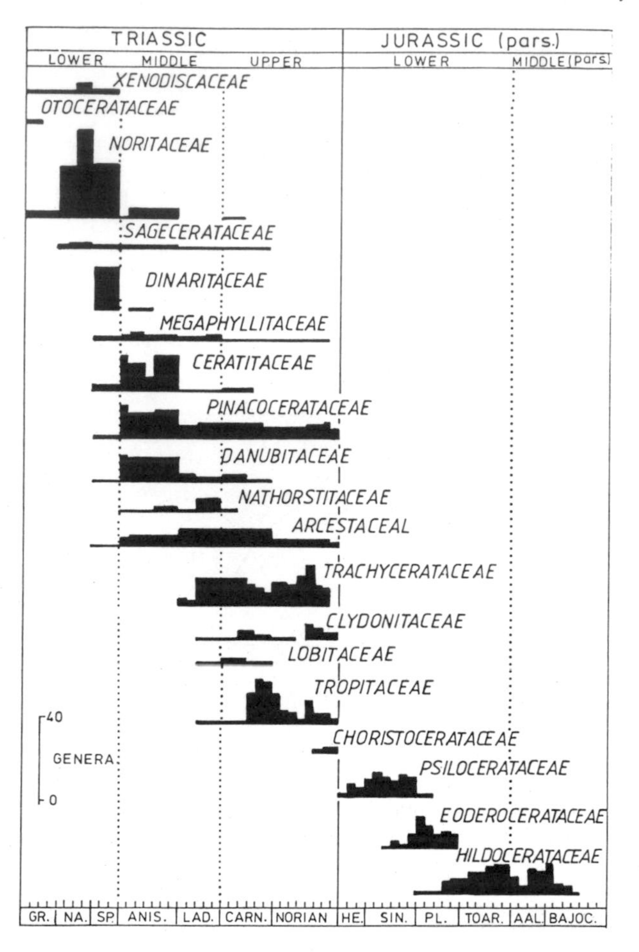

Figure 9.2. The succession of Triassic and early to mid-Jurassic ammonite superfamilies, with diversity represented in histogram form. GR, Griesbachian; NA, Nammalian; SP, Spathian; ANIS, Anisian; LAD, Ladinian; CARN, Carnian; HE, Hettangian; SIN, Sinemurian; PL, Pliensbachian; TOAR, Toarcian; AAL, Aalenian; BAJOC, Bajocian. After House (1985); based on data from Tozer (1981) and Donovan, Callomon, and Howarth (1981).

ered, the degree of overlap is progressively less, which strongly favors the preemptive model. The point is not easy to illustrate diagrammatically in the absence of adequate quantitative data but is very apparent in field collecting and can be supported with a few examples. Thus the family Schlotheimiidae is shown in figure 9.3 to range from the late Hettangian to the end of the Sinemurian, hence overlapping with three other families. By far the most important genus in terms of numbers of individuals is *Schlotheimia*, which became extinct at the end of the Hettangian, being replaced immediately in the earliest Sinemurian by arietitids; the Sinemurian schlotheimiids are highly subordinate in numbers and rank as mere stragglers. The replacement of *Schlotheimia* by the arietitid genus *Vermiceras* is a strikingly sharp event throughout northern Europe, from which it is easy to locate precisely the Hettangian-Sinemurian boundary, but in southern Europe arietitids occurred abundantly in late Hettangian times as contemporaries of the

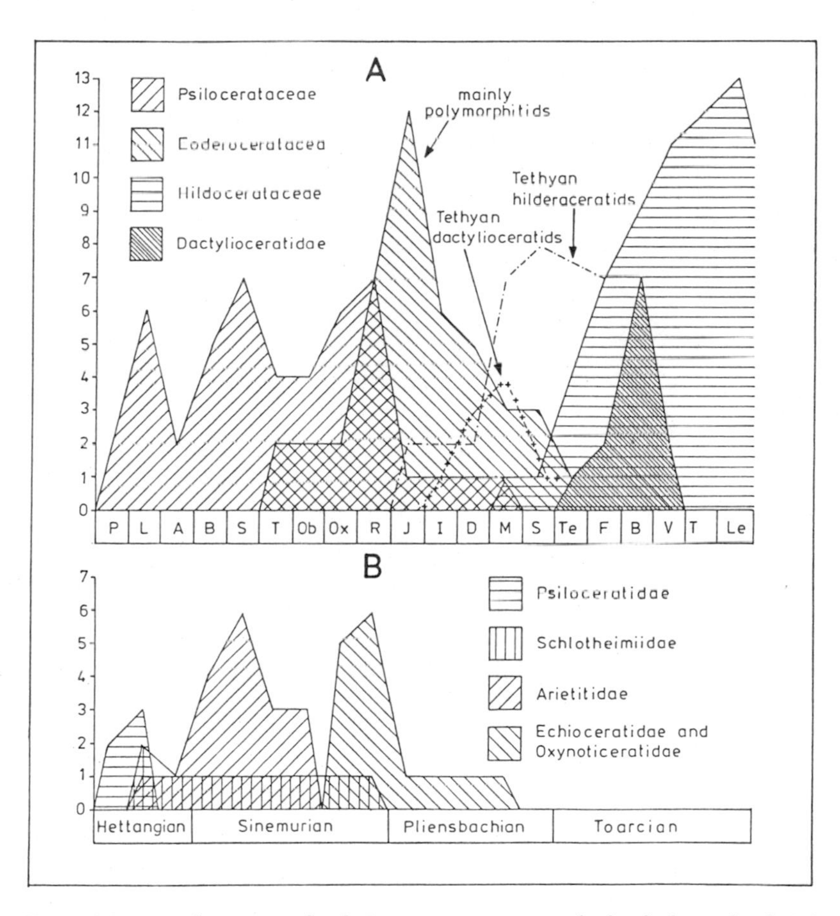

Figure 9.3. Faunal turnover of early Jurassic ammonites at the level of superfamily (**A**) and family (**B** for the Psilocerataceae). Numbers on ordinate refer to genera. Ammonite zones: P, Planorbis, L, Liasicus; A, Angulata; B, Bucklandi; S, Semicostatium; T, Turneri; Ob, Obtusum; Ox, Oxynotum; R, Raricostatum; J, Jamesoni; I, Ibex; D, Davoei; M, Margaritatus; S, Spinatum; Te, Tenuicostatum; B, Bifrons; V, Variabilis; T, Thouarsense; Le, Levesquei. Based on data from Donovan, Callomon, and Howarth (1981).

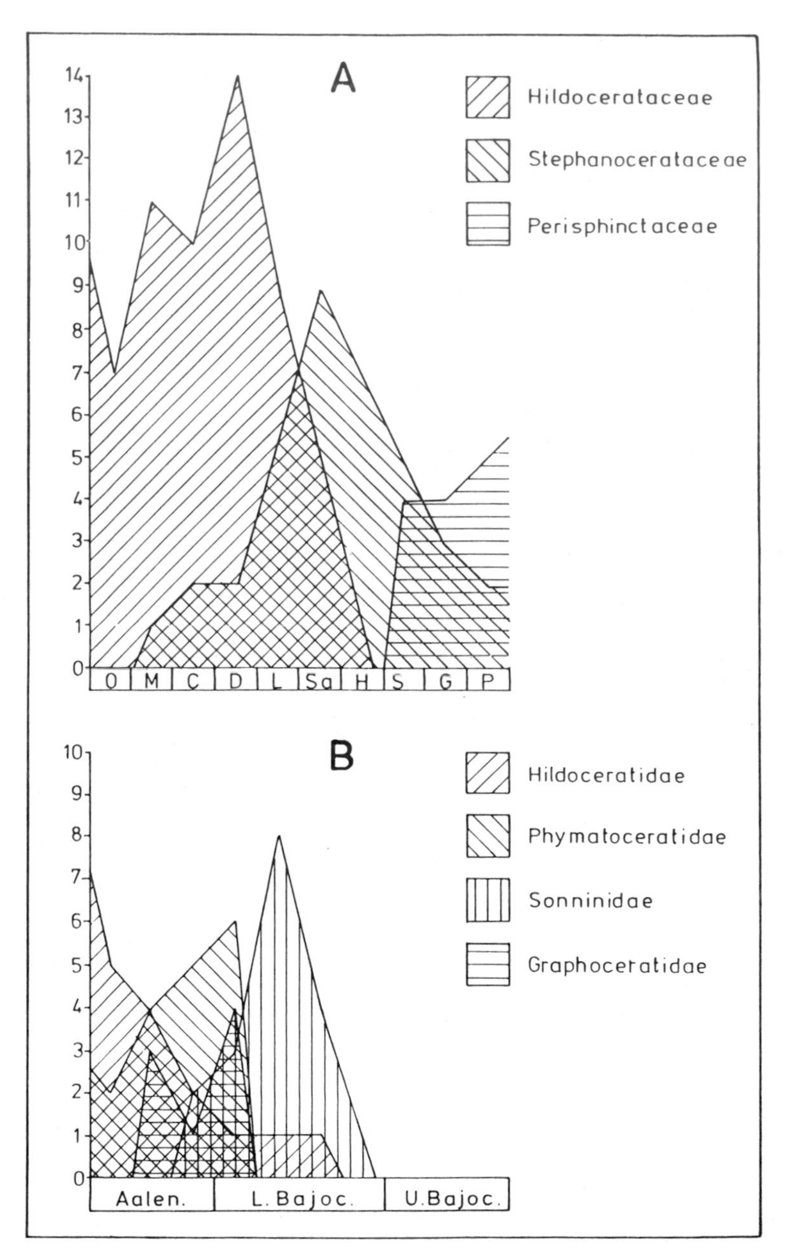

Figure 9.4. Faunal turnover of early mid-Jurassic ammonites at the level of superfamily (**A**) and family (**B** for the Hildocerataceae). Ammonite zones: O, Opalinum; M, Murchisonae; C, Concavum; D, Discites; L, Laevinscula; Sa, Sauzei; H, Humphriesianum; S, Subfurcation; G, Garantiana; P, Parkinsoni. Based on data from Donovan, Callomon, and Howarth (1981).

more boreal schlotheimiids (Donovan, Callomon, & Howarth 1981). It seems evident that the migration of arietitids into northern Europe took place after some event that caused the extinction of the schlotheimiids.

An even more striking example of northerly migration and radiation of ammonite groups following extinction of boreal contemporaries concerns the younger Liassic (fig. 9.3A). During the late Pliensbachian, hildoceratids and dactylioceratids flourished in southern Europe (Tethyan realm) and amaltheids in northern Europe (Boreal realm). Following extinction of the amaltheids at the end of the Pliensbachian (Spinatum zone) there was a northward migration of representatives of the Tethyan families, accompanied by significant evolutionary radiation. Once again, the change without any overlap in northern European sections from the latest Pliensbachian deposits rich in amaltheid individuals to earliest Toarcian deposits with prolific numbers of hildoceratids and dactylioceratids is an easily recognized phenomenon that is much more clear-cut than that which can be presented at the level of higher taxa. A third good example concerns the European appearance in large numbers of the genus *Sonninia* in early Bajocian times after extinction of the phymatoceratids and graphoceratids that dominated late Aalenian faunas (fig. 9.4B). *Sonninia* has no obvious ancestors in Europe and is thought to have immigrated from the eastern Pacific (Westermann and Riccardi 1985; Bayer and McGhee 1985).

Comparison of Major and Minor Extinction Episodes

For the time interval under consideration two events can be recognized that affected a high percentage of the total marine invertebrate fauna and that warrant the term mass extinction—one at the end of the Triassic and the other in the early Toarcian. The end-Triassic event is generally recognized as one of the five biggest in the Phanerozoic. Besides the almost complete faunal turnover among the ammonites, which has already been mentioned (see fig. 9.2) nearly half the bivalve genera and nearly all the European species became extinct (Hallam 1981). Although their more sporadic occurrence makes it more difficult to pin down precisely when they disappeared, it is apparent that the latest Triassic was also a time of crisis for the archeogastropods, with the Norian extinction of a number of groups important in the Paleozoic that had survived the end-Permian extinction event. These include all the Murchisonaceae and half the families of the Pleurotomariacea (Knight et al. 1960).

The early Toarcian mass extinction event, which is clearly recognizable in Europe if not yet elsewhere in the world, can be pinpointed with fine stratigraphic precision as the Exaratum subzone of the Falciferum zone and caused the extinction of the majority of benthic and nektobenthic species, including bivalves, ammonites, and belemnites; only the planktic coccolithophorids and pentacrinitids were unaffected (Hallam 1987).

The likeliest cause of the two extinction events was a severe reduction in benthic and nektobenthic habitat area caused by epicontinental sea regression or spread of anoxic bottom waters during the succeeding transgression. Both factors might have been operative at the end of the Triassic (Hallam 1981), but there is clear evidence that an anoxic event was at least the dominant factor for the early Toarcian extinctions (Hallam 1987). After the extensive niche vacations in Europe following the extinction events, recolonization took place from other parts of the world, as in Tethys and the Pacific margins (Hallam 1987).

In addition to these two major events a number of minor extinction events affecting only the ammonites are clearly recognizable in Europe. Characteristically these correlate with sedimentary cycles, for which facies analysis indicates a succession of shallowing and deepening phases recognizable widely across the continent (Hallam 1978a, 1989 and fig. 9.5). Whether or not these phases relate to regional epeirogeny or global eustasy, it is clear that ammonite extinctions correlate well with the shallowing phases and corresponding regressions and radiations with the deepening phases and corresponding transgressions (fig. 9.6 and table 9.1), as also inferred by Donovan (1985) and Bayer and McGhee (1985). Bayer, Altheimer, & Deutschle (1985) point out a pattern for the Aalenian that is more generally valid and widely recognizable in the Liassic. Major regressive cycles, when developed in thick stratal sequences, usually break down into a series of smaller cycles of similar character, with correspondingly smaller-scale ammonite extinction events. Broadly speaking, the large-scale regressions, exhibiting marked facies changes up the succession and traceable over a large region, tend to correspond to stage boundaries, intermediate-scale events to zonal boundaries and small-scale events, recognized only locally, to subzonal boundaries.

Evidently the ammonites were appreciably more stenotopic than other invertebrate groups and were affected by environmental changes of quite modest magnitude. This is evidently the cause of their high turnover rate, which in turn makes them good biostratigraphic indices, with generic longevity being generally shorter than species longevity of other groups (Hallam 1987).

Evidence also indicates variation of taxon longevity within the ammonites. It is well known that the deeper-water suborder Phylloceratina had higher generic longevity than the shallower-water Ammonitina and is hence much less useful stratigraphically. Based on the strength of their siphuncular tubes, Westermann (1971) inferred that adult representatives of the Ammonitina, unlike the Phylloceratina, would not have been able to withstand hydrostatic pressures at depths greater than 100 m. The difference in longevity must surely relate to this difference in habitat. The shallower water habitat of the Ammonitina, overwhelmingly the most dominant of the ammonites of the European epicontinental sea, would have been one of relatively low predictability in factors such as temperature, salinity, and oxygen content, with the corresponding environmental stress promoting high extinction rates at times of sea-level change on a regional or global scale. Speciation rates

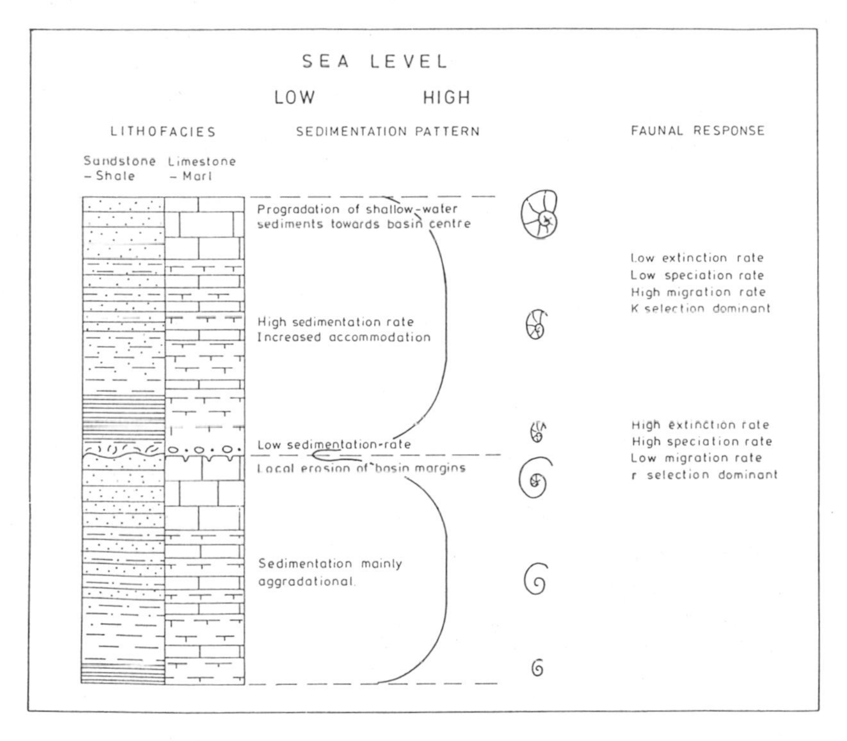

Figure 9.5. Evolution and extinction of ammonites in relation to sedimentary cycles. The lithofacies symbols are conventional. In the siliciclastic sequence the cycle begins with a condensed shell bed and is overlain by a laminated shale, but these deposits could equally well be found in the calcareous sequence. Two ammonite taxa are represented in simple diagrammatic form. Both exhibit phyletic size increase, and the younger ammonite has evolved from the older by means of paedomorphosis, with costae that only occur in the juvenile of the ancestor but throughout the life history of the descendant.

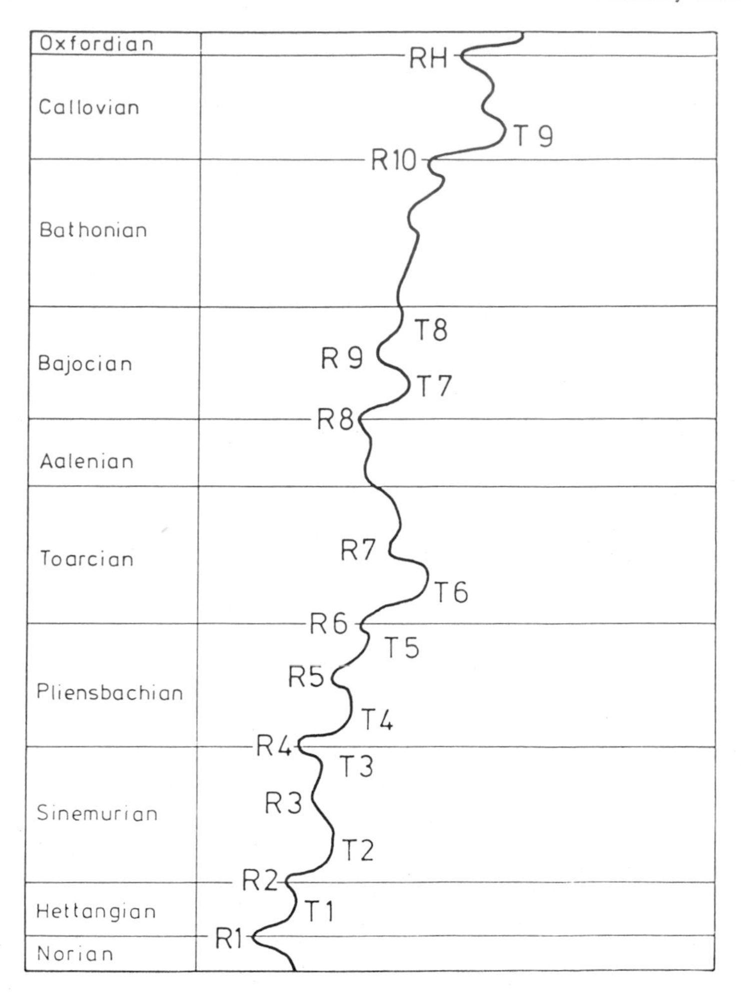

Figure 9.6. Sea-level changes from the end of the Triassic to the start of the late Jurassic. A series of deepening episodes/transgressions (T) and shallowing episodes/regressions (R) correlate respectively with ammonite radiations and extinctions; see table 9.1. Sea-level curve based on figure 10 of Hallam (1989).

are also likely to have increased as a result of increased allopatry caused by greater isolation of bodies of shallow water. Even within the Ammonitina depth-related differences may have existed. Thus Phelps (1982) found that within the Pliensbachian family Liparoceratidae the species longevity of *Liparoceras*, inferred on facies grounds to have lived in a deeper-water habitat, is greater than contemporary shallow-water *Androgynoceras*. I have recognized a similar pattern among Sinemurian Arictitidae. The large genera of this family, such as *Arietites* and *Coroniceras*, are common only in shallow-water facies and had only brief time spans, whereas the small *Arnioceras* is abundant only in deeper-shales facies and had an appreciably greater range in time.

Phyletic Size Increase

One of the most widely recognized and discussed patterns of evolutionary change discernible in the fossil record is phyletic size increase, commonly known as Cope's rule (Newell 1949; Stanley 1973b; La Barbera 1986). Phyletic size increase is a common phenomenon among

Table 9.1. Ammonite radiations and extinctions in relation to transgressions and regressions (See fig. 9.6)

		Radiations	Extinctions
R_{11}	End-Callovian		Kosmoceratids Reineckiids Tulitids
T_9	Early Callovian	Kosmoceratids Macrocephalitids	Last stephanoceratids (Cadomitinae)
R_{10}	End-Bathonian		Clydoniceratinae
T_8	Late Bajocian	Perisphinctids	
R_9	Mid-Bajocian		Sonniniids
T_7	Early Bajocian	Sonniniids	
R_8	End-Aalenian		Most hildoceratids and graphoceratids
R_7	Mid-Toarcian	Hildoceratids Dactylioceratids (plus migration into	Dactylioceratids
T_6	Early Toarcian	boreal realm)	
R_6	End-Pliensbachian		Amaltheids
T_5	Late Pleinsbachian	Amaltheids	
R_5	Mid-Pliensbachian		Liparoceratids
T_4	Early Pliensbachian	Polymorphitids etc.	
R_4	End-Sinemurian		Echioceratids Oxynoticeratids
T_3	Late Sinemurian	Echioceratids Oxynoticeratids	
R_3	Mid Sinemurian		Arietitids
T_2	Early Sinemurian	Arietitids	
R_2	End-Hettangian		Psiloceratids Most schlotheimiids
T_1	Hettangian	Psiloceratids Schlotheimiids	
R_1	End Triassic		Mass extinction

Jurassic ammonites and bivalves with the rate of increase in ammonites being appreciably faster than in bivalves, in correspondence with their higher rate of taxonomic turnover. Although small ammonites and bivalve taxa frequently evolved from larger ones, this almost invariably seems to have taken place at an appreciably more rapid rate than phyletic size increase, which is characteristically gradual. Indeed, evolutionary size decrease is normally abrupt in the stratigraphic record (Hallam 1975b, 1978b). In other words, there is apparently a temporal asymmetry to the phenomenon of evolutionary size change.

A clue to the interpretation of this asymmetry is provided by the study of ammonite size increase in conjunction with facies analysis of the sedimentary cycles characteristic of much of the European Jurassic. According to facies interpretation of environmental change, the times of abrupt size decrease in a given lineage correspond to times of marine shallowing and regression, with a corresponding restriction of habitat and consequent increased environmental stress. Such stressful environments favor, in ecologists' parlance, *r*-selection, with the premium being on early maturation and increased reproduction. At times of higher sea level and reduced stress, K-selection is favored, its most characteristic expression being delayed maturity and, consequently, larger maximum size of organisms. The most dramatic evolutionary changes take place at times of regression, with preferential extinction of larger taxa and survival of smaller opportunists, which may in turn become phyletically larger and become extinct during the next regressive event (fig. 9.5). I first tentatively put forward this model a decade ago (Hallam 1978b), and it has since received support from ammonoid workers on both the Jurassic (Bayer and McGhee 1985) and Carboniferous (Ramsbottom 1981).

Since phyletic size increase is such a widespread trend in the animal kingdom, there must be manifestly one or more selective advantages of larger size. Among those proposed are an improved ability to capture prey or ward off predators; greater reproductive success; increased regulation of the internal environment; and increased heat regulation per unit volume (Newell 1949; Rensch 1959; Odum 1971; Stanley 1973a; Calder 1984). A price must be paid, however, for such adaptive advantages. Because food resources are likely to remain approximately constant, population sizes must decrease, thereby increasing the probability of extinction (Hallam 1975b; Stanley, this volume). The correlative later maturation and slower reproductive rate reinforce the extinction vulnerability.

Recently Brown and Maurer (1986) have presented data for a variety of vertebrates and plants that indicate that large species use a disproportionately large share of resources within local ecosystems. Small species tend to have higher population densities; however, this does not compensate for their lower rate of energy use per individual. Thus selection pressures give rise to a situation exemplifying Cope's rule, which describes the effects of two opposing forces, one operating at the level of individuals within a population, the other at the level of species within ecosystems. Individuals of large size are favored by intraspecific selection because they can dominate resource use and consequently leave more offspring than their smaller relatives. However, the size increase is, as pointed out previously, accompanied by a higher probability of extinction because of smaller population size, lower population density, and slower population growth.

The pattern illustrated in figure 9.5, therefore appears to result from a fascinating interaction of biotic and abiotic factors. During times of relatively low environmental stress, the biotic factors predominate and ammonite taxa tend to increase their size more or less gradually. During times of increased stress, such as produced by regression, abiotic factors predominate and the extinction rate increases, with the survival of smaller taxa being favored. The model is based primarily on data from ammonites but appears to be valid also for at least some bivalve species, such as those of *Gryphaea*. Detailed analysis of *Gryphaea* size increase suggests that it may not be uniformly gradual, as earlier work suggested, but, in some cases at least, may decline with time from a higher to a lower rate (Hallam 1982).

Conclusions

Within the early Mesozoic ammonite and bivalve faunas of western Europe, abiotic factors appeared to have predominated over biotic factors in the promotion of evolutionary change. This is most evident for the two mass extinction events at the end of the Triassic and in the early Toarcian but is also clear for the more frequent ammonite extinction episodes at stage and zonal boundaries. For the longer-ranging bivalves the results are less clear, with stratigraphic and geographical variations of facies obscuring evolutionary change to a much greater extent. Nevertheless, detailed studies such as those of Hallam (1982) on *Gryphaea* and Johnson (1985) on *Radulopecten* suggest patterns of evolution and extinction in relation to physical changes in the environment comparable to the ammonites.

Where biotic factors play a role, as at times when sea level is relatively high and the marine environment relatively uniform, competition appears to be preemptive rather than displacive, with faunal change requiring the prior vacation of ecological niches. Phyletic size change within given taxa can be accounted for by an intriguing interplay between biotic and abiotic factors. During periods of low environmental stress the operation of modest selection pressures favors size increase, but vulnerability to extinction as a result of change in the physical environment is increased. Episodes of high environmental stress increase the extinction rate among K-related organisms, with *r*-selected opportunists, characteristically small, being favored. Heterochronic changes such as paedomorphosis and peramorphosis (McNamara 1982, 1986) can, as a result of progenesis and subsequent size increase, promote the evolution of new species within lineages (fig. 9.5). Admittedly the *r*-K dichotomy has been criticized as simplistic (Stearns 1976) but, nevertheless, Calder (1984) has summarized results that indicate a strong association between K-selecting environments and large body size, delayed reproduction, slowed development, and longer life span and the converse for *r*-selecting regimes. McKinney (1986) has also discussed the possible relationships between size changes and heterochrony. Such relationships are perceived most clearly in the ammonites, though they are also well exemplified by *Gryphaea* (Hallam 1982). Stanley (1973a) suggested that ammonites exhibit a higher rate of turnover than bivalves because of greater competitive interactions between related species that drive them to extinction more quickly. This interesting version of the Red Queen model is not supported by evidence presented here, which favors the idea that the shorter species durations of ammonites were the result of greater stenotopy and, hence, vulnerability to physical disturbance.

It is beyond the scope of this article to consider the more general question of speciation and diversification except to point out that substantial diversification due to in situ cladogenesis, either directly after extinction episodes or at any other time, has not yet been satisfactorily documented and is indeed difficult to recognize.

Literature Cited

Bayer, U., and G. R. McGhee, 1985. Evolution in marginal epicontinental basins: the role of phylogenetic and ecological factors (ammonite replacements in the German Lower and Middle Jurassic). In *Sedimentary and evolutionary cycles*, ed. U. Bayer and A. Seilacher, 164–220. Berlin: Springer-Verlag.

Bayer, U., E. Altheimer, and W. Deutschle. 1985. Environmental evolution in shallow epicontinental seas: Sedimentary cycles and bed formation. In *Sedimentary and evolutionary cycles*, ed. U. Bayer and A. Seilacher, 347–81. Berlin: Springer-Verlag.

Brown, J. H., and B. A. Maurer. 1986. Body size, ecological dominance and Cope's rule. *Nature* 324: 248–50.

Calder, W. A. 1984. *Size, function and life history.* Cambridge, Mass.: Harvard University Press.

Callomon, J. H. 1985. The evolution of the Jurassic ammonite family Cardioceratidae. The Palaeontological Association, *Spec. Papers Palaeontol.* 33: 49–90.

Darwin, C. R. 1859. *On the origin of species by means of natural selection, or the preservation of favoured races in the struggle for life.* London: John Murray.

Donovan, D. T. 1985. Ammonite shell form and transgression in the British Lower Jurassic. In *Sedimentary and evolutionary cycles*, ed. U. Bayer and A. Seilacher, 48–57. Berlin: Springer-Verlag.

Donovan, D. T., J. H. Callomon, and M. K. Howarth. 1981. Classification of the Jurassic Ammonitina. In, *The Ammonoidea*, System Assoc. spec. vol. 18, ed. M. R. House and J. R. Senior, 101–55. London: Academic Press.

Hallam, A. 1975a. *Jurassic environments.* Cambridge: Cambridge University Press.

———1975b. Evolutionary size increase and longevity in Jurassic bivalves and ammonites. *Nature* 258: 193–6.

———1976. Stratigraphic distribution and ecology of European Jurassic bivalves. *Lethaia* 9: 245–59.

———1978a. Eustatic cycles in the Jurassic. *Palaeogeog. Palaeoclimatol. Palaeoecol.* 23, 1–32.

———1978b. How rare is phyletic gradualism? Evidence from the Jurassic bivalves. *Paleobiology* 4: 16–25.

———1981. The end-Triassic bivalve extinction event. *Palaeogeog. Palaeoclimatol. Palaeoecol.* 35: 1–44.

———1982. Patterns of speciation in Jurassic *Gryphaea. Paleobiology* 8: 354–66.

———1987. Radiations and extinctions in relation to environmental change in the marine Lower Jurassic of north west Europe. *Paleobiology* 13: 152–68.

———1989. A re-evaluation of Jurassic eustasy in the light of new data and the revised Exxon curve. In Sea level changes—an integrated approach, ed. C. K. Wilgus et al. *Soc. Econ. Paleontol. Min. Spec. Publ.* 42: 261–276.

House, M. R. 1985. The ammonoid time-scale and ammonoid evolution. In *The chronology of the geological record*, Geol. Soc. ed. N. J. Snelling, 273–83. Lond. Mem. no. 10, Oxford: Blackwell Scientific Publications.

Howarth, M. K. 1973. The stratigraphy and ammonite fauna of the Upper Liassic Grey Shale of the Yorkshire coast. *Bull. Br. Mus. (Nat. Hist.) Geol.* 24: 237–77.

Jackson, J. B. C. 1972. The ecology of the molluscs of *Thalassia* communities, Jamaica, West Indies. II. Molluscan population variability along an environmental stress gradient. *Marine Biol.* 14: 304–37.

Johnson, A. L. A. 1984. The paleobiology of the bivalve families Pectinidae and Propeamussiidae in the Jurassic of Europe. *Zitteliana* 11: 1–235.

———1985. The rate of evolutionary change in European Jurassic scallops. *Spec. Papers Palaeontol.* 33: 91–102.

Knight, J. B. et al. 1960 Mollusca I. *Treatise Invert. Paleontol.*, part I. Lawrence: University of Kansas Press.

La Barbera, M. 1986. The evolution and ecology of body size. In *Patterns and processes in the history of life*, ed. D. M. Raup and D. Jablonski, 69–98. Berlin: Springer-Verlag.

McNamara K. J. 1982. Heterochrony and phylogenetic trends. *Paleobiology* 8: 130–42.

McNamara, K. J. 1986. A guide to the nomenclature of heterochrony. *J. Paleontol.* 60: 4–13.

McKinney, M. L. 1986. Ecological causation of heterochrony: a test and implications for evolutionary theory. *Paleobiology* 12: 282–9.

Miller, W. 1986. Paleoecology of benthic community replacement. *Lethaia* 19: 225–31.

Newell, N. D. 1949. Phyletic size increase—an important trend illustrated by fossil invertebrates. *Evolution* 3: 103–24.

Odum, E. P. 1971. *Fundamentals of ecology.* Philadelphia: W. B. Saunders Co.

Phelps, M. C. 1982. A facies and faunal analysis of the Carixian-Domerian boundary beds in north-west Europe. Ph.D. diss., University of Birmingham, England.

Ramsbottom, W. H. C. 1981. Eustatic control in Carboniferous ammonoid biostratigraphy. In *The Ammonoidea*, System. Ass. Spec. vol. 18, ed. M. R. House and J. R. Senior, 369–88. London: Academic Press.

Reeside, J. B., and Cobban, W. A. 1960. Studies of the Mowry Shale (Cretaceous) and contemporary formations in the United States and Canada, *U.S. Geol. Surv. Prof. Pap.* 355: 1–126.

Rensch, B. 1959. *Evolution above the species level.* New York: Columbia University Press.

Stanley, S. M. 1970. Relation of shell form to life habits of the Bivalvia. *Geol. Soc. Am. Mem. 125*: 1–296.

Stanley, S. M. 1973a. Effects of competition on rates of evolution, with special reference to bivalve mollusks and mammals. *System. Zool.* 22: 486–506.

Stanley, S. M. 1973b. An explanation for Cope's rule. *Evolution* 27: 1–26.

Stearns, S. C. 1976. Life history tactics: a review of the ideas. *Q. Rev. Biol.* 51: 3–47.

Stenseth, N. C., and J. Maynard Smith. 1984. Coevolution in ecosystems: Red Queen or stasis? *Evolution* 38: 870–80.

Tozer, E. T. 1981. Triassic Ammonoidea: Classification, evolution and relationship with Peruvian and Jurassic forms. In *The Ammonoidea*, System Assoc. Spec. vol. 18, ed. M. R. House and J. R. Senior, 66–100. London: Academic Press.

Van Valen, L. 1973. A new evolutionary law. *Evol. Theory* 1: 1–30.
Vermeij, G. J. 1977. The Mesozoic marine revolution: Evidence from snails, predators and grazers. *Paleobiology* 3: 245–58.
Westermann, G. E. G. 1966. Covariation and taxonomy of the Jurassic ammonite *Sonninia adicra* (Waagen). *Neues Jahrbuch für Geologie und Paläntologie, Abhandlungen* 124: 289–312.
Westermann, G. E. G. 1971. Form, structure and funciton of shell and siphuncle in coiled Mesozoic ammonoids. *Life Sci. Contrib. Roy. Ontario Mus.* no. 78, 1–39.
Westermann, G. E. G. and A. C. Riccardi. 1981. Middle Jurassic ammonite evolution in the Andean Province and emigration to Tethys. In *Sedimentary and evolutionary cycles*, ed. U. Bayer and A. Seilacher, 6–34. Berlin: Springer-Verlag.

10

Unoccupied Morphospace and the Coiled Geometry of Gastropods: Architectural Constraint or Geometric Covariation?

David E. Schindel

One picture is worth more than a thousand words.
Chinese Proverb

If we knew nothing of the biological particulars of a species, our uninformed expectation would be that evolution could proceed in all directions with equal probability. The ecological character of the species, its genetic composition, and its morphological makeup would be expected to change in any and all directions if we had no specific knowledge of the taxon. But we know that, in general, evolution does not send descendant taxa radiating out from their ancestors like the streaks from a skyrocket exploding in midair. A variety of forces act to keep the path of evolutionary descent constrained, most notably development, the physical environment, and interactions among species. Seilacher (1970) and Hickman (1980), among others, have characterized the kinds of constraints that might separate the forms we have seen from those that have not, and may never come to be. The descendants from a common ancestor comprise a nonrandom subset of the descendants that might have been. Each ancestral skyrocket gives off only a few streaks, and each of those follows its own crooked trajectory.

How can we evaluate and appreciate the degree to which a group's evolutionary potential has been constrained? Visualizing and measuring these evolutionary trajectories is no easy task, because every specimen of every taxon has more quantifiable dimensions than we can measure, perceive, or even imagine. Each genomic, morphological, and behavioral dimension can give a different view, a novel perspective on how descendants have been constrained during modification. Nevertheless, the study of evolutionary constraint requires a metric—a map for visualizing the occupied and unoccupied evolutionary pathways that are theoretically possible. Two such maps are the often cited "adaptive landscape" of genetic frequencies (from Sewall Wright) and David M. Raup's "morphospace" of coiled shells.

The study of evolutionary constraint comprises three phases: (1) choosing the axes of the coordinate system on which real or simulated organisms can be plotted, (2) mapping diversity onto the coordinate system, and (3) pondering the reasons why the map of morphospace is so full or so empty and why groups occur where they do on the map. The map's axes must be chosen to accommodate all the variability within the group (that is, the axes should be usable as a measurement and/or simulation scheme), produce clear separations among forms that may differ only subtly, and avoid using those characters on which the group's classification is based (thereby creating a circular argument by ordaining the location and extent of each group on the map). Dwelling on the choice of morphospace (or "genospace" or "ethospace") axes may seem like methodological hair splitting, but the distributions of groups on the map, and the inferences regarding contraint that are based on these distributions, all derive from this choice of axes.

I will focus on Raup's (1961, 1966) use of coiling parameters as axes of mollusc and brachiopod morphospace in order to present three points: (1) Our impression of realized versus unrealized evolutionary potential is dictated not only by the things we measure but also by the measurements we choose and the way we scale those measurements; (2) a group may display tremendous variability along several axes of the map and flexible covariation among these measures, yet produce forms that appear outwardly uniform and generalized; and (3) a group may display only limited variability along particular axes and tight covariation among measures, producing forms that are distinctive and specialized. Despite our intuition that "advanced" groups should be more widely distributed on a map of evolutionary potential, they may be narrowly distributed in morphospace. It is possible that "primitive" groups are developmentally more flexible and therefore cover wider areas of the map.

MEASURING GASTROPOD MORPHOLOGY

The morphology of gastropod shells historically has been approached in three ways. A pictorial approach, relying entirely on qualitative description, has been most widely used in systematic and evolutionary studies of snails. Features have been codified into a generally applicable but impressionistic system of terminology for characterizing shell geometry (Cox 1955). Overall shell shape could be "high-spired," whorl shape might be "inflated," sculptural ornament "weakly developed," but characterizations such as these can not be used objectively as the axes of a map.

Second, the biometric approach is based on shell measurements that are sufficiently detailed to characterize species- and even population-level distinctions. Systematic distinctions at these levels can require 10 or more variables that are tailored to the group under study (e.g., Kohn and Riggs 1975 on the genus *Conus;* Gould, Woodruff, & Martin 1974; Gould 1969 and other studies of *Cerion*). No single set of measurements has been found optimal or even sufficient for studying the full range of gastropods. This reflects the common trade-off between generality and precision. Biometric schemes designed for a particular group define a map with too many axes on which too few taxa can be located.

The third is the geometrical simulation approach, in which the fundamental geometry of growth in a group is distilled down to a few morphogenetic parameters. If done correctly, these growth parameters should be capable of simulating the spectrum of known variations within the group. Simple growth parameters have been used to study bryozoa (McKinney and Raup 1982), plants (Niklas 1982) and "triloboids" (Raup and Gould 1974). Surprisingly few investigators have followed up these simulation studies with empirical studies that used simulation parameters as metric characters (Cheetham, Hayek, & Thomsen 1981 on cheilostome bryozoa).

D'Arcy Thompson (1947) developed such a system of three "coiling angles," and D. M. Raup (1961, 1966) formalized and expanded them into a measure of apertural shape and three "coiling parameters." Raup described the shape index of the generating curve (S), its per whorl rate of expansion (W), and the curve's migration parallel and normal to the coiling axis (T and D, respectively). These parameters were established originally for the empirical analysis of morphological variation among living and fossil forms (Raup 1961), but they were then used as variables in a computer simulation (Raup and Michelson 1964; Raup 1966 and 1972). Changes in one or more parameter would produce different versions of shells displaying perfect equiangular (i.e., isometric) growth. Allometric growth was also simulated by varying one or more parameters during the growth process.

Raup used the three coiling parameters W, T, and D as axes of a three-dimensional "morphospace," within which he generated the spectrum of theoretically possible coiled shells, and mapped the approximate distribution of major coiling types (e.g., planispiral and conispiral) and higher taxonomic groups (pelecypods, gastropods, ammonoids, brachiopods, etc.) Similarities between simulated and real morphologies suggested that the geometric parameters chosen were "correct," or at least sufficient to encompass variation among many higher taxa.

Like Thompson before him, Raup was struck by the sufficiency of these few parameters in simulating most known coiling types. But unlike Thompson, he noted the vast regions of the theoretically possible morphospace that lacked any known representatives. Raup argued that these vacancies were mechanically unworkable (e.g., bivalves with interpenetrating umbones); functionally maladaptive (nonoverlapping "disjunct" whorls prone to breakage); or simply hadn't yet evolved. Without calling into question these alternate explanations, I argue that the apparent proportion of unoccupied morphospace results more from Raup's algebra than from the biology of taxa with coiled shells.

APPLICATION OF RAUP'S PARAMETERS IN EMPIRICAL STUDIES

Raup's coiling parameters have been used most effectively for surveys of planispiral groups (Ward 1980 and Saunders and Swan 1984 on coiled cephalopods; McGhee 1980 on biconvex brachiopods). The cephalopod studies generally establish whorl expansion, whorl displacement, and aperture shape (W, D, and S) using a few measurements made on the final whorl rather than an ontogenetic series of measurements. Without such repeated measurements, the possibility of allometric growth went unexplored in these studies. McGhee's (1980) study of biconvex brachiopods has been the only carefully documented analysis to date of ontogenetic variations in coiling parameters. He (1) measured the shell radius of the dorsal and ventral valves (umbo to the commissure along the sagittal plane) at 10° intervals; (2) calculated the instantaneous values of W (the regression slopes between adjacent radius measurements); and (3) reduced the ontogenetic series of expansion rates to a single average. McGhee found departures from simple logarithmic spiral growth that spanned nearly 10 orders of magnitude on a single individual, and he recorded whether whorl expansion rate increased or decreased during growth. It is worth noting that the variation in average expansion rates among major taxonomic groups can be less than the ontogenetic variation in expansion rate in a single individual. McGhee's findings have important implications for research on gastropods. Their reputation for geometric regularity notwithstanding, there are numerous gastropods that display striking midgrowth departures from ideal logarithmic coiling and common changes in coiling style during the final whorl if growth is determinant (especially in pulmonates). These allometric departures from the logarithmic spiral may prove to be valuable and sensitive taxonomic discriminators and should be considered in mapping evolutionary pathways.

Raup's coiling parameters have been used in few studies of gastropods, and these have paid little attention to allometric growth. Only Williamson (1981) and Kohn and Riggs (1975) have attempted to measure growth trajectories on individual specimens, from which allometric changes in coiling parameters could be established. Williamson presented plots of ontogenetic changes in W and T, but his analytical methods were not presented in detail. Kohn and Riggs established "rate of spire translation" and "rate of whorl expansion" (downward and outward migration of the suture line, respectively) by (1) measuring spire height and width on a whorl-by-whorl basis, (2) calculating the series of instantaneous growth rates, and (3) reducing each series to a single average. Any allometric patterns of growth that existed were thereby obscured. Vermeij (1971) did not attempt to establish allometry: Raup's T was approximated using the cotangent of half the apical angle and S with a single length/width ratio of the final aperture.

Subsequent Geometric Models of Gastropod Coiling. Since Raup's proposal, a few investigators have presented empirical studies of gastropod shape based on new simulation models of coiling geometry. These recent studies have attempted to model only one or two aspects of coiling, not the entire shell. Goodfriend (1983) established a measure for the peripheral angle and for whorl ontogeny but not for the shape and orientation of the true aperture. Heath (1985) explored the costs and benefits of whorl overlap using the cross-sectional areas of the aperture as it projects onto the midplane. Ekartane and Crisp (1983) proposed a compact mathematical representation of turbinate shells but one that is insensitive to the shape and orientation of the true aperture and to patterns of covariation and allometry during growth. In each of these three studies, a few related or morphologically similar taxa were compared. No wider taxonomic surveys were attempted, and so the general applicability of these measurement systems remains in doubt.

RAUP'S COILING PARAMETERS RECONSIDERED

Why haven't Raup's (1966) coiling parameters (or any such system of simulation parameters) been applied more widely and successfully to gastropods? Gastropod specialists commonly tout the value of studying the group "because they carry their ontogenetic history around with them." Yet the coiling parameters designed to simulate these ontogenies have not been used in their analyses. I will discuss briefly three possible reasons for the lack of such follow-up studies.

Lack of Biological Detail. The answer may lie in the fundamental differences between simulation and analysis as approaches to morphology.

Raup's simulation scheme is purposely free of specific advice regarding anatomical features to be used as measurement landmarks. In their economy, simulation parameters show how little information is needed to generate diverse forms. Measurement parameters require some added (but still minimal) specifications that will add biological realism, as the following examples illustrate.

Raup's coiling parameters were sufficient to produce axial cross-sections, not external views of shells (but see Savazzi 1985). In nature, however, two shells may be identical in cross-section and differ geometrically in other ways, such as the inclination of the aperture. Further, Raup calculated T using the center of the generating curve. Since the center of real apertures cannot be located on inspection, a widely recognizable conchological landmark would be a more practical point of reference. These added specifications would help in empirical studies, but they detract from the simplicity of a simulation model without adding insight. It is possible that no single scheme can accomplish both goals.

Logistical Difficulties. Several more concrete problems may have blocked wider use of Raup's simulation framework. Complete measurement of a shell's ontogeny requires cutting axial sections of shells, a laborious and partially destructive process. Perhaps the need for axial sections explains the lack of attempts (1) to study ontogenetic changes in gastropod coiling (Williamson used radiographs in his 1981 study, not axial sections) and (2) to measure Raup's whorl displacement rate D in gastropods. This stands in contrast to a considerable body of data on D in planispiral taxa such as ammonoids (Ward 1980; Saunders and Swan 1984) and brachiopods (McGhee 1980), which require no preparation before making the necessary measurements.

Lack of Satisfying Results. Proving the lack of something is a difficult task, so I will make the unsupported assertion that anyone calculating Raup's coiling parameters for gastropods has been disappointed to find how little variation there is among taxa. Values of W are generally concentrated below 2.0, and variation in D and T is anything but surprising: tall shells have high values of T, and shells with wide open umbilici have high values of D. As Raup showed in his original text (1966: 1184; fig. 4), gastropods occupy a relatively restricted flat ellipsoid of morphospace close to the W = 1 plane.

In the following section, I argue that this apparent restriction of gastropod coiling to a relatively small region of Raup's morphospace results from the algebraic construction of the coiling parameters. I then propose a modification of the parameters that removes the restriction and present a pilot survey of gastropod taxa with which to compare the two sets of parameters.

RAUP'S PARAMETERS REVIEWED

Whorl expansion rate, W - The general formula for an equiangular spiral is

$$r = a^{\Theta} \tag{1}$$

where r is the radius from the coiling axis to a point on the spiral, Θ is the angular position of the point (expressed throughout this article as whorl number, not radians or degrees), and a is a constant greater than 1. This produces a close facsimile of the three-dimensional spiral suture visible as a two-dimensional projection when viewed apically (down onto the apex of the spire). Both Thompson and Raup modified this form to add the scale factor r_O, the radius at whatever stage is defined to be $\Theta = O$. The general equation presented by Thompson and Raup was:

$$r_{\Theta} = r_O W^{\Theta} \tag{2}$$

Any two radius measurements were sufficient to estimate W, using "any point on the generating curve where a known angular distance separates it from the initial generating curve" (Raup 1966: 1180). Equation (2) reduces to:

$$W = \frac{r_{\Theta}}{r_O} \tag{2a}$$

when the two radius measurements are made one whorl apart. This has led to the practice of calculating W and other coiling parameters as a simple ratio of any two measurements taken one whorl apart (i.e., the slope of the regression line that connects only two points). Whorl expansion rate is a dimensionless number that reflects the percentage increase in radius per whorl rather than the absolute increase in radius. Defining $W = e^a$ and $b = \ln r_O$ leads to the transformation of (2) into:

$$\ln r_{\Theta} = b + a\,\Theta \tag{3}$$

which has the familiar form of a log-linear bivariate regression. Whorl expansion rate can be calculated as the antilog of a, the regression slope of $\ln r$ versus Θ.

Whorl translation rate, T, and displacement rate, D. To make the presentation of coiling parameter values as straightforward as possible, Raup scaled two coiling parameters (D and T) in terms of a third (W).

That is, he expressed the downward and outward migration of the whorl *relative* to the rate of whorl expansion. Rather than expressing T and D in terms of distances migrated per whorl, Raup made them dimensionless numbers that were scaled according to W.

Raup defined T in his equation for the vertical position of a point on the generating curve as it migrates parallel to the coiling axis.

$$Y_\Theta = Y_O W^\Theta + r_c T (W^\Theta - 1) \tag{4}$$

where the radius from the coiling axis to the center of the whorl is r_c when $\Theta = O$. When T is zero whorl translation contributes nothing to y, and vertical migration of a point on the generating curve is due entirely to the whorl expansion term. When T is not zero, its contribution to vertical migration is scaled according to W (fig. 10.1A).

Whorl displacement rate D originally was described as "position of the generating curve in relation to the axis" (Raup 1966: 1179) and "is defined as the ratio between the *r*-value [radius out from the coiling axis] of the axial margin of the generating curve and that of the outer margin. D is thus zero when the generating curve is in contact with the coiling axis (as in many gastropods) and increases as the generating curve moves away from the axis (as in serpenticone ammonoids)" (Raup 1966: 1181; comment in brackets added). Expressed this way, D is not a rate of change among whorls but a static ratio between measurements made on a single whorl (fig. 10.1B). Like whorl translation rate, displacement rate will be affected by variations in whorl expansion rate. In other words, the axial margin of two shells may each move away from the axis at the same rate (measured as mm/whorl),

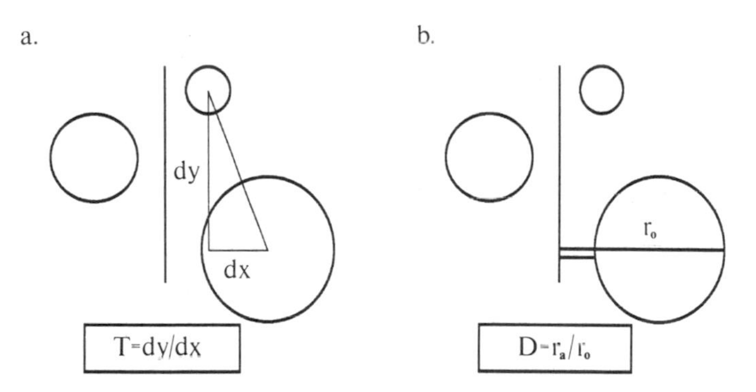

Figure 10.1. Schematic representation of two of Raup's coiling parameters. A, T, whorl translation rate, defined as the per whorl motion parallel to the coiling axis divided by its motion away from the axis. B, D, whorl displacement rate, defined as the ratio between r_a (radius from the coiling axis to the axial margin of the whorl) and r_o (radius to the outer margin).

but if they have different whorl expansion rates, their D values will differ.

Are the algebraic details of Raup's parameters important for interpreting the occupation of morphospace? Even Raup recognized the importance of their construction when he said: "the geometric model and the computer constructions based on it can be used to visualize the total spectrum of geometrically possible shell forms. This spectrum takes the form of a four-dimensional space with one dimension for each of the basic parameters. *It is assumed that each parameter is independent of the others*" (Raup 1966: 1181; emphasis added).

PROPOSAL FOR NEW COILING PARAMETERS

The proposed coiling parameters depart from Raup's scheme in one fundamental way. Each of these parameters independently describes the behavior of a different, easily recognizable point on each whorl. Each parameter is a log-linear regression slope of a distance associated with each reference point versus whorl number. The effects of specimen size on these distances are thereby normalized by the logarithmic transformation, rather than by using the measure (shell radius) on which one parameter (Raup's W) is based as a "size correction" for the other parameters. I propose a set of apertural measures in a later section, which, together with the proposed coiling parameters, comprise a generalized system for measuring gastropod architecture.

Standard orientations described. All the measurements needed to calculate coiling parameters can be taken with the shell in a single orientation—the *axial plane medial view* (figs. 10.2 and 10.3). Unfortunately, this involves cutting the shell through the coiling axis. Radiographs provide internal views without damaging the shell, but I am not convinced that the measurements they provide are as accurate as those made on medial shell sections. The primary problem is that secondary deposits on the inside of the apical whorls obscure the original surface of the whorl's interior on radiographs. The original shell layers can be distinguished from apical infilling on cut sections, however. Williamson (1981) used radiographs for establishing estimates of W and T but not D, perhaps because of this inability to locate the inner margin of the whorl on radiographs.

Previous investigators have avoided shell sectioning by measuring the motion of the suture line as seen in apical view (looking down on the apex along the coiling axis; fig. 10.4) or lateral view (looking at the side of the spire perpendicular to the coiling axis, e.g., Kohn and Riggs, 1975; see their fig. 2D). There are four drawbacks to this approach. First, using a single landmark (in this case the suture) to estab-

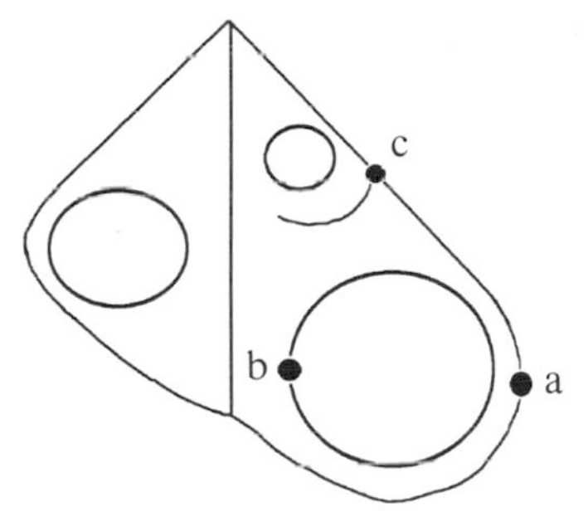

Figure 10.2. Landmarks used in measuring distances from which new coiling parameters were calculated. a, Periphery, the point on the outer surface of the shell farthest from the coiling axis. b, Axial margin, the point on the inner surface of the whorl closest to the coiling axis. c, Point of adherence, the point on the outside of the shell marking the suture between whorls.

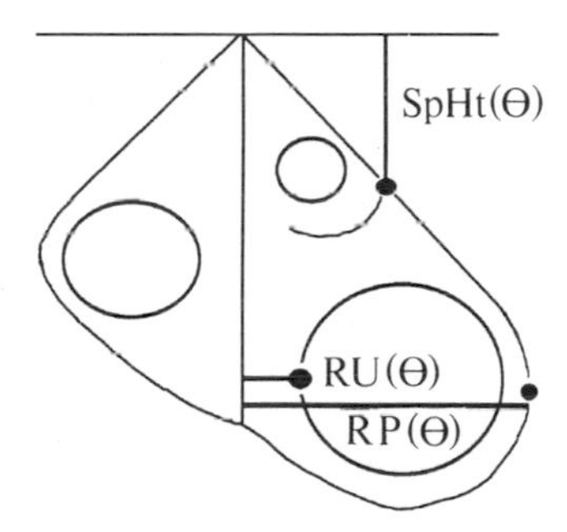

Figure 10.3. Distances used in calculating new coiling parameters. RP(Θ), radius to the periphery, measured as a perpendicular distance from the coiling axis; RU(Θ), umbilical radius, measured as the perpendicular distance from the coiling axis to the axial margin; SpHt(Θ), spire height, measured parallel to the coiling axis from the projection of the apex to the point of adherence. Θ is whorl number.

lish two coiling parameters (W and T) makes them nonindependent, as described previously. Second, apical and lateral views provide no data on whorl displacement (expansion of the umbilicus or columella). Third, measurements made on the shell exterior force the investigator to locate the apex (in apical view) or the coiling axis (in lateral view) in order to measure distances from each to the suture line. In practice, the locations of both the apex and the coiling axis are not obvious from the outside of the shell, and investigators usually just approximate their locations arbitrarily. I have constructed five such arbitrary "rules" for locating the apex (see fig. 10.4). Each of these definitions of the apex resulted in radius measurements with an oscillating error term, due to

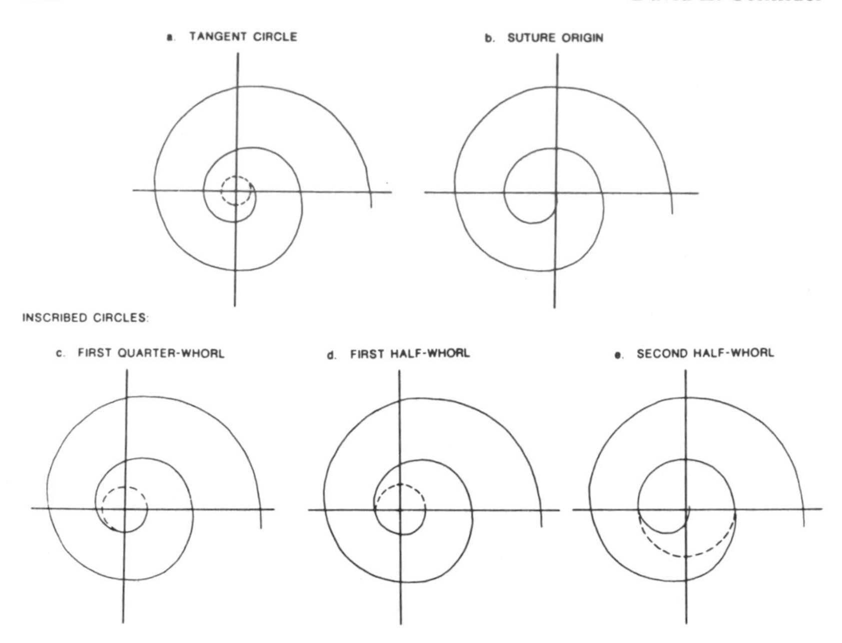

Figure 10.4. Five methods for estimating the position of the coiling axis in apical view. A, Center of a circle tangential to the suture where it first appears (radius of the circle equal to one-third the distance from the suture's origin to the suture half a whorl later). B, First appearance of the suture line. C, Center of a circle superimposed on the first quarter-whorl. D, center of a circle superimposed on the first half-whorl. E, Center of a circle superimposed on the second half-whorl.

mislocation of the apex (fig. 10.5). Fourth, the suture line is not usually the outermost point on the aperture. Whorl expansion rate based on the suture line will therefore discount any whorl expansion taking place peripheral to the suture line or unseen due to whorl overlap (except for the final whorl).

Reference points specified. Truly anatomical landmarks (developmentally homologous points) are difficult to find on the aperture. I have resisted the temptation to use reference points that are difficult to locate (such as the whorl's center of gravity) or too limited to particular taxa (e.g., elements of ornamentation). Instead, I have opted for what Bookstein (1978) termed "extremal landmarks." These are (1) the *periphery* (fig. 10.2, point a), being the point on the *outside* of the shell that is the greatest perpendicular distance from the coiling axis; (2) the *axial margin* (fig. 10.2, point b), being the point on the *inner* surface of the original whorl (ignoring subsequent internal deposits) closest to the coiling axis; and (3) the *point of adherence* (fig. 10.2, point c), being the point on the outside of the shell that marks the adherence between whorls (the suture traces the ontogeny of this landmark).

These reference points are not proposed as landmarks that are homologous among taxa, or even among the individuals of a population. These points may even represent different anatomical points on the mantle edge at different whorls of an individual if the shape and/or orientation of the aperture changes through growth. They are proposed as landmarks for an architectural, not an anatomical, analysis.

Measurements Described. The landmarks specified previously give rise to the following distances: (1) RP(Θ), radius to the periphery, is mea-

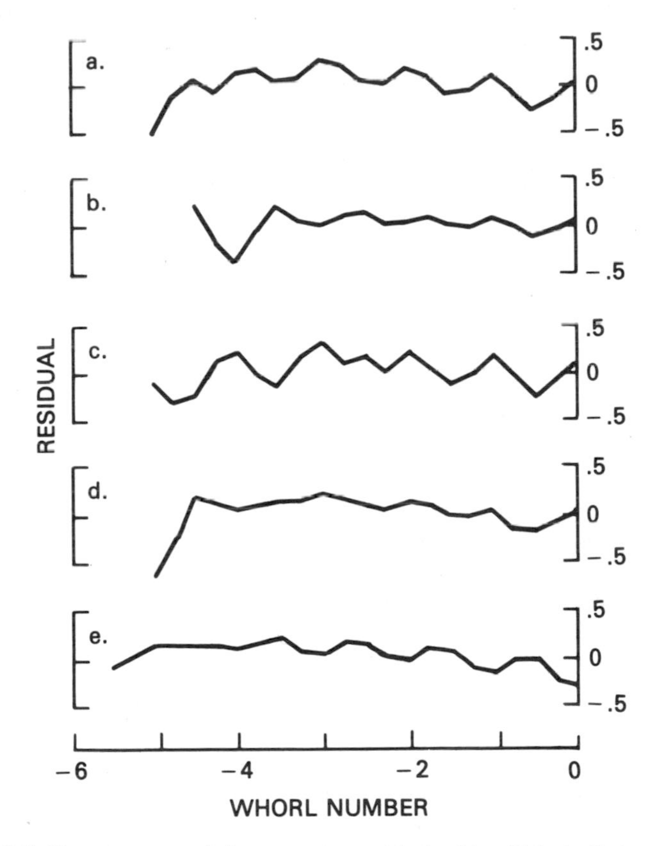

Figure 10.5. Error terms revealed as regression residuals of A to E, ln (radius), on whorl number. A to D, Different methods for locating the coiling axis in apical view (see fig. 10.4) used on a single camera lucida drawing. A, Tangent circle. B, Suture origin. C, Circle inscribed on the first half-whorl. D, Circle inscribed on the second half-whorl. E, Residuals for radius measurements made on a slightly misaligned shell. All data from *Trochus intextus*.

sured from the coiling axis to the *periphery* of whorl Θ, perpendicular to the axis; (2) RU(Θ), umbilical radius, is measured from the coiling axis to the *axial margin* of whorl Θ, measured perpendicular to the axis, and (3) SpHt(Θ), spire height, is measured from the apex to the *point of adherence* between whorls Θ and its predecessor, measured parallel to the coiling axis. Figure 10.3 illustrates these distances.

Whorl number Θ can be designated in two ways. The most general method is also arbitrary: number the periphery of the final whorl as Θ = O, with *n* decreasing into negative numbers toward the apex. This will result in Y-intercepts (expected measurement values at Θ = O; see following discussion of regression methods) that are highly variable among shells of different size. The second method is to designate as Θ = O a standard developmental stage, such as the protoconch/teleoconch boundary. Y-intercepts will have comparable meanings among samples, but it will be more difficult to identify such a comparable stage on all shells.

Coiling parameters defined. I propose the following replacement set of coiling parameters: (1) W, *shell expansion rate*, the antilog of the regression slope between log RP(Θ) and whorl number, Θ; (2) *M*, *suture migration rate*, the antilog of the regression slope between log SpHt(Θ) and whorl number, Θ; and (3) *U*, *umbilical expansion rate*, the antilog of the regression slope between log RU(Θ) and whorl number, Θ. A fourth coiling parameter, A, *apertural expansion rate*, may also prove useful. It can be calculated as the antilog of the regression slope between the logarithm of the width of the aperture, RP(Θ) – RU(Θ) (this distance will include the thickness of the outer shell margin) and whorl number, Θ. Alternatively, the cross-sectional area of the aperture can be used in place of apertural width for calculating A.

Measurement procedures. To reveal a medial view of the shell in axial plane orientation, each specimen needs to be embedded, cut, and polished. Attaining this view involves the following steps: (1) creating a bioplastic or epoxy "stub" in a greased cardboard box tall enough to accommodate the specimen's height; (2) removing the stub from the box and drilling a pit in the center of the stub large enough for the apex of the shell; (3) gluing the specimen into the pit with (a) the coiling axis perpendicular to the base of the stub and (b) the shell rotated so that a line drawn between the apex and the reference whorl Θ = O is parallel to one side of the stub; (4) returning the stub and specimen to the box and, with the apex down, filling the box with enough embedding medium to cover the aperture partially; (5) placing the box under vacuum before the medium hardens in order to fill the interior of the whorls;

(6) allowing the block to harden and then adding more embedding medium until the shell is encased; (7) sectioning the specimen and grinding the block to reveal a plane through the coiling axis and the point on the shell that defines $\Theta = O$. RP(Θ), RU(Θ), and SpHt(Θ) can now be measured at half-whorl increments.

Regression techniques for estimating coiling parameters. Equations 2, 2a, and 3 represent three alternate methods for estimating Raup's W, or any of the above four coiling parameters, by substituting the appropriate distance for r. For instance, any two values for umbilical radius RU(Θ) are sufficient for calculating U, as long as the number of whorls separating them is known (use equation 2). If the two measurements happen to be exactly one whorl apart, a simple ratio estimates U (equation 2a). Use of equations 2 and 2a assumes that growth is isometric, so two measures at any two different growth stages will give identical coiling parameters.

In the most general case, a log-linear regression can be calculated (using equation 3 as the regression model) from measurements made at any known whorl numbers. Least-squares regression is a reasonable method in the present case, because the independent variable, whorl number Θ, is measured without error. The coiling parameter would be calculated as the antilog of the slope a, and the antilog of b would be the predicted value of the shell measure (such as RU[Θ]) at $\Theta = O$.

Detecting allometric growth. Systematic departures from the ideal logarithmic spiral would constitute allometry, and these would be visible on plots of regression residuals versus whorl number. Polynomial expansion of (3) would be required to estimate these departures in the form of higher-order regression coefficients. Departures from ideal logarithmic growth will be established using the orthogonal polynomial regression model

$$\ln Y = b + a_1\Theta + a_2(\Theta - \bar{\Theta})^2 \quad (5)$$

where Y is a whorl-specific shell measure (either RP[Θ], RU[Θ], or SpHt[Θ]), the intercept b is its predicted value at $\Theta = O$, a_1 is the linear slope coefficient (and the logarithm of W, U, or M, respectively), and a_2 is the second-order coefficient representing degree of curvilinearity (concave-up if allometry is positive, concave-down if negative, and zero if growth is isometric); $\bar{\Theta}$ is the average whorl number for the measurements taken (see Sokal and Rohlf 1981: 677–83). The rate of change in the coiling parameter itself (rather than its logarithm) is the antilog of a_2.

I have chosen the orthogonal form of the second-order term shown in equation (5) rather than the simpler form

$$\ln Y = b + a_1\Theta + a_2\Theta^2 \tag{6}$$

which is akin to the one used by McGhee (1980: 76; equation 15) and McKinney (1984). In this form, the addition of the second-order term affects the first-order slope, because Θ and Θ^2 are not statistically independent.

In interpreting the results of regression analyses, it is common to use the *t*-statistic for testing the first- and second-order coefficients for departures from zero slope and zero curvilinearity, respectively. This would be inappropriate here, because measurements made in a growth series are not independent trials. Nevertheless, *t*-values can be used as an inverse metric for the total among-whorl departure from the regression model. Data that have near-perfect correspondence to a model would have large *t*-values. In contrast, correlation coefficients would always be very close to 1.0 for growing systems such as shell expansion. The nonlinear distribution of the correlation coefficient, *r*, makes this small range of variation less informative than *t*. It may be desirable to test for differences in slope or curvilinearity between samples. Analysis of covariance would be the appropriate method (see Sokal and Rohlf 1981: 509–30).

A COMPARISON OF COILING PARAMETERS

A survey of prosobranch gastropod form was undertaken for three reasons: (1) to compare the properties of Raup's coiling parameters with the properties of the parameters proposed here, (2) to explore covariation among parameters as expressed in a variety of underived (i.e., nonsiphonate) taxa, and (3) to explore constraints within a more derived taxonomic clade. Archaeogastropods and nonsiphonate caenogastropods were sampled to cover the greatest possible range of shapes (haliotids to turritellids) expressed by nonsiphonate prosobranchs. Cerithiids were sampled as a coherent taxonomic group, again with the goal of representing its full (though more limited) morphological diversity. I make no claim that this survey provides exhaustive coverage of all gastropod superfamilies. For example, it excludes pulmonates and opisthobranchs and a vast array of siphonate caenogastropods.

This survey therefore has two distinct flavors. Archaeogastropods and nonsiphonate caenogastropods represent a set of taxa that are ecologically less specialized and anatomically less derived: most of these nonsiphonate taxa are grazers with a preponderance of low- to medium-spired shells. Cerithiids are probably a monophyletic clade with a narrower range of form: all taxa are tall and siphonate. The survey of archaeogastropods and nonsiphonate caenogastropods explores the patterns of covariation among a wide range of taxa. The survey of cerithiids explores the nature of constraint within a more coherent and stereotyped taxon. Rather than discussing particular taxonomic groups, the samples have been separated into three "evolutionary grades" during the first part of the analysis, in which only coiling parameters are considered): (1) archaeogastropods, (2) caenogastropods that lack anterior siphons, and (3) siphonate caenogastropods, represented only by cerithiids. Current phylogenetic hypotheses for prosobranchs are not sufficiently robust to support a true evolutionary separation of species into clades (Golikov and Starobogatov 1975).

In the second part of the analysis (in which coiling parameters and apertural features are discussed), I distinguish four architectural "forms" of shells in the following discussion: (1) nonsiphonate shells with low spires (spire index equal to or less than 0.75), (2) nonsiphonate shells with moderate spires (spire index greater than 0.75 and less than 1.25), (3) nonsiphonate shells with tall spires (spire index equal to or greater than 1.25), and (4) siphonate shells (all of which are cerithiids and have spire indices greater than 1.18 in this study). Spire index is defined as the total shell height parallel to coiling axis divided by the maximum shell width, measured perpendicular to the coiling axis (Cain 1977).

Modern shells (no fossils) were sampled from the collections of the Yale Peabody Museum's Division of Invertebrate Zoology (listed with their localities in table 10.1, data listings in tables 10.1 and 10.2). Most taxa are represented by a single specimen, but two morphological extremes were sampled from specimen lots that displayed obvious variation in coiling.

Results

Sixty specimens representing 53 species from dozens of superfamilies are displayed on three-dimensional perspective plots of Raup's coiling parameters W, D, and T (fig. 10.6A) and the revised parameters W, U, and M proposed here (Figure 10.6B). Data for these figures are presented in table 10.2. Although I will not use the results of higher-order regression coefficients, it is worth noting that nearly all shells showed a slight negative allometry in W, U, and M (i.e., plots of RP, RU, and SpHt are slightly but consistently concave-down). Nerites and haliotids were the only exceptions, being consistently concave-up, suggesting a pervasive difference in coiling parameters related to their abilities to resorb and reshape the interior of the spire.

Table 10.1 Taxonomic, locality, and morphological data

Taxon	Locality	Spire Index	Form	Apertural Features		
				InAn	ApSh	ApAn
Grade 1: Archaeogastropods						
Astraea longispina (sh)	St. Croix	0.47	1	63.03	1.19	−15.82
Antraea longispina (tall)	St. Croix	0.56	1	47.02	0.94	−12.82
Bankivia sp.	Philippines	2.32	3	18.01	1.50	17.01
Calliostoma conulus (lge)	Mediterranean	0.94	2	57.03	0.78	34.02
Calliostoma conulus (med)	Mediterranean	1.12	2	52.03	0.75	20.01
Calliostoma conulus (sm)	Mediterranean	1.06	2	43.02	0.85	19.01
Clanculus sp.	?	0.71	1	55.03	1.16	53.03
Euchelus canaliculatus	Calapan, Philippines	1.06	2	40.02	1.09	4.00
Gibbula umbelicaris	Malta	0.88	2	34.02	1.03	−7.82
Haliotis tuberculata	Greece	0.34	1	43.00	1.32	−65.00
Liotia radiata	Bahama	0.69	1	38.02	1.00	0.00
Lischkeia ottoi	USFC Sta 2687	1.00	2	32.02	0.97	−25.83
Margarites costalis	Eastport, Maine	0.82	2	23.01	1.18	9.00
Margarites pupillus	Puget Sound	0.86	2	42.02	1.07	5.00
Monodonta canalifera	Philippines	0.89	2	53.03	0.87	−36.84
Nerita albicella	Philippines	0.74	1	5.00	1.55	8.00
Nerita latissima	Nicaragua	1.13	2	20.00	1.49	5.00
Nerita lineata	Ceylon	0.97	2	33.02	1.76	3.00
Norrisia norrissii	Catalina Is.	0.78	2	39.02	1.18	−19.83
Phasianella australia	Australia	1.70	3	23.01	1.60	19.01
Stomatia phymotis	Philippines	1.06	2	23.00	1.49	−22.00
Tegula excavata	West Indies	0.82	2	66.03	1.20	−61.85
Tegula gallina	San Pedro, Calif.	0.78	2	60.03	0.97	−27.83
Tricolia speciosa	Mediterranean	1.82	3	19.01	1.34	8.00
Trochus coelatus	St. Croix	0.89	2	54.03	1.08	−55.85
Trochus intextus	Pearl City, Hawaii	0.91	2	46.02	1.00	15.01
Turbo cinerea	Mindoro, Philippines	0.83	2	36.02	1.04	−37.84
Turbo coronata	East Indies	0.70	1	29.01	0.97	−12.82
Turbo fluctuosus	Gulf of California	1.09	2	24.01	1.11	10.01
Tubo niger	Callao, Peru	0.88	2	48.02	1.16	−22.83

Species	Locality	Spire index	Form	InAn	ApAn	ApSh
Turbo petholatus	Solomon Islands	0.91	2	39.02	0.95	−25.83
Umbonium vestarium	Philippines	0.61	1	35.02	1.01	−20.83
Grade 2: Nonsiphonate Caenogastropods						
Architectonica nobilis	Gulf of California	0.57	1	27.01	0.67	16.01
Bithynia tentaculata	England	1.55	3	23.01	1.41	15.01
Heliacus sp. (high)	Red Sea	0.75	1	24.01	1.03	21.01
Heliacus sp. (low)	Red Sea	0.61	1	22.01	1.02	13.01
Lacuna divaricata	Olga, Washington	1.47	3	27.01	1.34	17.01
Littorina irrorata	Dunedin, Florida	1.35	3	30.02	1.45	22.01
Littorina littorea	Squantum, Mass.	1.16	2	22.01	1.22	10.01
Littorina obtusata (sh)	Birch Is., Maine	1.05	2	24.01	1.36	8.00
Littorina obtusata (tall)	Birch Is., Maine	1.25	2	27.01	1.31	10.01
Littorina saxatilis	Quincy, Mass.	1.27	2	26.01	1.37	14.01
Littorina varia	Panama	1.58	3	43.02	1.41	13.01
Littorina ziczac (dark)	St. Kitts	1.46	3	30.02	1.34	15.01
Littorina ziczac (light)	St. Kitts	1.63	3	28.01	1.55	14.01
Modulus modulus	Captiva, Florida	0.82	2	47.02	1.09	−9.82
Nodolittorina tuberculata	West Indies	1.29	2	24.01	1.20	10.01
Rissella melanostoma	Yeppon Island	0.93	2	66.03	1.01	18.01
Tectarius muricatus	West Indies	1.36	3	34.02	1.26	11.01
Truncatella valida	Mindoro, Philippines	2.96	3	—	1.50	29.01
Turritella banksi	Panama	2.89	3	14.01	1.11	31.02
Turritella communis	Catalonia	3.99	3	9.00	1.24	20.01
Grade 3: Cerithiids (Siphonate Caenogastropods)						
Cerithidea obtusa	Shark's Bay, Aus.	2.25	4	2.00	1.11	34.02
Cerithium gemmatum	Panama	2.87	4	−3.82	1.91	40.02
Cerithium morus	Ceylon	1.65	4	12.01	1.50	33.02
Cerithium muscarum	Florida	2.66	4	7.00	1.64	34.02
Planaxis nucleus	St. Martin's, W.I.	1.75	4	25.01	1.43	18.01
Planaxis undulata	Ceylon	1.57	4	23.01	1.43	21.01
Potamis(?) *fuseatum*	Senegal	2.29	4	−26.83	1.22	40.02
Rhinoclavis asper	Friendly Islands	2.69	4	−1.82	1.68	32.02
Rhinocoryne humboldti	Pearl Is., Panama	1.73	4	—	1.30	21.01
Telescopium telescopium	N. Queensland, Aus.	1.64	4	27.01	1.14	68.03

Note: Spire index is total shell height/width; form is a classification based on spire index; InAn, inclination angle; ApAn, aperture angle; ApSh, aperture shape. Names and localities are reported exactly as they appear on Yale Peabody Museum Division of Invertebrate Zoology labels.

Table 10.2. Values for coiling parameters

Taxon	Proposed Parameters			Raup's Parameters		
	W	U	M	W	D	T
Grade 1: Archaeogastropods						
Astraea longispina (sh)	2.02	1.70	3.88	2.14	0.30	0.36
Astraea longispina (tall)	2.10	2.19	3.31	1.90	0.21	0.72
Bankivia sp.	1.40	1.08	1.84	1.43	0.07	3.39
Calliostoma conulus (lge)	1.63	1.19	1.53	1.61	0.01	0.83
Calliostoma conulus (med)	1.48	1.44	1.50	1.43	0.13	1.43
Calliostoma conulus (sm)	1.47	1.54	1.61	1.44	0.11	1.21
Clanculus sp.	1.93	1.60	2.20	1.63	0.08	0.77
Euchelus canaliculatus	2.29	1.58	2.73	2.12	0.07	0.79
Gibbula umbelicaris	1.84	2.05	1.82	1.64	0.22	0.96
Haliotis tuberculata	5.95	1.00	1.35	—	—	—
Liotia radiata	1.86	2.04	3.64	1.46	0.32	1.47
Lischkeia ottoi	1.82	1.05	1.94	1.97	0.06	0.77
Margarites costalis	1.77	1.87	2.74	1.62	0.15	1.12
Margarites pupillus	1.94	1.42	2.37	1.77	0.15	0.98
Monodonta canalifera	1.72	1.40	2.21	1.58	0.04	1.10
Nerita albicella	—	—	—	—	—	—
Nerita latissima	3.90	1.00	5.50	3.90	−0.30	0.71
Nerita lineata	3.04	1.00	3.69	3.04	0.00	0.34
Norrisia norrissii	2.50	2.27	2.28	2.32	0.20	0.23
Phasianella australia	1.76	0.83	1.99	1.68	0.03	1.53
Stomatia phymotis	—	—	—	—	—	—
Tegula excavata	1.90	1.42	2.63	1.63	0.10	1.09
Tegula gallina	1.99	1.62	2.19	1.83	0.14	0.58
Tricolia speciosa	2.08	0.84	3.18	2.06	0.00	1.72
Trochus coelatus	1.99	3.20	2.37	1.97	0.13	0.92
Trochus intextus	1.61	1.08	1.84	1.74	0.02	0.96
Turbo cinerea	2.64	2.64	3.21	2.69	0.23	0.23
Turbo coronata	2.41	2.02	3.67	2.36	0.15	0.17
Turbo fluctuosus	2.20	1.85	2.07	2.34	0.13	0.54
Turbo niger	2.34	1.28	3.25	—	—	—
Turbo petholatus	2.71	0.67	2.72	2.59	0.00	0.43
Umbonium vestarium	1.80	1.73	1.85	1.78	0.26	0.38

Grade 2: Nonsiphonate Caenogastropods						
Architectonica nobilis	1.63	1.63	2.05	1.51	0.35	0.74
Bithynia tentaculata	1.59	0.99	2.17	1.33	0.02	3.22
Heliacus sp. (high)	1.75	1.74	2.41	1.67	0.28	0.58
Heliacus sp. (low)	1.61	1.44	2.54	1.48	0.29	1.05
Lacuna divaricata	1.88	1.61	2.37	1.67	0.07	1.57
Littorina irrorata	1.76	1.92	1.89	1.59	0.09	1.10
Littorina littorea	1.82	1.80	2.17	1.65	0.21	1.04
Littorina obtusata (sh)	2.10	2.27	2.27	1.96	0.18	0.54
Littorina obtusata (tall)	1.70	1.69	2.52	1.35	0.07	1.31
Littorina saxatilis	1.83	1.63	2.32	1.71	0.12	1.28
Littorina varia	1.97	1.79	1.99	1.72	0.06	1.55
Littorina ziczac (dark)	1.71	2.03	2.13	1.82	0.14	1.34
Littorina ziczac (light)	1.75	1.44	1.87	1.70	0.05	1.62
Modulus modulus	2.00	2.19	2.20	1.88	0.15	0.74
Nodolittorina tuberculata	1.85	1.07	1.73	2.05	−0.18	0.90
Rissella melanostoma	1.71	0.93	2.04	1.25	0.05	2.35
Tectarius muricatus	1.57	1.34	1.95	1.39	0.12	2.15
Truncatella valida	1.14	1.09	1.66	1.07	0.07	22.92
Turritella banksi	1.24	1.13	1.27	1.50	0.00	2.61
Turritella communis	1.19	0.96	1.19	1.32	0.08	4.33
Grade 3: Cerithiids (Siphonate Caenogastropods)						
Cerithidea obtusa	1.24	1.01	1.39	1.63	0.02	1.48
Cerithium gemmatum	1.18	1.19	1.22	1.16	−0.28	3.01
Cerithium morus	1.39	1.23	1.68	1.41	0.06	2.45
Cerithium muscarum	1.21	1.15	1.41	1.36	−0.42	3.27
Planaxis nucleus	1.73	1.46	2.17	1.77	0.13	1.58
Planaxis undulata	1.74	1.18	1.89	2.13	0.02	1.05
Potamis(?) *fuseatum*	1.19	1.21	1.25	1.41	−0.09	2.09
Rhinoclavis asper	1.26	1.24	1.35	1.47	−0.24	2.30
Rhinocoryne humboldti	1.36	1.50	1.45	1.54	0.16	1.86
Telescopium telescopium	1.21	1.20	1.20	1.29	−0.23	1.98

Note: Newly proposed coiling parameters W, shell expansion rate; U, umbilical expansion rate; M, suture migration rate. Raup's (1966) coiling parameters W, whorl expansion rate; D, whorl displacement rate; T, whorl translation rate.

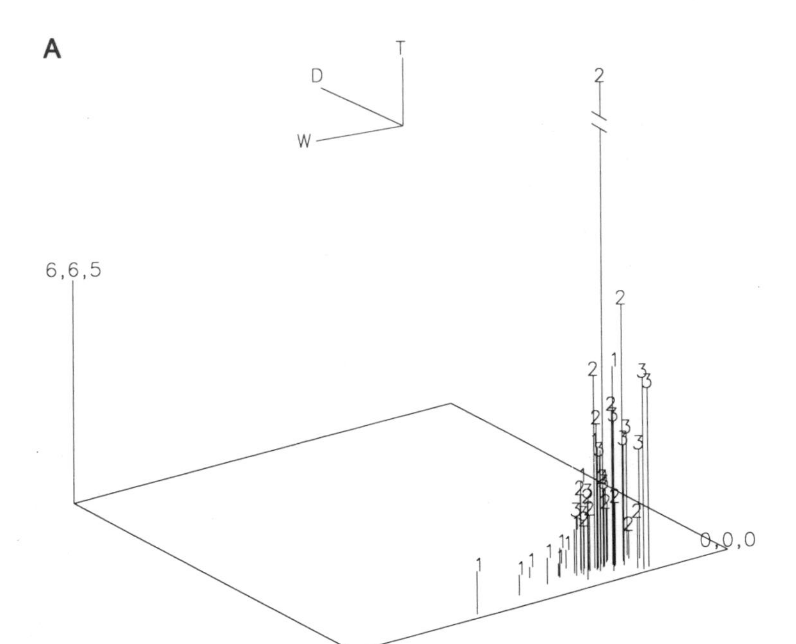

A
T
D
W
6,6,5
0,0,0

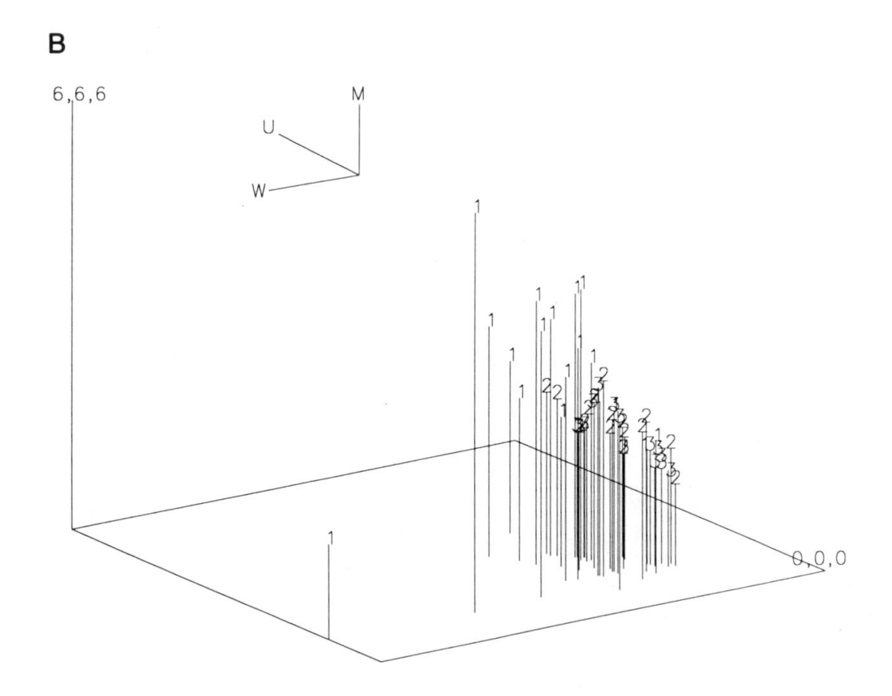

B
6,6,6
M
U
W
0,0,0

Values of Raup's parameters are confined to a relatively small region close to the plane of low D values (the maximum was 0.42) and an upper limit of W and T values shaped like a hyperbolic function (that is, W and T limits were inversely related). For instance, *Truncatella valida* had the highest *T*-value (22.9), partly because it also had the lowest W (1.07). This is the same pattern described by Raup (1966) as the distribution of gastropods in morphospace. Nearly all taxa have low D; high-spired caenogastopods (grades 2 and 3 as defined previously) have low W and high T, and low-spired archaeogastropods (grade 1) have low T and high W.

In contrast, the W, U, and M values calculated from the same data are more widely distributed in the interior of the cube defined by the three parameters. For instance, values of umbilical expansion rate U (the replacement proposed for Raup's D) varies from 0.67 to 3.2 (values of U below 1.0 indicate a decrease in RU(Θ) as growth proceeds, corresponding to a sinuate columella that "expands backwards" across the coiling axis onto the opposite side of the shell). Instead of the inverse relationship seem among Raup's parameters, there appears to be a direct relationship among the new parameters. Shells that are rapidly expanding at the periphery (high W) frequently have high rates of suture migration (M) and umbilical expansion (U).

The greatest variation in figure 10.6B thus separates slowly versus rapidly expanding shells rather than low- versus high-spired shells. Archaeogastropods (grade 1) are most widely distributed grade in figure 10.6B, including slow and rapid expanders. Nonsiphonate caenogastropods (grade 2) have a smaller distribution and are more concentrated in the region of slow expansion. Siphonate caenogastropods (grade 3) are obviously restricted to a very small region because they are repre-

◀ Figure 10.6. Plots of 62 shells representing 53 species and three evolutionary grades plotted on axes corresponding to (A) three coiling parameters defined by Raup (1966) and (B) three new coiling parameters proposed here. On each plot, the origin (0,0,0) is in the right-hand foreground and reference points (6,6,5) and (6,6,6) are added at the lefthand background. Specimens are labeled according to their grade: 1, Archaeogastropod. 2, Nonsiphonate caenogastropod. 3, Cerithiids (siphonate caenogastropod). A, Whorl expansion rate (W), displacement rate (D), and translation rate (T) were calculated from a pair of values of radius at the periphery (RP), radius of the umbilicus (RU) and spire height (SpHt), respectively. Each pair of measurements was made one whorl apart at the end of growth, as revealed on medial sections. B, Shell expansion rate (W), umbilical expansion rate (U), and suture migration rate (M) were calculated as the antilogs of migration slopes between whorl number and log(RP), log(RU), and log(SpHt), respectively. The vertical scale is broken on A for *Truncatella valida*, for which T = 22.9. The shell interiors of *Nerita* were resorbed, making calculation of U impossible. *Haliotis* was too rapidly expanding to provide sufficient whorl interiors for regression analysis. In these cases, U was defined arbitrarily as 1.0 (no expansion).

sented only by cerithiids, all of which display slowly expanding shells.

This finding suggests some interaction among the proposed coiling parameters that cannot be dismissed as an algebraic artifact. High-spired shells (siphonate and nonsiphonate) appear to be consistently slow in their expansion, whereas low-spired shells appear to include a variety of slow and rapid expanders. Before pursuing this pattern further, I will propose additional procedures for measuring other important aspects of the gastropod shell, especially those relating to the aperture. These will allow a more complete treatment of covariation and constraint among shell measures.

A DILEMMA IN THE ANALYSIS OF GASTROPOD COILING

Two components produce the geometry of a snail's shell: the aperture, being the growing margin through which the animal faces the environment, and the coiling behavior that describes how the aperture expands and moves through space. The coiling parameters established previously describe this expansion and motion but say nothing about the aperture itself. Only Linsley (1977 and 1978) and Vermeij (1971) have commented on covariation between coiling parameters and apertural morphology.

Students of gastropod shell geometry are faced with a dilemma. The shape, orientation, and coiling behavior of the aperture are, of course, ontogenetic trajectories, not static measurements. The true shape of the aperture at any point during growth can only be measured in the plane of the aperture, which in nearly all snails lies at some angle to the coiling axis. In contrast, the ontogenetic history of an aperture's coiling behavior can be measured most completely on an axial plane medial view, cut through the coiling axis (see fig. 10.3).

Herein lies the dilemma. Axial sections expose a measurable ontogenetic series of outlines at half-whorl intervals, but in most taxa they are distorted projections of the true aperture onto an axial plane. It is therefore impossible to establish both coiling parameters and the ontogenetic history of the true aperture unless the coiling axis lies on the plane of the aperture (i.e., the aperture is radial, *sensu* Linsley 1977). Specimens at different growth stages could be measured to construct a composite ontogenetic trajectory of the aperture within a population, but such a trajectory would also represent individual variation.

PROPOSAL FOR APERTURAL MEASURES

The problem of relating the aperture to its coiling history can be solved as follows: (1) measure the true shape of the aperture in the plane of the aperture on the final whorl, (2) establish the orientation of the true aperture relative to the coiling axis using several external views, and then (3) embed and section the shell to reveal an axial plane medial view for measuring the ontogenetic history of the projected aperture's shape, orientation, and coiling behavior. The shape and orientation of the true aperture and the projected apertures will have the same relationship throughout growth if the angle between the coiling axis and the apertural plane is constant throughout growth. This assumption can be tested directly by studying growth lines and indirectly by measuring the angles on separate shells at different growth stages.

Standard orientations described. Three shell orientations are proposed for characterizing the shape and orientation of the aperture (fig. 10.7). These are the *apertural plane view* normal to the plane of the aperture (figs. 10.7C and F), as suggested by Linsley (1977) and McNair et al. (1981); the *axial plane apertural view* (figs. 10.7A and D) normal to a plane defined by the coiling axis and the periphery; and the *axial plane side view* (figs. 10.7B and E) along a plane defined by the coiling axis and the periphery. These last two orientations are separated by 90°, and I have found that positioning a shell on a small wooden cube using modeling clay facilitates switching between these views.

Measurements described. I will describe the proposed variables in the order in which I normally measure them. Two variables are measured with the shell in apertural view (refer to fig. 10.7C and F): *MaxApHt*, maximum aperture height, and *MaxApWd*, maximum apertural width. MaxApWd is measured from below the point of adherence to the farthest point on the aperture. This second point has no distinction other than its position opposite the point of adherence on the aperture. I mark its position with a pencil for use as a landmark in another orientation. In shells with thick walls and a beveled apertural lip, this distance is measured from the slope break at the inside of the bevel (fig. 10.7F), because this inner outline is the meaningful one into which the animal withdraws ("clamping aperture" of McNair et al. 1981). These points define a standardized apertural long axis, though not necessarily the longest dimension of the aperture or a homologous anatomical axis. In most mesogastropods this axis will be oriented antero-posterior because the anal tube passes under the point of adherence. This axis will be transverse to the anatomical antero-posterior axis in many archaeogastropods.

MaxApWd is the greatest dimension perpendicular to the apertural long axis along which MaxApHt is measured. These two measures are

AXIAL PLANE APERTURAL VIEW

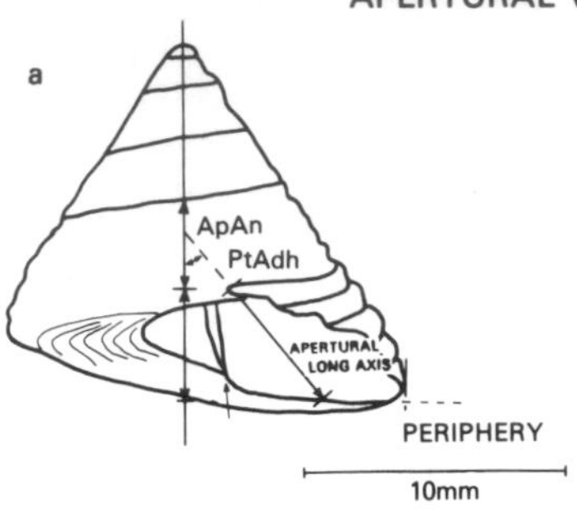

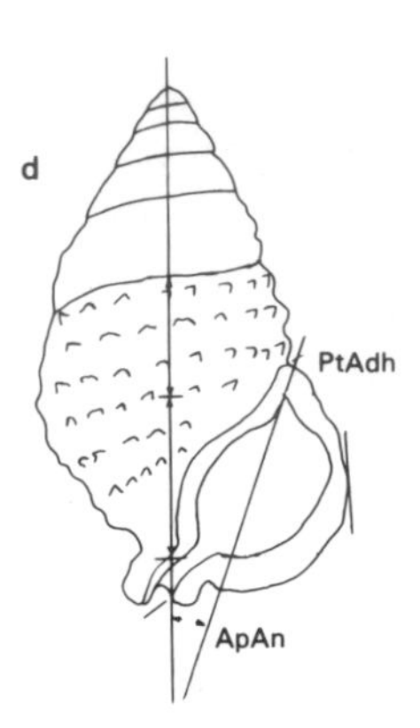

AXIAL PLANE SIDE VIEW

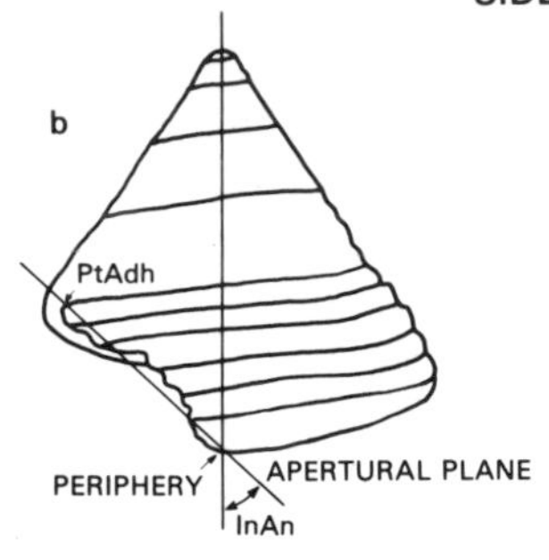

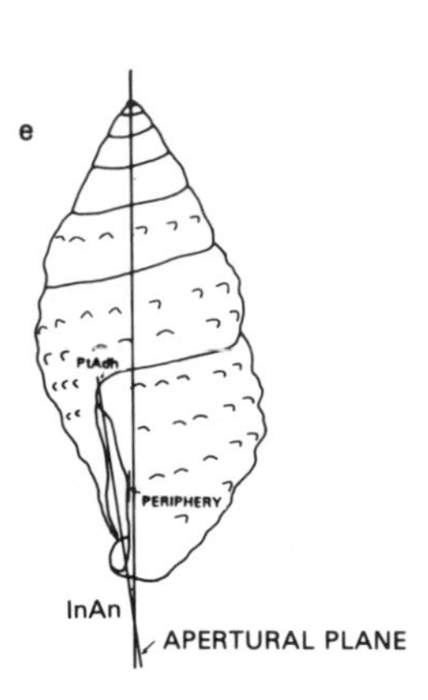

APERTURAL PLANE VIEW

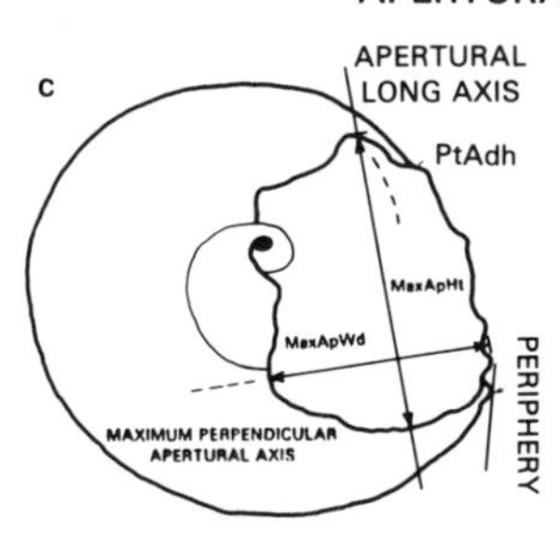

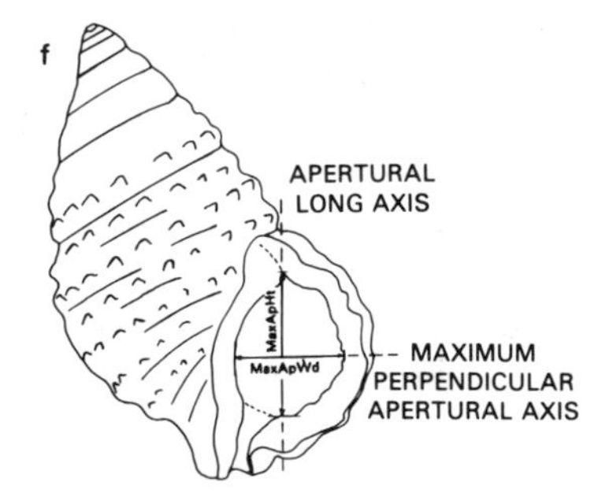

a crude estimate of apertural shape, and future studies will be more ambitious attempts to portray the apertural outline as a complex ellipse, perhaps using a harmonic series. The apertural shape index *ApSh* is calculated as MaxApHt/MaxApWd.

A single measurement is made with the shell in axial plane apertural view (figs. 7A and D): ApAn, apertural angle, is the angle between the coiling axis and the long axis of the aperture as defined by the point of adherence and its opposite point, as marked in apertural view. ApAn = 0 if the long axis (as seen in this view) is parallel to the coiling axis, positive if the long axis points away from the coiling axis as it passes through the point of adherence, and negative if it points toward the coiling axis. Apertural angle records the degree to which the longest dimension of the aperture (relative to the point of adherence) is "swung out" from the coiling axis (ApAn is negative, as in fig. 10.10H) or "tucked under" the whorl (ApAn is positive; see fig. 10.10J). There is no necessary relationship between ApAn and "angle D," defined by McNair et al. (1981) as the angle between the long axis of the foot and the long axis of the aperture, because the latter axis is not confined to pass through any standard landmark.

The only variable to be measured in axial plane side view is *InAn*, the inclination angle between the plane of the aperture and the coiling axis. This variable is equivalent to Vermeij's (1971) elevation angle "E" and Linsley's (1977) "angle of inclination."

APERTURAL FEATURES AND COILING PARAMETERS: COVARIATION AND CONSTRAINT

We now have six morphological axes along which we can view the diversity of gastropod architecture: three coiling parameters (W, U, and M), a crude but standardized measure of aperture shape (ApSh), and two angles that relate aperture shape to the coiling axis (ApAn and InAn). Each measure reflects a separate aspect of shell geometry, even

◀ Figure 10.7. Standard orientations and measurements for characterizing the aperture before sectioning. Each of these three orientations is shown for a shell with a highly inclined, nonsiphonate aperture (*Trochus intextus*, A to C) and another with a minimally inclined siphonate aperture with a thickened and beveled lip (*Cerithium morus*, D to F). ApAn, aperture angle; InAn, inclination angle; MaxApHt, maximum apertural height, measured in apertural plane view as the greatest distance through the point of adherence to the opposite lip of the aperture (on shells with a beveled aperture, the distance is measured between the "slope breaks" as in F); MaxApWd, maximum apertural width, measured as the greatest distance across the aperture (or the outline of the inner whorl on beveled apertures) perpendicular to MaxApHt; PtAdh, point of adherence.

though several of them are tied to the same landmark (e.g., the point of adherence).

In a preceding discussion I suggested that high-spired shells, especially cerithiids with anterior siphons, are confined to a relatively confined region of covariation among coiling parameters, specifically to regions of slow shell expansion (see fig. 10.6B). How do coiling parameters covary with aperture shape and orientation for a generalized group and for a more derived group? Figure 10.8 shows the broad distribution of low- and medium-spired shells with respect to shell expansion (W), suture migration (M), and inclination angle (InAn), and the more restricted distribution of tall nonsiphonate shells and cerithiids (see corresponding data, tables 10.1 and 10.2). These three axes provide no clear separation between siphonate and nonsiphonate high-spired shells. However, the cerithiids sampled (labeled as form group 4, fig. 10.8) generally have lower values for coiling parameter values (W and M generally below 1.5) and apertural planes that are less inclined than high spires without siphons.

Figure 10.9 also shows the smaller range of variation among high-spired shells with respect to aperture shape (ApSh) and orientation

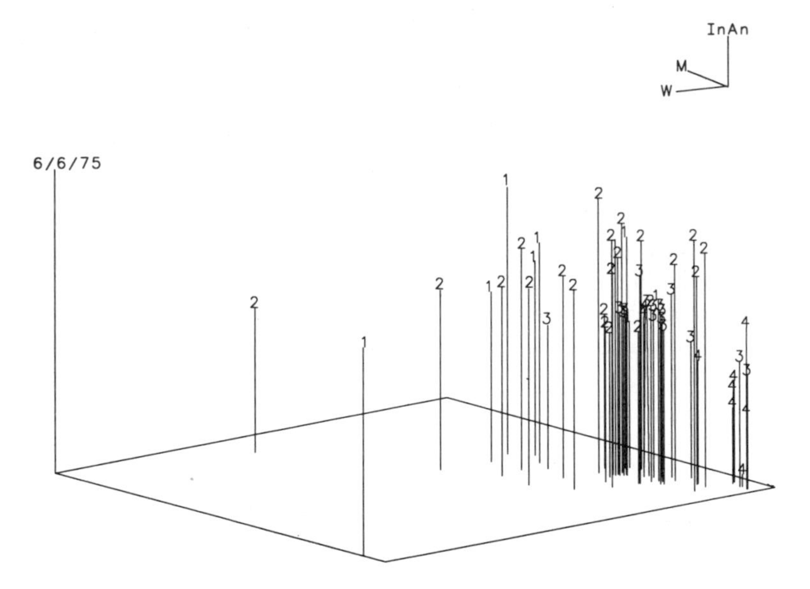

Figure 10.8. Plots of 62 shells on axes corresponding to W, shell expansion rate; M, suture migration rate; and InAn, inclination angle. The origin at the right-hand side of the plot is (W = 0, M = 0, InAn = −30°), and the reference point on the extreme left is (6,6,75). Specimens are labeled according to their shell form: 1, low spired; 2, moderate spires; 3, nonsiphonate high spires; 4, cerithiids (siphonate high spires).

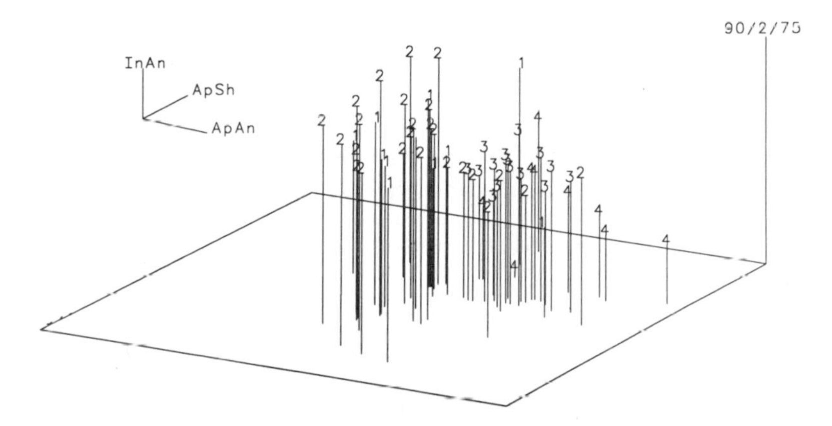

Figure 10.9. Plots of 62 shells on axes corresponding to ApAn, aperture angle; ApSh, aperture shape; and InAn, inclination angle. The origin at the left-hand side of the plot is (ApAn = −90°, ApSh = 0, InAn = −30°), and the reference point on the extreme right is (90,2,75). Specimens are labeled according to their shell form: 1, low spires; 2, moderate spires; 3, nonsiphonate high spires; 4, cerithiids (siphonate high spires).

(ApAn and InAn) and a clearer separation between tall nonsiphonate shells and cerithiids. Cerithiid shells (labeled as 4 on fig. 10.9) are confined to a separate and smaller range of inclination angles (roughly 0° to 20° with a single opisthoclinal outlier at InAn = −26°) than are other high-spired shells that lack siphons (InAn ranges from 10° to 40° in form group 3). The average inclination angle for the cerithiids is 7° and 25° for the nonsiphonate group. Shells with low and moderate spires (form groups 1 and 2) have much wider ranges of inclination angle (5° to 63° and 20° to 66°, respectively). Cerithiids also show the most extreme combination of aperture shape and angle (see fig. 10.9). They typically have elongate apertures (ApSh greater than 1.3 for the most part) that are slightly but consistently "tucked under" the whorl (ApAn is almost always between 20° and 40°). Nonsiphonate shells with high spires also have elongate apertures and a fairly narrow range of apertural angles, though the values are offset from the cerithiids (ApAn mostly 10° to 30°). This different pattern of tight covariation suggests a constraint among cerithiids that is absent from nonsiphonate taxa.

Shells with low and moderate spires have much wider ranges of apertural shape and angle (ApSh ranges from 0.7 to 1.6 for low spires, and 0.8 to 1.8 for moderate spires; ranges for ApAn are −65° to 53° and −62° to 34°, respectively). This result underscores Vermeij's (1971) and Linsley's (1977) assertions that the highest spires involve elongate apertures that are less inclined. It further suggests that low expansion rates may be an inherent characteristic of tall shells. Figure 10.8 suggests that among higher-spired shells (form groups 3 and 4), there is a positive covariation between expansion rates W and M and inclination angle. That is, more rapidly expanding shells must have a slightly more inclined aperture. In contrast, the distribution of form groups 1 and 2 on figure 10.9 denies any simple pattern of covariation.

HYPOTHESIS: TALL SPIRES AND SIPHONS CONSTRAIN COILING FLEXIBILITY

These patterns indicate that even though shells with low spires appear outwardly uniform, they are inwardly diverse. The same appears to be true for shells with moderate spires but less true for nonsiphonate shells with high spires, and simply wrong for the few cerithiids studied here. Shells with low and moderate spires exhibit a flexible system of internal trade-offs. Variations in expansion rates (W, U, and M) correspond to adjustments in the shape, inclination, and angle of the aperture (or some combination of the three). The outward result is a group of uniform-looking shells (like the archaeogastropods in general).

In contrast, high-spired taxa (form groups 3 and 4) display narrow ranges of variation for several parameters, and may have less ability to make trade-offs among parameters while maintaining overall shell shape. These patterns lead me to suggest that tall shells in general have a reduced range of morphogenetic pathways open to them. A more complete survey of taxa will reveal whether each high-spired group shows the full range of trade-offs among parameters or whether the range of trade-offs is expressed by different clades that are confined to different regions of the limited high-spired morphospace.

The very small representation of siphonate forms in this study precludes any firm conclusions. Nevertheless, these results suggest that siphonate groups like the cerithiids may each have very narrow ranges of coiling parameters and apertural characteristics and very tight patterns of covariation. The gastropod siphon was undoubtedly a critical novelty in their evolutionary success, but it may have carried with it a cost in terms of morphological flexibility. The siphon is usually described as anterior, but architecturally it is a part of the columella in nearly all cases. This projection would be too large an obstacle to coil over if it were located anywhere but on the coiling axis. What are the consequences of having the siphon restricted to an axial position?

The presence of a siphon makes the previous whorl a more complicated obstacle with which the growing apertural edge must contend. As seen in figure 10.9, inclination and apertural angles seem very constrained in siphonate shells (form 4). Any deviation from InAn = 0°

and ApAn = 30° would move the siphon from the columella. Too great an apertural angle would point the siphon across the base, obstructing coiling a half-whorl later unless the siphon were very short; too small an apertural angle would move the siphon from the columella, creating an obstacle one whorl away. This is the case for the caenogastropods *Ecphora* and *Rapana*, whose long siphons are reasonably far from the coiling axis. This results in an open umbilicus, not as a result of high umbilical expansion rate but as an exception to the covariation pattern that requires long apertures to lie alongside the previous whorl, not parallel to the coiling axis.

Illustrations of the proposed constraints. For example, the four low-spired shells shown in figure 10.10 (A to D) have essentially the same spire index and the same equant apertures (ApSh = 1.0), yet they vary widely as to coiling parameters (see data corresponding to figure 10.10; table 10.3). All four shells have inclined apertures, allowing the aperture (several of which have flared lips) to reside below the previous whorl. This degree of apertural inclination apparently accommodates wide variations in expansion rates. Another nonobvious pattern ap-

Table 10.3. Data for shells illustrated in figure 10.10

	W	U	M	InAn	ApSh	ApAn
A. Turbo coronata	2.42	2.02	3.67	29	1.0	−13
B. Liotia radiata	1.86	2.04	3.64	38	1.0	0
C. Heliacus	1.61	1.44	2.54	22	1.0	13
D. Umbonium vestarium	1.80	1.73	1.85	35	1.0	−21
E. Calliostoma conulus	1.57	1.54	1.61	43	0.0	19
F. Littorina obtusata	2.10	2.10	2.27	24	1.4	8
G. Euchelus canaliculatus	2.29	1.58	2.73	40	1.1	4
H. Stomatia phymatis	high	high	<0	23	1.5	−22
I. Telescopium telescopium	1.21	1.20	1.20	27	1.1	68
J. Rhinocoryne humboldti	1.36	1.50	1.45	>0	1.3	21
K. Phasianella australia	1.76	0.83	1.99	23	1.6	19

Note: W, shell espansion rate; U, umbilical expansion rate; M, suture migration rate; InAn, inclination angle; ApSh, aperture shape; ApAn, aperture angle.

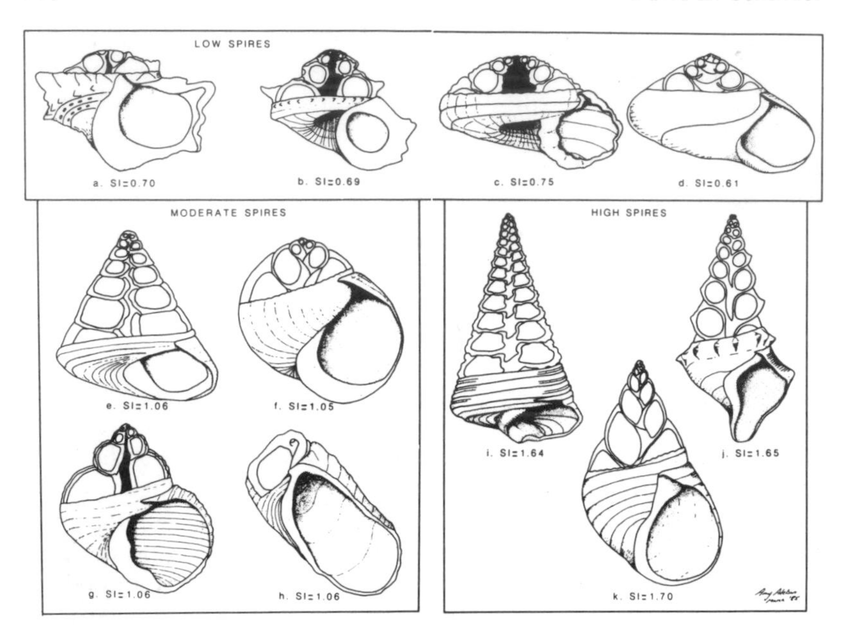

Figure 10.10. Partially cut-away specimens in axial-apertural view. Upper portion of spires reveal axial sections on which measurements for coiling parameters were made; lower portions are shown intact to exhibit covariation patterns between apertural features and coiling parameters. Specimens are grouped according to SI, spire index. Each grouping includes representatives of several grades. These shells were chosen and arranged to show the internal variability that occurs in the face of outward similarity (see text).

Topmost panel includes four low-spired shells (form = 1): A, *Turbo coronata* (an archaeogastropod, grade = 1). B, *Liotia radiata* (grade = 1). C, *Heliacus* (*Torinia*) sp. (a nonsiphonate caenogastropod, grade = 2. Robertson (1973) has suggested it is an opisthobranch). D, *Umbonium vestarium* (grade = 1). Left-hand panel includes four shells with moderate spires (form = 2). E, *Calliostoma conulus* (grade = 1). F, *Littorina obtusata* (grade = 2). G, *Euchelus canaliculatus* (grade = 1); H, *Stomatia phymotis* (grade = 1). The right-hand panel includes three high-spired shells (forms 3 and 4). I, *Telescopium telescopium* (a siphonate caenogastropod, grade = 3, form = 4). J, *Rhinocryne humboldti* (grade = 3, form = 4). K, *Phasianella australia* (grade = 1, form = 3).

pears: *Turbo* (fig. 10.10A) has the highest value of U but the narrowest umbilicus, whereas *Heliacus* and *Umbonium* have wide umbilici (though the latter's is filled by an inductural callus) and the lowest value of U. Clearly, parameters cannot be evaluated one at a time, because they covary strongly to form a basic morphogenetic field, one that determines whether the shell will be slowly or rapidly expanding in all respects.

The four shells with moderate spires show a striking pattern of covariation. As discussed previously, the three coiling parameters W, U, and M show a positive correlation (nearly all of them increase from E to H). *Haliotis*, the extreme case in the set, expands so rapidly that its whorls are too few to perform regressions. As the shell expands more rapidly, the previous whorl becomes more of an obstacle to coiling. The alternate solutions to the problem shown here either (1) have an elongate aperture that is "swung under" the previous whorl (higher ApSh and lower or negative ApAn, e.g., fig. 10.10F and G) or (2) have a more equant but inclined aperture that will reside below the bulge of the previous whorl (fig. 10.10G).

The three high-spired shells in figure 10.10 show the same pattern of covariation but much more restricted in scope. One might expect the shells in the tall spire grouping to have high values of Raup's T. In restructuring T as an independent parameter M, this expectation is lost, and the highest spires are found to have the lowest values of M. These low values are consistent with their low values for W and U—these shells are slowly expanding in every perspective. Their tall spires can only be understood, then, by establishing how this slow expansion harmonizes with other characters to arrange the many resulting whorls into a tall shell.

The shells are arranged in order of increasing expansion rate from figure 10.10I to J and correspondingly decreasing apertural angle. *Telescopium* (fig. 10.10I) has an extremely wide aperture, the most inclined aperture of any siphonate shell measured and the correspondingly highest aperture angle. This extreme degree of inclination could not be workable if *Telescopium* had a siphon of any significant length. It contends with the previous whorl by stacking its short, wide whorl entirely below its predecessor. *Rhinocoryne* and *Phasianella* (figs. 10J and K) show the more typical solution: have an elongate aperture that is swung under the previous whorl with minimal inclination. *Phasianella* is an unusually tall archaeogastropod (nonsiphonate, of course), suggesting that this pattern of covariation is tied to the architecture of the shells, not their evolutionary grade or taxonomic position.

Conclusions

The key to understanding the geometry of shell coiling lies in the integration of coiling behavior with the shape and orientation of the aperture. The proposed revision of Raup's coiling parameters seeks to remove their enforced algebraic relationship in order to reveal their architectural interactions. The primary pattern among coiling parameters is their strong three-way positive covariation: if a shell is rapidly expanding in one aspect (migration of the periphery, or the inner whorl margin, or the suture), it is likely to be rapidly expanding in the others as well.

The fact that the whorls are in contact in most snails produces covariation among coiling parameters and apertural shape and orientation. The archaeogastropods produce shells that seem fairly uniform in external aspect, but they do so through a great variety of compensating changes in coiling and apertural aspects. This suggests that there are selective forces acting on the overall shell shape and that the evolutionary history of archaeogastropods has been relatively unconstrained by specific architectural problems tied to individual aspects of the aperture or coiling. That is, sorting of any individual aspect of coiling seems weaker than sorting of the final shell shape.

In contrast, high-spired shells result from a smaller repertoire of coiling and apertural variations. The internal anatomy of gastropods may not provide the ability to produce tall spires by compensating for wide variation in one aspect of coiling with covariation in others. Certain features like the anterior siphon may leave no room for benign variation, and their "sequelae" (sensu Gould 1984) may reverberate through a host of morphological features. Sorting of these coiling characteristics might be more influential than sorting of overall shell shape, or they might be the invariant synapomorphies that unite a more derived group.

Thus, diversity of form can be understood on two levels: the product as it appears outwardly, or the flexible or rigid patterns of covariation that shape it inwardly. Any change in one aspect of coiling or the aperture would normally result in a novel shell form unless other aspects of shell geometry are free to compensate for the change. The distinctive morphologies of individual clades may correspond to particular *inabilities* to compensate for these changes. The "specialized" shells of advanced snails may not be specific answers to adaptive challenges, but rather they may be the unavoidable products of morphogenesis "locked in" by evolutionary novelty.

Acknowledgments

I gratefully acknowledge the work done by Amy Adelson in the course of this research, including specimen preparation, camera lucida drawing, measuring, data entry, and illustrating specimens. Many ideas for improving an earlier draft were offered by Warren Allmon, Fred Bookstein, Tim Collins, Douglas Erwin, Norman Gilinsky, Steve Gould, Carole Hickman, Michael Kellogg, Robert M. Linsley, Timothy Pearce, Geerat Vermeij, Peter Williamson, and Ellis Yochelson. Independent research leave time was provided by the Division of Biotic Systems and Resources, National Science Foundation.

Literature Cited

Bookstein, Fred L. 1978. *The measurement of biological shape and shape change. Lecture Notes in Biomathematic*, vol. 24. New York: Springer-Verlag.

Cain, A. J. 1977. Variation in the spire index of some coiled gastropod shells, and its evolutionary significance. *Proc. R. Soc. London* 277:377–428.

Cheetham, A. H., L. C. Hayek, and E. Thomsen. 1981. Growth models in fossil arborescent cheilostome bryozoans. *Paleobiology* 7:68–86.

Cox, L. R. 1955. Observations on gastropod descriptive terminology. *Proc. Malacolog. Soc.* 31:190–202.

Ekartane, S. U. K., and D. J. Crisp. 1983. A geometrical analysis of growth in gastropod shells, with particular reference to turbinate forms. *J. Mar. Biol. Assoc. U. K.* 63:777–97.

Golikov, A. N., and Y. I. Starobogatov. 1975. Systematics of prosobranch gastropods. *Malacologia* 15:185–232.

Goodfriend, G. A. 1983. Some new methods for morphometric analysis of gastropod shells. *Malac. Rev.* 16:79–86.

Gould, S. J. 1969. Character variation in two land snails from the Dutch Leeward Islands: Geography, environment and evolution. *Syst. Zool.* 18:185–200.

——— 1984. Covariance sets and ordered geographic variation in *Cerion* from Aruba, Bonaire, and Curacao: a way of studying nonadaptation. *Syst. Zool.* 33:217–37.

Gould, S. J., D. S. Woodruff, and J. P. Martin. 1974. Genetics and morphometrics of *Cerion* at Pongo Carpet: a new systematic approach to this enigmatic land snail. *Syst. Zool.* 23:518–35.

Heath, D. J. 1985. Whorl overlap and the economical construction of the gastropod shell. *Biol. J. Linn. Soc.* 24:165–74.

Hickman, C. S. 1980. Gastropod radulae and the assessment of form in evolutionary paleontology. *Paleobiology* 6:276–94.

Kohn A. J., and A. C. Riggs. 1975. Morphometry of the *Conus* shell. *Syst. Zool.* 24:346–59.

Linsley, R. M. 1977. Some "laws" of gastropod form. *Paleobiology* 3:196–206.

——— 1978. Locomotion rates and shell form in the Gastropoda. *Malacologia* 17:193–206.

Linsley, R. M., and W. M. Kier. 1984. The Paragastropoda: a proposal for a new class of Paleozoic Mollusca. *Malacalogia* 25:241–54.

McGhee, G. R., Jr. 1980. Shell form in the biconvex articulate Brachiopoda: a geometric analysis. *Paleobiology* 6:57–76.

McKinney, F. K., and D. M. Raup. 1982. A turn in the right direction: simulation of erect spiral growth in the bryozoans *Archimedes* and *Bugula*. *Paleobiology* 8:101–12.

McKinney, M. L. 1984. Allometry and heterochrony in an Eocene echinoid lineage: Morphological change as a by-product of size selection. *Paleobiology* 10:407–19.

McNair, C. G., W. M. Kier, P. D. LaCroix, and R. M. Linsley. 1981. The functional significance of aperture form in gastropods. *Lethaia* 14:63–70.

Niklas, K. J. 1982. Computer simulations of early land plant branching morphologies: Canalization of patterns during evolution? *Paleobiology* 8:196–210.

Raup, D. M. 1961. The geometry of coiling in gastropods. *Proc. Natl. Acad. Sci. U.S.A.* 24:602–9.

——— 1966. Geometric analysis of shell coiling: General problems. *J. Paleontol.* 40:1178–90.

——— 1972. Approaches to morphologic analysis. In *Models in paleobiology,* ed. T. J. M. Schopf, 28–44. Freeman, San Francisco: Cooper and Co.

Raup, D. M., and S. J. Gould. 1974. Stochastic simulation and evolution of morphology—towards a nomothetic paleontology. *Syst. Zool.* 23:305–22.

Raup, D. M., and A. Michelson. 1964. Theoretical morphology of the coiled shell. *Science* 147:1294–5.

Robertson, R. 1973. The biology of the Architectonicidae Gastropods combining Prosobranch and Opisthobranch Traits. *Malacologia* 14:215–20.

Saunders, W. B., and A. R. H. Swan. 1984. Morphology and morphologic diversity of mid-Carboniferous (Namurian) ammonoids in time and space. *Paleobiology* 10:195–228.

Savazzi, E. 1985. SHELLGEN: A Basic Program for the modeling of molluscan shell ontogeny and morphogenesis. *Computers Geosci.* 11(5):521–30.

Seilacher, A. 1970. Arbeitskonzept zur Konstruktionsmorphologie. *Lethaia* 3:393–6.

Sokal, R. R., and F. J. Rohlf. 1981. *Biometry,* 2d ed. San Francisco: W. H. Freeman and Co.

Thompson, D. W. 1947. *On growth and forms.* New York: Cambridge University Press.

Vermeij, G. J. 1971. Gastropod evolution and morphological diversity in relation to shell geometry. *J. Zool. London* 163:15–23.

Ward, P. D. 1980. Comparative shell shape distributions in Jurassic-Cretaceous ammonites and Jurassic-Tertiary nautilids. *Paleobiology* 6:32–43.

Williamson, P. G. 1981. Palaeontological documentation of speciation in Cenozoic molluscs from Turkana Basin. *Nature* 293:437–43.

11

Exploring the Roles of Intrinsic and Extrinsic Factors in the Evolutionary Radiation of *Melanopsis*

Dana H. Geary

The relationship of organic change to environmental change has long been an important focus of evolutionary inquiry. The debate has often been reformulated or refocused (Gould 1977), but the basic issue remains: What are the roles of intrinsic and extrinsic factors in the evolution of life? For instance, when a fossil species persists unchanged for some millions of years, does this indicate a lack of external impetus for change in the form of constant environmental conditions? Or does a long period without change indicate that organisms are homeostatic systems subject to developmental constraint? When we see rapid evolutionary change, is it necessarily preceded by some sort of environmental or ecological change, or might intrinsic factors, properties of the organisms themselves, determine rates of change? What is the degree to which organisms themselves can channel variation and in what kinds of ways?

Admittedly, dividing the world strictly into organismal (intrinsic) and environmental (extrinsic) components risks oversimplification. Rather than being two distinct entities, the organism and its environment "actively co-determine each other" (Levins and Lewontin 1985; see also Lewontin 1982). The organism plays a large role in determining what its environment actually is; most organisms seek out in some way the habitat appropriate for them, a sort of internal mechanism for controlling environment. Additionally, the organisms themselves determine which parts of the environment are relevant and which sets of conditions are tolerable. On the other hand, an organism cannot be separated from its history. Part of this history undoubtedly involves the process of adaptation to local surroundings; thus, the environment has shaped all organisms. Organism and environment are intricately linked, each with considerable influence on the other.

Despite its general fuzziness, the intrinsic/extrinsic categorization provides a useful framework for investigating evolutionary mechanisms. Consideration of important factors in evolution necessarily involves these two end members, even if the connections between them are intricate and sometimes unresolvable. Because of these connections, we should anticipate that many explanations will involve the *interaction* between organism and environment rather than either type of factor in isolation.

Explanations of evolutionary radiations must involve both intrinsic and extrinsic components. In a given situation, one or a few taxa may radiate while others remain at low diversity. For example, Cohen and Johnston (1987) identified several freshwater taxa that frequently give rise to species swarms (referred to as "species-flock prone") and others that "demonstrate little propensity towards speciation regardless of the lacustrine setting." Thus, there is undeniably a biological, or intrinsic, component to evolutionary radiations. On the other hand, the timing and location of many radiations suggests the importance of extrinsic factors: islands or isolated basins, for instance, have often provided the appropriate setting for rapid diversification.

In this paper, I explore a variety of methods of investigation into the relative contributions of intrinsic and extrinsic factors during evolution. I apply these methods to the evolutionary radiation of the mesogastropod *Melanopsis* (Ferrusac) in the Pannonian basin of central and eastern Europe (Geary 1986, 1988, in press). I focus on (1) the rate and timing of morphological change, (2) the direction of morphological change; and (3) total diversity patterns.

A Natural Experiment: the Evolution of *Melanopsis* in the Pannonian Basin

The Pannonian basin system provides an excellent setting for studying evolutionary mechanisms. It is one of a series of paratethyan basins in eastern and central Europe whose increased isolation during the Late Miocene led to the development of brackish and then freshwater conditions. A series of endemic radiations of brackish and freshwater organisms coincided with the isolation and freshening of these basins, including the well-known radiation of "Pontian cockles" in the Caspian basin (Gillet 1946; Stanley 1979; see following discussion).

The Pannonian basin system was isolated from the Late Sarmatian stage on (Nagymarosy 1981; Steininger and Rögl 1979; Rögl and Steininger 1983). The environmental history of the Pannonian Basin region has been well studied, and the physical changes that took place are generally well understood (Papp, Marinescu, and Senes 1974; Papp, Jámbor, and Steininger 1985; Jiřiček and Tomak 1981; Papp 1951, 1954; Royden, Horváth, and Burchfiel 1982; Royden, Horváth, and

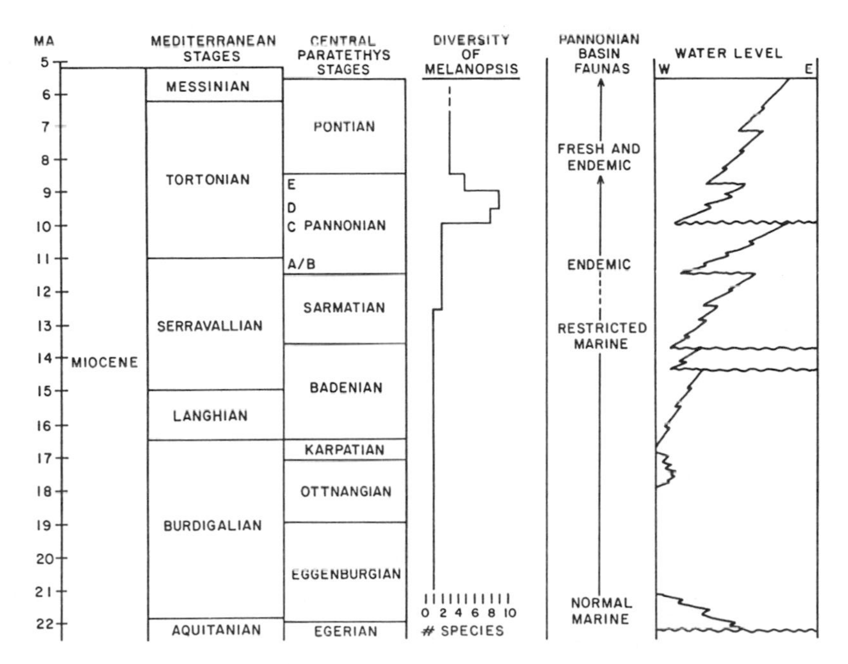

Figure 11.1. Correlation of central paratethys stages with those of the Mediterranean and relative timing of Miocene events in the Pannonian basin. Stage terminology based on Steininger, Müller, and Rögl 1988; diversity of *Melanopsis* based upon Geary 1986; Pannonian basin faunas based upon Papp 1951, 1954; Papp, Marinescu, and Senes 1974; Papp, Jámbor, and Steininger 1985; and Steininger, Müller, and Rögl 1988; water levels are after Rögl and Steininger 1983; Steininger, Müller, and Rögl 1988. Areas to the right of the curve indicate amount of inundated area. In general, regressive intervals affected the eastern parts of the basin less than they did the western parts. Figure modified from Geary et al. 1989.

Rumpler 1983; Rudinec, Tomek, and Jiřiček 1981; Bérczi et al. 1988; Nagymarosy and Müller 1988; Royden and Horváth 1988; and preceding references). Through the sedimentary record, we can approach such environmental factors as salinity, flow regime, turbidity, and overall water level in the basin.

With respect to *Melanopsis*, patterns of variation and change within and among species have been examined closely, and the phylogenetic relationships among many species are known (Geary 1986, 1988, in press). A sequence of Plio-Pleistocene melanopsids from Greece has been studied by Willman (1985) and provides an interesting basis for comparison. Finally, information on living species in this genus can contribute to our understanding of fossil patterns.

The changing patterns of faunal types and diversity within *Melanopsis* are summarized in figure 11.1. The melanopsid species living contemporaneously with normal marine faunas in the Early and Middle Miocene were *not* marine organisms. These early melanopsids lived only on the freshwater margins of the basin, and never co-occurred with the marine taxa. Key events during the Late Miocene include the final elimination of even euryhaline marine taxa by the end of the Sarmatian stage and the burst of diversification of *Melanopsis* in the Middle Pannonian stage zone C. Melanopsids exhibit a variety of evolutionary tempos, including an interval of stasis lasting at least 7 my, an interval of gradual change lasting approximately 2 my, and several instances of relatively abrupt cladogenesis (Geary in press).

The Rate and Timing of Morphological Change

THE TIMING OF DIVERSIFICATION

One of the striking features of the basic pattern of melanopsid diversity is the sudden appearance of the majority of species in the early part of Middle Pannonian zone C (fig. 11.1). These first appearances are not immigration events; the forms that arise in the Pannonian basin are never found elsewhere. The concentration of these cladogenic events in time (less than 0.5 my) prompts one to ask whether diversification can be linked to any known environmental changes or events. Possible intrinsic influences on the timing of diversification, although in general more difficult to document, must also be considered.

At first glance, one might imagine that the melanopsids are responding to a lack of competition from the marine organisms that had previously inhabited the basin. The extinction of the marine fauna occurred approximately 1.5 my earlier than the burst of diversification, however, so it is difficult to argue that reduced competition precipitated the radiation.

One potentially important environmental change was an increase in the overall water level in the basin at the base of zone C (Rögl and Steininger 1983; Steininger, Müller and Rögl 1988; Steininger pers. comm. 1983, 1989; Marinescu pers. comm. 1984, 1989 and unpubl. maps). Water levels were under tectonic control and independent of sea level changes by this time. At least two scenarios may be invoked to explain the correspondence of first appearances and increased water level. The Pannonian basin was structurally complex and topographically variable. A drop in water level would certainly have resulted in the fragmentation of formerly continuous populations; the increased isolation may have allowed populations to differentiate. When water levels rose again, newly derived forms could then spread across the basin, successful ones expanding their ranges and increasing their chances of being preserved. Alternatively, if the record is read literally, the interpretation is that speciation coincided with the increase in habitat associated with the transgression.

Elements of both of these scenarios may have contributed to the high rates of diversification. A fractionated population structure alone probably cannot explain the pattern; populations were presumably even more fragmented in the Sarmatian stage and earlier, when melanopsids were confined to only the most marginal habitats. Previous isolation without the opportunity of increased habitat did not result in diversification.

An explanation based entirely on the zone C increase in water level is not totally satisfactory, however, because water levels also rose at the base of zone A/B (Rögl and Steininger 1983; Steininger, Müller, and Rögl 1988), and no burst of diversification occurred then. An additional environmental consideration is the salinity of the basin's waters. Stable isotopic ratios of melanopsid shells from the Late Sarmatian through the Pannonian stage indicate that no major changes in salinity occurred during this interval (Geary et al. 1989). However, F. Marinescu (pers. comm. 1989) has suggested that salinities were lower during the Lower Pannonian zones A/B, and then increased in zones C/D, based upon the accompanying paleofauna (in Romania) during these two intervals. Our stable isotopic data do not rule out this possibility, because sampling in zones A/B is very sparse. Thus, a change in salinity (or some other aspect of water chemistry) at zone C may have helped facilitate the burst of diversification in *Melanopsis*.

The alternative to environmental explanations involves some intrinsic property of the melanopsids themselves: perhaps the delay between diversification simply represents the response time of the melanopsids. What sort of evidence would indicate an intrinsically mediated time lag? Aside from the negative evidence offered by a lack of correlation with known environmental factors, different patterns of response in different lineages might suggest a degree of intrinsic input. Such a difference exists in these data.

All of the diversity of Pannonian basin melanopsids can be traced to two stem species present in the Sarmatian stage: *M. impressa* and *M. bouei*. The descendants of each of these species define two clades (Geary 1986, in press), which I will refer to as the *M. impressa* clade and the *M. bouei* clade (fig. 11.2). The majority of Pannonian basin species belong to the *M. bouei* clade, and it is in this clade that diversification is concentrated during Pannonian zone C. Before this interval, there are no changes evident in *M. bouei*, nor does it give rise to any descendant species. The pattern in the *M. impressa* lineage is dif-

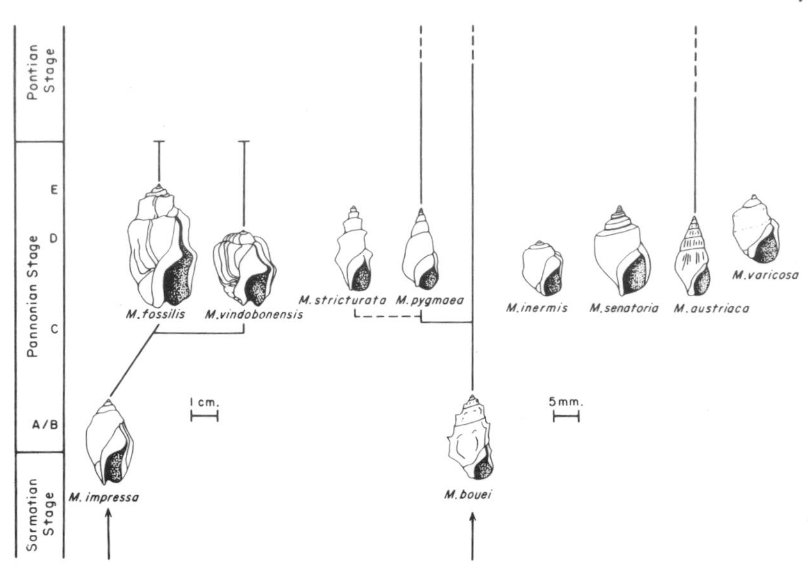

Figure 11.2. Total diversity pattern for all of the melanopsid species. Note different scales for the two clades (*M. impressa*, *M. fossilis*, and *M. vindobonensis* constitute the *M. impressa* clade; all other species belong to the *M. bouei* clade). Relative size within clades is approximate; maximum height for each species is *M. impressa*, 3.8 cm; *M. fossilis*, 5.6 cm; *M. vindobonensis*, 3.2 cm; *M. stricturata*, 1.4 cm; *M. pygmaea*, 1.4 cm; *M. bouei*, 1.9 cm; *M. inermis*, 1.3 cm; *M. senatoria*, 1.7 cm; *M. austriaca*, 1.1 cm; *M. varicosa*, 1.5 cm. Species ranges are as follows: *M. impressa* is present since at least the Eggenburgian stage of the Early Miocene, *M. fossilis* and *M. vindobonensis* range only from Pannonian zones C to E. *M. bouei* appears in the Sarmatian stage. All other species in the *M. bouei* clade arise in Pannonian zone C, except for *M. varicosa*, which appears in zone D only. *M. stricturata*, *M. inermis*, and *M. senatoria* persist only through zone D. *M. bouei*, *M. pygmaea*, and *M. austriaca* persist into the Pontian stage.

ferent; *M. impressa* exhibits a long sequence of gradual change (leading to *M. fossilis*) that begins by the latest Sarmatian. There is apparently no lag time between the Late Sarmatian change in environmental conditions and change in the *M. impressa* clade. (The only other event in the history of the *M. impressa* clade is a branching event that occurs in the Middle Pannonian zone C; this event parallels changes in the *M. bouei* clade.) Thus, the two melanopsid lineages differ in the timing (and style) of their evolutionary changes; the initiation of gradual change in *M. impressa* is approximately coincident with the extinction of the basin's marine fauna, whereas no change happens in the *M. bouei* clade until approximately 1.5 my later.

Thus, there are suggestions of both extrinsic and intrinsic contributions to the timing of diversity in *Melanopsis*. The correspondence of diversification with changes in water level in the basin is highly suggestive of at least some degree of environmental control. The absence of change during an earlier rise in water level, and the difference in "response time" between the two lineages, suggest some degree of intrinsic input to the pattern as well.

EPISODES OF STASIS

Discussion of the mechanisms responsible for stasis has focused on two general categories: developmental constraint and related factors, and stabilizing selection (selection for constant morphology mediated by the environment) (see Williamson 1981a, 1981b, 1987; Charlesworth, Lande, and Slatkin 1982; Van Valen 1982). For the most part, these two categories correspond to intrinsic and extrinsic mechanisms, respectively (though see Williamson 1987, for an intrinsically mediated form of stabilizing selection).

The most striking example of evolutionary stasis among Pannonian basin melanopsids is in *M. impressa*. Late Sarmatian samples of this species closely resemble those of the Karpatian and Eggenburgian stages, approximately 5 and 7.5 my earlier, respectively (Geary in press). This period of stasis ends with the onset of gradual changes, which lead eventually to *M. fossilis*. The period of stasis in *M. impressa* coincides with the presence of marine taxa in the basin; the onset of gradual change closely follows the extinction of the last remaining marine organisms. Thus, it appears that the change that precipitated the end of stasis in *M. impressa* was not a change in "melanopsidness," but rather a change in the Pannonian basin environment (either reduced competition or predation by euryhaline marine organisms or expansion of the brackish habitat melanopsids preferred). Indications are, then, that there was an extrinsic component to the maintenance of stasis in *M. impressa*.

The Direction of Morphological Change

Morphological change, be it driven by natural selection or genetic drift, is rooted in the variation found in populations. Potential intrinsic effects on the direction of morphological change involve the kinds of variation produced; organisms themselves may contribute to the directionality of morphological change by producing a limited range of variability in certain traits or by repeatedly developing similar kinds of variation. In this section, I investigate the contributions of intrinsic factors toward the directionality of morphological change in two ways: first, with respect to a character that arises repeatedly in Pannonian basin melanopsids and second, through examples of heterochrony.

SHOULDERING

Shouldering arises independently in at least three species of Pannonian basin melanopsids: *M. fossilis*, *M. stricturata*, and *M. senatoria* (fig. 11.2). (*M. varicosa* is also shouldered, but because its immediate ancestor is uncertain, it is not known if its shouldering is independently derived. The immediate ancestor of *M. senatoria* is also uncertain, but *M. fossilis*, *M. stricturata*, and *M. varicosa* are not considered as possibilities; its shouldering must then have been independently derived.)

It is important to investigate how shouldering arises in these species. If construction of the shouldering is achieved by different means, we might look for a selective explanation to explain the convergence. If, on the other hand, shouldering is homologous in all melanopsids, we might consider the importance of intrinsic mechanisms. As emphasized previously, the implication of intrinsic mechanisms does not mean that selection had no role in the process; the importance of intrinsic mechanisms lies in their ability to channel the variation available for selection.

With minor variation, shouldering is "built" by Pannonian basin melanopsids in a similar way. In each case, an inflection at midheight of the aperture forms a ridge, which, as growth continues, comes to lie immediately under the suture, where the initial bulge is reinforced by another bulge at the top of the aperture. Differences between the species involve the relative importance of these two elements. For instance, the initial bulge is very faint in *M. senatoria* but more pronounced in *M. fossilis* and *M. stricturata*. Interestingly, the nodose ornament of *M. bouei*, and the subtle inflations characteristic of *M. inermis*, originate from the same two locations on the shell margin (just below the suture and at half-height of the whorl).

Shouldering is also present in other melanopsids, from different places or different times. The living species *M. parresyi* (eastern Europe) has an inflection at the top of the aperture just below the suture, which is augmented by thickening of axial costae just below the suture. Shouldering in *M. costata* (Recent of Israel) is entirely a result of thickening of the costae just below the suture. *Melanopsis gorceixi heldreichi* (figured in Willmann 1985; from the Plio-Pleistocene of Kos) appears to have constructed its shouldering in the same way that *M. costata* does. In Willmann's sequence from Kos, contemporaneous lineages of *Viviparus* and *Theodoxus* also develop shouldering.

Thus, there are at least two basic ways that melanopsids develop shouldering (which may occur in combination): as apertural inflections (at half-height of the whorl and just below the suture) or by thickening of axial costae. The distribution of shouldering types in melanopsids does not give us an unambiguous answer to the question of why this trait appears so frequently, probably because the answer involves both selective forces and a degree of intrinsic directionality (Allmon and Geary 1986). Among the examples just described, the simultaneous and independent development of shouldering in *Melanopsis*, *Viviparus*, and *Theodoxus* provides the strongest evidence that in some environments there is selective pressure for this trait. On the other hand, the development of both shouldering and ornament from the same two homologous points in many Pannonian basin melanopsids suggests an intrinsic component to the directionality of morphological change.

HETEROCHRONY

Close examination of intraspecific variation and interspecific change among melanopsid species has revealed two examples of heterochrony (Geary 1986, 1988). *Melanopsis pygmaea* is the paedomorphic descendant of *M. bouei* ("*M. sturii*" of Geary 1988 should be *M. bouei*). Juvenile shells of both species are smooth and narrow. *Melanopsis pygmaea* retains this pattern of growth into adulthood, whereas *M. bouei* develops nodes and a more inflated shell. Thus, *M. pygmaea* is described as paedomorphic because its adult form resembles the juvenile stage of its ancestor. The particular process or processes involved (neoteny, progenesis, postdisplacement) cannot be specified without information on the ontogenetic age of individuals (Jones 1988; McKinney 1988). The crucial difference between these two species is the presence or absence of ornament; *M. bouei* "turns on" its ornament generally between the fifth and seventh whorls, whereas *M. pygmaea* stays with the juvenile program and never develops ornament. With respect to this important character, measured as the height of onset of ornamentation, *M. bouei* exhibits two interesting patterns of variation during its 5-my history. First, there are scattered occurrences of individuals ornamented only on the final whorl or two. These specimens appear intermediate in form between *M. bouei* and *M. pygmaea* and in fact were assigned to a "subspecies," *M. pygmaea turrita*, by Papp (1951). (I do not recognize Papp's subspecies of *M. pygmaea* because they are often sympatric [Geary 1986].) These "intermediate" individuals are always associated with samples of *M. bouei* and appear to be rather extreme variants of this species. Second, the height of onset of ornamentation varies considerably among samples of *M. bouei*. Of 66 possible pairwise comparisons between 12 measured samples, 43 (>65%) differed significantly in the height of onset of ornamentation (using ANOVA; significance at the .05 level). Thus, intraspecific variation in *M. bouei*, as expressed within and among samples, parallels the differences that distinguish it from its descendant species, suggesting an intrinsically defined pathway of variation and change.

The second instance of heterochrony involves the *M. impressa-fossilis-vindobonensis* sequence. In essence, the ontogenetic modification that produced *M. fossilis* from *M. impressa* is accelerated and exaggerated in *M. fossilis*' descendant species *M. vindobonensis* (fig. 11.3). This modification is an early increase followed by a decrease in the rate of shell expansion relative to translation, giving both *M. fossilis* and *M. vindobonensis* a three-phase ontogeny (an early phase of high translation, middle phase of rapid expansion, and a final phase of mostly translational growth). The resultant "*fossilis*" and "*vindobonensis*" morphologies are quite distinct, but plotted ontogenetic trajectories reveal an underlying similarity. Again, this example of heterochrony in *Melanopsis* indicates that the changes between species are often channeled by intrinsically defined pathways. Selective (extrinsic) forces are probably required to move populations along these pathways, but it is important to recognize the intrinsic component to the directionality of change.

Diversity Patterns

As described previously, all Pannonian basin melanopsids belong to one of two clades. A striking difference in diversity exists between these two clades, particularly during the Middle Pannonian stage. During this interval, the *M. impressa* clade includes only two species, whereas the *M. bouei* clade is represented by at least seven species. (fig. 11.2 gives a conservative estimate of species diversity in the *M. bouei* clade. Many other species have been described from this interval [see especially Jekelius 1944 and Papp 1954]. I do not have sufficient material to confirm the species-level status of these forms, but all would appear more closely related to the *M. bouei* clade.)

The difference in diversity between clades cannot be ascribed to different amounts of time available for the accumulation of species: each clade was represented by only one species in the Late Sarmatian. Furthermore, the differences cannot be attributed to purely environmental factors because both clades occupied the same general habitats at the same time. Thus, this diversity pattern is most likely due to some biological difference between the clades.

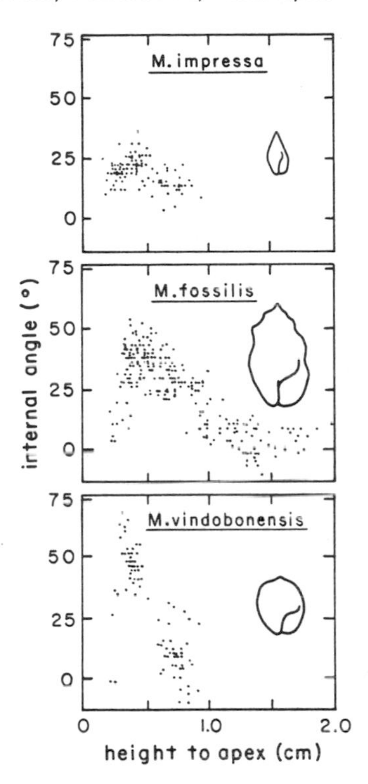

Figure 11.3. Internal angles for all individuals in single samples of *M. impressa*, *M. fossilis*, and *M. vindobonensis*. Internal angles were measured from x-radiographs of the shells. For each individual, a series of points marking an inflection at the upper right corner of each whorl interior was recorded; angles were measured between successive points and parallel to the coiling axis (see Geary, 1988). Low angles represent mostly translational growth; high angles represent expansional growth away from the coiling axis. Moving from left to right across each diagram corresponds to moving down the axis of coiling or through the collective ontogenies of each sample.

A comparison of species distributional data reveals contrasting patterns of variation in the two clades. From the time of their first appearances on, *M. fossilis* and *M. vindobonensis* (*M. impressa* clade) are numerous across the basin. Samples exhibit little or no geographic variation, and only minor temporal variation. Species in the *M. bouei* clade exhibit quite different patterns of distribution and variability. Most species are rare, geographically restricted, or both. Even the most abundant and widespread species, *M. bouei*, exhibits considerable temporal and geographical variability (Geary 1986, in press).

One possible explanation for these patterns is that the two clades differed in their dispersal abilities. In such a scenario, the poorly dispersing species in the *M. bouei* clade would have been more prone to isolation, hence their greater geographical and temporal variability and higher rates of speciation. Species in the *M. impressa* clade, assumed to be better dispersers, exhibit less variability, less diversity (only one branching event), and even an interval of gradual change. To account for the contrasting dispersal abilities, one must call upon a difference between clades in some aspect of their reproductive biology or mobility.

Unfortunately, very little is known about reproduction in *Melanopsis*, even with respect to extant species. It is recognized that fertilization is internal, probably through transfer of sperm as sperm balls (Bilgin 1973). It is presumed that eggs are laid, but none have been identified in nature. Freshwater prosobranchs are not known to have swimming veliger larvae, so calling upon planktotrophic versus nonplanktotrophic larval strategies to explain the diversity difference between clades is not reasonable for these snails.

Johnston and Cohen (1987) report on a similar pattern of different amounts of variability in two endemic genera of thiarid gastropods in Lake Tanganyika (freshwater mesogastropods, as is *Melanopsis*). These different patterns of variability led Johnston and Cohen to question their initial assumption of equal dispersal potential in the two genera and focus on a possible difference between brooding and nonbrooding. Their work illustrates the difficulties involved in specifying the proximal cause of differences in variability and diversity, even in extant groups.

As an alternative to the dispersal explanation, we might consider the possibility that some facet of the organisms' interactions with their environment differed between the two clades. By virtue of their considerably smaller size, for example, members of the *M. bouei* clade may in fact have experienced the same general environment in a different way (see size data in fig. 11.2 caption). A tabulation of gastropod diversity in three size classes in the Pannonian basin during the Sarmatian stage (summarized from Papp 1951) reveals that 65% of the species present (37 of 57 total species) are less than 1 cm in height (height of largest figured adult or largest height cited in description), 14% are between 1 and 2 cm, and 21% are larger than 2 cm. Thus, in the Pannonian basin region, several million years before the melanopsid radiation, prosobranch species diversity was concentrated in the smaller size ranges. This is clearly a rough approach to the question of size and diversity; the point is simply that an organism's basic perception of and reaction to a given environment will depend on many aspects of its biology, including its size. This size-environment explanation is interaction-based, rather than being exclusively intrinsic or extrinsic.

Discussion

The radiation of Pannonian basin melanopsids bears similarities to two other molluscan radiations. The common European cockle *Cerastoderma* underwent a spectacular radiation in the Pontian stage of the Late Miocene (and probably into the Pliocene) in the Caspian basin (Gillet 1946; Stanley 1979). More than 30 endemic genera comprising four subfamilies have been described from this time interval. The Caspian and Pannonian basins were both part of the paratethyan system and no doubt experienced many of the same environmental changes, although those of the Caspian generally took place several million years later than changes in the Pannonian basin. The second example comes from Runnegar and Newell (1971), who describe a remarkable radiation of bivalves in the Permian Paraná basin of eastern South America. Their revised taxonomy includes at least 12 endemic genera in two families. Each of these intrabasinal molluscan radiations (in the Pannonian, Caspian, and Paraná basins) took place in brackish water, and in all three cases accompanying faunas are sparse and taxonomically restricted.

In his 1987 review of rates of evolution in molluscs, Runnegar discusses the Pontian cockles and the bivalves of the Paraná basin. He contends that the most obvious explanations for the unusually rapid rates of diversification are "reduced population size, unstable conditions and an absence of competition from organisms excluded because of their intolerance to lowered or variable salinities" (Runnegar 1987). With respect to the Pannonian basin melanopsids, it is difficult to see how unstable conditions or reduced population size could have precipitated the radiation. Certainly the situation for *Melanopsis* before its radiation, when populations were living in the freshwater habitats marginal to the otherwise marine basin, would frequently have involved unstable conditions and reduced population sizes. Runnegar's third explanation, reduced competition from normal marine organisms, assumes that the organisms involved can tolerate marine conditions. Reduced competition may have been less important than simply an increase in appropriate habitat; brackish organisms prevail in brackish habitats, marine organisms prevail in marine habitats. More specifically, the melanopsid radiation does not occur immediately after elimination of the marine taxa but some time later, as discussed previously.

The Paraná, Caspian, and Pannonian basins differed in at least one important respect. The original size of the Permian Paraná basin and the size of the Ponto-Caspian Sea are both believed to have been approximately three times the current size of the Caspian Sea (Runnegar and Newell 1971). Runnegar and Newell suggest that a water body of this order of magnitude is needed "to provide sufficient space for effective radiation within a geographically restricted environment." The water body in the Pannonian basin was roughly half the size of those in the ancient Caspian or Paraná basins. The radiation of *Melanopsis* in this smaller basin may support Runnegar and Newell's basic contention, however, in that it is a radiation of considerably less taxonomic magnitude (intrageneric, rather than many new genera).

Summary

Complete explanations for the timing or direction of morphological change probably involve a combination of intrinsic and extrinsic factors, but some aspects of the observed patterns can be identified with one or the other set of influences. For instance, a variety of extrinsic factors appear to have facilitated the radiation of *Melanopsis*. Rapid diversification occurred as the water level in the basin rose, suggesting that the increase in available habitat associated with the transgression and/or the preceding period of relative isolation and fragmented population structures facilitated speciation. Extrinsic factors are also implicated in the maintenance of stasis in *M. impressa;* after more than 7 my of stasis, gradual change begins in this species at the same time that the last of the normal marine fauna are eliminated from the basin. Intrinsic influence on the direction of morphological change is indicated by examples of heterochrony and through the homologous development of several kinds of ornament in several melanopsid species.

The difference in species diversity between the two clades cannot be wholly explained by extrinsic factors. Possible explanations include differences in dispersal ability (an idea supported by differing patterns of variation between clades; but for which no reproductive data are available), or different perceptions of environment based upon the characteristically different sizes of individuals in the two clades. This latter explanation represents neither an intrinsic nor an extrinsic hypothesis but instead relies on the interaction between organismal and environmental factors.

Acknowledgments

My study of Pannonian basin melanopsids would not have been possible without the generous help of Fritz Steininger, Margit Bohn, Florian Marinescu, Kresimir Sakač, and Zlata Jirišić-Polšak. I would also like to thank Rudolf Jiřiček in Czechoslovakia; Áron Jámbor, Tamás Baldi, Geza Hámor, Pàl

Müller, Gitta and Laszlo Korpas, István Jánkovich, Ferenc Gàspar, and János Szabó in Hungary; Virgil Ghiurca, Eugen Nicorici, and Petru Bănărescu in Romania; Mato Pikija, Mato Brkić, Mme. Kochanski, and Zlatan Bajraktarević in Croatia; the late Adolf Papp, F. Sauerzopf, A. Schimatzek, Werner Pillar, Ortwin Schultz, Karl Schütz, and Karl Kleeman in Vienna; and Yossi Heller and Eitan Tchernov in Israel. I thank Steve Gould, Peter Williamson, and Ruth Turner for their assistance throughout this project. Robert Bleiweiss, Geerat Vermeij, and an anonymous reviewer made helpful comments on the manuscript. My work in Eastern Europe was funded and facilitated by the International Research and Exchanges Board (IREX) in Princeton, New Jersey. Work in Israel on living melanopsids was funded by the Geological Society of America and Sigma Xi.

Literature Cited

Allmon, W. D., and D. H. Geary. 1986. A pattern of homeomorphy in diverse lineages of gastropods. *Fourth North American Paleontological Convention Abstracts:* A1.

Bérczi, I., G. Hámor, Á. Jámbor, and K. Szentgyörgyi. 1988. Neogene sedimentation in Hungary. In *The Pannonian Basin: a study in basin evolution,* American Association of Petroleum Geologists Memoir 45, ed. L. H. Royden and F. Horváth, 57–67. American Association of Petroleum Geologists and Hungarian Geological Society. Tulsa, Oklahoma and Budapest, Hungary.

Bilgin, F. H. 1973. Studies on the functional anatomy of *Melanopsis praemorsa* (L.) and *Zemelanopsis trifasciata* (Gray). Proc. *Malacol. Soc. London.* 40:379–93.

Charlesworth, B., R. Lande, and M. Slatkin. 1982. A neo-Darwinian commentary on macroevolution. *Evolution* 36:474–98.

Cohen, A. S., and M. R. Johnston. 1987. Speciation in brooding and poorly dispersing lacustrine organisms. *Palaios* 2:426–35.

Foote, M., and R. H. Cowie. 1988. Developmental buffering as a mechanism for stasis: Evidence from the pulmonate *Theba pisana. Evolution* 42: 396–9.

Geary, D. H. 1986. The evolutionary radiation of melanopsid gastropods in the Pannonian Basin (Late Miocene, Eastern Europe). Ph.D. diss., Harvard University, Cambridge, Mass.

Geary, D. H. 1988. Heterochrony in gastropods: a paleontological view. In *Heterochrony in evolution: a multidisciplinary approach,* ed. M. L. McKinney, 183–96. New York: Plenum Press.

Geary, D. H. in press. Patterns of evolutionary tempo and mode in the radiation of *Melanopsis* (Gastropoda; Melanopsidae). *Paleobiology*

Geary, D. H., J. A. Rich, J. W. Valley, and K. Baker. 1989. Isotopic evidence for salinity changes in the Late Miocene Pannonian Basin: Effects on the evolutionary radiation of melanopsid gastropods. *Geology* 17:981–85.

Gillet, S. 1946. Lamellibranches dulcicoles, les limnocardiides. *Revue Scientifique* (Paris) 84:343–53.

Gingerich, P. D. 1985. Species in the fossil record: Concepts, trends, and transitions. *Paleobiology* 11:27–41.

Gould, S. J. 1977. Eternal metaphors of palaeontology. In *Patterns of evolution, as illustrated by the fossil record*, ed. A. Hallam, 1–26. Amsterdam: Elsevier.

Jekelius, E. 1944. *Sarmat und Pont von Soceni (Banat)*. Bucureşti: Imprimeria Naţională.

Jiřiček, R., and Č. Tomek. 1981. Sedimentary and structural evolution of the Vienna Basin. *Earth Evolution Sci.* 1:195–204.

Johnston, M. R., and A. S. Cohen. 1987. Morphological divergence in endemic gastropods from Lake Tanganyika: Implications for models of species flock formation. *Palaios* 2:413–25.

Jones, D. S. 1988. Sclerochronology and the size versus age problem. In *Heterochrony in evolution: a multidisciplinary approach*, ed. M. L. McKinney, 93–108. New York: Plenum Press.

Levins, R., and R. C. Lewontin. 1985. *The dialectical biologist*. Cambridge, Mass.: Harvard University Press.

Lewontin, R. C. 1982. Prospectives, perspectives and retrospectives. *Paleobiology* 8:309–13.

McKinney, M. L. 1988. Classifying heterochrony: Allometry, size, and time. In *Heterochrony in evolution: a multidisciplinary approach*, ed. M. L. McKinney, 17–34. New York: Plenum Press.

Nagymarosy, A. 1981. Chrono- and biostratigraphy of the Pannonian Basin: a review based mainly on data from Hungary. *Earth Evolution Sci.* 1:183–94.

Nagymarosy, A., and P. Müller. 1988. Some aspects of Neogene biostratigraphy in the Pannonian Basin. In *The Pannonian Basin: a study in basin evolution*, American Association of Petroleum Geologists Memoir 45, ed. L. H. Royden and F. Horváth, 69–78. American Association of Petroleum Geologists and Hungarian Geological Society.

Papp, A. 1951. Das Pannon des Wiener Beckens. *Mitteilungen der Geologischen Gesellschaft in Wien* 39:39–41, 99–193.

Papp, A. 1954. Die Molluskenfauna im Sarmat des Wiener Beckens. *Mitteilungen der Geologischen Gesellschaft in Wien* 45:1–112.

Papp, A., F. Marinescu, and J. Senes. 1974. *Chronostratigraphie und Neostratotypen, Miozan der Zentralen Paratethys*, Bd. IV, M5 Sarmatien. Bratislava: VEDA, Verlag der Slowakischen Akademie der Wissenschaften.

Papp, A., Á. Jámbor, and F. F. Steininger. 1984. *Chronostratigraphie und Neostratotypen, Miozan der Zentralen Paratethys, Bd. VI, M6 Pannonien (Slavonien und Serbien)*. Budapest: Akademiai Kiado.

Rögl, F., and F. F. Steininger. 1983. Vom Zerfall der Tethys zu Mediterran und Paratethys. Die neogene Palaogeographie und Palinspastik des Zirkum-Mediterranen Raumes. *Annals der Naturhistoriches Museum Wien* 85/A:135–63.

Royden, L. H., and F. Horváth, eds. 1988. *The Pannonian Basin, a study in basin evolution*, American Association of Petroleum Geologists Memoir 45. American Association of Petroleum Geologists and Hungarian Geological Society, Tulsa, Oklahoma and Budapest, Hungary.

Royden, L. H., F. Horváth, and B. C. Burchfiel. 1982. Transform faulting, extension, and subduction in the Carpathian Pannonian region. *Geological Soc. Am. Bull.* 93:717–25.

Royden, L. H., F. Horváth, and J. Rumpler. 1983. Evolution of the Pannonian Basin System. 1. Tectonics. *Tectonics* 2:63–90.

Rudenic, R., Č. Tomek, and R. Jiřićck. 1981. Sedimentary and structural evolution of the Transcarpathian Depression. *Earth Evolution Sci.* 1:205–11.

Runnegar, B., and N. D. Newell. 1971. Caspian-like relict molluscan fauna in the South American Permian. *Bull. Am. Mus. Natural Hist* 146:1–66.

Runnegar, B. 1987. Rates and modes of evolution in the Mollusca. In *Rates of evolution*, ed. K. S. W. Campbell and M. F. Day, 39–60. London: Allen and Unwin.

Stanley, S. M. 1979. *Macroevolution*. San Francisco: W. H. Freeman.

Steininger, F. F., C. Müller, and F. Rögl. 1988. Correlation of central paratethys, eastern paratethys, and Mediterranean Neogene stages. In *The Pannonian Basin, a study in basin evolution*, American Association of Petroleum Geologists Memoir 45, ed. L. H. Royden, and F. Horváth, 79–87. American Association of Petroleum Geologists and Hungarian Geological Society, Tulsa, Oklahoma and Budapest, Hungary.

Steininger, F. F., and F. Rögl. 1979. The Paratethys history—a contribution towards the Neogene geodynamics of the Alpine orogene (an abstract). *Annalles Geologique Pays Hellenes* 1979, 3:1153–65.

Van Valen, L. M. 1982. Integration of species: Stasis and biogeography. *Evolutionary Theory* 6:99–112.

Williamson, P. G. 1981a. Paleontological documentation of speciation in Cenozoic molluscs from Turkana Basin. *Nature* 293:437–43.

Williamson, P. G. 1981b. Morphological stasis and developmental constraint: Real problems for neo-Darwinism. *Nature* 294:214–5.

Williamson, P. G. 1987. Selection or constraint? a proposal on the mechanism for stasis. In *Rates of evolution*, ed. K. S. W. Campbell and M. F. Day, 129–42. London: Allen and Unwin.

Willmann, R. 1985. Responses of the Plio-Pleistocene freshwater gastropods of Kos (Greece, Aegean Sea) to environmental changes. In *Sedimentary and evolutionary cycles*, ed. U. Bayer and A. Seilacher, 295–321. New York: Springer-Verlag.

12

The Correlates of High Diversity in Lake Victoria Haplochromine Cichlids: a Neontological Perspective

Robert L. Dorit

The haplochromine cichlids of the East African Rift lakes are the fastest radiating and the most species-rich extant vertebrate taxon. The diversity and endemism that characterize the ichthyofaunas of many of the African Rift lakes—Malawi, Tanganyika, and Victoria—taunt the student of diversity. Might the forces underlying speciation be more easily glimpsed in these exaggerated natural experiments?

This chapter deals principally with the results of molecular and morphological studies undertaken on the haplochromine cichlid radiation in Lake Victoria. Within the context of the Victorian radiation, it addresses four interrelated questions:

1. Are current explanations for high-diversity clades conflating the *origin* of high diversity with the *maintenance* of high diversity?
2. What are the intrinsic features of haplochromine cichlids that might underlie their rapid speciation?
3. How do the apparent morphological trends in skull shape exhibited by certain haplochromine lineages come about?
4. To what extent do our systematic and evolutionary reconstructions of the haplochromine radiation depend on neontological data?

The Victorian basin, including two now separate lakes (Kivu and Edward) is the youngest of the African Rift lakes. Estimated to be approximately 750,000 years old, it was formed by rifting events in the Pleistocene (Beadle 1981). Modern Lake Victoria is both the shallowest (88 m maximum depth) and the largest (surface area, 69,000 km^2) of the Rift lakes (fig. 12.1).

From the earliest fish collections undertaken in Lake Victoria (Graham 1929; Worthington 1929, 1937), the unique nature of its ichtyofauna was obvious. Three features of this assemblage stand out: first, the sheer species diversity present in the lake; second, the complete dominance of the fauna by members of the family Cichlidae; and finally, the proportion of cichlid species endemic to the Victoria basin (>98%) (Greenwood 1984).

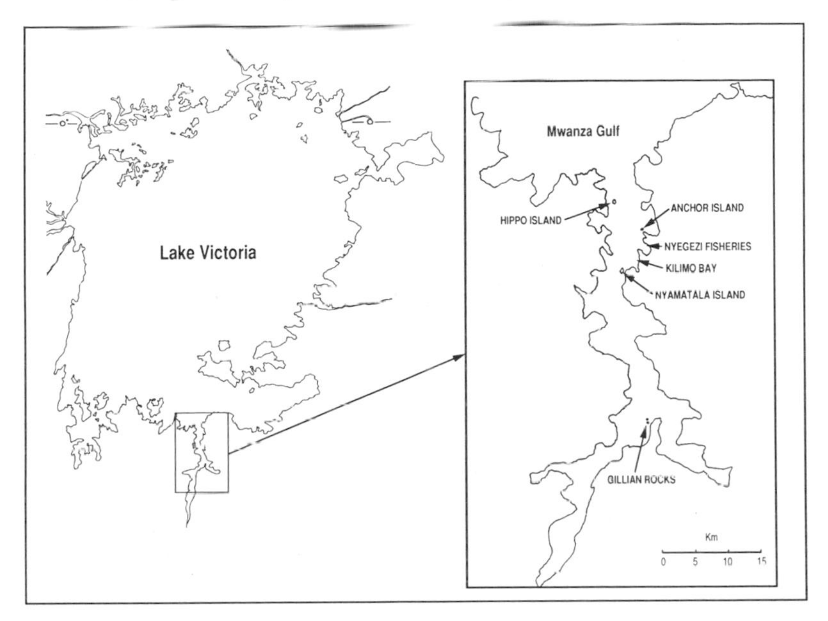

Figure 12.1. Present outline of Lake Victoria. Inset shows detailed shoreline of the Mwanza Gulf and the localities sampled in this study.

A conservative estimate of the number of cichlid species in Lake Victoria puts the figure at "200+" (Greenwood 1974); more recent work suggests it may be closer to 300 species (Barel et al. 1977). All but four of these species were previously grouped in the genus *Haplochromis*. A more recent cladistic revision has divided the assemblage into over 20 phylogenctic clusters, all of which, following cladistic convention, have been given generic status (Greenwood and Barel 1978; Greenwood 1979a, 1980).

The original haplochromine stock is likely derived from the riverine faunas inhabiting the eastward flowing rivers that once drained into the Zaire basin (Temple 1969; Greenwood 1974, 1984). However, although the monophyly of the Cichlidae has been established (Stiassny 1981), the monophyly of the Victorian haplochromines is far less certain and the sister group or groups remain unknown. Greenwood (1983) has suggested that the faunas of each of the lakes (Malawi, Tanganyika, and Victoria) should not be treated indcpendently, since a given monophyletic unit may have representatives in more than one lake.

Over the past 60 years, a number of mechanisms have been proposed to account for this exaggeratedly species-rich fauna. The majority of the explanations for the haplochromine radiations focus on two main themes: the ecological plasticity and consequent fine-scale resource partitioning of haplochromine cichlid communities (Fryer and Iles 1972; Greenwood 1974; Barel et al. 1977; Witte 1984a, 1984b) and the geological and climatic history of the Rift lakes (Bishop 1965, 1969; Kendall 1969; Beadle 1981). These themes reflect prevailing notions in much of the literature about diversity, where extinction is seen as the main regulator of diversity, and where speciation is discussed primarily as a passive consequence of (frequently abiotic) extrinsic events. I will argue that both of these assumptions deserve closer scrutiny and that they represent—at best—only part of a complete theory of diversity.

Accounting for High Diversity: a Nonequilibrium View

Species diversity is a composite measure, the balance between inputs (speciation and immigration) and outputs (extinction and emmigration). High diversity per se tells us little about the factors that underlie it. Despite this, most discussions concerning the haplochromine radiations have focused on the "output" side of the diversity equation (for an exception, see Cracraft 1982). Attention centers on those aspects of haplochromine biology, ecology, or plasticity that reduce the probability of extinction, permitting the coexistence of such large numbers of species.

The elegant work of several investigators supports the notion that certain derived features of the Cichlidae permit a degree of ecological accommodation (or niche partitioning, to use the older term) far greater than can be achieved by other teleost families. How this accommodation is brought about remains a controversial issue. Liem (1973, 1978, 1979; Liem and Osse 1975) argues that that the fusion of the lower pharyngeal jaw and the decoupling of the premaxilla from the maxilla represent "key innovations" of the Cichlidae, features that confer functional flexibility. These features, in short, allow the Cichlidae to escape the tyranny of form and provide the flexibility and trophic versatility necessary for the fine-scale partitioning of resources in a species-rich assemblage. Conversely, work carried out by members of the Leiden school of constructional morphology suggests a far closer fit between form and function (van Oijen 1982; Hoogerhoud, Witte, and Barel 1983; Barel 1983, 1984). Resource partitioning, under this rubric, is brought about by subtle morphological and habitat differences between species.

Yet another tantalizing element in the search for mechanisms underlying ecological accommodation comes from work on ontogenetic plasticity, which shows that a remarkable degree of phenotypic change can be induced by manipulating diet in both African and Neotropical cichlids (Greenwood 1965; Witte 1984a; Hoogerhoud 1986; Meyer 1987). Finally, the intraspecific trophic polymorphism exhibited by *Cichlasoma minckleyi* in the Cuatro Cienagas basin complicates the relationship between form and ecological specialization even further (Kornfield, Smith, and Gagnon 1982; Kornfield and Taylor 1983, Liem and Kaufman 1984).

Crucial as the search for the causes of coexistence is, it is not de facto an explanation for high diversity. What is being conflated here are the *origins* of diversity with the *maintenance* of diversity. The focus on mechanisms preventing competitive exclusion as the explanation for high diversity assumes that the inputs to the diversity equation are seldom limiting. Speciation is seen in diversity models much the way mutation is seen in population genetic models—important, but always sufficient.

Strong support for the view that interspecific competition regulates community membership certainly exists (Williams 1983; Roughgarden, Heckel and Fuentes 1983; Schoener 1984; Werner 1986; Roughgarden 1986). Particularly detailed work has been carried out on certain herpetofaunas of the Virgin Islands to show that their diversity and composition are clearly regulated by competition. In the Virgin Islands, the species diversity and composition of each island can be predicted by a set of simple, deterministic rules of assembly, mediated by habitat characteristics and ecological interaction (Mayer 1989). But this may be an unusual case. In contrast to the situation in the Greater Antilles, the Virgin Islands were once connected—the entire potential species pool could be found (or could reach) all the islands. The development of isolated islands and the consequent reduction in area and habitat diversity imply that each island decayed from its initial diversity toward a lower-diversity equilibrium. Under these circumstances, rates of speciation are essentially unimportant (or not limiting) in determining final diversity. Equilibria are quickly achieved without extensive speciation; preexisting in situ species pools form the basis for the eventual equilibrium diversity.

The situation in the Rift lakes, however, may be qualitatively different. Here, the levels of endemism suggest that the faunas diversify primarily by cladogenetic events, not by the decay of higher diversity species pools. Furthermore, although the stability of this complex community has not been analyzed in any detail, Lake Victoria may represent an icthyological assemblage far from equilibrium. It is a young lake, subject to frequent fluctuations in lake level and topography. In addition, theoretical models suggest the difficulty of maintaining stability in ecosystems of such high species diversity (Cohen and Newman 1985). In short, our static picture of current haplochromine diversity cannot tell us whether the system is at equilibrium. Although the older lakes in the Rift complex are similarly characterized by exceedingly diverse cichlid faunas, the cichlid species of Lake Malawi and Lake Tanganyika are far more differentiated, perhaps reflecting the longer-term consequences of ecological competition (Fryer and Iles 1972).

I do not wish to argue that competition is not a powerful regulator of community diversity, but instead to suggest that it is responsible primarily—through the agency of niche partitioning—for the *maintenance* of high species diversity in equilibrium situations. As such, it is a necessary, but not sufficient, explanation. We must still account for the *origins* of diversity, for the input terms of the diversity equation. Mechanisms responsible for ecological accommodation and partitioning may not become the main determinants of diversity until a sufficient number of ecological players has been placed on the stage. Clearly, there is a danger of drawing a false dichotomy: high diversity does not solely come about *either* because of high rates of origination *or* because of low rates of extinction (see Stanley, this volume). Nonetheless, traditional accounts have neglected the possibility that haplochromines currently dominate the Victorian icthyofauna, not through competitive superiority or ecological plasticity, but by virtue of the high rates at which new haplochromine species are produced.

Community ecologists have recently expressed a similar plea for increased attention to the factors controlling the input of new species (or propagules) into a given ecosystem. At least in some cases, community structure appears to depend not only on competitive interactions, but also on the mechanisms responsible for the introduction of new colonists into a given community (Ricklefs 1987; Roughgarden, Gaines, and Pacala 1987; Roughgarden 1989). At both the ecological and the evolutionary scales, inputs, and not just interactions, may be crucial regulators of overall diversity.

The high endemic diversity and relative recency of the Victorian haplochromine fauna suggest that the *origination* component of the diversity equation might still be discerned in this ecosystem. If so, could features *intrinsic* to haplochromines account for their higher rates of origination and hence for their dominance relative to the other teleost families in the lake?

Intrinsic Features: Intraspecific Variation and Population Structure in *Neochromis "velvet-black"*

The neglect of the origination term in the diversity equation stems in part from the perception that speciation is a passive consequence of chance environmental events. If, after all, speciation depends principally on chance abiotic phenomena, then differential rates of speciation arise stochastically. These different rates may still have important evolutionary consequences, but they do not come about as a consequence of biological features of the organisms involved. We gain little by seeking particularistic explanations for the dominance (achieved through high origination rates) of a given clade within a larger ecosystem.

A model that emphasizes origination as a component of evolutionary success assumes an interplay between the biotic and abiotic components of speciation. The study of speciation must consist of more than the identification of the geological or geographical barriers that initiate peripheral isolation. The genetic, ecological, and behavioral features that make different taxa differentially susceptible to extrinsic barriers need to be identified.

Early work on the teleost diversity in Lake Victoria focused almost exclusively on the search for extrinsic barriers. In part, this work sought to address a paradox: How could such a high degree of endemism exist in what appeared to be a large, continuous, relatively stable habitat? The evidence for frequent regressions in lake level during the Pleistocene provided at least a partial answer. These drops in water level would result in a series of small, peripherally isolated lakes fringing the main water body (Fryer 1959a, 1959b; Fryer and Iles 1972; Greenwood 1974).

This reconstruction of environmental events is likely correct, but it provides no explanation for one of the main features of the Victoria fauna: the dominance of haplochromines. Lake level fluctuations would presumably affect all the families of fishes living in the lake, resulting in small, polyspecific, peripheral isolates—yet only for the haplochromine cichlids have these external events led to an observable increase in diversity. The fossil and distributional evidence suggest that several teleost families were trapped in proto-Lake Victoria and were thus subjected to the same set of large-scale vicariance events (Greenwood 1959b). Could haplochromine propagules be more likely to result in viable, reproductively isolated populations after short periods of geographical isolation?

If haplochromine species are arrayed as a series of small, partially isolated demes, then environmental fracturing could have far more profound effects than it would upon a large panmictic population. Differentiation between populations (and the onset of reproductive isolation) would be far easier to achieve if environmental fluctuation were *superimposed* on a strongly deme-structured species (Wright 1941, 1945; Lande 1979; Templeton 1980, 1981; Patton and Smith 1989).

The extensive conservatism (or convergence) of morphological features in Victorian haplochromines makes it extremely difficult to analyze population structure using morphological characters. Genetic structure is more easily approached using molecular characters: this analysis focuses on restriction site variation in mitochondrial DNA (mtDNA). Mitochondrial DNA is a matrilineally inherited, circular piece of DNA that codes for certain of the proteins and tRNA's required by the eukaryotic mitochondrion (Clayton 1982, 1984; Wilson et al. 1985). Its function, however, need not concern us in this context. The main virtue of mtDNA is its ability to serve as a marker of relatedness between individuals or lineages, a marker that, in principle, is fully independent of morphology (Avise and Lansman 1983; Avise and Saunders 1984). To add to its appeal, it is a rapidly evolving molecule and hence a fast-ticking metronome of evolutionary change, ideal for reconstructing recent events (Brown, George, and Wilson 1979; Aquadro, Kaplan, and Risko 1984; Bermingham and Avise 1986).

The patterns of intraspecific variation in mtDNA provide important information about a key question: How are populations of a given haplochromine species arrayed over space? This study analyzes the population structure of a near-shore haplochromine species, provisionally named *Neochromis "velvet-black"* (Witte 1984b). This algal-grazing species can be collected at a number of localities throughout the lake without resorting to methods (such as trawling) that might artificially coalesce different populations.

I analyzed fine-scale population structuring by comparing mitochondrial variation in six population samples collected at localities no more than 10 km apart [fig. 12.1]. Intraspecific mitochondrial variation was established by examining the restriction patterns obtained with 17 different restriction enzymes. Each specimen yielded a particular pattern of fragments for each of the enzymes; these fragment patterns were then combined to produce a composite mitochondrial haplotype for each individual (Dorit 1986). The analysis of mitochondrial haplotype frequencies at each locality revealed that although all but one of the haplotypes was present in all of the populations sampled, the frequency distributions of the haplotypes showed significant heterogeneity. Even over the short distances separating the different localities, we are not simply sampling from the same underlying distribution [fig. 12.2].

Are these heterogeneity data sufficient to conclude that low levels of effective gene flow occur between populations, despite the short distances between localities? The haplotype distributions are not statistically independent, but we can determine the "minimum heterogeneous set," indicating the localities that must be included for the overall

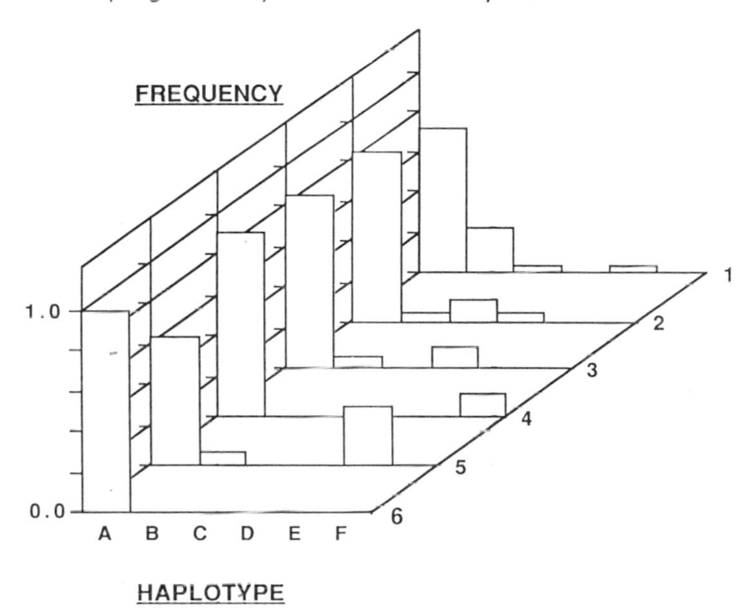

Figure 12.2. *Neochromis "velvet-black"* mitochondrial haplotype frequencies at six localities in the Mwanza Gulf, Lake Victoria, Tanzania. Populations are as follows: (1) Hippo Island, (2) Nyamatala Island, (3) Anchor Island, (4) Gillian Rocks, (5) Nyegezi Fisheries, (6) Kilimo Bay. Haplotypes are composites derived from the restriction patterns of 17 restriction enzymes (Dorit 1986).

Table 12.1. Heterogeneity G-test (a posteriori) of haplotype frequencies

Locality Set	G-value
[4, 5]	11.95 (n.s.)
[4, 5, 6]	15.48 (n.s.)
[2, 3, 4, 5, 6]	35.42 (n.s.)
[1, 2, 3, 4, 5]	43.04 (*)
[1, 2, 4, 5, 6]	39.01 (*)

Note: Localities are grouped into sets shown for heterogeneity analysis. Critical G-value derived from analysis using all localities: 37.652 (p = .05; d.f. = 25).

sample to remain significantly heterogeneous [table 12.1]. No clear relationship exists between the geographical position of a given population and overall heterogeneity. Thus, although the inclusion of the southernmost population (Gillian Rocks) is a necessary condition for significant heterogeneity when all haplotypes are considered, it is not a sufficient condition, as some subsets that include the Gillian Rocks population do not show significant G-values. Population I (Hippo Island) is not always critical for significant heterogeneity, although it is the geographical outlier. Only the population at Nyegezi (5), was present in all heterogeneous subsets, a fact unexpected from the geographical position of that population.

We are most likely observing in situ differentiation between populations of N. "*velvet-black*" and not just an artifact of advancing colonization or a haplotype frequency cline in the Mwanza Gulf. Although we cannot ascertain the composition of the original source population that gives rise to all the sampled populations, it probably included haplotypes A to E. The alternative, that identical haplotypes appeared independently in several of the populations, appears less parsimonious. The rare haplotypes (B to F) appear to have been previously generated from haplotype A by one to four restriction site changes.

We are clearly not sampling within a continuous and fully outbreeding population over the six localities in the study. The existence of heterogeneity in the haplotype distributions of the six populations suggests instead that a strong degree of population structuring exists over relatively short geographical ranges (Takahata and Slatkin 1984; Slatkin 1985). Furthermore, the structuring has remained sufficiently stable over evolutionary time to permit the development of observable between-locality frequency differences in mitochondrial haplotypes. The presence of variant haplotypes in low frequencies in each of the populations suggests that these populations have not undergone strong bottlenecking in population size (Avise, Neigel, and Arnold 1984b; Wilson et al. 1985).

This last aspect of the data is surprising in view of recent work on the geological and limnological history of Lake Victoria and other African Rift lakes (Stager 1984; Stager, Reinthal and Livingstone 1986). Cores from the northern regions of the lake reveal a strong sedimentary discontinuity at approximately 13,000 years B.P., thought to indicate a drop of as much as 66 m in the level of the lake, with a consequent 20% reduction in surface area and 45% reduction in shoreline. Given that the maximal depth of the Mwanza Gulf is probably less than 12 m., this scenario would indicate a complete drying up of the entire Gulf and a subsequent reinvasion less than 12,000 years ago. If the newly colonizing populations were homogeneous and monomorphic upon reinvasion, they are unlikely to have differentiated so appreciably in 12,000 years. This suggests either that the discontinuity seen in the cores of the northern section of the lake may not reflect a lakewide phenomenon or that swamps, streams, and other river drainages may have served as refugia during the period of aridity (Mann 1964; Stager, Reinthal, and Livingstone 1986). One additional scenario reconciles the paleoecological and neontological data: the population recolonizing the Mwanza Gulf (presumably from the more central and deeper regions of Lake Victoria) could have contained several haplotypes (A through E). The differentiation of haplotype *frequencies* that we observe has occurred since the recolonization; the appearance of multiple haplotypes precedes it.

MtDNA, by virtue of its maternal mode of inheritance, accurately mirrors the patterns of female dispersal. Infrequent male dispersal, though of profound relevance to the evolutionary fate of populations, is unlikely to be picked up by the analysis of mtDNA restriction patterns. Unless strong differences exist between the successful movement of males and that of females, the mtDNA data sketch a portrait of a highly structured species with limited effective dispersal of individuals, even over short distances.

What aspects of haplochromine biology underlie the strong degree of population structuring and the extremely rare dispersal of individuals seen in N. *"velvet black"*? Differentiation in lacustrine cichlids is frequently attributed to the ecological or habitat specialization of rock-dwelling forms (Witte 1981, 1984b). Rock-frequenting cichlids are never found over sandy or muddy habitats. They have been described as "reluctant" to cross stretches of open water, an observation often used to explain in situ differentiation (Fryer 1959a). Although the habitat fidelity of these fishes is certainly suggestive, the levels of migration required to prevent local differentiation are low enough to be missed by all but the most intensive of direct sampling methods. Furthermore, habitat restriction should not be confused with site fidelity: the fact that certain species are only found in the vicinity of rocks says nothing about their dispersal from one rocky habitat to another. Evidence on the colonization of experimental reefs should make us cautious about assuming the inability of these forms to migrate and colonize new habitats under appropriate circumstances (McKaye and Gray 1984).

The mouth-brooding habit that characterizes all of the Victoria haplochromines may be an important contributor to the observed differentiation, a point which has been raised by other cichlid workers (Fryer and Iles 1972; Dominey 1984; Dominey and Bulmer 1984). In particular, the fact that the fry remain tightly bound to their parent for several weeks has the unexpected consequence of precluding passive dispersal or active migration at a stage of the life cycle during which many freshwater teleosts tend to disperse. Although no direct evidence suggests that haplochromines do not disperse significantly at later stages in their ontogeny, it is clear that mouth brooding brings about, at least initially, a degree of philopatry not often found in other teleosts, including the far less speciose tilapiine cichlids (Lowe-McConnell 1969; Fryer and Iles 1969). A number of authors have coupled reduced dispersal, breeding structure, and complex courtship in haplochromines to argue for the importance of runaway sexual selection in establishing reproductive isolation among haplochromine species (McKaye 1983; Dominey 1986).

Differentiation in mtDNA frequencies, suggestive as it is, is of course *not* speciation. These results should not be construed as implicit support for sympatric models of speciation—the establishment of reproductive isolation undoubtedly requires extrinsic barriers. Nevertheless, the evidence for strong population structuring and incipient differentiation must be incorporated in any complete explanation of haplochromine diversity. Population structure clearly plays a crucial role in genetic differentiation, drift, and speciation (Wright 1932, 1941, 1982).

Molecular Phylogenies and the Interpretation of Trends: the Psammochromis-Macropleurodus Superlineage

A phylogeny of the Lake Victoria haplochromines is crucial to understanding this radiation. This phylogeny has proven elusive because of the absence of clear morphological synapomorphies on which to base a cladogram. Molecular analyses supply a different set of characters for phylogenetic reconstruction.

I constructed a branching diagram, based on mtDNA restriction sites, for the species in the *Psammochromis-Macropleurodus* superlineage. Greenwood previously defined this assemblage on morphological grounds (Greenwood 1980). It includes a variety of "trophic types," as well as the three monotypic genera (defined on the basis of their overall morphological divergence) present in Lake Victoria (Greenwood 1959a, 1974). MtDNA analysis offered the possibility of establishing a cladogram independent of morphology, which could then be compared to the morphological cladogram. Convergences, homologies, and parallelisms in form and function could then be identified (Kessler and Avise 1984; Neigel and Avise 1985).

Eight taxa in the *Psammochromis-Macropleurodus* superlineage were used (Fig. 12.4). Variable restriction sites were divided into unique, shared-primitive, and shared-derived sites; only shared-derived sites were used to establish a branching diagram of the species. The data were analyzed using the PHYSIS programs (Mickevich and Farris 1984), which generated the most parsimonious networks uniting the lineages. Two tree topologies are equally consistent with the data [fig.

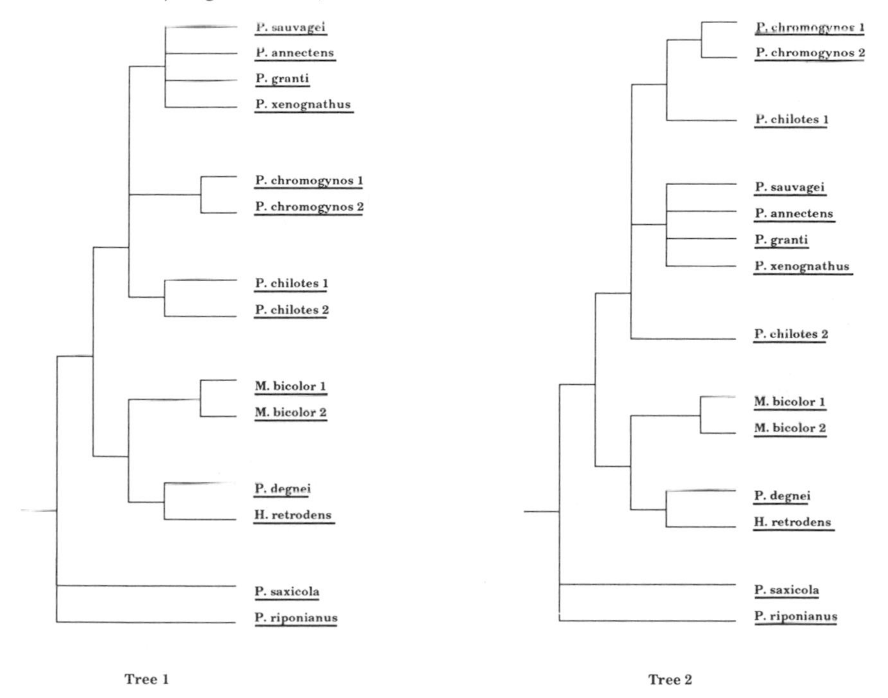

Figure 12.3. Alternate minimum path length trees for the *Psammochromis-Macropleurodus* superlineage, obtained from phylogenetically informative mtDNA restriction sites. For both topologies, the path length = 25; consistency index = 56.00.

12.3]. The consistency index of the two trees, a measure of the extent to which the synapomorphic characters are fully congruent with the tree, indicates that 60% of the sites support both of these trees. The remaining 40% of the sites fit the tree if parallel site gains or losses have occurred.

A number of tentative conclusions can be drawn:

1. The molecular data support several, but not all, of Greenwood's generic assignments made on the basis of morphology. These include the genera *Paralabidochromis* and *Psammochromis*. The remaining genera are neither supported nor contradicted by these data.

2. Not all species show characteristic (species-specific) mtDNA haplotypes. MtDNA haplotype differences accumulate as a function of time and are more likely to be preserved given reproductive isolation. The absence of differentiation between legitimate species may simply reflect the recency of the speciation events (Neigel and Avise 1985; Avise 1986) or, possibly, the existence of leaky reproductive barriers, which lead to introgression of extraspecific haplotypes (Powell 1983;

Spolsky and Uzzell 1984). Intraspecific hybridization in aquarium-kept Victorian haplochromines has been well documented (Capron de Caprona and Fritzsch 1984).

The interspecific differentiation in mtDNA haplotypes described here stands in contrast with the very low levels of differentiation in allozyme types or frequencies found in Lake Victoria cichlids (Sage et al. 1984; Verheyen, Van Rompaey, and Selens 1985) and in those of other lakes (Kornfield 1978; Kornfield, Smith, and Gagnon 1982). Low levels of mtDNA differentiation may also characterize certain Malawi cichlids (Kornfield, pers. comm. 1985).

3. The three monotypic genera emerge as sister groups in both of the tree topologies. This arrangement is supported by two synapomorphic site changes and contrasts with the arrangement proposed by Greenwood. Because the monotypic genera are considered the end points of a morphological trend in skull shape, the discordance between the molecular and morphological cladograms is more than a mere curiosity. It forces us to reconsider the explosion of forms that accompanies the diversification of Victorian haplochromines. The differing pathways and stages of morphological specialization embodied in each of these haplochromine species may shed light on the ways in which morphological trends arise.

In the context of evolutionary reconstruction, the existence of trends—orderly, steady progressions of form—always commands our attention. The hope has always been that the real forces of evolution will show their hand through the steady transformation of organisms in a single direction or through the repeated occurrence of similar trends in unrelated lineages. If there are any laws of evolution, trends are our best hope of identifying them.

The dense packing of haplochromine species in morphological space has led cichlid workers (Fryer and Iles 1972; Greenwood 1974, 1980, 1981, 1984; Fryer, Greenwood and Peake 1985) to identify series of extant species that appear to represent a sort of living trend—from a hypothesized primitive, generalized ancestor to a derived and more complex (specialized) form. The fossil record of the African Cichlidae consists primarily of a few scattered specimens from the Miocene of Rusinga Island (Greenwood 1959b; van Couvering and Miller 1969), and offers no clue about the order of appearance of the different forms. Extant species, however, can be thought of as the neontological equiv-

Figure 12.4. Representative neurocranial shapes arrayed on mtDNA-derived consensus cladogram. Neurocranial shape "trend" is assumed to run from the more generalized *P. degnei* or *H. retrodens* → *P. chilotes* → *P. annectens* → *P. xenognatus* → *M. bicolor* (in order of increasing derivation). ▶

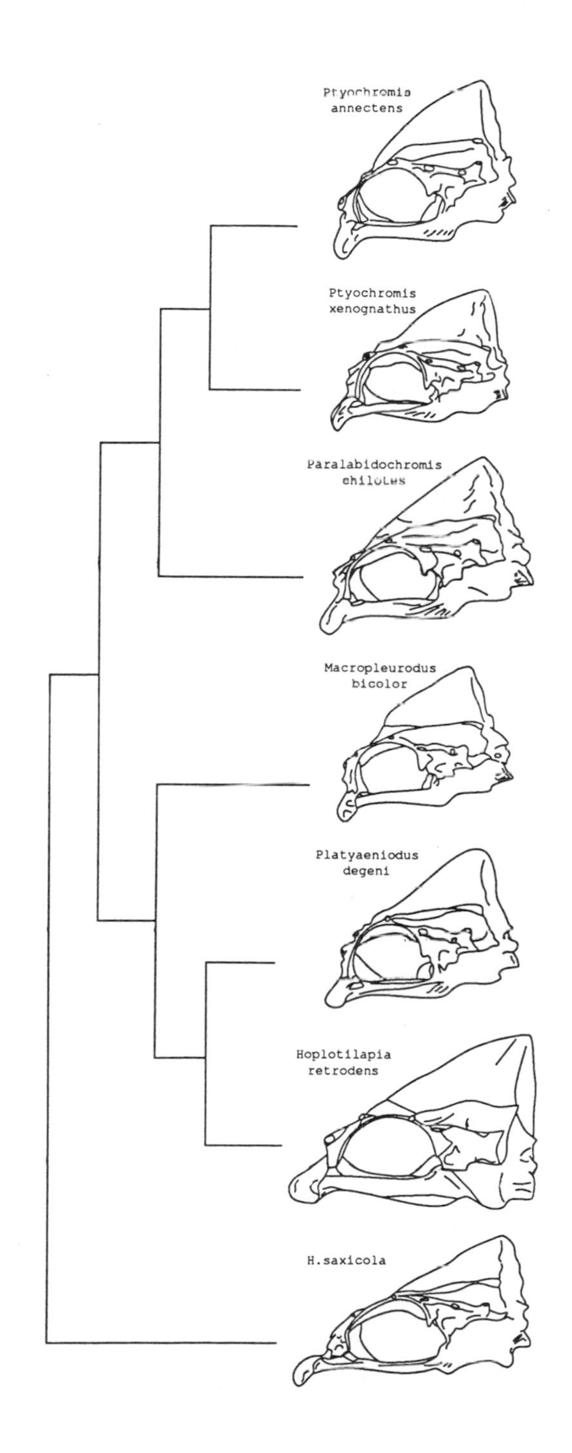
Ptyochromis annectens
Ptyochromis xenognathus
Paralabidochromis chilotes
Macropleurodus bicolor
Platyaeniodus degeni
Hoplotilapia retrodens
H.saxicola

1 cm

alent of a trend in the fossil record: rather than a series of successive fossil forms, each extant species connects primitive and derived morphologies. Greenwood (1981:70) has suggested the term "cladistic gradualism," which he introduces thus: "Admittedly, within lineages of the Lake Victoria flock one can see many examples of gradual change in a particular character or suite of characters. But, each point on the grade is a species, and the different species are contemporaneous, not successive elements in a temporal and phyletic continuum as they would be in the case of true phyletic gradualism. Perhaps one should call this phenomenon 'cladistic gradualism'?"

The *Psammochromis-Macropleurodus* superlineage, the object of much of this study, displays trends in at least three different character complexes: the proportions of the dentary, the attendant tooth implantation patterns, and the overall shape of the neurocranium (Greenwood 1980). We focus on the trend in neurocranial shape.

Although several species in the superlineage appear to share the same "degree" of derivation in neurocranial shape, it is nonetheless possible to establish a sequence of increasingly derived forms, based on the discussions in Greenwood (1980, 1981, 1984) (fig. 12.4).

The extent to which this series represents a sequence of ancestral and descendant forms can only be ascertained by erecting a classification independent of these characters. It is in this context that the tentative phylogenetic arrangement derived from the molecular data is particularly useful. The molecular cladogram suggests that the trend in neurocranial shape purportedly exhibited by the *Psammochromis-Macropleurodus* superlineage is more apparent than real [fig. 12.4]. The "trend"—namely the existence of intermediate forms linking apparent end points—exists only as a *static* description of the various morphologies present in the lineage. But the *evolutionary history* giving rise to this apparent trend is likely to be far more complex. The notion that cladistic gradualism is the process responsible for the trend may suffer from the same potential weakness as the claim of phyletic gradualism in the fossil record (Eldredge and Gould 1972; Gould and Eldredge 1977). Rather than a correct rendering of the history of morphological change in a lineage, gradual trends may simply reflect our ability to arrange forms in a sequence. Cladistic gradualism assumes that ancestral forms give rise at (or following) speciation to descendant forms that differ only slightly from their parent species. A number of such splitting events, coupled with the ability of ancestors to resist extinction in the face of descendant taxa, would yield a trend, a phylogenetic morphocline of increasing derivation.

The molecular cladogram, however, argues that the end points of the trend (*Platytaeniodus*, *Macropleurodus*, and *Hoplotilapia*) are sister groups, and the intermediate forms, as far as can be discerned, form the apomorphic sister group to the monotypic genera. If phylogenetic sister groups are not phenetically adjacent, as is the case in the *Psammochromis-Macropleurodus* superlineage, a different explanation for the apparent trend in neurocranial shape is required.

The molecular cladogram emphasizes that there is more than one way to skin a trend. In effect, the phylogenetic arrangement indicates that the "intermediate" phenotypes (intermediate neurocranial shapes) fill in the gap in morphospace left by the original branching events. The limits of the morphological envelope have been explored first, and subsequent branching events have filled in the available or accessible morphospace. Intermediate forms do indeed exist, but they are not being generated sequentially; extreme morphologies are not respectively the first and the most recent members of the radiation. Just as establishing a true gradualistic trend in the fossil record requires more than the connecting of two end points and the extrapolation of events in between, the apparent trend in neurocranial shape does not de facto assure us of how it came about.

Would a Fossil Record Suffice to Reconstruct the Cichlid Radiations?

The overall theme of this volume raises a crucial issue concerning our analyses of cichlid diversity: To what extent do our conclusions depend on neontological data? How much of the history of the Victorian radiation could we reconstruct using only the kinds of information available in the fossil record?

A truly pessimistic answer to this question has been set forth by Fryer et al. (1985), who suggest that much of the radiation would be missed if we had to rely on skeletal morphology alone. Characters unavailable in the record, such as live coloration and ecological habit, are often crucial in distinguishing haplochromine species, particularly in Lake Victoria (Greenwood 1965; Greenwood and Peake 1979b; Lewis 1982; Witte 1981). Nonetheless, if only post hoc, we can explore the extent to which morphometric information reflects the patterns and processes we can discern using molecular or other neontological evidence. I suggest that both the population structuring in *Neochromis "velvet-black"* and the radiation of the *Psammochromis-Macropleurodus* superlineage can be discerned by a sufficiently detailed morphometric analysis using characters available in a relatively complete fossil record that contained well-localized samples. The following sections describe the results of both intra- and interspecific morphometric analyses carried out on the specimens used in the previous molecular studies.

MORPHOMETRIC VARIATION IN *NEOCHROMIS "VELVET-BLACK"*

To what extent are the molecular interpopulation differences echoed at the morphological level? We performed a detailed morphometric analysis of skull shape on specimens of N. "*velvet-black*" in the six populations previously described. We took 40 measurements, using as reference points homologous landmarks that could be unambiguously discerned in all specimens (appendix 12.1). These linear measures formed a grid of partially redundant morphometric information (Bookstein et al. 1985): even slight changes in the shape or proportions of the head can thus be detected.

I used discriminant analysis to find differences between the six populations in the analysis, defined a priori. The resulting discriminant functions accurately classified all the N. "*velvet-black*" specimens [fig. 12.5]. This suggests that the morphometric differences between populations are consistent and geographically concordant. Nineteen additional specimens of N. "*velvet-black*," drawn from three of the original

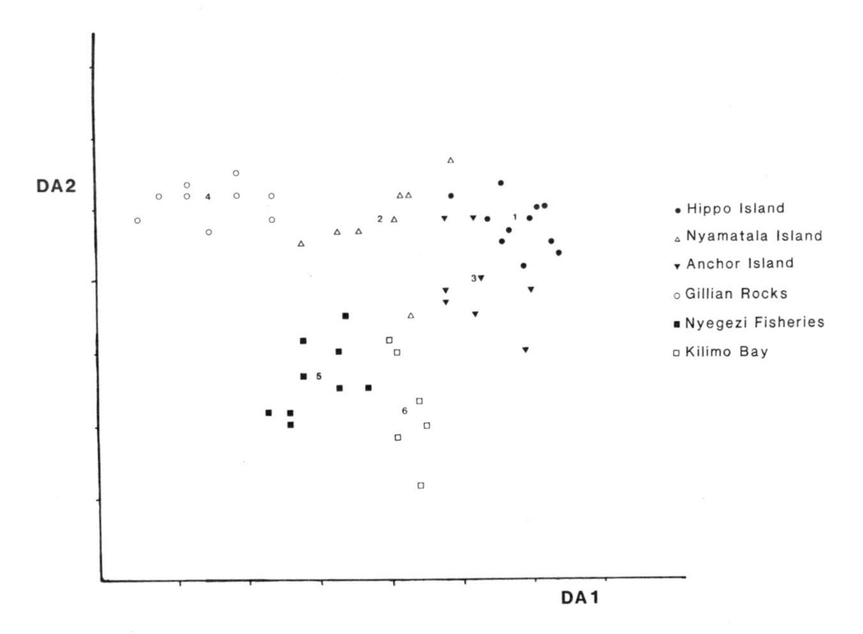

Figure 12.5. Positions in discriminant space of specimens of *Neochromis "velvet-black"* drawn from six localities. Numbers indicate locality centroids.

Table 12.2. Percent discriminant function classification efficiency using only pharyngeal jaw measurements

Actual Locality	Predicted Locality					
	1	2	3	4	5	6
1	46	18	18	9	0	9
2	34	55	11	0	0	0
3	13	0	61	0	13	13
4	22	22	0	34	11	11
5	11	11	0	11	56	11
6	0	0	0	0	43	57

Note: Discriminant analysis was carried out using only pharyngeal jaw measures (measures 25–31 in appendix 12.1). Localities for *N. "velvet-black"* as previously described.

localities (but not used to generate the discriminant functions), were also correctly assigned using the original discriminant functions. Given the relatively short distances separating the populations, the success of the discriminant analysis was unexpected: it suggests that subtle but consistent differences exist between populations. These populations do not differ significantly in the size distributions of the specimens. All pairwise Mahalanobis distance comparisons between populations are significant, except between the populations at Nyegezi Fisheries (5) and Kilimo Bay (6) ($p = 0.064$).

Oral jaw, opercular, and neurocranial measures must all be used to discriminate completely among all six localities. Analyses using only subsets of measurements substantially reduce the classifying accuracy of the discriminant functions (an example is shown in table 12.2). Using only characters of the pharyngeal jaw and branchial basket results in 50% misclassification (a random assignment of specimens to groups produces, on average, a 15% to 20% correct assignment). Despite the high correlations between the 40 measurements, each adds important information (Dorit 1986). The issue here is not the *magnitude* of the difference between localities but the existence of interpopulational differences detectable from information present in an idealized fossil sequence.

How would such differentiation, if encountered in the fossil record, be interpreted? One possibility is to describe it as adaptation to the subtly different resources at each locality (van Oijen 1982; Witte 1984b). Interlocality differences in the availability and distribution of resources exist, but these differences must be consistent and stable over evolutionary time if heritable differentiation is to develop as a result of adaptation. This account also assumes an extremely precise and detailed fit between the morphology and the resource utilization of these fishes, an assumption frequently challenged (Liem 1980a, 1980b). If the observed differentiation is considered simply ecophenotypic, this nonetheless requires that the resources at each site be distinct enough to produce the divergence. No indication of that degree of habitat heterogeneity exists for neighboring rocky shores in Lake Victoria.

Conversely, the interlocality differentiation may simply reflect the shallow fitness function that characterizes morphometric traits within a certain range. Rather than representing differential adaptation to differing ecologies, the different morphologies may be functionally equivalent. The morphometric differences between localities may be no more than evolutionary noise brought about by differences in the available genetic variation sequestered at each locality by the founding population or by slight variations in the epigenetic context in which development occurs. The stability over time of these morphometric differences has not been addressed and requires further neontological work. The remarkable phenotypic plasticity of certain cichlid species (Greenwood 1965; Witte 1984a; Meyer 1987) further complicates the interpretation of these results.

Whatever the basis of the interlocality differentiation, these six populations have not been sampled from a single underlying distribution. A discontinuity between the populations has existed long enough for morphometric differentiation to have occurred. Similar results are reported in N. "*velvet-black*" by Witte-Maas (cited in Witte 1984b), as well as for other Victorian haplochromines (Witte 1987).

MORPHOMETRIC VARIATION IN THE *PSAMMOCHROMIS-MACROPLEURODUS* COMPLEX

We analyzed interspecific differences in cranial morphology using the measurements described in the previous section. Species in the *Psammochromis-Macropleurodus* superlineage could not be distinguished unequivocally using principal components analysis (PCA). This technique uncovers the underlying "simple structure" of the multivariate data by collapsing the variance for all characters into a handful of orthogonal axes. The inability of this method to discriminate suggests a common *Baüplane* shared by all the species in the superlineage. In contrast, a broader analysis of more differentiated haplochromine taxa suggests that several of neurocranial shapes result from allometric growth (Strauss 1984).

The species in this superlineage can, however, be distinguished consistently by using discriminant analysis [fig. 12.6]. Sixty-seven of 69

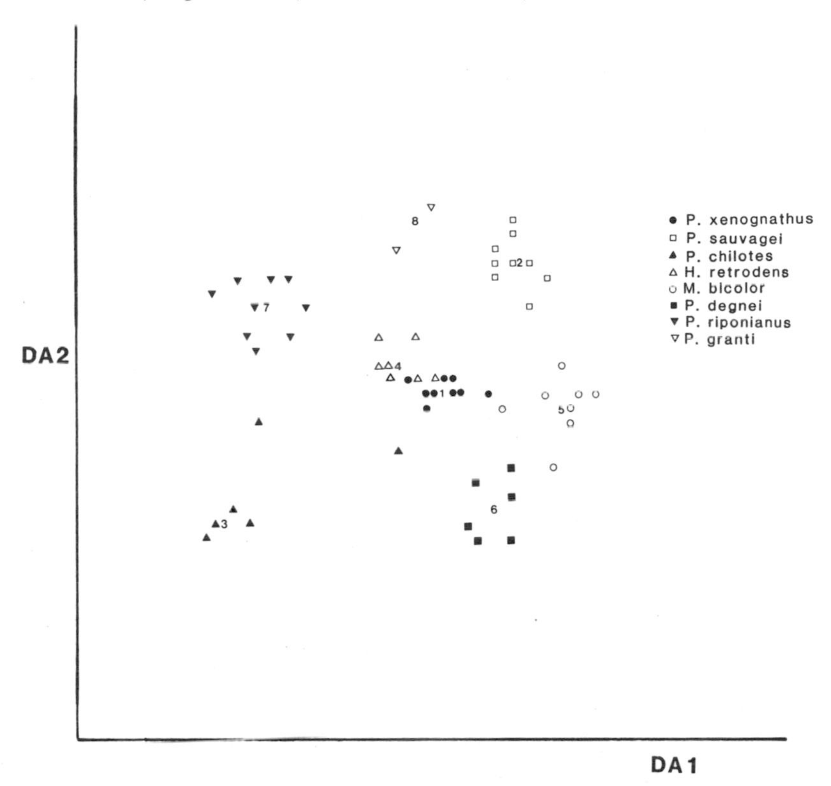

Figure 12.6. Positions in discriminant space of specimens of representative species in the *Psammochromis-Macropleurodus* superlineage. Numbers indicate species centroids.

specimens in this study were classified correctly (97.1%), indicating that the measurements used effectively describe the species involved. Two specimens of *Paralabidochromis chilotes* are misclassified: one is grouped with *Ptyochromis xenognathus*, the other with *Psammochromis riponianus*. A palentologist encountering a similar level of differentiation in a fossil assemblage would likely not confer specific rank to all of the currently recognized species.

Trends Revisited: High Diversity and the Filling of Morphological Space

If we assume the molecular phylogeny is correct, how might the apparent trend in neurocranial shape arise? How might such trends appear in the fossil record?

The molecular data suggest that the neurocranial shape trend in the

Psammochromis-Macropleurodus lineage has not arisen by the steady progression from generalized to specialized morphology. Instead, extreme morphologies appear to arise first, and later branching events give rise to shapes that fill the gap.

Two scenarios are possible. The monotypic genera may now be the "extreme" forms because they are the oldest, and they have slowly diverged over time (driven by competitive interactions) to occupy their positions as the extreme forms within the trend. Conversely, the branching events that gave rise to *Hoplotilapia*, *Macropleurodus*, and *Platytaeniodus* resulted in widely divergent morphologies that have since remained relatively static.

At present, we cannot distinguish between these two reconstructions. Given the recency of the Victorian radiation, neither the order of appearance of the taxa in the superlineage nor their individual transformational histories would likely be resolvable in the fossil record. However, in cases where apparent trends are generated over longer time spans, the fossil record remains the sole source of information about the order of appearance of related taxa. Embedded in the record is the only information that might distinguish between cladogenetic and anagenetic explanations for phenetic packing. Evidence of intraspecific stasis for the characters involved in the trend would argue that successful speciation events intercalate "intermediate" morphologies. Conversely, if phenetic packing arises as a consequence of anagenetic transformations following the appearance of new taxa, some form of character displacement must be invoked.

The discordance between the molecular and morphological cladograms of the *Psammochromis-Macropleurodus* superlineage sharply emphasizes the risks of using morphological (or stratigraphic) series for phylogenetic reconstruction. The phylogenetic utility of trends in the fossil record has been championed by systematists who frequently have had to rely on the geological sequence of appearance of forms to establish phylogenetic hypotheses (Bock 1970, 1979). But the potential for the evolutionary misinterpretation of phenetic series has been a quietly recurring theme of a number of studies (Lauder 1981; Eldredge and Cracraft 1980; Cracraft 1982; Vrba 1980; Larson 1983a, 1983b, 1989). Many of these critiques emphasize that an evolutionary trend is a historical hypothesis and thus requires some prior knowledge of ancestor-descendant (or sister group) relations. Clearly, a trend within a monophyletic group can only be identified post facto once a phylogeny (derived independently of the trend) has been erected. As Lauder (1981:432) warns: "the morphological series . . . , may bear no relation to the historical sequence of structural change."

Regardless of the exact way in which the phenetic packing in this superlineage came about, it is still paradoxical. Phenetic packing may be the price paid for the rapid and extensive speciation that has taken place in Lake Victoria. The monotypic genera may represent the accessible limits of the haplochromine *Baüplane*. Strictly speaking, these are not "constraints" (sensu Alberch 1982) but historically contingent limits. More extreme neurocranial shapes can be found in the cichlids from other Rift lakes and are therefore functionally possible; but these may only come about in far older radiations, as the edges of the morphological envelope are pushed out. This does not explain, however, why the "extreme" morphologies are explored first. Alternatively, phenetic packing could reflect the limits of available genetic variation, some form of canalization, or developmental buffering (Waddington 1957; Rendel 1979).

Finally, the packing may reflect an underlying adaptive topography: the neurocranial forms we find represent selective peaks, several of which converge near the intermediate neurocranial morphology. Once again, this selective scenario accounts for the forms encountered but not for their peculiar order of appearance.

The "cladistic gradualism" explanation for the neurocranial shape trend in the *Psammochromis-Macropleurodus* complex depends on two central assumptions. First, morphological evolution in this group is seen to proceed toward increased morphological specialization. Second, it assumes that the small morphological differences between species reflect gradual sequential change at branching events.

Both of these assumptions bear closer scrutiny. Liem (1978, 1979, 1980a, 1980b; Liem and Osse 1975) has questioned the concept of "specialization" as an explanation of morphological change in teleosts. In Victorian haplochromines, it is clear that evolutionary success, at least as defined by diversity, has been achieved by both so-called specialized (e.g., piscivorous) and generalized (e.g., detritivore or planktivore) lineages.

The peculiar way in which the neurocranial "trend" comes about—end points first, intermediate forms later—emphasizes the important distinction between phylogenetic and phenetic adjacency. The observation that a cluster of related species is tightly packed in phenetic space is not in itself evidence of gradual change. The morphological changes accompanying speciation can indeed be small (or nonexistent, as with sibling species). But unless and until branching order can be determined, continuous, gradualistic change at branching events remains a hypothesis to be tested, not an axiom to be relied upon. The temptation to use phenetic trends as a guide to evolutionary relationship is

ularly strong when only fossil evidence is available. As the neurocranial shape trend illustrates, however, morphological similarity between related taxa *need not* have arisen because morphological changes accompanying speciation are always small. The fact that ontogeny is continuous does not necessarily mean that transformational series are continuous themselves.

The fundamental causes underlying the haplochromine radiation in Lake Victoria, and in the other Rift lakes, are still poorly understood. As in all evolutionary reconstruction, we seek to infer dynamic mechanisms from static observations. The use of molecular tools, coupled with the detailed morphometric analysis of extant and fossil radiations, may allow for a more general theory of the origins of high diversity to emerge. Such a theory is, however, not yet at hand.

Literature Cited

Alberch, P. 1982. Developmental constraints in evolutionary processes. In *Evolution and development*, ed. J. Bonner, 313–32. Dahlem Konferenzen. Berlin: Springer-Verlag.

Aquadro, C. F., N. Kaplan, and K. J. Risko. 1984. An analysis of the dynamics of mammalian mitochondrial DNA sequence evolution. *Molec. Biol. Evol.* 1:423–34.

Avise, J. C. 1986. Mitochondrial DNA and the evolutionary genetics of higher animals. *Phil. Trans. R. Soc. Lond.* B312:325–42.

Avise, J. C., and R. A. Lansman. 1983. Polymorphism of mitochondrial DNA in populations of higher animals. In *Evolution of genes and proteins*, ed. M. Nei and R. K. Koehn, 147–64. Sunderland, Mass.: Sinauer Associates.

Avise, J. C., and N. C. Saunders. 1984. Hybridization and introgression among species of sunfish (*Lepomis*): Analysis by mitochondrial DNA and allozyme markers. *Genetics* 108:237–55.

Avise, J. C., J. E. Neigel, and J. Arnold. 1984b. Demographic influences on mitochondrial DNA lineage survivorship in animal populations. *J. Molec. Evol.* 20:99–105.

Barel, C. D. N., F. Witte, and M. J. P. van Oijen. 1976. The shape and the skeletal elements in the head of a generalized *Haplochromis* species: *H. elegans* Trevawas 1933 (Pisces, Cichlidae). *Neth. J. Zool.* 26:163–265.

Barel, C. D. N., M. J. P. van Oijen, F. Witte, and E. Witte-Maas. 1977. An introduction to the taxonomy and morphology of the haplochromine cichlidae from Lake Victoria. *Neth. J. Zool.* 27:333–89.

Barel, C. D. N. 1983. Towards a constructional morphology of cichlid fishes. *Neth. J. Zool.* 33:357–424.

Barel, C. D. N. 1984. From relations in the context of constructional morphology: the eye and suspensorium of lacustrine Cichlidae (Pisces, Teleostei). *Neth. J. Zool.* 34:439–502.

Beadle, L. C. 1981. *The inland waters of tropical Africa: an introduction to tropical limnology,* 2d ed. New York: Longman Press.

Bermingham, E., and J. Avise. 1986. Molecular zoogeography of freshwater fishes in southeastern United States. *Genetics* 113:939–65.

Bishop, W. W. 1965. Quaternary geology and geomorphology in the Albertine Rift Valley, Uganda. *Geol. Soc. Am. Sp. Paper* 84:293–321.

Bishop, W. W. 1969. *Pleistocene Stratigraphy in Uganda* Entebbe: Geological Survey of Uganda, Mem. X. Gov. Printer, 128 pp.

Bock, W. J. 1970. Microevolutionary sequences as a fundamental concept in macroevolutionary models. *Evolution* 24:704–22.

Bock, W. J. 1979. The synthetic explanation of macroevolutionary change—a reductionistic approach. In *Major patterns in vertebrate evolution,* ed. M. K. Hecht, P. C. Goody, B. M. Hecht, 851–95. New York: Plenum Press.

Bookstein, F., B. Chernoff, R. Elder, J. Humphries, G. Smith, and R. Strauss. 1985. *Morphometrics in evolutionary biology,* Special Publication 15. Philadelphia: Academy of National Sciences.

Brown, W. M., M. George, Jr., and A. C. Wilson. 1979. Rapid evolution of animal mitochondrial DNA. *Proc. Natl. Acad. Sci. U.S.A.* 76:1967–71.

Clayton, D. A. 1982. Replication of animal mitochondrial DNA. *Cell* 28:693–705.

Clayton, D. A. 1984. Transcription of the mammal mitochondrial genome. *Ann. Rev. Biochem.* 53:573–94.

Cohen, J., and C. M. Newman. 1985. When will a large and complex system be stable? *J. Theor. Biol.* 113:153–6.

Couvering, J. A. van, and J. A. Miller. 1969. Miocene stratigraphy and age determinations, Rusinga Island, Kenya. *Nature* 221:628–32.

Cracraft, J. 1982. A non-equilibrium theory for the rate control of speciation and extinction and the origin of macroevolutionary patterns. *Syst. Zool.* 31:348–65.

Crapon de Caprona, M. D., and B. Fritzsch. 1984. Interspecific fertile hybrids of haplochromine Cichlidae (Teleostei) and their possible importance for speciation. *Neth. J. Zool.* 34(4):503–38.

Dominey, W. J. 1984. Effects of species selection and life history on speciation: Species flocks in African cichlids and Hawaiian *Drosophila.* In *Evolution of fish species flocks,* ed. A. A. Echelle and I. Kornfield, 231–50. Orono: University of Maine Press.

Dominey, W. J., and L. S. Bulmer. 1984. Cannibalism of early life stages in fishes. In *Infanticide: Comparative and evolutionary perspectives,* ed. G. Hausfater and S. Blaffer Hrdy. Hawthorne, N.Y.: Aldine.

Dorit, R. L. 1986. Molecular and morphological variation in Lake Victoria haplochromine cichlids (Perciformes: Cichlidae). Ph.D. diss. Harvard University, Cambridge, Mass.

Eldredge, N., and S. J. Gould. 1972. Punctuated equilibria: an alternative to phyletic gradualism. In *Models in paleobiology,* ed. T. J. M. Schopf, 82–115. San Francisco: Freeman Cooper and Co.

Eldredge, N., and J. Cracraft. 1980. *Phylogenetic patterns and the evolutionary process*. New York: Columbia University Press.

Fryer, G. 1959a. The trophic interrelationships and ecology of some littoral communities of lake Nyasa with special reference to the fishes, and a discussion of the evolution of a group of rock-frequenting *Cichlidae*. *Proc. Zool. Soc. Lond*. 132:153–281.

Fryer, G. 1959b. Some aspects of evolution in Lake Nyasa. *Evolution* 13:440–51.

Fryer, G., and T. D. Iles. 1969. Alternatives routes to evolutionary success as exhibited by african cichlid fishes of the genus Tilapia and the species flocks of the Great Lakes. *Evolution* 23:359–69.

Fryer, G., and T. D. Iles. 1972. *The cichlid fishes of the Great African lakes*, Edinburgh: Oliver and Boyd.

Fryer, G., P. H. Greenwood, and J. F. Peake. 1985. The demonstration of speciation in fossil molluscs and living fishes. *Biol. J. Linn. Soc.* 26:325–36.

Gould, S. J., and N. Eldredge. 1977. Punctuated equilibria: the tempo and mode of evolution reconsidered. *Paleobiology* 3:115–51.

Graham, M. 1929. *The Victoria Nyanza and its fisheries*. London: Crown Agents for the Colonies.

Greenwood, P. H. 1951. Fish remains from the Miocene deposits of Rusinga Island and Kavirondo Province, Kenya. *Ann. Mag. Nat. Hist.* (12)4:192–201.

Greenwood, P. H. 1959a. The monotypic genera of cichlid fishes of Lake Victoria. II. *Bull. Br. Mus. Nat. Hist. (Zool.)* 5(7):163–77.

Greenwood, P. H. 1959b. Quaternary fish fossils. *Explor. du Parc Nat. Albert*. Fasc 4, no. 1.

Greenwood, P. H. 1965. Environmental effects on the pharyngeal mill of a cichlid fish, *Astatoreochromis alluaudi* and their taxonomic implications. *Proc. Linn. Soc. Lond.* 176:1–10.

Greenwood, P. H. 1974. *Cichlid fishes of Lake Victoria, East Africa: the biology and evolution of a species flock. Bull. Br. Mus. Nat. Hist. (Zool.)* 6 (suppl.):1–134.

Greenwood, P. H., and C. D. N. Barel. 1978. A revision of the Lake Victoria *Haplochromis* species (Pisces, Cichlidae). VIII. *Bull. Br. Mus. Nat. Hist. (Zool.)* 33(2):141–92.

Greenwood, P. H. 1979a. Towards a phyletic classification of the 'genus' *Haplochromis* (Pisces, Cichlidae) and related taxa. I. *Bull. Br. Mus. Nat. Hist. (Zool.)* 34(4):265–322.

Greenwood, P. H. 1979b. Macroevolution—myth or reality? *Biol. J. Lin. Soc.* 12:293–304.

Greenwood, P. H. 1980. Towards a phyletic classification of the 'genus' *Haplochromis* (Pisces, Cichlidae) and related taxa. II. The species from Lake Victoria, Nabugabo, Edward, George and Kivu. *Bull. Br. Mus. Nat. Hist. (Zool.)* 39(1):1–101.

Greenwood, P. H. 1981. Species flocks and explosive evolution. In *Chance, change and challenge—the evolving biosphere*, ed. P. H. Greenwood and P. L. Forey, 61–74. London: Cambridge University Press and British Museum (Nat. Hist.)

Greenwood, P. H. 1983. Notes on the anatomy and phyletic relationships of hemichromis Peters, 1858. *Bull. Br. Mus. Nat. Hist.* 48:131–71.

Greenwood, P. H. 1984. African cichlids and evolutionary theories. In *Evolution of fish species flocks*, ed. A. A. Echelle and I. Kornfield, 141–54. Orono: University of Maine Press.

Hoogerhoud, R. J. C., F. Witte, and C. D. N. Barel. 1983. The ecological differentiation of two closely resembling *Haplochromis* species from Lake Victoria (*H. iris* and *H. hiatus;* Pisces, Cichlidae). *Neth. J. Zoology* 33:283–305.

Hoogerhoud, R. J. C. 1986. Taxonomic and ecological aspects of morphological plasticity in molluscivorous haplochromines (Pisces, Cichlidae). *Ann. Mus. R. Afr. Centre Sci. Zool.* 251:131–4.

Kendall, R. L. 1969. An ecological history of the Lake Victoria basin. *Ecol. Monog.* 39:121–76.

Kessler, L. G., and J. C. Avise. 1984. Systematic relationships among waterfowl (Anatidae) inferred from restriction endonuclease analysis of mitochondrial DNA. *Syst. Zool.* 33:370–80.

Kornfield, I. 1978. Evidence for rapid speciation in African cichlid fishes. *Experientia* 34:335–6.

Kornfield, I., D. C. Smith, and P. S. Gagnon. 1982. The cichlid fish of Cuatro Cienagas, Mexico: Direct evidence of conspecificity among distinct trophic morphs. *Evolution* 36(4):658–64.

Kornfield, I. L., and J. N. Taylor. 1983. A new species of polymorphic fish, *Cichlasoma minckleyi* from Cuatro Cienagas, Mexico (Teleostei: Cichlidae). *Proc. Biol. Soc. Wash.* 96:253–69.

Lande, R. 1979. Effective deme sizes during long-term evolution estimated from rates of chromosomal rearrangement. *Evolution* 33:234–51.

Larson, A. 1983. A molecular phylogenetic perspective on the origins of a tropical lowland salamander fauna. I. Phylogenetic inferences from protein comparisons. *Herpetologica* 39:85–99.

Larson, A. 1983. A molecular phylogenetic perspective on the origins of a lowland tropical salamander fauna. II. Patterns of morphological evolution. *Evolution* 37(6):1141–53.

Larson, A. 1989. The relationship between speciation and morphological evolution. In *Speciation and its consequences*, ed. D. Otte and J. A. Endler, 579–98. Sunderland, Mass.: Sinauer Associates.

Lauder, G. V. 1981. Form and function: structural analysis in evolutionary morphology. *Paleobiology* 7(4):430–2.

Lewis, D. S. C. 1982. Problems of species definition in Lake Malawi cichlid fishes (Pisces: Cichlidae). *J. L. B. Smith Institute of Ichthyology Spec. Pub.* 23:1–5.

Liem, K. F. 1973. Evolutionary strategies and morphological innovations: Cichlid pharyngeal jaws. *Syst. Zool.* 22:425–41.

Liem, K. F., and J. W. M. Osse. 1975. Biological versatility, evolution and food resource exploitation in African cichlid fishes. *Am. Zool.* 15:427–54.

Liem, K. F. 1978. Modulatory multiplicity in the functional repertoire of the feeding mechanism in cichlid fishes. *J. Morphol.* 158(3):323–60.

Liem, K. F. 1979. Modulatory multiplicity in the feeding mechanism in cichlid fishes, as exemplified by the invertebrate pickers of Lake Tanganyika. *J. Zool. Lond.* 189:93–125.

Liem, K. F. 1980a. Acquisition of energy by teleosts: Adaptive mechanisms and evolutionary patterns. In *Environmental physiology of fishes*, ed. M. A. Ali, 299–333. New York: Plenum Publishers.

Liem, K. F. 1980b. Adaptive significance of intra- and interspecific differences in the feeding repertoires of cichlid fishes. *Am. Zool.* 20:295–314.

Liem, K. F., and L. S. Kaufman. 1984. Intraspecific macroevolution: functional biology of the polymorphic cichlid species *Cichalsoma minckleyi*. In *Evolution of Fish Species Flocks*, ed. A. A. Echelle and I. Kornfield, 155–67. Orono: University of Maine Press.

Lowe-McConnell, R. H. 1969. Speciation in Tropical freshwater fishes. *Biol. Jour. Linn. Soc* 1:51–75.

Mann, M. J. 1964. *Report on a fisheries survey of Lake Rukwa*. Report of the East African Freshwater Fisheries Research Organization, 1962–1963, 42–52.

Mayer, G. C. 1989. Deterministic aspects of community structure in West Indian amphibians and reptiles. Ph.D. diss., Harvard University, Cambridge, Mass.

Meyer, A. 1987. Phenotypic plasticity and heterochrony in *Cichlasoma managuense* (Pisces: Cichlidae) and their implications for speciation in cichlid fishes. *Evolution* 41:1357–69.

Mickevich, M. F. and J. S. Farris. 1984. *PHYSIS (Phylogenetic Analysis System)*. Documentation for VAX/VMS installation at Harvard University, Cambridge, Mass.

McKaye, K. R. 1983. Ecology and Breeding Behavior of a Cichlid Fish, *Cyrtocara Eucinostonus* on a Large Lek in Lake Malawi, Africa. *Envir. Biol. Fish.* 8:81–96.

McKaye, K. R. and W. N. Gray. 1984. Extrinsic barriers to gene flow in rock-dwelling cichlids of Lake Malawi: macrohabitat heterogeneity and reef colonization. In *Evolution of fish species flocks*, ed. A. A. Echelle and I. Kornfield, 169–84. Orono: University of Maine Press.

Neigel, J. E. and J. C. Avise. 1985. Phylogenetic relationships of mitochondrial DNA under various demographic models of speciation. In *Evolutionary processes and theory*, ed. E. Nevo and S. Karlin, 515–34. New York: Academic Press.

Oijen, M. J. P. van. 1982. Ecological differentiation among the haplochromine piscivorous species of lake Victoria. *Neth. J. Zool.* 32:336–63.

Patton, J. L., and M. F. Smith. 1989. Population structure and genetic and morphologic divergence among pocket gopher species (Genus *Thomomys*). In *Speciation and its consequences*, ed. D. Otte and J. A. Endler, 579–98. Sunderland, Mass.: Sinauer Associates.

Powell, J. R. 1983. Interspecific cytoplasmic gene flow in the absence of nuclear gene flow: Evidence from *Drosophila*. *Proc. Natl. Acad. Sci. U.S.A.* 80:492–5.

Rendel, J. M. 1979. Canalisation and selection. In *Quantitative genetic variation*, ed. J. N. Thomson and J. M. Thoday, 139–56. New York: Academic Press.

Ricklefs, R. E. 1987. Community diversity: Relative roles of local and regional processes. *Science* 235:167–71.

Roughgarden, J., D. Heckel, and E. Fuentes. 1983. Coevolutionary theory and the biogeography and community structure of *Anolis*. In *Lizard ecology: Studies on a model organism*, ed. R. Huey, E. Pianka and T. Schoener, 371–410. Cambridge, Mass.: Harvard University Press.

Roughgarden, J. 1986. A comparison of food-limited and space-limited animal competition communities. In *Community ecology*, ed. J. Diamond and T. Case, 492–516. New York: Harper and Row.

Roughgarden, J., S. Gaines, and S. Pacala. 1987. Supply-side ecology: the role of physical transport processes. In *Organization of communities: Past and present*, ed. P. Giller and J. Gee, 459–86. London: Blackwell Scientific.

Roughgarden, J. 1989. The structure and assembly of communities. In *Perspectives in ecological theory*, ed. J. Roughgarden, R. M. May, S. Levin, 203–26. Princeton, N.J.: Princeton University Press.

Sage, R. D., P. V. Loiselle, P. Basasibwaki, and A. C. Wilson. 1984. Molecular versus morphological change among cichlid fishes of Lake Victoria. In *Evolution of fish species flocks*, ed. A. A. Echelle and I. Kornfield, 185–201. Orono: University of Maine Press.

Schoener, T. 1984. Size differences among sympatric bird-eating hawks: a worldwide survey. In *Ecological communities: Conceptual issues and the evidence*, ed. D. Strong, D. Simberloff, L. Abele and A. Thistle, 254–81. Princeton, N.J.: Princeton University Press.

Slatkin, M. 1985. Gene flow in natural populations. *Ann. Rev. Ecol. Syst.* 16:393–430.

Spolsky, C., and T. Uzzell. 1984. Natural interspecies transfer of mitochondrial DNA in amphibians. *Proc. Nat. Acad. Sci. U.S.A.* 81:5802–5.

Stager, J. C. 1984. The diatom record of Lake Victoria, East Africa: the last 17,000 years. *Proc. VIIth Int. Diat. Symp., Philadelphia, 1982*, ed. D. G. Mann, 455–76. Koenigstein, Germany: Strauss and Cramer.

Stager, J. C., P. N. Reinthal, and D. A. Livingstone. 1986. A 25,000 year history for Lake Victoria, East Africa, and some comments on its significance for the evolution of cichlid fishes. *Fresh. Biol.* 16:15–19.

Stiassny, M. L. J. 1981. The phyletic status of the family Cichlidae (Pisces, Perciformes): a comparative anatomical investigation. *Neth. J. Zool.* 31:275–314.

Strauss, R. A. 1984. Allometry and functional feeding morphology in haplochromine cichlids. In *Evolution of fish species flocks*, ed. A. A. Echelle and I. Kornfield, 217–30. Orono: University of Maine Press.

Takahata, N., and M. Slatkin. 1984. Mitochondrial gene flow. *Proc. Nat. Acad. Sci. U.S.A.* 81:1764–7.

Temple, P. H. 1969. Some biological implications of a revised geological history of Lake Victoria. *Biol. J. Linn. Soc.* 1:363–71.

Templeton, A. R. 1980. The theory of speciation via the founder principle. *Genetics* 94:1011–38.

Templeton, A. R. 1981. Mechanisms of speciation—a population genetic approach. *Ann. Rev. Ecol. Syst.* 12:23–48.

Verheyen, E., J. van Rompaey, and M. Selens. 1985. Enzyme variation in haplochromine cichlid fishes from Lake Victoria. *Neth. J. Zool.* 35(3):469–78.

Vrba, E. S. 1980. Evolution, species and fossils: How does life evolve? *South Afr. J. Sci.* 76:61–84.

Waddington, C. H. 1957. *The strategy of the genes*. London: Allen Unwin.

Werner, E. 1986. Species interactions in freshwater fish communities. In *Community ecology*, ed. J. Diamond and T. Case, 344–58. New York: Harper and Row.

Williams, E. 1983. Ecomorphs, faunas, island size and diverse endpoints. In *Lizard ecology: Studies on a model organism*, ed. R. Huey, E. Pianka and T. Schoener, 326–70. Cambridge, Mass.: Harvard University Press.

Wilson, A. C., R. L. Cann, S. M. Carr, M. George, U. B. Gyllensten, K. M. Helm-Bychowski, R. G. Higuchi, S. R. Palumbi, E. M. Prager, R. D. Sage, and M. Stoneking. 1985. Mitochondrial DNA and two perspectives on evolutionary genetics. *Biol. J. Linn. Soc.* 26(4):375–400.

Witte, F. 1981. Initial results of the ecological survey of the haplochromine cichlid fishes from the Mwanza Gulf of Lake Victoria (Tanzania): Breeding patterns, trophic and species distribution. *Neth. J. Zool.* 31:175–202.

Witte, F. 1984a. Consistency and functional significance of morphological differences between wild-caught and domestic *Haplochromis squamipinnis*. *Neth. J. Zool.* 34(4):613–21.

Witte, F. 1984b. Ecological differentiation in Lake Victoria haplochromines: comparison of cichlid species flocks in African Lakes. In *Evolution of fish species flocks*, ed. A. A. Echelle and I. Kornfield, 155–67. Orono: University of Maine.

Witte, F. 1987. From form to fishery. Ph.D. diss. Staatsuniversiteit Leiden, Leiden, The Netherlands.

Worthington, E. B. 1929. The life of Lake Albert and Lake Kioga. *Geogr. J.* 74:109–32.

Wright, S. 1932. The roles of mutation, inbreeding, crossbreeding and selection in evolution. *Proc. VI Int. Cong. Genet.* 1:356–66.

Wright, S. 1941. On the probability of fixation of reciprocal translations. *Am. Nat.* 74:513–22.

Wright, S. 1945. Tempo and mode in evolution: a critical review. *Ecology* 26:415–9.

Wright, S. 1982. Character change, speciation and the higher taxa. *Evolution* 36:427–43.

Appendix

A12.1. Measurements employed in inter- and intraspecific morphometric analyses.

1. Standard length
2. Ascending arm of the premaxilla
3. Dentigerous arm of the premaxilla
4. Maxilla length: neurocranial condyle to shank protuberance
5. Dentary length: posterior tip of ascending wing to point of implantation of anteriormost tooth
6. Dentary length: point of anteriormost tooth implantation to dorsal, posterior margin of dentary
7. Lower jaw length (to posterior articulation of anguloarticular)
8. Anguloarticular, tip of primordial process to posterior articulation
9. Anguloarticular, posterior articular to point of anterior reentry into dentary
10. Preoperculum length: slf1 to slf4, posterior vertical margin
11. Preoperculum length: slf4 to slf7, ventral margin
12. Preoperculum length: slf7 to slf1
13. Lower margin of interoperculum
14. Dorsal margin of suboperculum
15. Dorsal margin of operculum
16. Posterior margin of operculum
17. Anterior margin of operculum
18. Hyomandibular length: width at articulation, neural condyle to operculum condyle
19. Hyomandibular length from condyle to lower (ventral) tip of bone
20. Interoperculum, length of upper (dorsal) margin
21. Lacrymal (L), length of anterior margin
22. Lacrymal (L), length of dorsal margin
23. Lacrymal, length of posterior margin
24. Lacrymal (L), length of ventral margin
25. Lower pharyngeal element, right to left horn
26. LPE, right horn to tip, dorsal view, of keel
27. LPE, width of dentigerous area, at posterior margin
28. LPE, Length of dentigerous area, at posterior margin
29. Length of branchial basket, from below central cartilage to anterior edge of glossohyoid
30. Width of branchial basket
31. Right margin of branchial basket, tip of R in.-hy. to tip of glossohysid
32. Posterior tip of neurocranial crest to articulation facet with first vertebra

A12.1. (*continued*)

33. Posterior tip of neurocranial crest to maxillary articulation zone, anterior margin of neurocranium
34. Tip of neurocranial crest to pharyngobranchial apophysis
35. Ventral margin, first vertebral articulation to vomerine notch
36. Vertebral articulation to pharyngobranchial apophysis
37. Ventral margin, orbit, taken immediately above vomer
38. Pharyngobranchial apophysis to maximum articulation, anterior tip of neurocranium
39. Pharyngobranchial apophysis to vomerine notch
40. Pharyngobranchial apophysis to protruberance, ventral crest of parasphenoid.

Note: Figures illustrate the actual measurements on an idealized haplochromine skull. Drawings and measurement designations are derived from Barel, Witte, and van Oijen (1976).

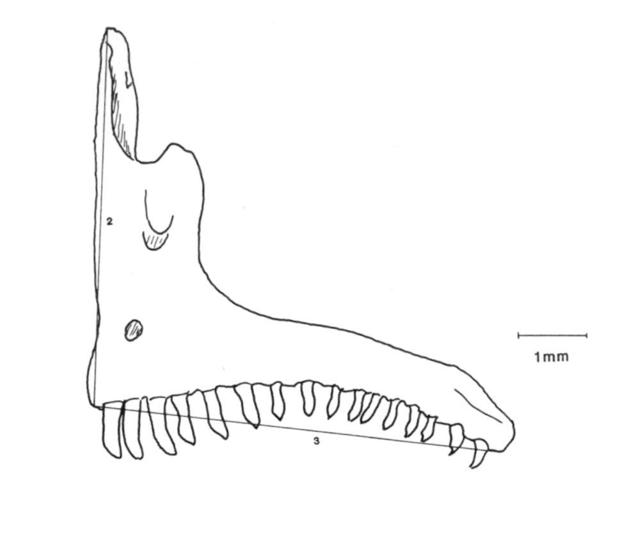

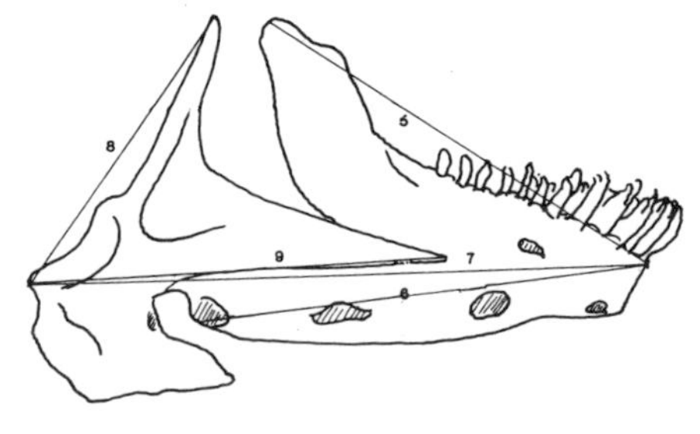

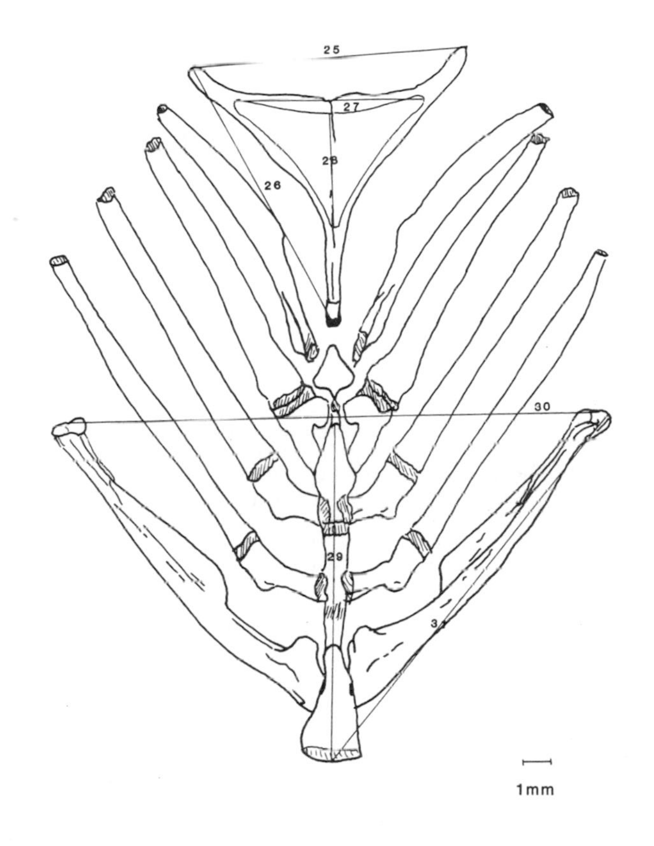
25
27
28
26
30
29
1mm

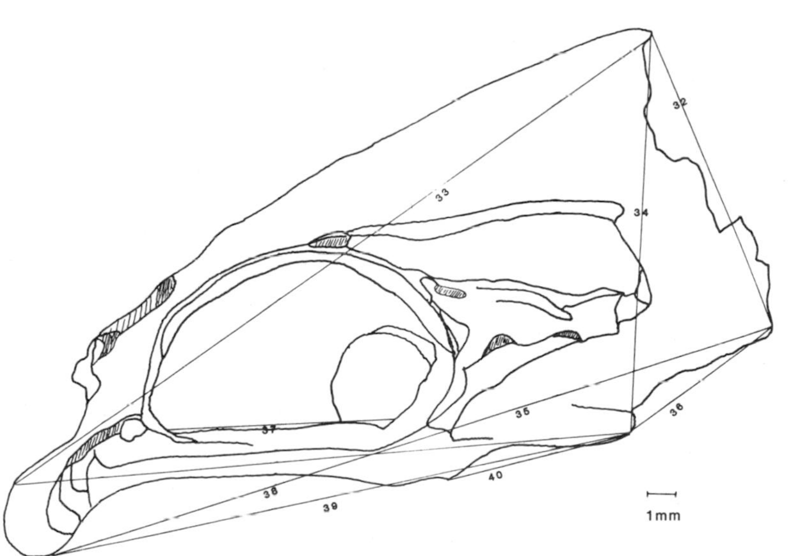
32
33
34
35
36
37
38
39
40
1mm

13

Ecological Causes of Clade Diversity in Hummingbirds: A Neontological Perspective on the Generation of Diversity

Robert Bleiweiss

A fundamental question about the origin of biological diversity is the extent to which the intrinsic attributes of organisms pace evolutionary radiations. Intrinsic control implies that diversification is governed by biological properties of organisms rather than by idiosyncratic environmental variation (Mayr 1963; Jackson 1974; Vrba 1983, 1984; Stenseth and Maynard Smith 1984). Indeed, differences in ecological specialization (breadth of a species' niche) and dispersal ability (vagility) are often associated with differences in diversity among related lineages, with higher diversity being positively correlated with increased ecological specialization and reduced vagility. Examples include drosophilid flies in Hawaii (Templeton 1979), cichlid fishes in African Rift lakes (Echelle and Kornfield 1984; Dorit, this volume), antelopes in sub-Saharan Africa (Vrba 1980, 1983, 1984, 1987), and several invertebrate groups (Hansen, 1983; Valentine and Jablonski 1983; Cohen and Johnston 1987). Paleontologists in particular have used the correlation as the empirical foundation for recent theories of macroevolution (Eldredge 1979; Vrba 1984).

In view of these speculations, hummingbirds (family Trochilidae) are an important group in the examination of intrinsic controls on diversification. This monophyletic New World family of specialized nectar feeders consists of two subfamilies, the Phaethornithinae (approximately 35 species) and Trochilinae (approximately 295 species). The ninefold difference in species diversity between subfamilies is remarkable because the depauperate Phaethornithinae are both less vagile *and* more ecologically specialized than are most species in the Trochilinae.

The hummingbird pattern is a striking counterexample to the prevailing views of macroevolutionary theory. In this paper, I conclude that intrinsic factors alone cannot account for the diversity difference between hummingbird subfamilies, largely because the effects of intrinsic factors on diversification cannot be divorced from the context of the environment.

Hummingbirds as a Natural Test of Macroevolutionary Hypotheses

Current evidence strongly suggests that the two hummingbird subfamilies are sister groups—clades (strictly monophyletic groups)—that share a more recent common ancestor with each other than either does with any other taxon. This phylogenetic hypothesis appears corroborated by both anatomical (Zusi 1980; Zusi and Bentz 1982), and biochemical (Sibley, Ahlquist, and Monroe 1988; Bleiweiss et al., in prep.) data. Such a pattern of diverse and depauperate sister clades within a certainly monophyletic family is widely recognized as essential to testing evolutionary modes during diversification (Mayden 1986; Raikow 1986; Fitzpatrick 1988).

The two hummingbird subfamilies meet several restrictive conditions necessary for examining the role of intrinsic and extrinsic factors in the generation of diversity. As sister groups, the two hummingbird subfamilies are the same age. The subfamilies' ecologies are understood to an extraordinary degree, and it is known that they exhibit striking differences in ecological specialization and in vagility. Moreover, the vast majority of trochilines (93%) breed and winter within the geographical range occupied by phaethornithines (Mexico to Argentina; Olrog 1959; Lack 1973; Johnsgard 1983). In particular, both subfamilies occur in mountainous areas, where many extrinsic barriers may promote speciation (Mayr 1963; Futuyma 1986; Graves 1985). Given these advantages, the relative importance of intrinsic and extrinsic factors can be adequately considered.

Differences in Intrinsic Attributes

Even among extant avian groups, the behavior and ecology of hummingbirds is documented and understood to an extraordinary degree. My review of these data indicates that the prime differences between subfamilies relate to feeding specializations within the nectar-feeding niche and to vagility.

REPRODUCTION

Among intrinsic factors, different reproductive biologies can probably be dismissed as an explanation for the different number of species in the two hummingbird subfamilies. There is no evidence for differences in life span, generation time, or clutch size among hummingbirds (Perrins and Middleton 1985). All hummingbird species studied to date are polygynous (Wolf and Stiles 1970; Stiles and Wolf 1979), a mating system generally thought to enhance speciation (Futuyma 1986). The available data suggest that, if anything, sexual selection is more intense among phaethornithines; a greater percentage of phaethornithines ex-

Table 13.1. The characterization of hummingbird subfamilies

	Number of Species (%)	
	Phaethornithinae (specialist)	Trochilinae (generalist)
Breeding System†		
Lekking	15 (83)	16 (55)
Nonlekking	3 (17)	13 (45)
Bills†		
Decurved	25 (73.5)	9 (3)
Straight‡	7 (20.5)	250 (95)
Other§	2 (6)	4 (2)
Vagility†		
Sedentary	11 (85)	37 (37)
Altitudinal migrants	2 (15)	46 (45)
Long-distance migrants	0 (0)	18 (18)
Island		
Oceanic island	1 (3)	22 (7)
Endemic island species	0 (0)	15 (5)

*Based on broad survey of ecological and behavioral characteristics from among 297 of the approximately 330 known species and most genera of hummingbird. Hummingbird mating systems vary from nectar-centered territories held by males to lek-mating systems in which the male does not defend a resource. I considered any species with some lekking behavior to be a lek-mating species. The higher proportion of lekkers in the Phaethornithinae must be viewed as a provisional trend until more data are obtained; breeding biologies are documented for relatively few trochilines, and the breeding system of species in both subfamilies may vary geographically from a nectar-centered territorial to a lek-type one. Bill structure indicates the degree of ecological specialization in nectar-foraging (see text). Dispersal tendencies were divided into three categories, sedentary (includes local movements among habitat patches), attitudinal migrants, and long-distance (latitudinal or longitudinal) migrants. Oceanic islands are defined as those that have never been connected to the mainland. These islands must be reached by dispersal, and a groups incidence on oceanic islands is a measure of its dispersal abilities. Literature citations and data for tables 13.1 to 13.3 are provided in appendices 13.1 to 13.3. All statistical analyses were performed with the *Statistical Package for the Social Sciences* (Nie et al. 1983).

†Significant at $p < 0.01$ by chi square with Yates correction

‡Includes species with slightly decurved bills

§Recurved or with a strong hook at the tip

hibit lek-type mating systems (table 13.1), a form of polygyny in which sexual selection may be the strongest. In lek-type mating systems members of one sex, typically males, display at traditional sites visited by the other sex. Relatively few males garner the majority of matings, thereby skewing reproductive success and increasing the strength of sexual selection on males.

ECOLOGICAL SPECIALIZATION

Most trochilines studied have generalized feeding ecologies and exploit a wide range of flower taxa and morphological types (Feinsinger 1976; Feinsinger and Colwell 1978; Stiles 1985). Morphological specialization need not correlate with ecological specialization, but it appears to do so in hummingbirds. Hummingbirds with straight bills of moderate length exploit a greater taxonomic and structural diversity of flowers than do species with other bill shapes (Feinsinger and Colwell 1978). Although the ecology of many trochilines is unknown, the fact that most have generalized bill shapes indicates that the subfamily as a whole can be characterized as having generalized feeding ecologies.

Because many trochiline species often compete for nectar from the same flowers, behavioral dominance plays an important role in how they partition plants; different species employ territorial or nonterritorial feeding strategies, and sexual differences in territoriality are common. Foraging behaviors both among and within species, and even within individuals, are remarkably plastic (Ewald and Carpenter 1978; Snow and Snow 1980; Feinsinger, Swarm, and Wolfe 1985). For instance, individuals may opportunistically switch between territorial and nonterritorial strategies. In addition to hovering and probing directly through the corolla opening, they may perch while feeding, often piercing the base of long flowers whose nectar they otherwise cannot reach (Feinsinger and Colwell 1978).

Phaethornithine feeding ecologies are more specialized in every respect. All species have long (usually curved) bills that closely match the long tubular corollas of their preferred food plants, giant monocotyledonous forest herbs of the order Scitamineae (table 13.1, fig. 13.1). Most species hover at their food plants and probe directly into the corolla. Thus, differences in bill structures permit phaethornithine species to partition nectar sources based on floral morphology. Because this gives each species preferential access to a particular set of floral resources, phaethornithines usually do not defend their food plants. Instead, they follow regular routes between dispersed clumps of flow-

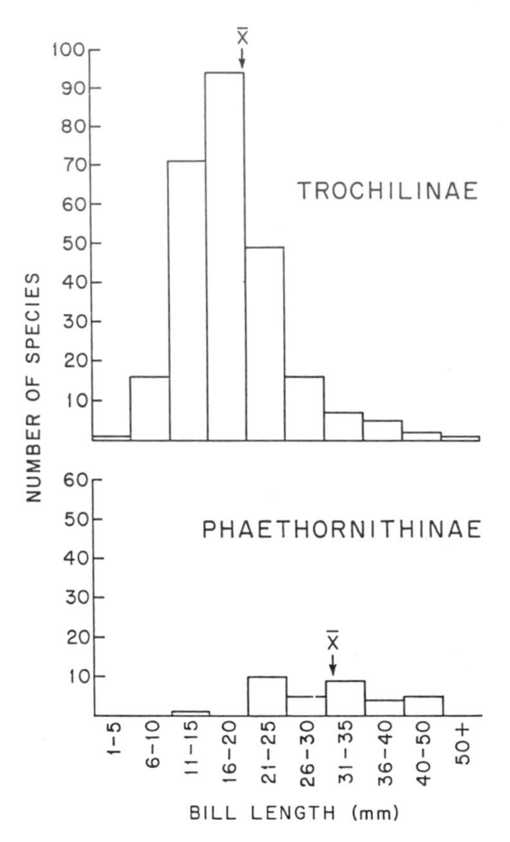

Figure 13.1. Distribution of bill sizes between hummingbird subfamilies. Those of the trochilinae are significantly shorter, $\bar{X}$ = 19.5 mm, than are those in the phaethornithinae, $\bar{X}$ = 31.7 mm (Mann-Whitney U-test, two-tailed probability $p < 0.0001$). Some trochilines with more specialized feeding ecologies have long straight bills (> 30 mm, n = 15), but they are greatly outnumbered by shorter-billed species.

ers, a foraging pattern called "trap-lining" (Stiles 1975, 1978; Feinsinger and Colwell 1978; Stiles and Wolf 1979).

VAGILITY

By any criterion, trochilines are more vagile than phaethornithines. Many trochilines move among different habitats or undertake seasonal altitudinal or long-distance migrations (Feinsinger 1980; Feinsinger, Swarm, and Wolfe 1985; Stiles 1985); similar movements are infrequent among phaethornithines (table 13.1; Skutch 1964; Hilty 1975; Stiles 1977; Stiles and Wolf 1979; Feinsinger 1980). Trochilines are often seen crossing large bodies of water, and many species occur on oceanic islands (Lack 1973). Bodies of water are apparently effective barriers to phaethornithines; only one species occurs on an oceanic island (table 13.1).

Explanation of the Hummingbird Pattern

The opposing ecological (generalist, specialist) and dispersal (vagile or sedentary) characteristics may be combined in four ways (fig. 13.2). Current macroevolutionary theory has focused on predicting and explaining the diversity of low-dispersing specialists and their opposites, the high-dispersing generalists. The reason for this is clear; current microevolutionary theory leads one to expect that both specialization and low dispersal ability will augment the process of diversification by increasing the rate of speciation: the former by increasing the likelihood that a population will evolve in response to competition or environmental change, the latter by decreasing gene flow between populations that may then diverge. Conversely, one expects ecological generalists with high dispersal ability to be much less diverse (Eldredge 1979; Stanley 1979; Vrba 1980, 1983).

The hummingbird case clearly contradicts current macroevolutionary theories because the extremely diverse trochilines are vagile ecological generalists, whereas the depauperate phaethornithines are sedentary ecological specialists. How can this contradiction be explained? I suggest that the environmental context provides the missing factor, without which the effects of intrinsic attributes cannot be understood. Specifically, I propose that the propensity to colonize new habitats, exploit new resources, or both can be the limiting factor for diversification when an evolutionary radiation takes place in a topographically complex region.

Most hummingbirds live on continents that harbor a great diversity

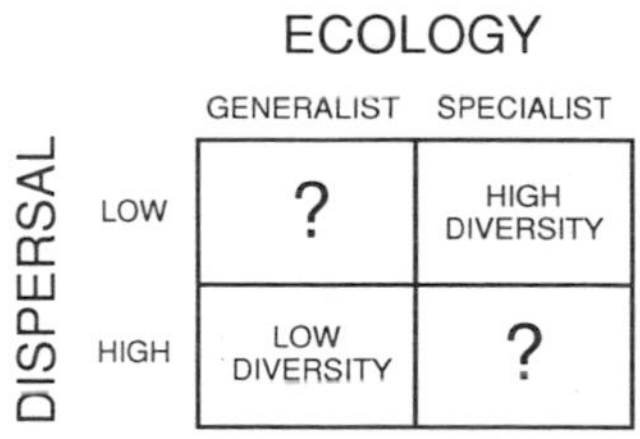

ECOLOGY

DISPERSAL	GENERALIST	SPECIALIST
LOW	?	HIGH DIVERSITY
HIGH	LOW DIVERSITY	?

Figure 13.2. Predicted species diversity for pairwise character combinations of ecological niche (high specialization versus low specialization = generalization) and dispersal ability (low dispersal = sedentary, high dispersal = vagile). See text for further discussion.

Table 13.2. Relative diversity of phaethornithines and trochilines within different habitats*

Ecology	Phaethornithinae (sedentary specialist)	Trochilinae (vagile generalist)
Habitats/subfamily	17	34
Tropical forest habitats/subfamily	13	13
Elevation range (m)/species ($\bar{x} \pm$ S.E.)†	1098 ± 128	1490 ± 53
Habitats/species ($\bar{x} \pm$ S.E.)‡	3.30 ± 0.42	3.81 ± 0.11
Species/habitat ($\bar{x} \pm$ S.E.)†·§	5.2 ± 1.5	27.8 ± 7.1
Sympatric species/habitat ($\bar{x}$)†	2.6	12.2

*See appendix 13.2. Habitats of the tropical wet forest region are followed by (F). Wet forest (F), humid forest (F), cloud forest (F), elfin forest (F), *terra firme* forest (F), *várzea* forest (F), swamp forest (F), river edge forest (F), sandy-belt forest (F), mangrove (F), savanna woodland (F), dry forest (F), gallery forest, lighter thinned woodland, second-growth woodland, forest or woodland border (F), shrubby areas and clearings, river island shrubbery, scrubby areas, arid/desert scrub, coastal zone, marshes, savanna, paramo, cultivated lands, pine-oak woodland, deciduous woodland, coniferous forest, open coniferous forest, lowland pine savanna, pinyon-juniper association, willow-alder thickets, chaparral, meadows. Habitats are those used in the American Ornithologists; Union's *Check-list of North American Birds* (1983) and those in Hilty and Brown (1986). Data for "sympatric species" collected from the following sources: DesGranges and Grant (1980); Feinsinger (1980); Snow and Snow (1980); Stiles (1980, 1985).

†Two-tailed $p < 0.02$ by Mann-Whitney U-test.

‡Two-tailed $p > 0.1$ by Mann-Whitney U-test.

§Average based on only those habitats in which at least one species occurs.

of habitats. The radiation of any group with a continental distribution depends in part on its ability to exploit the available habitats. In this regard, organisms with generalized ecologies and high vagility are superior. Their high vagility gives them access to new habitats, and their generalized ecologies permit them to utilize the often novel resources encountered there. In a complex topography, colonizing ability should actually augment the production of geographical isolates and their speciation because the physical environment provides adequate barriers to gene flow among different populations. In contrast, diversification in a more uniform habitat depends on a group's ability to subdivide resources and on biological attributes such as low vagility that lead to genetic divergence among local populations in the absence of physical environmental barriers. Therefore, the effect of intrinsic attributes on diversity depends on the environmental setting.

The importance of the environmental context is illustrated by dramatic differences in the range of habitats that each of the two subfamilies occupies (table 13.2). Trochilines occur in almost every vegetated habitat from sea level to snow line, whereas the phaethornithines are limited to lowland and montane tropical rain forests. This difference can be attributed directly to the subfamilies' different ecologies and vagilities. The greater vagility of trochilines has given them greater access to a variety of environments in the Andes, on islands, and in temperate North America. The more generalized feeding ecologies of trochilines has enabled each species, and the subfamily as a whole, to utilize a wider spectrum of flowers and hence habitats (with respect to rainfall, vegetation structure, and elevation). In fact, trochilines have adapted to the rigors of life in habitats at or near the snow line through remarkable behavioral adaptations (Carpenter 1976) rather than through anatomical ones (Wolf and Gill 1986). In contrast, the phaethornithines' specializations apparently limit them to forested habitats because this is where their preferred food plants occur (Weske and Terborgh 1977; Feinsinger et al. 1979; Stiles 1981).

Generalized feeding ecologies certainly favor colonization of the montane habitats where trochilines attain their greatest diversity (table 13.3). Flowers in these temperate environs are typically less specialized and more sporadic (Snow and Snow 1980; Stiles 1981, 1985) than those in lowland tropical forests, where phaethornithines are most diverse (Stiles 1977; Stiles and Wolf 1979; Gill, Mack, and Ray 1982). Moreover, most montane hummingbird-pollinated plants have been derived from insect-pollinated forms (Grant and Grant 1968; Kennedy 1977; Snow and Snow 1980). Presumably, ancestral generalist feeders were favored to initiate coevolutionary relationships with these plants.

The existence of only a few montane trochilines with phaethornithinelike ecologies and decurved bills reinforces my explanation that ecological specialization has prevented phaethornithines from diversifying. Montane trochilines with decurved bills include three species of *Campylopterus*, *Lafresnaya lafresnayi*, and *Rhodopis vesper*. Like phaethornithines, many trochilines with long and/or decurved bills may trap-line (Feinsinger and Colwell 1978; Snow and Snow 1980). The

Table 13.3. Relative proportions of trochilines (vagile generalists) and phaethornithines (sedentary specialists) in lowland and montane habitats

	Number of species (%)	
	Phaethornithinae	Trochilinae
Continent		
Lowland (< 2,000 m)*	21 (84)	94 (45)
Montane† (> 2,000 m)	4 (16)	115 (55)

*Significant at $p < 0.01$ by chi square with Yates correction.

†Nine other phaethornithines occur above 1,000 m.

paucity of species with these characteristics outside lowland forested habitats implies a genuine lack of ecological opportunities for them elsewhere.

Patterns of ecological coexistence in each subfamily, another important factor in the production and maintenance of diversity, also appear to depend on vagility. In any habitat where both subfamilies occur, trochilines are 5 times more diverse on average than are phaethornithines (table 13.2). As emphasized earlier, competition for nectar has produced patterns of ecological dominance among hummingbird species. The evolution of socially subordinate species in the trochilinae is facilitated by their high vagility because such species often move to other habitats when the bet nectar resources are controlled by behaviorally dominant species (Feinsinger 1976; Feinsinger, Swarm, and Wolfe 1985). Without the ability to emigrate to other habitats, the number of ecological generalists that could coexist would be greatly reduced.

Discussion

Other than the explanation I have provided, the greater number of trochilines might be due to coincidence or to other characters that distinguish the two subfamilies. Based on present evidence, however, it seems likely that attributes associated with feeding ecology (structural and behavioral) have played the dominant role in the generation and maintenance of hummingbird species diversity. Competition for nectar has been demonstrated repeatedly in these birds and is a prime determinant of species coexistence (Feinsinger, Swarm, and Wolfe 1985).

Information on the ecology and dispersal capabilities of other extant groups supports the importance of these factors in promoting evolutionary radiations. The superior colonizing ability of high dispersing ecological generalists is widely documented (MacArthur and Wilson 1967; Williams 1969; Williamson 1981). Such groups are more likely to reach oceanic islands where they may, if ecological opportunities permit, undergo dramatic evolutionary radiations. In addition, forest organisms in general, not just the phaethornithines, are notoriously poor colonizers. In New Guinea, for example, forest birds are much less likely to colonize offshore islands than are birds of disturbed habitats and ecotones (Diamond 1974; Diamond and Marshall 1977). The same factors should operate in insular regions on continents.

The hummingbird case is not the only example of high diversity in a clade of vagile organisms. Marine taxa with planktonic larvae are believed to disperse widely because they spend a long period of time in ocean currents, which can transport them over great distances (but see Johannessen 1988). However, Vermeij (1987) has noted that contrary to the conventional view, many molluscan families with planktonically dispersing larvae are very rich in extant species, including the Cypraeidae, Mitridae, Costellariidae, Strombidae, and Nassariidae. Moreover, Vrba (1980) has shown for antelopes that the vagile Alcelaphini are more diverse than their low-dispersing sister clade, the Aepycerotini.

Significantly, in the case of the antelopes, the diverse Alcelaphini are more specialized ecologically than are the Aepycerotini (the Alcelaphini would fall in the lower right hand cell of fig. 13.2). Thus, ecological specialization and low vagility are decoupled in antelopes, in contrast to hummingbirds and other organisms. Several other groups of highly mobile large land mammals also seem to have undergone rapid diversification (Kurten 1968; Bush et al. 1977). Like hummingbirds, all of these vertebrate groups have continental distributions. Thus, dispersal ability alone may limit diversification on continents.

The role of ecological generalization and high dispersal ability in generating diversity may be underestimated because of modifications in these characteristics through time. Many of the classic examples of a correlation between high diversity and ecological specialization and/or low dispersal ability are of radiations on oceanic islands (for birds, the Hawaiian honeycreepers [Fringillidae, Drepanidinae] and Galapagos finches [Emberizidae, Geospizinae]). In many cases, the ancestors to these radiations were probably ecologically generalized and highly vagile (Amadon 1986; Grant 1986); these characteristics favor successful island colonization. Ecological specializations may therefore be more the products of radiations than their causes.

In light of neontological evidence for the positive association of ecological generalization and vagility with diversity, one is led to ask why paleontological examples have failed to reveal the pattern. Several reasons may be suggested. First, many insular ecosystems, whether they are oceanic islands, isolated mountains, or bodies of water, are ephemeral; this may cause most insular radiations to be poorly documented in the fossil record. Therefore, the actual diversity of ecologically generalized, high-dispersing clades may be underestimated by sampling error; their diverse descendants on "islands" are never documented.

Second, ecological specialization leads to morphological specialization, which in turn makes assessment of taxonomic diversity in the fossil record relatively easy. In contrast, ecologically generalized forms, if trochiline hummingbirds are any indication, utilize behavioral differences to segregate ecologically. Certainly, the number of species in many trochiline genera would be difficult to assess from their fossil remains. For example, the approximately 30 species in the trochiline genus *Amazilia* are all very similar in bill and body size and shape and most would be cryptic as fossils. Because most phaethornithines have specialized bills and differ in size, their fossil remains would provide a good estimate of their taxonomic diversity. A paleontologist working with only hard-part morphology might therefore underestimate diversity in groups with generalized ecologies.

Finally, among terrestrial forms at least, the energetic costs of dispersal mandate that vagile forms be lightly constructed. For this reason, good dispersers may be less preservable and so rarer as fossils. Therefore, we may be able to say little about fossil diversity patterns for organisms with certain kinds of ecologies (birds in general and birds with good dispersal capabilities in particular). This creates yet another bias in the fossil record.

A combination of neontological and paleontological studies will provide the best picture of macroevolution. The hummingbird example emphasizes that neontological examples are indispensable for formulating general macroevolutionary laws because so much of nature's complexity is lost in the fossil record. Neontological examples also demonstrate the unique value of the fossil record for estimating rates of speciation and extinction, which ultimately determine the diversity of a clade. Without this information, the picture of diversity is a static one that limits what one can say about its *generation*.

The hummingbird example makes two contributions to macroevolutionary theory. First, it suggests that the effects of vagility and ecological specialization on diversification depend on the physiographic context in which the radiation takes place. If environmental heterogeneity and topographic complexity severely restrict access to patches of habitat, and conditions in each patch vary, then the propensity to colonize (generalist ecology, high vagility) will enhance diversification. Under more homogeneous physiographic settings, the opposite attributes will favor diversification. Second, hummingbirds serve as a reminder that the formulation of macroevolutionary hypotheses must include explicit reference to the environmental context. The intrinsic biological attributes of organisms cannot be treated as separate from the extrinsic environment in which they live.

Summary

A common sense tenet of macroevolutionary theory asserts that specialized ecologies and low dispersal ability enhance diversification by increasing the number of speciation events. Among clades, these attributes often correlate positively with diversity. I report an anomalous case involving the two sister subfamilies of hummingbirds (Trochilidae). The subfamily with less specialized ecologies and greater vagility (Trochilinae) is nine times more diverse than its sister taxon (Phaethornithinae). Analysis of this case suggests that the biological attributes of trochilines make them superior colonizers, which has favored their diversification in a complex physiographic setting. The biological basis for evolutionary laws therefore appears to be contingent on the environmental and historical context of the organisms.

Acknowledgments

I thank Kurt Fristrup, Dana Geary, Tom Givnish, John Kirsch, Karen Steudel, and Geerat Vermeij for helpful comments on the manuscript. Cheryle Hughes drafted and Don Chandler photographed the figures. This work was funded in part by the National Science Foundation Grant No. BSR 8817330.

Literature Cited

Amadon, D. 1986. The Hawaiian honeycreepers revisited. *Elepaio* 46:83–84.

American Ornithologists' Union. 1983. *Check-list of North American birds*, 6th ed. Committee on classification and nomenclature of the A. O. U. Lawrence, Kans.: Allen Press.

Bush, G. L., S. M. Case, A. C. Wilson, and J. L. Patton. 1977. Rapid speciation and chromosomal evolution in mammals. *Proc. National Acad. Sci. U.S.A.* 74:3942–6.

Carpenter, F. L. 1976. Ecology and evolution of an Andean hummingbird (*Oreotrochilus estella*). *Univ. Calif. Publ. Zool.* 106:1–74.

Cohen, A. S., and M. R. Johnston. 1987. Speciation in brooding and poorly dispersing lacustrine organisms. *Palaios* 2:426–35.

Diamond, J. M. 1974. Colonization of exploded volcanic islands by birds: the supertramp strategy. *Science* 184:803–6.

Diamond, J. M., and A. G. Marshall. 1977. Distributional ecology of New Hebridean birds: a species kaleidoscope. *J. Animal Ecol.* 46:703–27.

Eldredge, N. 1979. Alternative approaches to evolutionary theory. *Bull. Carnegie Mus. Natural Hist.* 13:7–19.

Echelle, A. A., and I. Kornfield, eds. 1984. *Evolution of fish species flocks*. Orono: University of Maine Press.

Ewald, P. W., and F. L. Carpenter. 1978. Territorial responses to energy manipulations in the anna hummingbird. *Oecologia* 31:277–92.

Feinsinger, P. 1976. Organization of a tropical guild of nectarivorous birds. *Ecolog. Monogr.* 46:257–91.

Feinsinger, P. 1980. Asynchronous migration patterns and the coexistence of tropical hummingbirds. In *Migrant birds in the neotropics*, ed. A. Keast and E. S. Morton, 411–9. Washington, D.C.: Smithsonian Institution Press.

Feinsinger, P., and R. K. Colwell. 1978. Community organization among neotropical nectar-feeding birds. *Am. Zoologist* 18:779–95.

Feinsinger, P., R. K. Colwell, J. Terborgh, and S. B. Chaplin. 1979. Elevation and the morphology, flight energetics, and foraging ecology of tropical hummingbirds. *Amer. Naturalist* 113:481–97.

Feinsinger, P., L. A. Swarm, and J. A. Wolfe. 1985. Nectar-feeding birds on Trinidada and Tobago: Comparison of diverse and depauperate guilds. *Ecolog. Monogr.* 55:1–28.

Fitzpatrick, J. W. 1988. Why so many passerine birds? A response to Raikow. *Syst. Zool.* 37:71–76.

Futuyma, D. J. 1986. *Evolutionary biology*, 2d ed. Sunderland, Mass.: Sinauer Associates.

Gill, F. B., A. L. Mack, and R. T. Ray. 1982. Competition between hermit hummingbirds Phaethornithinae and insects for nectar in a Costa Rican rain forest. *Ibis*. 124:44–49.

Grant, K. A., and V. Grant. 1968. *Hummingbirds and their flowers*. New York: Columbia University Press.

Grant, P. R. 1986. *Ecology and evolution of Darwin's finches*. Princeton, N.J.: Princeton University Press.

Graves, G. R. 1985. Elevational correlates of speciation and intraspecific geographic variation in plumage in Andean forest birds. *Auk* 102:556–79.

Hansen, T. 1983. Modes of larval development and rates of speciation in early tertiary neogastropods. *Science* 220:501–2.

Hilty, S. 1975. Year-round attendance of white-whiskered and little hermits, *Phaethornis* spp., at singing assemblies in Colombia. *Ibis* 117:382–4.

Jackson, J. B. C. 1974. Biogeographic consequences of eurytopy and stenotopy among marine bivalves and their evolutionary significance. *Am. Naturalist* 108:541–60.

Johannesson, K. 1988. The paradox of Rockall: Why is a brooding gastropod (Littorina saxatalis) more widespread than one having a planktonic larval dispersal stage (L. littorea)? *Marine Biol.* 99:507–13.

Johnsgard, P. A. 1983. *The hummingbirds of North America*. Washington, D.C.: Smithsonian Institution Press.

Kennedy, H. 1977. Unusual floral morphology in a high altitude *Calathea* (Marantaceae). *Brenesia* 12:1–9.

Kurten, B. 1968. Pleistocene mammals of Europe. London: Weidenfeld and Nicolson.

Lack, D. 1973. The numbers of species of hummingbirds in the West Indies. *Evolution* 27:326–37.

MacArthur, R. H., and E. O. Wilson. 1967. *The theory of island biogeography. Monographs in Population Biology*, vol. I. Princeton, N.J.: Princeton University Press.

Mayden, R. L. 1986. Speciose and depauperate phylads and tests of punctuated and gradual evolution: Fact or artifact? *Syst. Zool.* 35:591–602.

Mayr, E. 1963. *Animal species and evolution.* Cambridge, Mass.: The Belknap Press of Harvard University Press.

Nie, N. H., C. H. Hull, J. G. Jenkins, K. Steinbrenner, D. H. Brent. 1983. *Statistical package for the social sciences, SPSSx.* New York: McGraw-Hill.

Olrog, C. 1959. *Las aves argentinas: una guia de campo.* Tucuman, Argentina: Universidad Nacional de Tucuman, Instituto "Miguel Lillo".

Perrins, F. S., and R. Middleton. 1985. *The encyclopedia of birds.* New York: Equinox Books.

Peters, J. L. 1945. *Check-list of birds of the world, vol. V.* Cambridge, Mass.: Harvard University Press.

Raikow, R. J. 1986. Why are there so many kinds of passerine birds? *Syst. Zool.* 35:255–9.

Sibley, C. G., J. E. Ahlquist, and B. L. Monroe, Jr. 1988. A classification of the living birds of the world based on DNA-DNA hybridization studies. *Auk* 105:409–23.

Skutch, A. F. 1964. Life histories of hermit hummingbirds. *Auk* 81:5–25.

Snow, D. W., and B. K. Snow. 1980. Relationships between hummingbirds and flowers in the Andes of Colombia. *Bull. Br. Mus. Natural Hist. (Zoology)* 38:105–39.

Stanley, S. M. 1979. *Macroevolution: Pattern and process.* San Francisco: W. H. Freeman.

Stenseth, N. C., and J. Maynard-Smith. 1984. Coevolution in ecosystems: Red queen evolution or stasis. *Evolution* 38:870–80.

Stiles, F. G. 1975. Ecology, flowering phenology, and hummingbird pollination of some Costa Rican *Heliconia* species. *Ecology* 56:285–301.

Stiles, F. G. 1977. Coadapted competitors: flowering seasons of hummingbird food plants in a tropical forest. *Science* 198:1177–8.

Stiles, F. G. 1978. Ecological and evolutionary implications of bird pollination. *Am. Zoologist* 18:715–27.

Stiles, F. G. 1980. The annual cycle in a tropical wet forest hummingbird community. *Ibis* 122:322–43.

Stiles, F. G. 1981. Geographical aspects of bird-flower coevolution, with particular reference to Central America. *Ann. Missouri Botan. Gardens* 68:323–51.

Stiles, F. G. 1985. Seasonal patterns and coevolution in the hummingbird-flower community of a Costa Rican subtropical forest. In *Neotropical ornithology,* A. O. U. Monograph no. 36, ed. P. A. Buckley, M. S. Foster, E. S. Morton, R. S. Ridgely, and F. G. Buckley, 757–87.

Stiles, F. G., and L. L. Wolf. 1979. *Ecology and evolution of lek mating behavior in the long-tailed hermit hummingbird,* A. O. U. Monograph no. 27.

Templeton, A. R. 1979. Once again, why 300 species of Hawaiian *Drosophila? Evolution* 33:513–7.

Valentine, J. W., and D. Jablonski. 1983. Larval adaptation and patterns of brachiopod diversity in space and time. *Evolution* 37:1052–61.

Vermeij, G. J. 1987. The dispersal barrier in the tropical Pacific: Implications for molluscan speciation and extinction. *Evolution* 41:1046–58.

Vrba, E. S. 1980. Evolution, species and fossils: How does life evolve? *South Afr. J. Sci.* 76:61–84.

Vrba, E. S. 1983. Macroevolutionary trends: New perspectives on the roles of adaptation and incidental effect. *Science* 221:387–9.

Vrba, E. S. 1984. Evolutionary pattern and process in the sister-group Alcelaphini-Aepycerotini (Mammalia: Bovidae). In *Living fossils*, ed. N. Eldredge and S. M. Stanley, 62–79. New York: Springer-Verlag.

Vrba, E. S. 1987. Ecology in relation to speciation rates: Some case histories of Miocene-Recent mammal clades. *Evolutionary Ecol.* 1:283–300.

Weske, J. S., and J. W. Terborgh. 1977. *Phaethornis koepckeae*, a new species of hummingbird from Peru. *Condor* 79:143–7.

Williams, E. E. 1969. The ecology of colonization as seen in the zoogeography of anoline lizards on small islands. *Q. Rev. Biol.*44:345–89.

Williamson, M. 1981. *Island populations*. New York: Oxford University Press.

Wolf, L. L., and F. B. Gill. 1986. Physiological and ecological adaptations of high montane sunbirds and hummingbirds. In *High altitude tropical biogeography*, ed. F. Vuilleumier and M. Monasterio, 103–19. New York: Oxford University Press.

Wolf, L. L., and F. G. Stiles. 1970. Evolution of pair cooperation in a tropical hummingbird. *Evolution* 24:759–73.

Zusi, R. L. 1980. Paper presented at the American Ornithologist's Union Conference, Fort Collins, Colo.

Zusi, R. L., and G. D. Bentz. 1982. Variation in a muscle in hummingbirds and swifts and its systematic implications. *Proc. Biological Soc. Wash.* 95:412–20.

Appendixes

A13.1. Ecological and morphological data for hummingbird species

		H	S	Z D	I	MIN	MAX
Phaethornithinae							
001	*Ramphodon naevius*	01	6	036 0	1	9999	9999
002	*Glaucis dohrnii*	01	6	025 0	1	9999	9999
003	*Glaucis aenea*	08	2	030 1	1	0000	0800
004	*Glaucis hirsuta*	07	2	033 1	3	0000	1000

H = number of habitats where found (see table 13.1 and appendix table A13.2 for list of habitats); S = bill shape (0 = no data, 1 = strongly decurved, 2 = decurved, 3 = slightly decurved, 4 = straight, 5 = upturned; 6 = other); Z = bill size (in millimeters; 0 = no data); D = vagility (0 = no data, 1 = sedentary, 2 = seasonal movements among habitats, 3 = altitudinal migrant, 4 = long distance latitudinal migrant); I = occurce on islands (0 = no data, 1 = absent, 2 = land bridge island, 3 = oceanic island); MIN and MAX = minimum and maximum extent of elevational range in meters (9999 = no data).

A13.1. *(continued)*

		H	S	Z	D	I	MIN	MAX
005	*Threnetes niger*	00	3	034	0	1	9999	9999
006	*Threnetes loehkeni*	00	3	029	0	1	9999	9999
007	*Threnetes leucurus*	03	2	030	0	1	0000	1000
008	*Threnetes ruckeri*	04	2	030	1	1	0000	1050
009	*Threnetes grzimicki*	01	0	000	0	1	9999	9999
010	*Phaethornis yaruqui*	01	2	046	1	1	0000	1500
011	*Phaethornis guy*	04	2	043	3	2	0900	2000
012	*Phaethornis syrmatophorus*	02	2	041	0	1	0800	2400
013	*Phaethornis superciliosus*	04	2	041	2	1	0000	1800
014	*Phaethornis malaris*	00	2	046	1	1	9999	9999
015	*Phaethornis margarettae*	01	2	039	0	1	9999	9999
016	*Phaethornis eurynome*	01	2	038	0	1	0076	2242
017	*Phaethornis nigrirostris*	01	2	035	1	1	0950	0950
018	*Phaethornis hispidus*	03	2	033	1	1	0000	1000
019	*Phaethornis anthophilus*	07	3	038	0	2	0000	1200
020	*Phaethornis bourcieri*	02	4	033	0	1	0000	0400
021	*Phaethornis philippi*	01	4	032	0	1	0000	0200
022	*Phaethornis koepckæ*	01	3	034	0	1	0500	1130
023	*Phaethornis squalidus*	04	2	025	0	1	0000	0500
024	*Phaethornis augusti*	04	2	033	0	1	0450	2500
025	*Phaethornis pretrei*	05	2	034	1	1	0364	1091
026	*Phaethornis subochraceus*	00	3	025	0	1	9999	9999
027	*Phaethornis nattereri*	00	2	030	0	1	9999	9999
028	*Phaethornis maranhaoensis*	00	2	024	0	1	9999	9999
029	*Phaethornis gounellei*	00	2	025	0	1	0455	0485
030	*Phaethornis ruber*	06	2	023	1	1	0000	0500
031	*Phaethornis griseogularis*	02	2	023	0	1	0300	1800
032	*Phaethornis longuemareus*	06	2	025	1	2	0000	1700
033	*Phaethornis idaliae*	00	2	015	0	1	0121	0121
034	*Eutoxeres aquila*	05	1	025	3	1	0000	2100
035	*Eutoxeres condamini*	04	1	025	0	1	0000	0700
Trochilinae								
036	*Androdon aequatorialis*	06	6	041	0	1	0000	1590
037	*Doryfera johannae*	02	4	030	0	1	0280	1800
038	*Doryfera ludovicae*	04	4	036	0	1	0900	2700
039	*Phaeochroa cuvierii*	05	4	018	2	2	0000	0350
040	*Campylopterus curvipennis*	02	0	000	0	1	0000	0350
041	*Campylopterus excellens*	00	0	000	0	0	9999	9999
042	*Campylopterus largipennis*	03	3	028	1	1	0000	0550
043	*Campylopterus rufus*	06	0	000	0	1	1250	1850
044	*Campylopterus hyperythrus*	02	4	020	0	1	1300	2600
045	*Campylopterus duidae*	04	4	020	0	1	1200	2400
046	*Campylopterus hemileucurus*	05	2	038	3	1	0000	2450

A13.1. *(continued)*

		H	S	Z	D	I	MIN	MAX
047	*Campylopterus ensipennis*	02	2	028	0	2	0700	2000
048	*Campylopterus falcatus*	03	2	028	0	1	0900	3000
049	*Campylopterus phainopeplus*	03	3	025	3	1	1200	4800
050	*Campylopterus villaviscensio*	00	4	000	0	1	0400	1500
051	*Eupetomena macroura*	00	3	020	0	1	9999	9999
052	*Florisuga mellivora*	05	3	020	2	2	0000	1600
053	*Melanotrochilus fuscus*	02	3	020	2	1	0825	0825
054	*Colibri delphinae*	08	4	018	3	2	0300	2800
055	*Colibri thalassinus*	05	3	020	3	1	0600	3350
056	*Colibri coruscans*	02	3	025	3	1	0600	3600
057	*Colibri serrirostris*	05	3	023	3	1	2075	2075
058	*Anthracothorax viridigula*	04	3	025	0	2	9999	9999
059	*Anthracothorax prevostii*	05	3	025	0	3	0000	1500
060	*Anthracothorax nigricollis*	05	3	025	4	2	0000	1750
061	*Anthracothorax veraguensis*	00	0	000	0	1	9999	9999
062	*Anthracothorax dominicus*	04	3	024	0	3	9999	9999
063	*Anthracothorax viridis*	03	3	024	0	3	9999	9999
064	*Anthracothorax mango*	03	3	025	0	3	9999	9999
065	*Avocettula recurvirostris*	01	5	018	0	1	9999	9999
066	*Eulampis jugularis*	04	3	027	1	3	9999	9999
067	*Sericotes holosericeus*	04	3	024	1	3	9999	9999
068	*Chrysolampis mosquitus*	04	3	015	4	3	0000	1750
069	*Orthorhynchus cristatus*	04	4	011	1	3	9999	9999
070	*Klais guimeti*	04	4	013	3	1	0400	1900
071	*Abeillia abeillia*	04	4	013	3	1	1000	1850
072	*Stephanoxis lalandi*	05	4	012	1	1	1150	2490
073	*Lophornis ornata*	05	4	010	0	2	0100	0950
074	*Lophornis gouldii*	00	4	010	0	1	9999	9999
075	*Lophornis magnifica*	04	4	010	3	1	9999	9999
076	*Lophornis delattrei*	04	4	010	3	1	0600	2000
077	*Lophornis stictolopha*	03	4	010	0	1	0000	1300
078	*Lophornis chalybea*	02	4	015	0	1	0100	0600
079	*Lophornis pavonina*	04	4	013	0	1	0500	2000
080	*Lophornis helenae*	07	4	014	3	1	0350	1450
081	*Lophornis adorabilis*	05	4	015	3	1	9999	9999
082	*Popelairia popelairii*	02	4	013	0	1	0500	1200
083	*Popelairia langsdorffi*	02	4	013	0	1	0000	0300
084	*Popelairia letitiae*	00	0	000	0	1	9999	9999
085	*Discosura conversii*	06	4	010	3	1	0000	1400
086	*Discosura longicauda*	03	4	012	0	1	0000	0200
087	*Chlorestes notatus*	04	4	018	0	2	0000	1000
088	*Chlorostilbon mellisugus*	02	4	013	0	2	0000	2200

A13.1. *(continued)*

		H	S	Z	D	I	MIN	MAX
089	*Chlorostilbon aureoventris*	03	4	014	1	1	1160	1160
090	*Chlorostilbon canivetii*	07	4	019	2	2	0000	1850
091	*Chlorostilbon assimilis*	00	4	019	0	0	9999	9999
092	*Chlorostilbon ricordii*	03	4	018	1	3	9999	9999
093	*Chlorostilbon swainsonii*	04	4	013	0	3	9999	9999
094	*Chlorostilbon maugaeus*	04	4	013	0	3	9999	9999
095	*Chlorostilbon gibsoni*	05	4	013	3	1	0000	2300
096	*Chlorostilbon russatus*	03	4	015	0	1	0600	2600
097	*Chlorostilbon stenura*	03	4	018	0	1	1000	3000
098	*Chlorostilbon alice*	03	4	018	0	1	0750	1800
099	*Chlorostilbon poortmani*	04	4	018	0	1	0500	2800
100	*Chlorostilbon auratus*	00	0	000	0	0	9999	9999
101	*Cynanthus sordidus*	03	4	021	0	1	9999	9999
102	*Cynanthus latirostris*	03	4	021	4	3	0150	3000
103	*Cyanophaia bicolor*	02	4	020	0	3	9999	9999
104	*Thalurania colombica*	06	3	023	3	1	0000	1900
105	*Thalurania furcata*	03	3	025	0	2	0000	1900
106	*Thalurania watertonii*	00	3	023	0	1	9999	9999
107	*Thalurania glaucopis*	02	3	017	1	1	0980	0980
108	*Panterpe insignis*	06	4	023	1	1	1800	3000
109	*Damophila julie*	04	4	013	3	2	0000	1750
110	*Lepidopyga coeruleogularis*	05	3	018	0	2	0000	0100
111	*Lepidopyga lilliae*	01	3	018	0	1	0000	0100
112	*Lepidopyga goudoti*	03	4	018	0	1	0000	1600
113	*Hylocharis xantusii*	04	4	017	0	2	9999	9999
114	*Hylocharis leucotis*	03	4	017	3	1	0900	3900
115	*Hylocharis eliciae*	05	4	015	3	2	0000	0350
116	*Hylocharis sapphirina*	04	3	020	0	1	0000	0500
117	*Hylocharis cyanus*	04	3	023	0	1	0000	1250
118	*Hylocharis chrysura*	03	3	020	4	1	9999	9999
119	*Hylocharis grayi*	07	3	020	3	1	0000	2600
120	*Chrysuronia oenone*	04	3	020	0	2	0000	1500
121	*Goldmania violiceps*	03	4	018	0	1	0600	1400
122	*Geothalsia bella*	01	4	013	0	1	0600	1650
123	*Trochilus polytmus*	03	3	019	0	3	9999	9999
124	*Leucochloris albicollis*	04	4	020	0	1	0825	2455
125	*Polytmus guainumbi*	03	2	025	2	2	0000	0600
126	*Polytmus milleri*	02	2	025	0	1	1300	2200
127	*Polytmus theresiae*	02	2	025	0	1	0000	0300
128	*Leucippus fallax*	02	3	020	0	2	0000	0800
129	*Leucippus baeri*	00	3	020	0	1	9999	9999
130	*Leucippus taczanowskii*	00	3	025	0	1	9999	9999

A13.1. *(continued)*

		H	S	Z	D	I	MIN	MAX
131	*Leucippus chlorocercus*	02	3	018	0	2	0000	0430
132	*Taphrospilus hypostictus*	00	0	000	0	1	0500	1200
133	*Amazilia chionogaster*	00	4	023	0	1	9999	9999
134	*Amazilia viridicauda*	00	4	023	0	1	9999	9999
135	*Amazilia candida*	07	0	000	0	1	0000	0800
136	*Amazilia chionopectus*	06	4	015	0	2	0000	0500
137	*Amazilia versicolor*	05	4	018	1	1	0000	1700
138	*Amazilia luciae*	00	0	000	0	1	9999	9999
139	*Amazilia fimbriata*	04	4	020	0	1	0000	1300
140	*Amazilia distans*	01	4	020	0	1	0300	0300
141	*Amazilia lactea*	01	4	020	0	1	0300	1400
142	*Amazilia amabilis*	05	4	018	1	1	0000	1580
143	*Amazilia decora*	00	4	018	0	1	9999	9999
144	*Amazilia rosenbergi*	03	4	018	0	1	0000	0200
145	*Amazilia boucardi*	02	0	000	0	1	9999	9999
146	*Amazilia franciae*	03	4	023	3	1	1000	2000
147	*Amazilia leucogaster*	03	4	023	0	1	0000	0250
148	*Amazilia cyanocephala*	05	4	020	0	1	0000	2550
149	*Amazilia cyanifrons*	06	4	018	0	1	0400	2000
150	*Amazilia beryllina*	10	3	020	3	1	0000	3000
151	*Amazilia cyanura*	06	0	000	0	1	0000	1050
152	*Amazilia saucerrottei*	07	4	018	0	1	0000	3000
153	*Amazilia tobaci*	10	4	020	2	2	0000	1800
154	*Amazilia viridigaster*	02	4	018	0	1	0000	2100
155	*Amazilia edward*	03	4	023	0	2	0000	1830
156	*Amazilia rutila*	07	4	022	2	3	0000	1350
157	*Amazilia yucatanensis*	05	4	021	4	1	0000	0600
158	*Amazilia tzacatl*	05	4	020	1	2	0000	1800
159	*Amazilia castaneiventris*	00	3	020	0	1	0150	2045
160	*Amazilia amazilia*	07	4	020	0	1	0000	2500
161	*Amazilia viridifrons*	04	0	000	0	1	9999	9999
162	*Amazilia violiceps*	07	3	023	4	1	0000	2250
163	*Eupherusa poliocerca*	04	0	000	0	1	9999	9999
164	*Eupherusa eximia*	05	4	023	2	1	0000	2500
165	*Eupherusa cyanophrys*	03	0	000	0	1	9999	9999
166	*Eupherusa nigriventris*	04	4	015	0	1	1370	2100
167	*Elvira chionura*	04	4	015	0	1	0750	1980
168	*Elvira cupreiceps*	05	0	019	2	1	9999	9999
169	*Microchera albocoronata*	05	4	010	3	1	9999	9999
170	*Chalybura buffonii*	07	3	025	0	1	0000	2000
171	*Chalybura urochrysia*	06	3	025	0	1	0000	0900
172	*Aphantochroa cirrochloris*	00	0	000	0	1	9999	9999

A13.1. *(continued)*

		H	S	Z	D	I	MIN	MAX
173	*Lampornis clemenciae*	07	4	026	3	1	0300	3900
174	*Lampornis amethystinus*	06	0	000	3	1	0900	3050
175	*Lampornis viridipallens*	05	0	000	0	1	0900	2200
176	*Lampornis sybillae*	04	0	000	0	0	9999	9999
177	*Lampornis hemileucus*	04	4	023	0	1	1320	9999
178	*Lampornis calolaema*	04	4	023	2	1	9999	9999
179	*Lampornis castaneoventris*	04	4	023	0	1	1220	3050
180	*Lampornis cinereicauda*	00	0	000	0	1	9999	9999
181	*Lamprolaima rhami*	05	0	000	0	1	0900	2950
182	*Adelomyia melanogenys*	02	4	013	0	1	1000	2500
183	*Anthocephala floriceps*	03	4	013	0	1	0600	2300
184	*Urosticte ruficrissa*	01	4	020	0	1	1600	2300
185	*Urosticte benjamini*	03	4	020	1	1	0700	1500
186	*Phlogophilus hemileucurus*	01	4	018	0	1	0400	1500
187	*Phlogophilus harterti*	00	4	015	0	1	9999	9999
188	*Clytolaema rubricauda*	05	3	018	0	1	0000	2410
189	*Polyplancta aurescens*	03	3	020	0	1	0000	0550
190	*Heliodoxa rubinoides*	02	3	023	0	1	1200	2600
191	*Heliodoxa leadbeateri*	05	3	023	0	1	0500	2400
192	*Heliodoxa jacula*	05	4	023	1	1	0500	2300
193	*Heliodoxa xanthogonys*	04	4	020	0	1	0700	2000
194	*Heliodoxa schreibersii*	02	4	023	0	1	0300	1000
195	*Heliodoxa gularis*	01	3	028	0	1	0600	1100
196	*Heliodoxa branickii*	00	4	023	0	1	9999	9999
197	*Heliodoxa imperatrix*	02	3	025	0	1	0400	2050
198	*Eugenes fulgens*	08	4	029	4	1	0900	3300
199	*Hylonympha macrocerca*	03	4	025	0	1	0900	1200
200	*Sternoclyta cyanopectus*	04	3	030	0	1	0000	1900
201	*Topaza pella*	01	3	022	1	1	0250	0500
202	*Topaza pyra*	03	3	025	0	1	0100	0300
203	*Oreotrochilus melanogaster*	00	3	015	0	1	9999	9999
204	*Oreotrochilus chimborazo*	01	3	017	1	1	5300	5300
205	*Oreotrochilus estella*	01	3	021	0	1	1830	4300
206	*Oreotrochilus leucopleurus*	00	3	019	3	1	1525	3660
207	*Oreotrochilus adela*	00	0	000	0	1	9999	9999
208	*Urochroa bougueri*	04	3	030	3	1	0500	2500
209	*Patagona gigas*	03	3	041	3	1	0000	3660
210	*Aglaeactis cupripennis*	04	4	018	0	1	2900	3400
211	*Aglaeactis aliciae*	00	4	018	0	1	9999	9999
212	*Aglaeactis castelnaudii*	00	4	018	0	1	9999	9999
213	*Aglaeactis pamela*	00	4	018	0	1	9999	9999
214	*Lafresnaya lafresnayi*	04	2	025	3	1	1500	3700

A13.1. *(continued)*

		H	S	Z	D	I	MIN	MAX
215	*Pterophanes cyanopterus*	03	4	030	0	1	2600	3600
216	*Coeligena coeligena*	03	4	036	0	1	1000	2600
217	*Coeligena wilsoni*	02	4	033	0	1	0700	2000
218	*Coeligena prunellei*	01	4	030	0	1	1400	2600
219	*Coeligena torquata*	03	4	033	0	1	1500	3000
220	*Coeligena phalerata*	04	4	030	3	1	1400	3300
221	*Coeligena bonapartei*	04	4	030	0	1	1400	3200
222	*Coeligena helianthea*	05	4	033	0	1	1900	3300
223	*Coeligena lutetiae*	05	4	033	0	1	2600	3600
224	*Coeligena violifer*	00	4	033	0	1	2000	9999
225	*Coeligena iris*	02	4	033	0	1	1700	3500
226	*Ensifera ensifera*	05	5	119	3	1	1700	3300
227	*Sephanoides sephanoides*	00	4	016	3	3	0000	2135
228	*Sephanoides fernandensis*	02	4	016	0	3	0000	0000
229	*Boissonneaua flavescens*	03	4	018	0	1	1400	2800
230	*Boissonneaua matthewsii*	03	4	018	0	1	1600	2200
231	*Boissonneaua jardini*	03	4	018	1	1	0350	2200
232	*Heliangelus mavors*	03	4	015	0	1	2000	3200
233	*Heliangelus spencei*	03	4	013	0	1	2000	3600
234	*Heliangelus clarisse*	03	4	018	0	1	1800	3000
235	*Heliangelus amethysticollis*	00	4	013	0	1	9999	9999
236	*Heliangelus strophianus*	04	4	015	0	1	1200	2800
237	*Heliangelus exortis*	03	4	015	0	1	1500	3400
238	*Heliangelus viola*	03	4	013	0	1	2000	3500
239	*Heliangelus micraster*	00	4	015	0	1	9999	9999
240	*Heliangelus regalis*	03	4	014	0	1	1950	2200
241	*Eriocnemis nigrivestis*	03	4	018	3	1	2750	4700
242	*Eriocnemis vestitus*	03	4	018	0	1	2250	3850
243	*Eriocnemis godini*	00	4	018	0	1	2100	2300
244	*Eriocnemis cupreoventris*	04	4	018	0	1	1950	3000
245	*Eriocnemis luciani*	04	4	020	0	1	2800	4800
246	*Eriocnemis mosquera*	03	4	020	3	1	1200	3600
247	*Eriocnemis glaucopoides*	00	0	000	0	1	9999	9999
248	*Eriocnemis mirabilis*	03	4	015	0	1	2200	2200
249	*Eriocnemis alinae*	02	4	015	0	1	2000	2800
250	*Eriocnemis derbyi*	02	4	020	0	1	2500	3600
251	*Haplophaedia aureliae*	04	4	020	3	1	1500	3100
252	*Haplophaedia lugens*	03	4	018	0	1	1100	2000
253	*Ocreatus underwoodii*	05	4	013	3	1	0850	3100
254	*Lesbia victoriae*	04	3	015	0	1	2600	4000
255	*Lesbia nuna*	04	4	010	0	1	2000	3800
256	*Sappho sparganura*	00	3	020	0	1	9999	9999

A13.1 *(continued)*

		H	S	Z	D	I	MIN	MAX
257	*Polyonymus caroli*	03	4	018	0	1	1500	9999
258	*Ramphomicron microrhynchum*	04	4	005	1	1	1700	3400
259	*Ramphomicron dorsale*	03	3	008	3	1	2000	4500
260	*Metallura phoebe*	00	4	015	0	1	1980	3050
261	*Metallura teresiae*	00	4	013	0	1	9999	9999
262	*Metallura aeneocauda*	00	4	018	0	1	9999	9999
263	*Metallura baroni*	00	4	013	0	1	1900	1900
264	*Metallura eupogon*	00	4	013	0	1	9999	9999
265	*Metallura williami*	03	4	015	0	1	2100	3800
266	*Metallura odomae*	00	4	013	0	1	9999	9999
267	*Metallura tyrianthina*	03	4	010	3	1	1700	3800
268	*Metallura iracunda*	03	4	010	0	1	1800	3100
269	*Chalcostigma ruficeps*	02	4	013	0	1	2100	2700
270	*Chalcostigma olivaceum*	00	4	015	0	1	9999	9999
271	*Chalcostigma stanleyi*	02	4	010	0	1	3500	4500
272	*Chalcostigma heteropogon*	04	4	013	0	1	2900	3500
273	*Chalcostigma herrani*	04	4	013	0	1	2700	3600
274	*Oxypogon guerinii*	02	4	008	3	1	3200	5200
275	*Opisthoprora euryptera*	04	5	013	0	1	2500	3600
276	*Taphrolesbia griseiventris*	00	4	018	0	1	9999	9999
277	*Aglaiocercus kingi*	05	4	013	1	1	0900	3000
278	*Aglaiocercus coelestis*	02	4	013	0	1	0300	2100
279	*Oreonympha nobilis*	00	4	025	0	1	9999	9999
280	*Augastes scutatus*	03	4	017	1	1	1000	2000
281	*Augastes lumachellus*	03	4	018	0	1	0950	1600
282	*Schistes geoffroyi*	03	4	015	1	1	1400	2500
283	*Heliothryx barroti*	07	4	015	1	1	0000	1830
284	*Heliothryx aurita*	03	4	015	1	1	0000	0950
285	*Heliactin cornuta*	03	4	013	3	1	9999	9999
286	*Loddigesia mirabilis*	00	4	013	0	1	9999	9999
287	*Heliomaster constantii*	06	4	036	3	1	0000	1500
288	*Heliomaster longirostris*	05	4	038	1	2	0000	1525
289	*Heliomaster squamosus*	00	4	030	0	1	9999	9999
290	*Heliomaster furcifer*	05	4	030	4	1	9999	9999
291	*Rhodopis vesper*	01	2	031	0	1	0000	3050
292	*Thaumastura cora*	01	4	010	0	1	9999	9999
293	*Philodice evelynae*	03	3	017	1	3	9999	9999
294	*Philodice bryantae*	03	4	020	3	1	1830	1830
295	*Philodice mitchellii*	04	4	015	3	1	0000	1900
296	*Doricha enicura*	04	0	000	0	1	1000	1750
297	*Doricha eliza*	04	0	000	0	2	9999	9999

A13.1. *(continued)*

		H	S	Z	D	I	MIN	MAX
298	*Tilmatura dupontii*	05	0	000	2	1	0500	1950
299	*Microstilbon burmeisteri*	00	0	000	0	1	9999	9999
300	*Calothorax lucifer*	05	2	021	4	1	1100	2250
301	*Calothorax pulcher*	02	0	000	0	1	9999	9999
302	*Archilochus colubris*	08	4	018	4	3	0000	2450
303	*Archilochus alexandri*	06	4	021	4	1	9999	9999
304	*Calliphlox amethystina*	05	4	013	4	1	0000	1500
305	*Mellisuga minima*	01	0	000	0	3	9999	9999
306	*Mellisuga helenae*	03	0	000	0	3	9999	9999
307	*Calypte anna*	06	4	019	4	2	0000	1800
308	*Calypte costae*	04	4	018	4	2	9999	9999
309	*Stellula calliope*	08	4	016	4	1	0180	3500
310	*Atthis heloisa*	05	4	012	2	1	1500	2900
311	*Atthis ellioti*	06	4	000	0	1	0900	3300
312	*Myrtis fanny*	02	3	018	0	1	1200	2800
313	*Eulidia yarrellii*	02	3	013	0	1	9999	9999
314	*Myrmia micrura*	02	3	013	0	1	0000	0100
315	*Acestrura mulsant*	02	4	018	0	1	1500	2800
316	*Acestrura decorata*	00	0	000	0	1	9999	9999
317	*Acestrura bombus*	00	4	010	0	1	9999	9999
318	*Acestrura heliodor*	06	4	013	0	1	0500	3000
319	*Acestrura astreans*	00	4	013	0	1	9999	9999
320	*Acestrura berlepschi*	00	4	013	0	1	9999	9999
321	*Acestrura harterti*	00	0	000	0	1	9999	9999
322	*Chaetocercus jourdanii*	05	4	010	0	2	0900	3000
323	*Selasphorus platycercus*	07	4	019	4	1	0900	3350
324	*Selasphorus rufus*	06	4	018	4	2	0000	2250
325	*Selasphorus sasin*	04	4	018	4	3	9999	9999
326	*Selasphorus flammula*	05	4	015	3	1	1700	3100
327	*Selasphorus torridus*	00	4	013	3	1	9999	9999
328	*Selasphorus simoni*	00	4	012	0	1	9999	9999
329	*Selasphorus ardens*	02	4	015	3	1	9999	9999
330	*Selasphorus scintilla*	05	4	015	3	1	1220	3000

Note: Recent studies suggest that some phaethornithine species included in my list may not be valid (Hinkelmann 1988, 1989). Thus, my tally may in fact underestimate the difference in species diversity between subfamilies.

A13.2. Species diversity of subfamilies in 34 selected habitats

Habitat Type	Number of Species Per Habitat Type	
	Phaethornithines	Trochilines
Wet forest	004	009
Humid forest	024	076
Cloud forest	003	070
Elfin forest/woodland	000	017
Terra firme forest	000	002
Várzea forest	002	002
Swamp forest	000	002
River edge forest	003	000
Sandy belt forest	002	005
Mangrove	001	009
Savanna woodland	003	008
Dry forest	004	022
Gallery forest	002	016
Lighter/thinned woodland	001	076
Second-growth woodland	012	063
Forest or woodland border	013	159
Shrubby areas and clearings	008	149
River island shrubbery	000	003
Scrubby areas	001	049
Arid/desert scrub	000	030
Coastal zone	000	002
Marshes	000	002
Savanna	001	015
Paramo	000	022
Cultivated lands	005	075
Pine-oak woodland	000	017
Deciduous woodland	000	002
Coniferous forest	000	001
Open coniferous forest	000	001
Lowland pine savanna	000	001
Pinyon-juniper association	000	001
Willow-alder thickets	000	001
Chaparral	000	005
Meadows	000	007

Note: Habitats are those designated in the American Ornithologists' Union's *Check-list of North American Birds* (1983) and those in Hilty and Brown (1986).

A13.3. Principal literature sources for A13.1 and A13.2

American Ornithologists' Union. 1983. *Check-list of North American birds*, 6th ed. Committee on classification and nomenclature of the A. O. U. Lawrence, Kan.: Allen Press.

Barash, D. P. 1972. Lek behavior in the broad-tailed hummingbird. *Wilson Bull.* 84:202–3.

Bond, J. 1971. *Birds of the West Indies*. Boston, Mass.: Houghton Mifflin Co.

Des Granges, J-L., and P. R. Grant. 1980. Migrant hummingbirds accommodation into tropical communities. In *Migrant birds in the neotropics: Ecology, behavior, distribution and conservation*, ed. A. Keast and E. S. Morton, 395–409. Washington, D.C.: Smithsonian Institution Press.

Feinsinger, P. 1977. Notes on the hummingbirds of Monteverde, Cordillera de Tilaran, Costa Rica. *Wilson Bull.* 89:159–64.

Feinsinger, P. 1980. Asynchronous migration patterns and the coexistence of tropical hummingbirds. In *Migrant birds in the neotropics: Ecology, behavior, distribution and conservation*, ed. A. Keast and E. S. Morton, 411–9. Washington, D.C.: Smithsonian Institution Press.

Feinsinger, P., and S. B. Chaplin. 1975. On the relationship between wing dic loading and foraging strategy in hummingbirds. *Am. Naturalist* 109:217–24.

Feinsinger, P., and R. K. Colwell. 1978. Community organization among neotropical nectar-feeding birds. *Am. Zoologist* 18:779–95.

Feinsinger, P., L. A. Swarm, and J. A. Wolfe. 1985. Nectar-feeding birds on Trinidad and Tobago: Comparison of diverse a depauperate guilds. *Ecological Monogr.* 55:1–28.

Fitzpatrick, J. W., D. E. Willard, and J. W. Terborg. 1979. A new species of hummingbird from Peru. *Wilson Bull.* 91:177–86.

French, R. 1973. A *guide to the birds of Trinidad and Tobago*. Newton Square, Pa.: Harrowood Books.

Gill, F. B. 1987. Ecological fitting: Use of floral nectar in *Heliconia stilesii* Daniels by three species of hermit hummingbirds. *Condor* 89:779–87.

Gill, F. B., A. L. Mack, and R. T. Ray. 1982. Competition between hermit hummingbirds Phaethorninae and insects for nectar in Costa Rican rain forest. *Ibis* 124:44–49.

Hilty, S. L., and W. W. Brown. 1986. *A guide to the birds of Colombia*. Princeton, N.J.: Princeton University Press.

Hinkelmann, C. 1988. On the identification of *Phaethornis maranhaoensis* Grantsav. 1968 (Trochilidae) *Bull. Brit. Orn. Club* 108:14–18.

Hinkelmann, C. 1989. Ammerkungen zu *Phaethronis margarettae* und anderen nicht validen nev beschriebenen Kolibriarten. *Trochilus* 10:64–70.

Holt, E. G. 1928. An ornithological survey of the Serra do Itatiaya, Brazil. *Bull. Am. Mus. Natural Hist.* 57:251–326.

Johnsgard, P. A. 1983. *The hummingbirds of North America*. Washington, D.C.: Smithsonian Institution Press.

Johnson, A. W. 1967. *The birds of Chile and adjacent regions of Argentina*,

Bolivia, and Peru, vol. 2. Buenos Aires, Argentina: Platt Establecimientos Graficos.

Koepcke, M. 1964. *The birds of the Department of Lima, Peru*. Newton Square, Pa.: Harrowood Books.

Land, H. C. 1970. *Birds of Guatemala*. Wynnewood, Pa.: Livingston Publishing Co.

Lyon, D. L. 1976. A montane hummingbird territorial system in Oaxaca, Mexico. *Wilson Bull.* 88:280–99.

Meyer de Schauensee, R., and W. H. Phelps, Jr. 1978. *A guide to the birds of Venezuela*. Princeton, N.J.: Princeton University Press.

Moore, R. T. 1939. Habitats of white-eared hummingbird in northwestern Mexico. *Auk* 56:422–6.

Nicholson, W. M. 1931. Field notes of the Guiana king hummingbird. *Ibis* 13:534–53.

Pitelka, F. A. 1942. Territoriality and related problems in North American hummingbirds. *Condor* 44:189–204.

Ridgely, R. S. 1976. *A guide to the birds of Panama*. Princeton, N.J.: Princeton University Press.

Ruschi, A. 1972. *Beija-flores (Hummingbirds)*. Museu de Biologia "Prof. Mello Leitão"; São Paulo Brazil.

Sazima, M. 1977. Hummingbird pollination of *Barbacenia flava* (Velloziaceae) in the Serra do Cipo, Minas Gerais, Brazil. *Flora Bild.* 166:239–47.

Schemske, D. W. 1975. Time budget and foraging site preference of the cinnamon hummingbird in Costa Rica. *Condor* 77:216–7.

Sick, H. 1986. Ornitologia Brasileira, Uma Introdução, 2d ed., vol. 1. Brasilia, Brazil: Editora Universidade de Brasilia.

Skutch, A. 1958. Life history of the violet-headed hummingbird. *Wilson Bull.* 70:5–19.

Skutch, A. 1961. Life history of the white-creasted coquette hummingbird. *Wilson Bull.* 73:5–10.

Snow, D. W., and B. K. Snow. 1980. Relationships between hummingbirds and flowers in the Andes of Colombia. *Bull. Br. Mus. Natural Hist. (Zoology)* 38:105–39.

Stiles, F. G. 1971. Time, energy and territoriality of the Anna hummingbird (*Calypte anna*) *Science* 173:818–21.

Stiles, F. G. 1975. Ecology, flowering phenology, and hummingbird pollination of some Costa Rican *Heliconia* species. *Ecology* 56:285–301.

Stiles, F. G. 1980. The annual cycle in a tropical wet forest hummingbird community. *Ibis* 122:322–43.

Stiles, F. G. 1983. Systematics of the southern forms of *Selasphorus* (Trochilidae). *Auk* 100:311–25.

Stiles, F. G. 1985. Seasonal patterns and coevolution in the hummingbird-flower community of a Costa Rican subtropical forest. In *American Ornithologists' Union Monograph no. 36*, ed. P. A. Buckley, M. S. Foster, E. S. Morton, R. S. Ridgely, and F. G. Buckley, 757–87.

Stiles, F. G., and L. L. Wolf. 1970. Hummingbird territoriality a a tropical flowering tree. *Auk* 87:467–91.

Stiles, F. G., and L. L. Wolf. 1979. Ecology and evolution of lek mating behavior in the long-tailed hermit hummingbird. *American Ornithologists' Union Ornithological Monographs* no. 27.

Weske, J. S., and J. W. Terborgh. 1977. *Phaethornis koepckeae*, a new species of hummingbird from Peru. *Condor* 79:143–7.

Wetmore, A. 1968. *The birds of the Republic of Panama, Part* 2. Smithsonian Miscellaneous Collections, vol. 150. Washington, D.C.: Smithsonian Institution Press.

Wolf, L. L. 1964. Nesting of the fork-tailed emerald in Oaxaca, Mexico. *Condor* 66:51–55.

Wolf, L. L. 1969. Female territoriality in a tropical hummingbird. *Auk* 86:490–504.

Wolf, L. L. 1975. "Prostitution" behavior in a tropical hummingbird. *Condor* 77:140–4.

Wolf, L. L., F. G. Stiles, and F. R. Hainsworth. 1976. Ecological organization of a tropical, highland hummingbird community. *J. Animal Ecol.* 45:349–79.

Wolf, L. L., and J. S. Wolf. 1971. Nesting of the purple-throated carib hummingbird. *Ibis* 113:306–15.

14

Faunal Diversity and Turnover in a Miocene Terrestrial Sequence

John C. Barry, Lawrence J. Flynn, and David R. Pilbeam

Recent discussions of biotic diversity and faunal change have focused on the relative importance of biotic and abiotic factors in evolution (e.g., Connell 1978; Hoffman and Kitchell 1984; Stenseth and Maynard Smith 1984). Using examples from the fossil record of Neogene mammals, Vrba (1980, 1985) in particular has emphasized a primary role for the physical environment in forcing evolutionary change in organisms. Beginning with the premise that organismic evolution is conservative and that most change in lineages is concentrated in abrupt speciation events and accumulates through successive replacement of closely related forms, she has argued that immigration, speciation, and extinction should only occur when forced by changes in the physical environment. The impact of environmental change is considered to be primarily through modification of habitat, which in the Neogene is seen as largely the result of climatic change, although tectonic and oceanographic events also had effects (Vrba 1985).

Vrba's turnover pulse hypothesis has implications for community evolution and succession. The difficulties of recognizing communities among assemblages of fossil vertebrates and their modification by various processes were discussed by Olson (1966, 1980) and Van Couvering (1980), among others. Noting that temporally successive fossil assemblages generally have a strong resemblance, Olson (1966, 1980) has observed that their most salient feature is maintenance of ecological structure through time, despite considerable species replacement between assemblages. For that reason he distinguished community succession from community evolution as the replacement of one community by another with a markedly different ecological structure. If species are static entities, then communities should evolve and succeed each other primarily through replacement of their constituent species, not through gradual anagenetic transformation of members. Thus the turnover hypothesis implies that community evolution, and not just the more radical community succession, should be driven primarily by environmental events.

In the fossil record, speciation, immigration, and extinction are manifested as first and last occurrences in local stratigraphic sequences. Therefore, temporal patterns of first and last occurrences should conform to the expectations of the hypothesis. As articulated by Vrba (1985) there are two principal expectations. First, evolutionary events (first or last occurrences) should be concentrated within brief time intervals to form turnover pulses and, secondly, turnover pulses should be closely synchronous with changes in the physical environment. As secondary expectations any turnover pulses should also be composed of both first and last occurrences and turnovers should affect different lineages synchronously and in like manner. These four expectations are in some degree testable, but the literature on terrestrial vertebrates offers contradictory conclusions. Webb (1977, 1984) and Schankler (1980) both found evidence for episodes of rapid faunal change during the Miocene and early Eocene of North America. These episodes contrasted to periods of little change and were interpreted by these authors as resulting from increased rates of immigration and extinction due to climatic or tectonic events. On the other hand Bernor (1985), Hill (1988), and Badgley and Gingerich (1988) have argued cogently that such episodes are likely to be artifacts of varying sample size or other accidental factors. Hill (1988) and Bernor (1985) conclude that pulses cannot be recognized in either the East African or Eurasian Miocene records, although they do not rule them out.

Schankler (1980) and Webb (1984) both noted that clusters of immigration events did not necessarily coincide with extinction events. Webb therefore distinguished between origination-type and extinction-type faunal turnovers and suggested they could have fundamentally different environmental causes—perhaps operating on different time scales. This was also implied by Olson (1966), but at present there is no clear understanding as to the time scale on which intervals, events, and pulses are to be recognized, nor of the time scale on which they are to be correlated to environmental changes. The late Miocene and Pliocene turnovers described by Webb occurred over time periods of 1 to 5 millions years duration, whereas the supposed early Eocene episodes where much shorter, lasting no more than 500,000 years. In either case, time scale is important in evaluating the extent to which episodes or pulses are biologically relevant and likely to be explained by discrete environmental change. Presumably, however, the relevant time scale is less than 10^6 years, and more probably on the order of 10^5 or 10^4 years.

In this paper we discuss the implications of new stratigraphic data for rodents and artiodactyls from the Neogene Siwalik formations of northern Pakistan. We use these data to describe patterns of faunal change, as evident in both changes in the number of species present and the timing and intensity of faunal turnover. We focus particularly on the pattern of turnover in these two groups between 16 and 7 Ma in order to determine how Siwalik faunal change relates to climatic and other environmental events and how closely it conforms to the expectations of Vrba's turnover pulse hypothesis. Our principal questions are: (1) What evidence is there for the presence of episodes or pulses of turnover in the Siwalik record? and (2) How convincing are potential correlations between turnover pulses and global or regional environmental events as documented in the isotopic or other proxy records? We have also examined the record for evidence that turnover pulses were composed of both first and last occurrences and whether both rodents and artiodactyls were affected synchronously. Because the timing of change in successive fossil assemblages is difficult to delineate and "causes" of change are not easily recognized, as both depend on the vagaries of the stratigraphic record (Gradstein and Agterberg 1985; Jablonski 1986; Hill 1988), we have attempted to analyze this record with attention to the completeness and quality of the data.

The Siwalik formations of the Indian Subcontinent are a classic terrestrial Neogene sequence, which has been of paleontological interest for over 150 years. They are remarkable in that vertebrate and invertebrate fossils are common throughout a nearly 18-million-year interval spanning the early Miocene through Pleistocene (N.M. Johnson et al. 1982, 1985; Opdyke et al. 1979), thus preserving fossil remains of terrestrial communities existing in one geographically restricted region for millions of years. In the past decade there has been renewed research interest in closely coordinated studies of the fossils and sediments and analysis of their age and the tectonic setting. As a result, potential exists for correlating the long Siwalik faunal record to sedimentary, tectonic, and climatic records, making is possible to examine with greater resolution than previously the timing of faunal turnovers and their relationship to climatic and other environmental events. As such, the fossil assemblages of the Siwaliks constitute a rare data set for vertebrates and terrestrial communities and are directly relevant to hypotheses of faunal change.

Methods

The terrestrial Miocene deposits of Pakistan and India occur as thick clastic wedges deposited by large river systems along the southern margins of the northern and western mountains between India and Asia.

The deposits of the Potwar Plateau of northern Pakistan include channel belt sands, complex levee and channel fills, and the soils and other fine-grained sediments of the floodplains (Behrensmeyer 1987) and span the interval between 18.3 and 0.7 Ma (N.M. Johnson et al. 1982, 1985). Exposures are extensive, and composite stratigraphic sections are readily constructed and dated, with fossil sites being traced into them. Superposition, radiometric dates on bentonites, and magnetostratigraphy combined give a high degree of chronological precision and resolution (Opdyke et al. 1979; G.D. Johnson et al. 1982; Tauxe and Opdyke 1982; N.M. Johnson et al. 1982, 1985). As a result the ages of fossil sites can, in general, be estimated to 200,000 or perhaps even 100,000 years, the exceptions being sites within long magnetozones that necessitate interpolation of age based on sediment accumulation rates (Badgley, Tauxe, and Bookstein 1986). Nevertheless, even when the absolute ages of sites are not known precisely, the stratigraphic order is unambiguous from superpositional relationships. For consistency with earlier publications of Siwalik faunas, in the following we use age estimates for the magnetic chrons that are based on the Mankinen and Dalrymple (1979) geomagnetic reversal time scale.

In the Potwar Plateau, the Miocene Siwalik Group rests unconformably on Eocene marine rocks with the base of the section on the northern flank of the Salt Range being approximately 18.3 Ma (N.M. Johnson et al. 1985). The oldest 2 millions years of this sequence are still inadequately sampled, as are Pliocene and younger horizons (Keller et al. 1977; Jacobs 1978; Opdyke et al. 1979). In contrast the 9 million years between 16 and 7 Ma include numerous and extensive fossil localities distributed throughout the section in a variety of lithologies. These includes sites with both large and small mammals. For this interval, knowledge of mammal faunas has progressed beyond simple description of faunal composition to recognition of community structure, faunal turnover, and trends in species abundance.

Some Siwalik intervals have localities on the order of every 100,000 years, but in general the sampling, especially for sites with rodents, only permits resolution on the order of every half million years. For this reason our analysis is based on a subdivision of the period between 16 and 7 Ma (more precisely, 16.25 to 6.75 Ma) into 19 intervals, each 500,000 years long. (By convention, intervals are centered at whole multiples of a half millions years: e.g., the boundaries of the 8 Ma interval are at 8.25 and 7.75 Ma.) There are indications, however, that some major faunal transitions in the Siwaliks have occurred over much shorter time intervals, perhaps only 100,000 years (Barry et al. 1985), so that if shorter intervals were used, the intensity of single episodes of change might in some cases be seen to be stronger. That is, the events might be concentrated within a single brief subinterval. Even so, the total amount of change within the longer interval would not be diminished and the patterns of change, as seen at that time scale, would not be altered. As previously noted, at present there is no understanding of the time scale on which pulses are biologically significant. Nevertheless, it must be realized that conclusions about faunal change depend on the time scale used.

Questions of faunal change are most informatively addressed at the level of the species, and we have attempted to identify all records at that level. In this preliminary analysis, we have concentrated on artiodactyls and rodents, because they are abundant and their taxonomic study is generally at an advanced stage. These are, of course, the dominant mammalian herbivores in modern communities and evidently also in the Miocene Siwaliks. The two groups have the potential to reflect climatic and habitat changes in complementary ways. Some other members of the Siwalik herbivore trophic level are the Proboscidea and Rhinocerotidae and, after 9.5 Ma, hipparionine horses. The biostratigraphy and, to some extent, the systematics of these groups are not yet well enough understood to include them in this study.

Identifications of the rodents are based on work in progress and on published studies (Jacobs 1978; Jacobs and Flynn 1981; Flynn and Jacobs 1982; Cheema, Sen, and Flynn 1983; Flynn, 1982, 1986; Flynn, Jacobs, and Sen 1983; Lindsay 1987, 1988; Jacobs, Flynn, and Downs 1989). Identifications for artiodactyls other than bovids are also from work in progress, whereas the identifications of the middle Miocene bovids are based on unpublished research by Raza and Solounias, supplemented by Thomas (1983). Detailed systematic and biostratigraphic analysis of Siwalik Bovidae younger than 10.5 Ma has not been completed. Age estimates are based on both work in progress and published studies, including N.M. Johnson et al. (1982, 1985), Tauxe and Opdyke (1982), Barry, Behrensmeyer, and Monaghan (1980) and Barry, Lindsay, and Jacobs (1982). Using the results of these taxonomic and stratigraphic studies, we have determined temporal ranges of species throughout the well-sampled 16 to 7 Ma interval of the Siwaliks; subsequently, from that compilation, we have derived the number of species of rodents and artiodactyls for each half-million-year subdivision. Due to the preliminary nature and complexity of the data and because many of the species names are in manuscript and unavailable for citation, the data are presented in tabular form (table 14.1).

Table 14.1 combines the number of species that were actually found in each interval with the number of species that might reasonably be

Table 14.1. Number of species and number of first and last occurrences

A. *Rodents*																
Intrv (Ma)	Q	Rhizomyids			Cricetids			Murids			Others			Total		
		#Sp	F	L	#Sp	F	L	#Sp	F	L	#Sp	F	L	#Sp	F	L
7.0	1	7	4	5	0	0	0	3	1	2	4	3	3	14	8	10
7.5	1	3	0	0	4	4	4	2	0	0	3	0	2	12	4	6
8.0	1	6	4	3	0	0	0	4	2	2	4	0	1	14	6	6
8.5	4	2	2	0	0	0	0	2	0	0	4	0	0	8	2	0
9.0	2	2	0	2	1	0	1	2	0	0	5	0	1	10	0	4
9.5	1	2	1	0	9	0	8	2	0	0	5	2	0	18	3	8
10.0	4	1	0	0	9	0	0	2	0	0	3	0	0	15	0	0
10.5	2	2	0	1	10	1	1	2	1	0	4	0	1	18	2	3
11.0	2	3	1	1	9	2	0	1	1	0	4	0	0	17	4	1
11.5	5	2	–	–	7	–	–	–	–	–	4	–	–	13	–	–
12.0	3	2	0	0	7	0	0	1	0	1	4	0	0	14	0	1
12.5	2	3	0	1	16	2	9	1	0	0	5	0	1	25	2	11
13.0	5	3	–	–	14	–	–	1	–	–	5	–	–	23	–	–
13.5	1	3	1	0	17	6	3	1	0	0	5	3	0	26	10	3
14.0	3	2	0	0	11	0	0	1	0	0	2	0	0	16	0	0
14.5	2	2	0	0	13	2	2	1	1	0	3	0	1	19	3	3
15.0	3	2	0	0	11	2	0	0	0	0	3	0	0	16	2	0
15.5	5	2	–	–	9	–	–	0	–	–	3	–	–	14	–	–
16.0	2	2	2	0	11	11	2	0	0	0	3	3	0	16	16	2

B. *Artiodactyls*

Intrv (Ma)	Q	Guoids			Tragulids			Giraffids			Bovids			Total		
		#Sp	F	L	#Sp	F	L	#Sp	F	L	#Sp	F	L	#Sp	F	L
7.0	2	4	0	0	3	0	2	2	0	1						
7.5	2	4	0	0	4	0	1	2	1	0						
8.0	1	7	1	3	4	0	0	1	0	0						
8.5	2	7	1	1	4	0	0	1	1	0						
9.0	1	7	5	1	4	1	0	1	0	1						
9.5	3	5	0	3	3	2	0	1	0	0						
10.0	4	5	0	0	1	0	0	1	1	0						
10.5	2	8	3	3	3	1	2	1	0	1	4	1	4	16	5	10
11.0	3	6	1	1	3	1	1	1	0	0	5	0	2	15	2	4
11.5	1	5	1	0	4	0	2	1	0	0	7	0	2	17	1	4
12.0	3	6	2	2	4	0	0	1	0	0	8	1	1	19	3	2
12.5	3	4	0	0	5	1	1	1	0	0	8	1	1	18	2	2
13.0	3	4	0	0	5	1	1	1	0	0	8	1	1	18	2	2
13.5	1	4	0	0	5	2	1	1	0	0	10	3	3	20	5	4
14.0	2	5	1	1	4	1	0	1	0	0	8	8	0	18	10	1
14.5	4	4	0	0	3	1	0	1	0	0	1	0	1	9	1	1
15.0	3	4	2	0	2	0	0	1	0	0	1	0	0	8	2	0
15.5	4	3	1	1	2	0	0	1	0	0	1	0	0	7	1	1
16.0	3	3	3	1	3	2	1	2	1	1	1	0	0	9	6	3

Note: Intrv (Ma), median age of interval; Q, quality class; #Sp, number of species found or inferred to be present in interval; F, number of first occurrences; L, number of last occurrences.

expected to be present, based on earlier and later records. This technique, the range-through method (e.g., Boltovsky 1988; originally the "range method" of Cheetham and Deboo 1963), augments the numbers of species but evidently without distorting findings, because plots of the observed numbers of species show the same trends as the augmented totals (see fig. 14.3 and following discussion). As an example, no sites with small mammals are presently known for the 13 Ma interval, but 23 species of rodents are found in both well-sampled intervals immediately preceding and following it. These 23 species are therefore assumed to have been present through the intervening interval as well. Since no data exist for this interval, table 14.1 shows no observed first or last occurrences. However, these absences should not be understood to meant that there were no events in the interval. Note also that "first" and "last" occurrences are observations that are not equivalent to, but only estimates of, local appearances (by either immigration or speciation) and extinctions. Some aspects of the problems of approximating local appearances and extinctions, and deriving event patterns, are dealt with later.

Sites or horizons with abundant fossils usually have more taxa than those with fewer fossils and as a result may give better estimates of diversity and greater confidence in constraining first and last occurrences of taxa. The abundance of fossils throughout the Siwalik sequence, however, is not uniform, and our analysis therefore begins with an appraisal of the quality of the Siwalik record, followed by a discussion of how variations in quality affect comparisons between intervals. Since this paper concerns the number of species and their stratigraphic ranges, by "quality" we mean the degree to which the known fossil assemblages of an interval might represent the actual fauna present during that interval; better-quality intervals give more confidence about presence or absence and community composition. Many factors, such as the number of localities, types of depositional environments, number of specimens, and fossil preservation, can influence interval quality. Also important, presumably, are attributes of individual species such as size, habitat preference, or ease of collection and identification. These factors may vary between sites and taxa (Badgley 1986a), and evaluation of their relative importance provides an assessment of quality for each interval. The species of each order are generally similar to each other, so that within each group taphonomical biases tend to be similar. However, because of the very different taphonomical processes to which small and large mammals are subjected, the quality of the rodent and artiodactyl records must be evaluated separately.

Surface finds of rodents are few in comparison to those of artiodactyls, but rodents commonly occur as isolated teeth that can be extracted

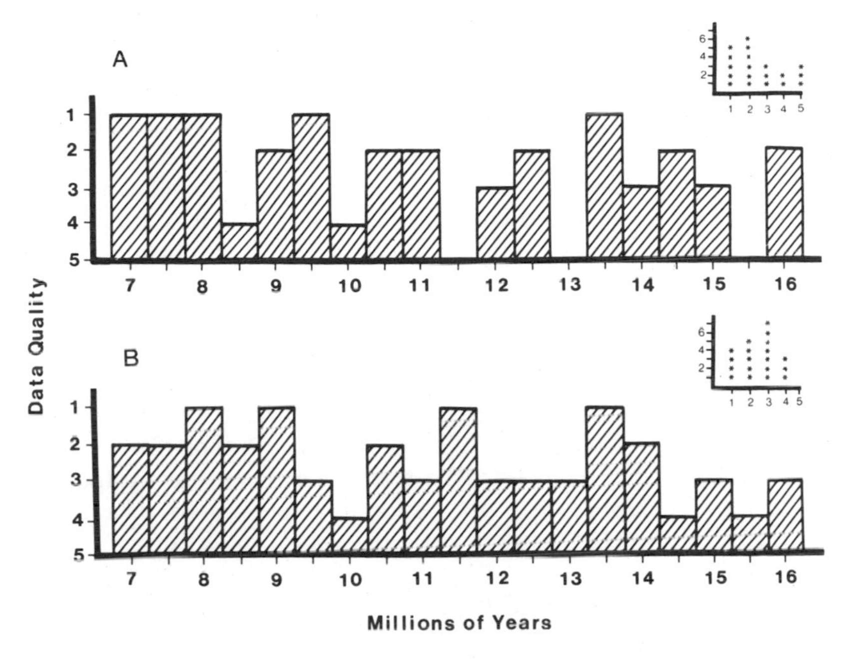

Figure 14.1. Quality of data. A; Rodents. B; Artiodactyls. Insets are histograms for five quality categories.

from bulk samples of fossiliferous sediments. Individual sites may be affected by different taphonomical processes, but the factors related to discovery and identification should be similar between sites and independent of taxon because the teeth are physically alike and have the same chance for recognition in the picked concentrate. Site-specific biases, such as hydraulic size sorting and predator selectivity, may be more important, but with typically only a few sites per interval it is difficult to judge how important. Hydraulic size sorting is probably not critical. Predator selectivity remains a concern, as fossil assemblages formed primarily from scats or pellets may give biased estimates of small mammal diversity within habitats. However, for intervals where we have more than one site, we find no evidence for a strong bias that can be attributed to predator selectivity. Thus, while we do not have quantitative data to evaluate taxonomic composition relative to site-related biases, our qualitative assessments suggest they are constant between intervals. Rodent "quality" is therefore considered to be a function of the number of specimens. The largest sites have in excess of 300 specimens, and five intervals with over 300 teeth are in class 1 (fig. 14.1a; abundance data from Jacobs et al., 1990; updated by W.R. Downs, June 1988, pers. comm.). Six well-sampled intervals with at least 150 specimens are given a rank of class 2. Three intervals with under 100 specimens are ranked as class 3, whereas two poorly represented intervals with fewer than 50 specimens are class 4. Three intervals without rodents are class 5.

Large mammals such as artiodactyls typically occur as surface finds within small areas of outcrop derived from one stratigraphic horizon but are rarely concentrated enough to quarry. In that they are irregularly distributed throughout the sequence, we have attempted to minimize the effects of low fossil productivity by intensive collecting at less fossiliferous levels and by basing identifications on as many body parts as possible. Intervals are ranked using several variables, including number of localities, number of specimens, and preservation. Nine intervals are considered well known and given a rank of class 1 or 2 (fig. 14.1B, whereas seven are class 3. Three intervals are poorly known and rank as class 4, but none lack data completely. Compared to the rodents, there are fewer class 1 and 2 intervals and no class 5. The mode for the rodents in class 2, whereas for the artiodactyls it is class 3 (fig. 14.1, insets).

The data on species ranges from which the values of table 14.1 were derived also provide estimates of species longevities. Apparent longevities range from ca. 100,000 years to about 6 my and average around 3 my. Two factors decrease confidence in such figures. First, many taxa of long duration cannot be used to compute longevities, as those first found at 16 Ma cannot be assumed to have originated at that time, whereas those last found at 7 Ma may not have become extinct then. Second, the level of systematic analysis is not yet uniform for all groups, and some apparently long-lived species may eventually be recognized as a succession of closely related species. Species longevity and validity of the half-life concept for Siwalik faunas (sensu Kurten 1972) will be subjects of future study.

Finally, most Siwalik species exhibit morphological stasis and are without apparent ancestors or descendants within the sequence. Many probably originated as immigrants (Barry et al. 1985). A few lineages (e.g., the rodent *Kanisamys*, the tragulid *Dorcatherium*, and the giraffe *Giraffokeryx*), however, do show evidence of in situ evolution by anagenesis or various forms of cladogenesis. We have deliberately minimized effects of phyletic terminations within these lineages by conservatively using few arbitrary cutoffs within lineages. In addition, for some taxa there appear to be changes in relative abundance, which may also indicate changes in biomass. If so, these changes are presumably related to environmental fluctuations and historical events, such as replacement of one taxon by another. A possible case has been discussed by Jacobs et al. (1990). Information on the relative abundance of taxa

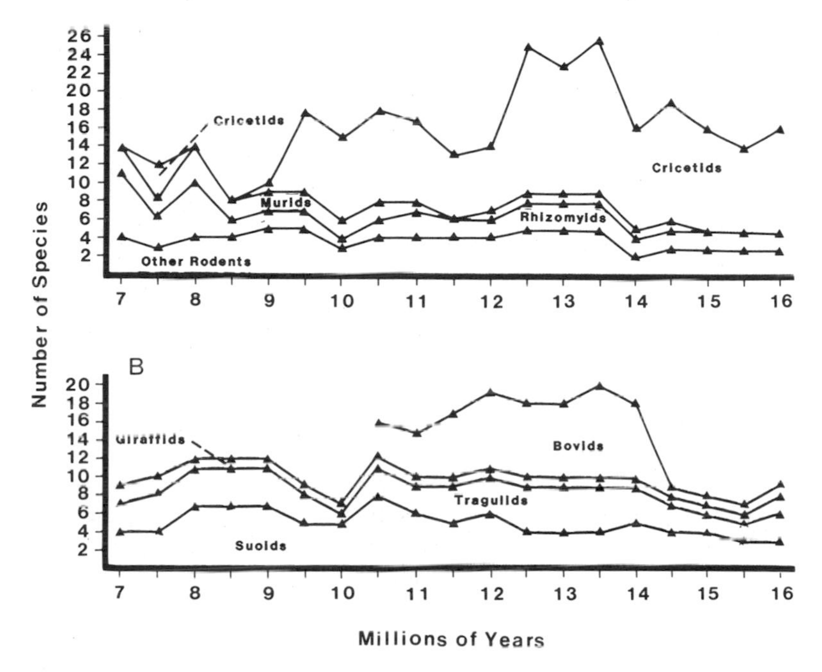

Figure 14.2. Number of species found or inferred to be present in each interval. A; Rodents. B; Artiodactyls.

is important but difficult to document (Badgley 1982, 1986a). It is potentially available for some taxa but not for all intervals. Therefore, we use only presence/absence data. Following Connell (1978) we use diversity to mean the total number of species present.

Analysis

NUMBER OF SPECIES AND DIVERSITY

Figure 14.2 shows the number of species known for each interval, with the cumulative totals for the rodents and artiodactyls kept separate to facilitate comparisons between them. Inspection of the figure suggests that during most intervals rodents, and particularly the muroid families, predominate. Further, while the number of species in most groups is nearly constant (and low), substantial change seems to occur within some groups. The total number of species increases during the middle Miocene and then decreases at the beginning of the late Miocene. Large increases in the number of bovids after 14.5 Ma and rhizomyids after 8.5 Ma are notable. Also notable are the irregular, steplike changes in the total numbers of rodents after 14.0, 12.5, and 9.5 Ma, which appear to be due largely to pronounced increases and decreases among cricetids.

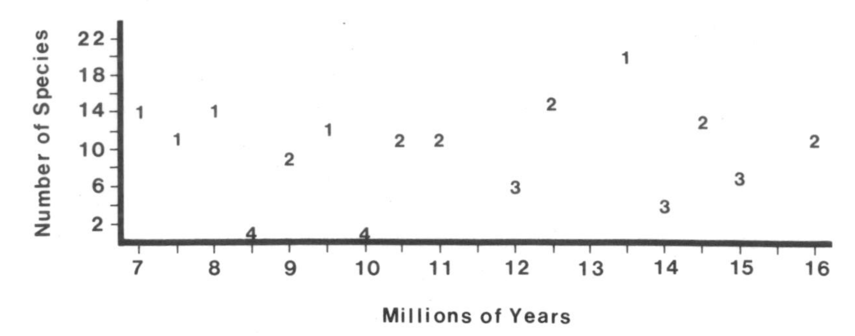

Figure 14.3. Number of species of rodents actually recovered in each interval; data points represented by numerals indicating class.

Before the significance of these observed changes among intervals can be assessed, it is necessary to judge how closely the fossil record estimates the original diversity. It is generally assumed that the number of species found is a good estimate of faunal diversity and that observed differences between intervals reflect genuine changes in diversity (see Andrews 1981). However, because of sampling effects, the number of species identified for an interval will underestimate true diversity (Nichols and Pollack 1983; Koch 1987) and sometimes markedly so. This is largely a consequence of some species being rare, restricted to very limited habitats, or having short temporal ranges. Finding rare or restricted species requires large numbers of specimens and sites sampling a range of habitats; finding short-lived species requires more sites differing in age within each interval. Sampling bias therefore includes the effects of both preservational processes, which limit the number of available specimens or localities, and collection effort, which limits the number of specimens or localities found.

Since the quality of an interval is a combination of both collecting effort and richness, quality should affect the number of species found. This is seen most clearly in figure 14.3, where the number of rodent species collected from each interval is categorized by our estimates of data quality. From the figure it is evident that intervals with the fewest species have the poorest-quality data. The effect is particularly apparent for classes 3 and 4 when compared to class 2, suggesting the class 3 and 4 intervals should not be compared directly to the class 1 and 2 intervals, although class 3 and 4 intervals do provide some useful data. As the difference between the number of species in class 1 and class 2 is less, the effect of data quality appears weaker for these two classes. A nonparametric Mann-Whitney U test of the rank order of the class 1 and class 2 values supports this observation, as there is no significant difference ($T = 28.5$, $p < 0.05$) in the distribution of values between the two classes. Therefore, not only do we consider the class 1 and class 2 data to be much more reliable than those of the other classes, but we also consider class 2 intervals to be directly comparable to those of class 1.

The effects of data quality complicate interpretation of figure 14.2, because differences in numbers of species observed are the result of differences in the quality of the record and of actual diversity. Although there is no way to completely isolate the effect of quality from diversity, comparisons using only intervals of high quality should more accurately reflect any underlying differences in diversity. The following discussion is therefore based principally on the two higher-quality classes.

Figure 14.4 shows the number of rodent and artiodactyl species in class 1 and 2 intervals. The basic features are the same as those of Figure 14.2, except for lack of information on the number of artiodactyls present before 14 Ma. Again, the number of species of artiodactyls in each interval is generally fewer than the number of rodents, and over time the number of artiodactyls is nearly constant. In contrast, the

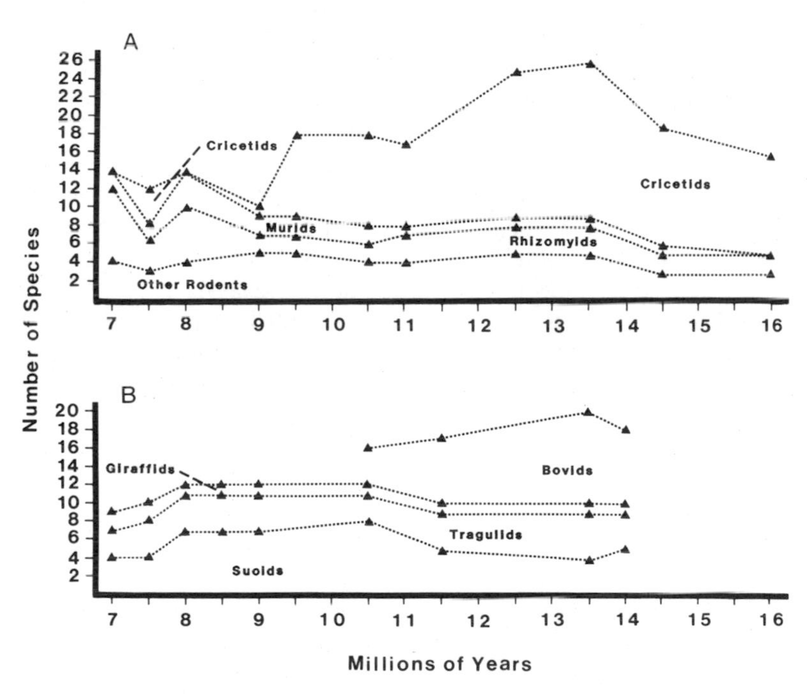

Figure 14.4. Number of species found or inferred to be present in quality class 1 and 2 intervals. A; Rodents. B; Artiodactyls.

number of rodents, and especially of cricetids, fluctuates more widely with increases in the cricetids after 14.5 Ma and the rhizomyids after 8.5 Ma. There is also a strong decreasing trend, possibly stepped, after 12.5 Ma, due primarily to loss of cricetids.

However, the significance of these apparent temporal trends is difficult to assess, because the observed differences in species number among the intervals must be considered relative to the unknown differences in diversity of successive faunas. Given high-quality data, increases or decreases in the number of species found in different intervals will reflect changes in diversity and, to some degree, sampling error. The value of the observed increase or decrease as an estimator of change in diversity clearly depends on how closely the number of species found approaches the number of species that were actually present. For example, if the fauna of an interval had about 50 species, then an increase or decrease of 10 or so species would be a substantial proportion of the total. However, if the diversity of rodents and artiodactyls were much larger—for example, more than 100—then the observed changes will be proportionately less important and more likely to be masked by sampling error.

Unfortunately, there is no reliable method for deriving diversity from the number of observed species without making one or more unrealistic assumptions (Nichols and Pollock 1983). Modern rodent faunas from tropical regions comparable in area to the Potwar Plateau (ca. 20,000 km^2) typically have numbers of species approximating those of our fossil faunas. For instance, Van Couvering (1980) and Van Couvering and Van Couvering (1976) cite 31 and 19 species of rodents from two species-rich forests in Zaire, 26 species for a Zaire forest-savanna site, and 20, 12, and 8 species for three African savanna parks. Rijksen (1978) lists 12 species of rodents for a 15,000 km^2 forest reserve on Sumatra, Kitchener et al. (1987) report 16 rodents from a smaller area near Kunming, China, and Roberts (1977) lists 11 extant species for the Potwar Plateau. By comparison the Siwalik class 1 and 2 intervals have between 10 and 26 species (table 14.1). These numbers compare favorably with modern diversities, and it therefore seems plausible that the class 1 and class 2 intervals record a large proportion of the rodents in the fossil communities.

Large mammals are probably more adversely affected by human disturbance; as a consequence, modern artiodactyls in most areas may be more depauperate than rodents. However, African and subtropical Asian reserves typically have between 6 and 28 species, with the more closed forests of Asia having fewer species and the woodlands and savannas of Africa having the most (Medway 1965; Rijksen 1978; Roberts 1977; Schaller 1967; Van Couvering 1980; Van Couvering and Van

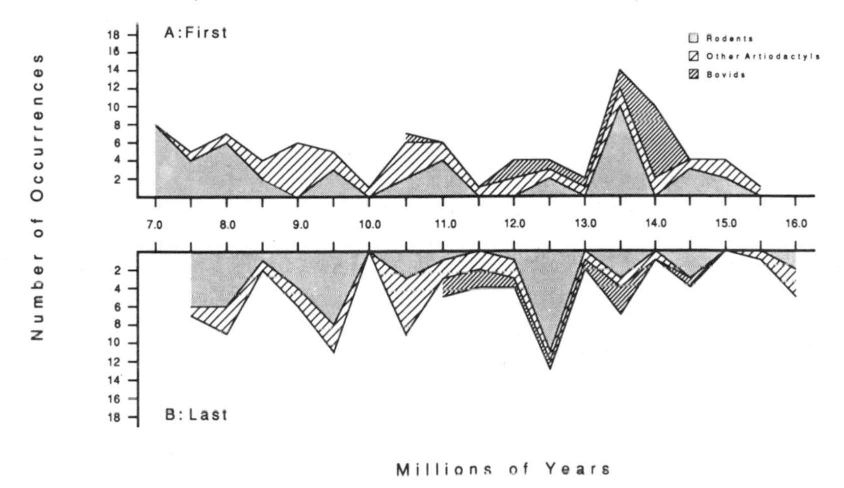

Figure 14.5. Number of first and last occurrence events recorded per interval. A; First occurrences. B; Last occurrences.

Couvering 1976). By comparison, Siwalik middle Miocene intervals have between 15 and 20 species. We therefore infer that for the class 1 and 2 intervals the number of rodents and artiodactyls found or assumed to be present approximates the diversity of the two groups in the interval. If this inference is correct, then differences in the number of species between intervals should reliably track changes in diversity. Furthermore, since the number of species present in a fossil assemblage will underestimate true diversity, it is of some interest that middle Miocene Siwalik assemblages have nearly as many species as modern faunas. This may indicate higher diversity in the Miocene than at present, as suggested by Van Couvering and Van Couvering (1976) for East Africa.

FIRST AND LAST OCCURRENCES

Figure 14.5 is a plot of the number of first (fig. 14.5A) and last (fig. 14.5B) occurrences of all rodent and artiodactyl species, using the values of table 14.1. The most obvious feature of this figure is that all intervals contain first or last occurrence events, but the number of events is variable among intervals. Some intervals, such as those centered on 10.0, 13.0, 15.0, and 15.5 Ma, have few events. By contrast, other intervals have more than the average number of events, which form peaks or maxima. The presence of these minima and maxima suggests that first and last occurrence events may have clustered as pulses, some of which were larger than others. But it is not clear how much of the variation in the number of events per interval is random scatter and at what level a maximum or minimum can be seen as being unusual or especially meaningful when compared to other intervals.

It is not possible to test individual maxima or minima for significance, but a chi-square goodness-of-fit test can be used to assess the significance of the departure of the observed distributions (table 14.1; exclusive of bovids) from an even distribution. The null hypothesis in both cases is that the 78 first occurrences and 79 last occurrences are distributed evenly among 18 intervals. The expected and observed values are shown in table 14.2, along with the individual contributions to the chi-square. With 16 degrees of freedom the chi-square values of 33.93 and 55.55 have very low probabilities (less than .01 for the first occurrences and less than .005 for the last occurrences), and we conclude it is very unlikely the two distributions are chance departures from an even distribution.

The large contributions to the chi-square of first occurrence maxima at 13.5 and 7.0 Ma and the maxima and minima in last occurrences at 15.0, 12.5, 10.5, 10.0, 9.5, and 8.0 Ma indicate that these intervals are largely responsible for the departure from a random distribution of the events. Notice should also be taken of the minima of first occurrences at 15.5, 13.0, 11.5, and 10.0 Ma, as collectively these contribute nearly a third of the total chi-square value. However, as we discuss later, some of these maxima and minima may be consequences of varying quality of the record from interval to interval and are not accurate descriptions of the pattern of Siwalik faunal change. The maxima and minima at 15.0, 10.5, and 10.0 Ma are especially suspect.

While some intervals have many first and last occurrences, maxima of first and last occurrences do not always coincide (fig. 14.5). For example, a first occurrence peak at 13.5 Ma does not have a complementary last occurrence maximum, and a peak of last occurrences at 12.5 Ma has no corresponding maximum of first occurrences. It is also apparent that maxima do not necessarily synchronously coincide in different taxonomic groups. This is best seen in the 8.0 Ma interval, where 12 of 14 rodents are first or last occurrences, while only 4 of 8 artiodactyls represent first or last records. It may be significant, however, that maxima and minima on each curve seem to follow one another closely.

Figure 14.5 is based on all known first and last occurrences but does not incorporate information about the varying quality of the data from individual intervals. It is therefore analogous to figure 14.2 as a depiction of the Siwalik record as it is actually observed. However, as the apparent ages of first and last occurrences are strongly biased by the quality of the fossil record, it is difficult to be confident about when a species first appears or disappears in even a local sequence (Hay 1972;

Table 14.2. Chi-square goodness-of-fit to an even distribution of the number of first and last occurrences of rodents and artiodactyls (excluding bovids) using actual record

Interval	Expected	Observed	$X^2 = (O - E)^2/E$
A. First Occurrences: p = 0.056, *n* = 78			
7.0	4.3	8	3.18
7.5	4.3	5	0.11
8.0	4.3	7	1.70
8.5	4.3	4	0.02
9.0	4.3	6	0.67
9.5	4.3	5	0.11
10.0	4.3	1	2.53
10.5	4.3	6	0.67
11.0	4.3	6	0.67
11.5	4.3	1	2.53
12.0	4.3	2	1.23
12.5	4.3	3	0.39
13.0	4.3	1	2.53
13.5	4.3	12	13.79
14.0	4.3	2	1.23
14.5	4.3	4	0.02
15.0	4.3	4	0.02
15.5	4.3	1	1.23
			X^2 = 33.92[a]
B. Last Occurrences: p = 0.056, *n* = 79			
7.5	4.4	7	1.54
8.0	4.4	9	4.81
8.5	4.4	2	1.31
9.0	4.4	6	0.58
9.5	4.4	11	9.90
10.0	4.4	0	4.40
10.5	4.4	9	4.81
11.0	4.4	3	0.45
11.5	4.4	2	1.31
12.0	4.4	3	0.45
12.5	4.4	12	13.13
13.0	4.4	1	2.63
13.5	4.4	4	0.04
14.0	4.4	1	2.63
14.5	4.4	3	0.45
15.0	4.4	0	4.40
15.5	4.4	1	2.63
16.0	4.4	5	0.08
			X^2 = 55.55[b]

[a]With d.f. = 16, the probability of $\chi^2 < .01$.
[b]With d.f. = 16, the probability of $X^2 < .005$.

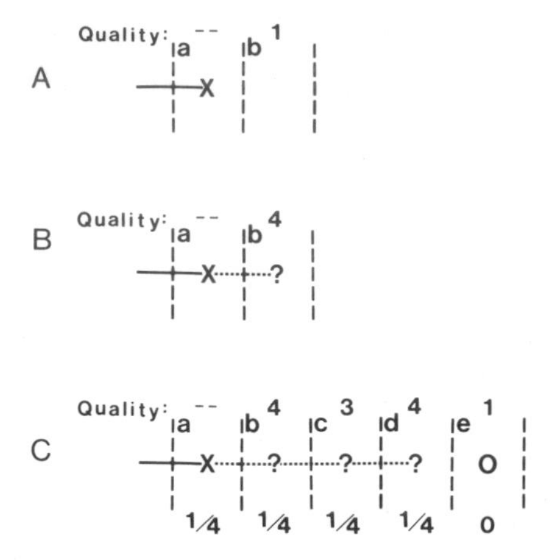

Figure 14.6. Method for using quality information to constrain the potential stratigraphic range of a species. A; With the adjacent interval B being a class 1, the first occurrence of species X is constrained to interval A. B; With the adjacent being a class 4, the first occurrence of species X could be in either interval A or interval B. C; With the consecutive intervals B, C, D, and E being class 4, 3, 4, and 1, the first occurrence of species X is constrained to intervals A, B, C, or D but could be in any one of these four.

Agterberg 1985; Koch 1987). Information about the expectation of finding a species can be used to infer absence from the negative evidence of nonoccurrence (Nichols and Pollock 1983), and such inferences can be used with information on varying data quality to constrain the stratigraphic ranges. As previously discussed, class 1 and 2 intervals probably record a large fraction of the rodents or artiodactyls that lived in the area, and what follows is based on this assumption. Therefore, if a species has not been found in a class 1 or 2 interval, it can be inferred that the species probably was not present rather than that the absence was the result of incomplete sampling. A hypothetical example is shown in figure 14.6A, where a species is recorded in interval A but not in the older adjacent interval B. If the quality of the older interval B is either class 1 or class 2, then we can be reasonably confident that the species was absent during the time represented by interval B. Its first occurrence therefore lies within interval A. If, however, the quality of the adjacent interval B is poorer than class 2, as in figure 14.6B, the absence of the species from the adjacent interval may only be due to incomplete sampling and the species might well have occurred within the interval. Similarly, if the quality of adjacent intervals is less than class 2, the species range might well have extended through several additional intervals. As shown in figure 14.6C, the potential range might span any number of poor-quality intervals before an interval of sufficient quality is reached to be reasonably confident that the species is absent. The sum of these intervals establishes the extent of the potential range but does not indicate where the first occurrence is within it. In the 16 to 7 Ma time period there are only two segments with more than two adjacent poor-quality intervals, and therefore most potential range extensions are 1 million years or less.

The same reasoning holds for last occurrences, although it is the quality of the adjacent younger intervals that is important, and extensions of last occurrences would be into younger horizons.

Figure 14.7 shows the resulting stratigraphic distribution of first and last occurrences if, using information on data quality, the ranges of all species are extended to the oldest (or youngest) reasonable interval. Because events shift toward the intervals with the highest-quality data, the general effect has been to make maxima more pronounced. Otherwise, figure 14.7 has much the same form as figure 14.5 and, in particular, the maxima originally at 12.5 and 13.5 Ma are still present. They are, however, centered on different ages, more widely separated, and neither is as prominent as before because of the growth of new maxima elsewhere in the sequence. Comparison of figures 14.5 and 14.7 emphasizes the uncertainty of the age estimates for maxima and, since the quality estimates for rodents and artiodactyls may differ for an interval (fig. 14.1), it particularly emphasizes the uncertainty of the relative ages of maxima in the two groups.

Figures 14.5 and 14.7 are only two nonarbitrary examples of many combinations of first and last occurrences that are permitted by the

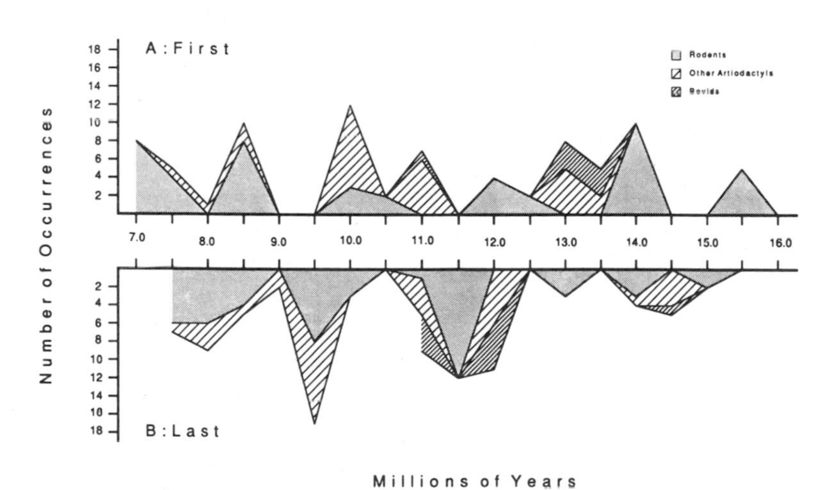

Figure 14.7. Number of first and last occurrence events with the range of each species maximally extended. A; First occurrences. B; Last occurrences.

data. In figure 14.7 the species' ranges are extended to the oldest and youngest possible intervals, whereas in figure 14.5 each species range is the minimum duration allowable (which is the range given by the observed record). However, for many species there are intermediate intervals that instead could contain the first or last occurrence (i.e., intervals B, C, or D on fig 14.6C). For each set of combinations a diagram similar to figures 14.5 and 14.7 can be constructed. Collectively all of these diagrams will be geometrically transitional between figures 14.5 and 14.7. There is no a priori reason to choose any one of the possibilities over the others. Nevertheless, figures 14.5 and 14.7 are of special interest because they are the two cases that bound the possibilities.

Figures 14.5 and 14.7 can be linked to the probability of finding a first or last occurrence event within a subdivision of a taxon's potential range. That probability is unknown, but if it could be inferred from the frequency of occurrences and data quality (McKinney 1986) or if plausible a priori assumptions could be made, then the range extensions can be weighted. A conservative initial a priori assumption is to assume the probability for occurrence of an event is the same for each interval within a species' potential range. In terms of calculating the resulting probability-weighted distribution of events, this corresponds to partitioning each event equally among all the intervals in which it might lie (including the one containing the observed record) so that each interval gets an equal fraction of one event. For example, in figure 14.6C the first occurrence of the species might be in any of the four intervals A, B, C, or D; each then would be given a value of one-fourth of a first occurrence.

The weighted distributions of first and last occurrences through time using equal probability partitioning (table 14.3) are shown in figure 14.8. Because the partitioning is equal for each interval, figure 14.8 is geometrically the intermediate between the extremes of figures 14.5 and 14.7. (Figures 14.5 and 14.7 also correspond to assumed probability distributions in which the probability is one for either the first or last interval and zero for the rest, which seems unlikely.) Figure 14.8 has two important qualities. First, it is the weighted distribution that most strongly underestimates, or averages out, the absolute magnitude of maxima and minima. Secondly, any maxima or minima will span the largest number of intervals, giving the best visual expression to the uncertainties of their ages. In these effects it is the distribution which is the least likely to show large maxima of short durations; for that reason we focus on it in the following discussion.

As before, we do not know what absolute number of events should distinguish a maximum as significant when compared to other values.

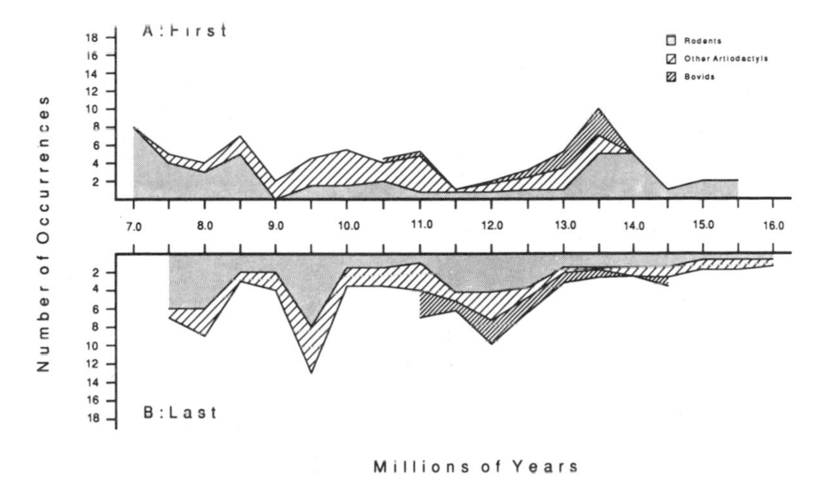

Figure 14.8. Number of first and last occurrence events using equal probability partitioning. A; First occurrences. B; Last occurrences.

A chi-square goodness-of-fit test can be used to judge the significance of the overall departure of the calculated distributions of first and last occurrences (table 14.3, exclusive of bovids) from an even distribution of the events. The null hypothesis in both cases is again that the first occurrences (or last occurrences) are distributed evenly among 18 intervals. The expected and observed values are shown in table 14.4, along with the individual contributions to the chi-square. With 16 degrees of freedom the chi-square value of 36.43 for last occurrences has a very low probability of less than .005, and we conclude it is very unlikely the distribution of last occurrences is due to chance. However, the chi-square of 19.84 for the first occurrences is only modestly significant, having with 16 degrees of freedom a probability slightly less than 0.25. This is a larger probability than the other values we have found, but it is still likely that the distribution of first occurrences is not due entirely to chance. Therefore, although the first occurrences are only marginally convincing, we conclude that both the first and last occurrence events do form pulses in the Siwalik record.

There are no very large single contributions to the chi-square value for the first occurrences, but most of the total comes from the maxima at 13.5, 8.5, and 7.0 Ma and the minima at 14.5 and 11.5 Ma. The 13.5 and 7.0 Ma maxima and the minimum at 11.5 Ma correspond to events on figure 14.5, but the other two are new features of figure 14.8. For the last occurrences, the largest contributions are from the maxima at 9.5 and 8.0 Ma, and these also correspond to maxima on figure 14.5. The very large maximum of last occurrences at 12.5 Ma on fig-

Table 14.3. Calculated number of first and last occurrence events, using equal probability partitioning*

Intrv (Ma)	Rodents				Other Artiodactyls				Bovids			
	Q	#Ev	#Sp	Rate	Q	#Ev	#Sp	Rate	Q	#Ev	#Sp	Rate
A. First Occurrences												
7.0	1	8	14	0.57	2	0	9	0.00				
7.5	1	4	12	0.33	2	1	10	0.10				
8.0	1	3	14	0.21	1	1	12	0.08				
8.5	4	5			2	2	12	0.17				
9.0	2	0	10	0.00	1	2	12	0.17				
9.5	1	1.5	18	0.08	3	3						
10.0	4	1.5			4	4						
10.5	2	2	18	0.11	2	2	12	0.17	2	0.5	4	0.13
11.0	2	0.8	17	0.05	3	4			3	0.5		
11.5	5	0.8			1	0.3	10	0.03	1	0	7	0.00
12.0	3	0.8			3	0.9			3	0.3		
12.5	2	1	25	0.04	3	1.4			3	0.8		
13.0	5	1			3	2.4			3	1.8		
13.5	1	5	26	0.19	1	2	10	0.20	1	3	10	0.30
14.0	3	5			2	0	10	0.00	2	0	8	0.00
14.5	2	1	19	0.05	4	0			4	0		
15.0	3	2			3	0			3	0		
15.5	5	2			4	0			4	0		
16.0	2	–	16		3	–			3	–		

B. Last Occurrences

7.0	1	–	14		2	–	9					
7.5	1	6	12	0.50	2	1	10	0.10				
8.0	1	6	14	0.43	1	3	12	0.25				
8.5	4	2			2	1	12	0.08				
9.0	2	2	10	0.20	1	2	12	0.17				
9.5	1	8	18	0.44	3	5						
10.0	4	1.5			4	2						
10.5	2	1.5	18	0.08	2	2	12	0.17	2	–	4	
11.0	2	1	17	0.06	3	3			3	3		
11.5	5	4.2			1	1	10	0.10	1	1	7	0.14
12.0	3	4.2			3	3.1			3	2.6		
12.5	2	3.7	25	0.15	3	1.1			3	1.6		
13.0	5	1.5			3	0.6			3	1.1		
13.5	1	1.5	26	0.06	1	0.3	10	0.03	1	1.8	10	0.08
14.0	3	1.5			2	1	10	0.10	2	0	8	0.00
14.5	2	1.5	19	0.08	4	1.1			4	1		
15.0	3	0.7			3	1.1			3	0		
15.5	5	0.7			4	1.1			4	0		
16.0	2	0.7	16	0.04	3	0.7			3	0		

Note: *Intrv (Ma), median age of interval; Q, quality class; #Ev, number of events calculated for interval; #Sp, number of species found or inferred to be present in interval; Rate, #Ev/#Sp (number of events per species per half million years).

Table 14.4. Chi-square goodness-of-fit to an even distribution of the number of first and last occurrences of rodents and artiodactyls (excluding bovids) using equal probability partitioning

Interval	Expected	Observed	$X^2 = (O - E)^2/E$
A. First Occurrences: p = 0.056, n = 70.2			
7.0	3.9	8	4.31
7.5	3.9	5	0.31
8.0	3.9	4	0.00
8.5	3.9	7	2.46
9.0	3.9	2	0.93
9.5	3.9	4.5	0.09
10.0	3.9	5.5	0.66
10.5	3.9	4	0.00
11.0	3.9	4.75	0.19
11.5	3.9	1	2.16
12.0	3.9	1.65	1.30
12.5	3.9	2.4	0.58
13.0	3.9	3.4	0.06
13.5	3.9	7	2.46
14.0	3.9	5	0.31
14.5	3.9	1	2.16
15.0	3.9	2	0.93
15.5	3.9	2	0.93
			X^2 = 19.84[a]
B. Last Occurrences: p = 0.056, n = 78.25			
7.5	4.3	7	1.70
8.0	4.3	9	5.14
8.5	4.3	3	0.39
9.0	4.3	4	0.02
9.5	4.3	13	17.60
10.0	4.3	3.5	0.15
10.5	4.3	3.5	0.15
11.0	4.3	4	0.02
11.5	4.3	5.2	0.19
12.0	4.3	7.3	2.09
12.5	4.3	4.8	0.06
13.0	4.3	2.1	1.13
13.5	4.3	1.75	1.51
14.0	4.3	2.5	0.75
14.5	4.3	2.6	0.67
15.0	4.3	1.8	1.45
15.5	4.3	1.8	1.45
16.0	4.3	1.4	1.96
			X^2 = 36.43[b]

[a]With d.f. = 16, X^2 has a probability between .25 and .1.

[b]With d.f. = 16, X^2 has a probability $< .005$.

ure 14.5 is composed of the events between 11.5 and 12.5 on figure 14.8, but none of the latter make an especially notable contribution to the chi-square. (Note that bovids are excluded.) The maxima and minima at 15.0, 10.5, and 10.0 Ma on figure 14.5 are not evident on figure 14.8.

On figure 14.8 the maxima centered at 13.5 Ma and 12.0 Ma span multiple intervals, but it is clear that they do not overlap and that neither they nor the maximum at 9.5 Ma have complementary peaks. This then demonstrates that maxima of first and last occurrences are not necessarily synchronous. Also, as only rodents show large numbers of events after 8.5 Ma, it appears that maxima do not necessarily include all taxonomic groups.

In our discussion of figures 14.5, 14.7, and 14.8, we have compared the magnitude of maxima and minima of one time interval to those of other intervals. However, these direct comparisons can be misleading. If the rates of first and last occurrences were constant regardless of the number of species, then more events should be expected in intervals with larger total faunas. Since it appears that diversity in the Siwalik faunas is not constant between intervals, this effect must be controlled. We have therefore calculated the relative change (Table 14.3), which is expressed as the number of events per species per half million years. The values used for the number of events per interval are those computed for figure 14.8.

Figure 14.9 shows the resulting plot of relative proportions of first and last occurrences. Only intervals with class 1 or class 2 data are plotted, since only these intervals have species numbers that we believe are adequate to closely reflect the diversity of the rodents and artiodactyls. Inspection of this figure indicates that some intervals are characterized by high rates of first or last occurrences, whereas others have low rates. Notable in this regard are the high first occurrence rates and low last occurrence rates at 13.5 Ma for both rodents and artiodactyls, the rather low rates of both first and last occurrences for artiodactyls at 14.0 and 11.5 Ma, the zero first occurrence rate for rodents at 9.0 Ma, and the very large values for rodents at 9.5, 8.0, 7.5, and 7.0 Ma.

The high rate of first occurrences at 13.5 Ma corresponds to the maximum seen on the preceding figures and, while all taxa are affected, there clearly is not a corresponding peak of high rates of last occurrences in this interval. The maximum of last occurrences centered on 12.0 Ma on figure 14.8 is missing from figure 14.9, but this may not be meaningful. Since the number of species is only poorly estimated, this interval could not be included in the analysis. Therefore, it is not valid to conclude that the maximum is absent, but only that we are unable to demonstrate that it is present. For the same rea-

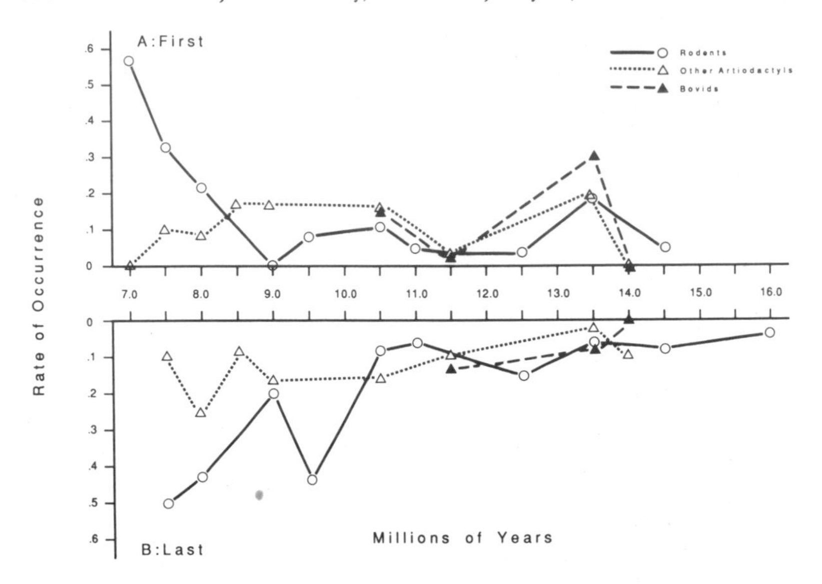

Figure 14.9. Relative number of first and last occurrence events for quality class 1 and 2 intervals. Rates as number of events per species per half million years. A; First occurrences. B; Last occurrences.

sons, poor data quality for artiodactyls in the 9.5 Ma interval precludes calculating a rate of last occurrences. We expect future collecting to provide evidence to decide these points.

In the preceding discussions we have apparently greatly underestimated the importance of the number of first and last occurrences for rodents after 9.0 Ma, as figure 14.9 shows that this period contains some of the highest rates of either first or last occurrences. Our previous conclusion, however, that only rodents are undergoing significant turnover, seems to be supported. Except for the events at 9.0 Ma, the rates for nonbovid artiodactyls are modest when compared to the values for rodents. Nevertheless, it should be noted that after 10.5 Ma the artiodactyl rates, and especially last occurrences, are generally higher than those typical of the more diverse assemblages of the middle Miocene.

OTHER SIWALIK TAXA

In some instances first and last occurrences in other groups are known to coincide with faunal episodes documented for the rodents and artiodactyls. Most notable are the appearance of hipparionine equids at 9.5 Ma and the appearances of leporids and cercopithecids after 7.0 Ma. Large hominoids of the *Sivapithecus* group first appear in the sequence at about 12.0 Ma and may become locally extinct shortly after

7.5 Ma. Hyenas, which are the most important large predators in the middle and late Miocene Siwalik communities, first occur at 12.7 Ma.

The Siwalik Rhinocerotidae may be somewhat exceptional, as it is a diverse family with long-lived species (Heissig 1972). The family shows some replacement of species at about 10 to 9 Ma, but not the rapid speciation and turnover characteristic of other large herbivores, particularly the bovids. The systematics of Elephantoidea have been revised by Tassy (1983), who showed that this group is also diverse in both the middle (at least six taxa) and late (about nine taxa) Miocene. Diversity does not diminish before 8 Ma, but at present the stratigraphic ranges of individual species are not well enough known to assess the timing and significance of turnover within the superfamily.

PALEOCOMMUNITIES

Siwalik paleocommunities and their depositional environments and ecology have been discussed by Badgley and Behrensmeyer (1980), Andrews (1983), Raza (1983), Badgley (1986a, 1986b), Kappelman (1986), and Behrensmeyer (1987), but only one analysis of the ecological structure of the Siwalik fossil assemblages has been published (Andrews 1983). That study indicates that the late Miocene assemblages had an ecological structure most similar to woodland or bushland communities. Our own preliminary analyses of older and younger assemblages suggest that, despite the apparent decrease in diversity and considerable turnover, the middle Miocene communities had very much the same structure as those of the late Miocene. In the latest Miocene, however, the animals may have lived in more open habitats.

CORRELATIONS TO GLOBAL EVENTS

Changes in apparent diversity and faunal composition of Siwalik terrestrial vertebrates show general, but weak, correlations to global environmental events. Although early Miocene climates appear to have been relatively equable and stable, several lines of evidence indicate that from 16 to 12 Ma there were substantial changes in the physical environment of the earth. Intensification of oceanic and atmospheric circulation, steepening of temperature gradients, and expansion of Antarctic glaciation seem particularly important (Kennett, Keller, and Svinivansan 1985; Kennett 1986; Savin et al. 1985; Woodruff, Savin, and Douglas 1981; Woodruff 1985). In addition, there were major oscillations in eustatic sea level near 16 Ma and two smaller falls between 14 and 12 Ma (Haq, Hardenbol, and Vail 1987).

Subsequently (ca. 12 to 8 Ma), isotopic studies suggest temperatures were variable but without significant trend (e.g., Kennett 1986; Savin et al. 1985; Woodruff, Savin, and Douglas 1981). Two cool episodes bracketing the middle/late Miocene boundary have been identified, one between 11.0 and 9.5 Ma and one between 9.0 and 8.0 Ma (Kennett 1986; Berggren et al. 1985). The second episode is one of the strongest isotopic excursions in the Miocene (Kennett 1986). In the same 12 to 8 Ma interval, the Haq, Hardenbol, and Vail (1987) eustatic curve indicates a progressive fall in sea level ending at about 10.5 Ma with one of the lowest stands of the entire Neogene. Keller and Barron (1983), Barron, Keller, and Dunn (1985), and Mayer, Shipley, and Winterer (1986) have also correlated middle Miocene deep sea hiatuses and regional seismic reflectors to isotopic and biostratigraphic events. They interpret the NH4 and NH5 hiatuses in particular as due to reorganization of oceanic circulation attendant on climate change. Although the correlation and dating of these various events is uncertain, the low stand of sea level at the end of the middle Miocene, the first of the isotopic excursions, and the NH4 hiatus may all record a single, major global climatic event around 11 to 9.5 Ma.

The remainder of the late Miocene (ca. 8 to 5 Ma) initially had somewhat warmer conditions (Kennett 1986), followed by very marked cooling after 6.5 Ma. The cooling is observed in both the faunal and oxygen isotope records (e.g. Burckle 1985; Kennett 1986) and can be related to a carbon isotopic event of the same age and to the eventual desiccation of the Mediterranean (Haq et al. 1980; Vincent, Killingley, and Berger 1985; Hodell, Elmstrom, and Kennett 1986). Latest Miocene deep sea hiatuses and sea level oscillations have also been noted (Keller and Barron 1983; Haq, Hardenbol, and Vail 1987). These paleoceanographic events are probably related to the widespread increasing seasonability and aridity noted in terrestrial environments during the late Miocene (Jacobs and Flynn 1981; Flynn and Jacobs 1982; Janecek and Rea 1983; Stein 1985; Van Zinderen Bakker and Mercer 1986).

Climatically induced habitat change, coupled with brief intervals of low sea level and northward movements of the Afro-Arabian and Indo-Australian plates, would have permitted migration of African mammals into Europe and Asia in the early Miocene. Corresponding faunal events have been noted in the vertebrate record (Adams, Gentry, and Whybrow 1983; Thomas 1985; Barry et al. 1985; Barry, Jacobs, and Kelley 1987; Bernor et al. 1988), and these profoundly affected Eurasian terrestrial communities. Because our analysis of Siwalik faunas begins at 16 Ma, these early appearances and events are not evaluated here. We note, however, that the early to middle Miocene diversification of muroids and ruminants may be related to these biotic and environmental events.

Our analysis of first and last occurrences of Siwalik rodents and artiodactyls highlights four major faunal episodes (figs. 14.5, 14.7, and 14.8). These are the peak of the first appearances between 14.0 and 13.5 Ma, the two peaks of last occurrences between 12.5 and 11.5 Ma and at 9.5 Ma, and finally the increased numbers of both first and last occurrences of rodents that characterize the post-8.5 Ma intervals. (These dates are based on the Mankinen and Dalrymple [1979] magnetic time scale. They convert to approximately 14 to 13.5, 13 to 12, 10, and 8.5 Ma on the Berggren et al. [1985] time scale.) In addition, after 12.5 Ma (approximately 13 Ma) there is a decreasing trend in the number of Siwalik rodents and artiodactyls, and this we have interpreted as a result of declining diversity in these two groups. There are suggestive correlations between the 14 to 13.5, 13 to 12, and 10 Ma Siwalik episodes and paleoceanographic events, including (1) the fall in sea level at about 14 to 13 Ma, (2) the end of the early middle Miocene ^{18}O enrichment and the second fall in sea level between 13 and 12 Ma, and (3) the oxygen isotope event, deep sea hiatus, and low stand of sea level between 11 to 9.5 Ma. However, because of the uncertainties associated with the ages of both the Siwalik episodes and the paleoceanographic events, it is not yet possible to determine how closely contemporaneous the two series of events may have been.

The apparent decreased diversity after 12.5 Ma and increasing rates of faunal turnover after 8.5 Ma may both have been due to a local, long-term trend of altered climate, with decreased precipitation and increased seasonality being likely causes (Jacobs and Flynn 1981; Flynn and Jacobs 1982; Barry et al. 1985). This local trend may be part of the more general increasing seasonality and aridity of late Miocene terrestrial environments, but understanding the causes of climatic change on the Indian Subcontinent is complicated by local factors. These include the northward movement of the subcontinent and the uplift of its northern and western mountains, as well as the development of the Indian monsoon system. These must have had strong local influences on the atmosphere-ocean climate system, possibly overriding or reversing more general trends. Although an isotopic excursion near 8.5 Ma is within the relevant time range, significant faunal change was already in progress in the Siwaliks before the major global climate perturbations at the end of the Miocene. Research in progress on stable isotopes in the Siwaliks (Quade, Cerling, and Bowman 1989) may eventually document local climate changes and show directly how they are correlated to the vertebrate faunal changes.

In summary, the Miocene history of Siwalik terrestrial vertebrates only approximates the oceanographic record. We recognize some coincident events but at this time are not able to establish a clear correlation between any climatic, tectonic, or eustatic event and Siwalik faunal turnover. Episodes of faunal change in the Siwaliks may have been related to global climate or tectonic events, but until more precise correlations can be established this remains only a strong possibility.

Discussion

Despite their preliminary nature, the following conclusions seem relatively secure.

1. The quality of the data varies from interval to interval. Differences in data quality must be considered whenever comparisons of numbers of species from different intervals are made or whenever the age and magnitude of episodes of faunal turnover are considered.

2. Observed diversity underestimates original diversity. However, for intervals with high-quality data, the number of species of rodents and artiodactyls found or inferred to be present appears to be close to original diversity. Our conclusions are based primarily on these intervals.

3. In most intervals more species of rodents than of artiodactyls are recovered, and the diversity of rodents may have been greater than that of artiodactyls, as in modern faunas. Among rodents, cricetids are most diverse. Bovids are more species rich than other artiodactyls.

4. The number of recorded species changes over time, with the most notable change being between 12.5 Ma and 10.5 Ma. That change is due primarily to a decrease in the number of species of cricetids. The number of species of other groups changes less, although rhizomyids increase notably after 9.0 Ma. If the number of species found reflects diversity, then the data suggest, but do not demonstrate, high diversity from 13.5 to 12.5 Ma, lower diversity from 12.5 until 9.5 Ma, and lowest diversity from 9.5 to 7.0 Ma.

5. First and last occurrence events are present in all intervals, but intervals vary considerably in the number of events recorded. Because of variation in data quality from interval to interval, the magnitude and age of the maxima and minima may be misleading when directly determined from the actual record.

6. We have attempted to incorporate the quality of data into our assessment of the patterns of first and last occurrences. Our objective was to specify limits for estimates of species ranges. These limits, together with the observed record, constrain the first or last occurrences to short stratigraphic intervals.

7. When the first or last occurrences are partitioned equally among all intervals within their potential ranges, a diagram is constructed that minimizes the amplitude of maxima. This diagram (fig. 14.8) is particularly useful in testing for the presence of pulses in the sequence, since

it is the theoretical distribution that least emphasizes pulses. Nevertheless, there are large maxima of first occurrences between 14.0 and 13.5 Ma, at 8.5 Ma, and at 7.0 Ma and minima between 12.5 and 11.5 Ma and at 9 Ma. There are maxima in last occurrences between 12.5 and 11.5 Ma, at 9.5 Ma, and after 8.0 Ma. There is a series of minima before 12.5 Ma, between 11.0 and 10.0 Ma, and between 9.0 and 8.5 Ma. Although it is not possible to test the significance of the individual maxima, we find that the observed distributions are not random departures from an even distribution. On that basis we conclude that faunal events probably form pulses in the Siwalik record.

8. The maxima between 14 and 13.5 Ma and between 12.5 and 11.5 Ma do not overlap and do not have corresponding complementary maxima. Maxima younger than 8.5 Ma are complementary but only for rodents. At 9.5 Ma most of the events are last occurrences, with nearly equal numbers for rodents and artiodactyls.

9. Rates of first and last occurrences can be calculated relative to the number of species in those intervals with the best data (fig. 14.9). The highest rates of appearances are at 13.5 and after 8.0 Ma. These affect all taxa in the first instance but only rodents in the second. The highest rates of last occurrences are at 9.5 and after 8.0 Ma and in the second instance involve only rodents. Noteworthy minima in these rates are at 14.0 and 11.5 Ma (first and last occurrences: artiodactyls), at 13.5 Ma (last occurrences: rodents and artiodactyls), and at 9.0 Ma (first occurrences: rodents). These correspond to maxima and minima in the numbers of events seen on figure 14.8, but the 12.0 Ma last occurrence peak has not been detected because the data quality is poor for the 12.0 Ma interval. The 9.5 Ma last occurrence peak for artiodactyls has not been detected for the same reason.

10. Therefore, it would appear that first and last occurrences do not necessarily coincide to form synchronous turnover pulses. Nor do they necessarily affect all taxonomic groups equally or synchronously.

11. The record of faunal change for Siwalik vertebrates correlates to the oceanographic record of environmental change, but at present we are not able to demonstrate a direct correspondence between any single climatic, tectonic, or eustatic event and a Siwalik faunal turnover. Early and middle Miocene diversification of muroids and ruminants occurred during an interval of significant global climate change. Potential correlations exist between the 14 to 13.5, 13 to 12, and 10 Ma Siwalik episodes and specific paleoceanographic events. These include the fall in sea level at about 14 to 13 Ma, the end of the early middle Miocene ^{18}O enrichment and a second fall in sea level between 12 and 13 Ma, and an isotopic excursion, low sea level, and deep sea hiatus near the middle to late Miocene boundary. The 8.5 to 7.0 Ma high

turnover rates briefly precede a period of marked cooling and increasing aridity.

IMPLICATIONS

In the detailed Siwalik record the observed patterns of faunal change only partly fulfill the expectations of Vrba's (1985) turnover pulse hypothesis. Overall, the evidence for pulses of turnover in the Siwalik record is strong, although not conclusive. Some intervals have substantially larger concentrations of events than others, but all intervals have some events and the distinction between intervals with many and those with few is not easily made. Nevertheless, events are not evenly distributed, and the intervals with the largest numbers of events could be considered to represent pulses of turnover in the sense of Vrba (1985). The evidence for synchroneity between turnover pulses and climatic, tectonic, or eustatic events is also inconclusive, due primarily to the difficulty of establishing temporal correlations that are precise enough to be compelling. The question of what causal relationships exist between specific environmental events and observed local terrestrial effects is a separate inferential process. However, episodes of faunal change and changes in the physical environment tend to occur at approximately the same time and, in some cases—as in the late Miocene—we can visualize how change in the environment could have produced the observed effect on the fauna. Finally, in contrast to the primary expectations, the secondary expectations of the hypothesis are clearly not met, as first and last occurrences are not necessarily synchronous and turnover pulses do not always affect different taxonomic groups to the same extent.

These conclusions depend on the quality of the record, but we believe that neither additional data nor greater precision in the correlations will substantially alter them. It is important, however, to recognize how our conclusions are conditioned by the temporal scale we have used. Our analysis is based on intervals of 0.5 my duration, and at this level of temporal resolution we have been able to detect concentrations of first and last occurrences that are distinct enough for us to argue that there are turnover pulses in the Siwalik record. If our analysis had been done on a coarser scale, it is unlikely that we would have detected any pulses, whereas an analysis using shorter intervals might possibly show an even more distinct clustering of events. That is, there may be subintervals of intense turnover hidden within longer intervals of normal or below normal turnover. Previous analyses have indicated that major faunal transitions may have occurred over such short time intervals (Barry et al. 1982, 1985). Failure to find them at one time scale does not preclude finding them with finer resolution. The maxima of first and last occurrences centered on 13.5 Ma and 12.0 Ma correspond to Webb's (1984) origination-type and extinction-type turnover episodes but are recognized on a time scale that is nearly an order of magnitude finer.

Faunal turnover is complex, influenced by many factors, including climate, tectonics, eustacy, and biological processes such as competition and predation. The potential correlations with pulses of turnover provide some evidence for the impact of environmental events on terrestrial faunas. Evidence for environmental effects is also found in the patterns of faunal turnover and may indicate the relative importance of the different types of events, such as extinction, speciation, or immigration. Previous analyses suggested that the first and last occurrences of the middle Miocene part of the Siwalik record were more episodic than those of the late Miocene part (Barry et al. 1985). That observation is supported by the present study, in that the record before 9 Ma is dominated by three distinct maxima, whereas after 9 Ma faunal turnover is consistently high, at least for the rodents. This pattern implies that the middle and late Miocene communities may have been perturbed in different ways, with episodic events or threshold effects being more important before 9 Ma and nonepisodic or continuous change afterward. Physical or ecological barriers that were suddenly breached by tectonic or eustatic events, or perhaps by the distant effects of climatic change, could have produced the episodic pattern.

The lack of synchroneity between maxima of first and last occurrences and the differential effects on taxa suggest that the biological attributes of the taxa were also important. This may have been especially true for the late Miocene turnover, which involved primarily small mammal groups (Flynn and Jacobs 1982) and possibly also the bovids, but not other artiodactyls. The changes among these taxa may reflect loss of permanent forest cover and probably altered precipitation patterns, perhaps involving both greater seasonality and a decrease in precipitation. In such circumstances the differing physiological thresholds and ecological relationships of different groups will probably ensure that changing environments do not affect all groups to the same extent simultaneously. As a result the local extinction or immigration of the various species are likely to come at different times as the local habitats are progressively altered.

Summary

It is useful to reiterate our main points and tentative conclusions. In this study we discuss changes in diversity and patterns of first and last occurrences among Neogene mammals, using a nine million year (16 to 7 Ma) detailed record of rodents and artiodactyls from the Siwalik formations of northern Pakistan. We use this record to investigate how faunal turnover in terrestrial ecosystems is related to climatic and other environmental events. Because large and small mammals have different taphonomic biases and may vary in abundance between stratigraphic horizons, the apparent patterns and changes among intervals are sensitive to the quality of the data. After adjustments are made for interval quality, it is apparent that the Siwalik sequence documents changes over time both in the number of species present and in the taxonomic composition of the rodent and artiodactyl assemblages, with the number of species first increasing and then decreasing after 12.5 Ma. First and last occurrences are not evenly distributed through time but form clusters of events. This distribution of events is not random, as some intervals have more events than would be expected by chance, whereas others have fewer. The clusters or maxima correspond to more intense periods of faunal change or pulses of turnover as discussed by Vrba (1985). A large maximum of first occurrence events is centered on 13.5 Ma, whereas two large maxima of last occurrences are at 12.0 and 9.5 Ma. After 8.5 Ma there are large numbers of both first and last occurrences. The maximum of first occurrences at 13.5 Ma is complemented by a minimum of last occurrences, whereas the maximum of last occurrences at 12.0 Ma is complemented by a minimum of first occurrences. After 8.5 Ma maxima of first and last occurrences coincide for the rodents but not for the artiodactyls. The inferred pulses of turnover are approximately synchronous with episodes of environmental change, but because temporal resolution is not precise, correlations are weak. The diversification of the muroids and ruminants at the beginning of the middle Miocene occurred during a period of global climatic cooling, while the faunal changes in the Siwaliks at the end of the Miocene occurred just before another period of marked cooling and increasing aridity. Potential correlations also exist between turnover pulses at 13.5, 12.0, and 9.5 Ma and paleoceanographic events. The presence of turnover pulses approximately synchronous with intervals of environmental change conforms to the primary expectations of Vrba's (1985) turnover hypothesis, but some of these observations are inconsistent with the secondary expectations. Middle Miocene faunal change differs from that of the late Miocene, suggesting that the underlying causes were different. Episodic environmental change may have been most important in the middle Miocene and nonepisodic change in the late Miocene. Although abiotic factors of climate, tectonics, and eustacy play critical roles in Siwalik faunal history, biotic factors are also important.

Acknowledgments

Our close collaboration with colleagues in the Geological Survey of Pakistan has made this research possible. Our work has benefited particularly from interaction with S. M. Ibrahim Shah and with S. M. Raza. We thank R. Strong and T. Disotell for assisting with statistical calculations and J. Kingston, M. Morgan, and A. Uchida for help in analyzing community structure. We also thank the preceding and C. Badgley, A. K. Behrensmeyer, B. Brown, A. Hill, L. L. Jacobs and E. H. Lindsay for careful reading and critique of earlier drafts. Lindsay, Jacobs, J. A. Baskin, R. Daams, and W. R. Downs generously made small mammal data available and assisted in many other ways. Financial support derives from NSF grants BSR 85–00145 and BNS-8812306 and SFCP grant 7087120000.

Literature Cited

Adams, C. G., A. W. Gentry, and P. J. Whybrow. 1983. Dating the terminal Tethyan event. *Utrecht Micropaleontolog. Bull.* 30:273–98.

Agterberg, F. P. 1985. Methods of scaling biostratigraphic events. In *Quantitative stratigraphy,* ed. F. M. Gradstein, F. P. Agterberg, J. C. Brower, and W. S. Schwarzacher, 195–241. Dordrecht, Netherlands: D. Reidel Publishing Co.

Andrews, P. 1981. Species diversity and diet in monkeys and apes during the Miocene. In *Aspects of human evolution,* ed. C. B. Stringer, 25–61. London: Taylor and Francis.

Andrews, P. 1983. The natural history of *Sivapithecus.* In *New interpretations of ape and human ancestry.* ed. R. L. Ciochon and R. S. Corruccini, 441–63. New York: Plenum.

Badgley, C. E. 1982. Community reconstruction of a Siwalik mammalian assemblage. Ph.D. diss., Yale University, New Haven, Conn.

Badgley, C. E. 1986a. Taphonomy of mammalian fossil remains from Siwalik rocks of Pakistan. *Paleobiology* 12:119–42.

Badgley, C. E. 1986b. Counting individuals in mammalian fossil assemblages from fluvial environments. *Palaios* 1:328–38.

Badgley, C. E., and A. K. Behrensmeyer. 1980. Paleoecology of middle Siwalik sediments and faunas, northern Pakistan. *Palaeogeog., Palaeoclimatol., Palaeoecol.,* 30:133–55.

Badgley, C. E., and P. D. Gingerich. 1988. Sampling and faunal turnover in early Eocene mammals. *Palaeogeog., Palaeoclimatol., Palaeoecol.,* 63:141–57.

Badgley, C. E., L. Tauxe, and F. L. Bookstein, 1986. Estimating error of age interpolation in sedimentary rocks: a bootstrap method. *Nature* 319:139–41.

Barron, J. A., G. Keller, and D. A. Dunn. 1985. A multiple microfossil biochronology for the Miocene. In *The Miocene ocean: Paleoceanography and biogeography*, ed. J. P. Kennett, 21–36, Geolog. Soc. Am. Memoir 163.

Barry, J. C., A. K. Behrensmeyer, and M. Monaghan. 1980. A geologic and biostratigraphic framework for Miocene sediments near Khaur Village, northern Pakistan. *Postilla* 183:1–19.

Barry, J. C., L. L. Jacobs, and J. Kelley. 1987. An early middle Miocene catarrhine from Pakistan with comments on the dispersal of catarrhines into Eurasia. *J. Human Evolution* 15:501–8.

Barry, J. C., N. M. Johnson, S. M. Raza, and L. L. Jacobs. 1985. Neogene mammalian faunal change in southern Asia: Correlations with climatic, tectonic, and eustatic events. *Geology* 13:637–40.

Barry, J. C., E. H. Lindsay, and L. L. Jacobs. 1982. A biostratigraphic zonation of the middle and upper Siwaliks of the Potwar Plateau of northern Pakistan. *Palaeogeog., Palaeoclimatol., Palaeoecol.*, 37:95–130.

Behrensmeyer, A. K. 1987. Miocene fluvial facies and vertebrate taphonomy in northern Pakistan. *Soc. Econ. Paleontologists* a *Mineralogists* Spec. Pub. 39:169–76.

Berggren, W. A., D. V. Kent, and J. A. Van Couvering. 1985. Neogene geochronology and chronostratigraphy. 1–6 The chronostratigraphy of the Geological Record. N. J. Snelling, ed., Blackwell Scientific Publications, Oxford, p. 211–60.

Bernor, R. L. 1985. Neogene paleoclimatic events and continental mammalian response: Is there global synchroneity? *South Af. J. Sci.* 81:261.

Bernor, R. L., L. J. Flynn, T. Harrison, S. T. Hussain, and J. Kelley. 1988. *Dionysopithecus* from southern Pakistan and the biochronology and biogeography of early Eurasian catarrhines. *J. Human Evolution* 17:339–58.

Boltovsky, D. 1988. The range-through method and first-last appearance data in paleontological surveys. *J. Paleontol.*, 62:157–9.

Burckle, L. H. 1985. Diatom evidence for Neogene palaeoclimate and palaeoceanographic events in the world ocean. *South Afr. J. Sci.*, 81:249.

Cheema, I. U., S. Sen, and L. J. Flynn. 1983. Early Vallesian small mammals from the Siwaliks of northern Pakistan. *Bulletin du Muséum national d'Histoire naturelle, Paris*, séries 4, section C, 5:267–80.

Cheetham, A. H., and P. B. Deboo. 1963. A numerical index for biostratigraphic zonation in the mid-Tertiary of the eastern Gulf. *Gulf Coast Assoc. Geolog. Sci. Trans.*, 13:139–47.

Connell, J. H. 1978. Diversity in tropical rain forests and coral reefs. *Science* 199:1302–10.

Flynn, L. J. 1982. Systematic revision of Siwalik Rhizomyidae (Rodentia). *Géobios* 15:327–89.

Flynn, L. J. 1986. Species longevity, stasis, and stairsteps in rhizomyid rodents. In *Vertebrates, phylogeny, and philosophy*, ed. K. M. Flanagan and J. A. Lillegraven, 273–85. Contributions to Geology, University of Wyoming, Spec. Paper 3.

Flynn, L. J., and L. L. Jacobs. 1982. Effects of changing environments on

Siwalik rodent faunas of northern Pakistan. *Palaeogeog. Palaeoclimatol. Palaeoecol.* 38:129–38.

Flynn, L. J., L. L. Jacobs, and S. Sen. 1983. La diversité de *Paraulacodus* (Thryonomyidae, Rodentia) et des groups apparantés pendant le Miocène. *Annales de Paléontologie* 69:355–66.

Gradstein, F. M., and F. P. Agterberg. 1985. Quantitative correlation in exploration micropaleontology. In *Quantitative stratigraphy*, ed. F. M. Gradstein, F. P. Agterberg, J. C. Brower, and W. S. Schwarzacher, 309–57. Dordrecht, Netherlands: D. Reidl Publishing Co.

Haq, B. U., J. Hardenbol, and P. R. Vail. 1987. Chronology of fluctuating sea levels since the Triassic. *Science* 235:1156–67.

Haq, B. U., T. R. Worsley, L. H. Burckle, R. G. Douglas, L. D. Keigwin, N. D. Opdyke, S. M. Savin, M. A. Sommer, E. Vincent, and F. Woodruff. 1980. Late Miocene Marine carbon-isotopic shift and synchroneity of some phytoplanktonic biostratigraphic events. *Geology* 8:427–31.

Hay, W. W. 1972. Probabilistic stratigraphy. *Eclogae Geologicae Helvetiae* 65:255–66.

Heissig, K. 1972. Paläontologische und geologische Untersuchungen im Teriär von Pakistan. 5. Rhinocerotidae (Mamm.) aus den unteren und mittleren Siwalik-Schicthen. *Bayerische Akademie der Wissenschaften* Math.-Naturwiss. Klasse, Abhandlungen 152:1–112.

Hill, A. 1988. Causes of perceived faunal change in the later Neogene of East Africa. *J. Human Evolution* 16:583–96.

Hodell, D. A., K. M. Elmstrom, and J. P. Kennett. 1986. Latest Miocene benthic d18O changes, global ice volume, sea level and the "Messinian salinity crisis." *Nature* 320:411–14.

Hoffman, A., and J. A. Kitchell. 1984. Evolution in a pelagic planktic system: a paleobiologic test of models of multispecies evolution. *Paleobiology* 10:9–33.

Jablonski, D. 1986. Causes and consequences of mass extinctions: a comparative approach. In *Dynamics of extinction*, ed. D. K. Elliott, 183–229. New York. John Wiley.

Jacobs, L. L. 1978. Fossil rodents (Rhizomyidae and Muridae) from Neogene Siwalik deposits, Pakistan. *Mus. Northern Ariz. Press Bull. Series* 52:1–103.

Jacobs, L. L., and L. J. Flynn. 1981. Development of the modern rodent fauna of the Potwar Plateau, northern Pakistan. Neogene-Quarternary boundary field conference, India, 1979. *Proceedings*, 79–81.

Jacobs, L. L, L. J. Flynn, and W. R. Downs. 1989. Neogene rodents of southern Asia. In *Papers on fossil rodents in honor of Albert E. Wood*, ed C. C. Black and M. R. Dawson, 157–77, Los Angeles County Museum Science Series 33.

Jacobs, L. L., L. J. Flynn, W. R. Downs, and J. C. Barry. 1990. *Quo vadis, Antemus?* The Siwalik muroid record. In *European Neogene mammal chronology*, ed. E. H. Lindsay, V. Fahlbusch and P. Mein, 573–86. New York: Plenum.

Janecek, T. R., and D. K. Rea. 1983. Eolian deposition in the northeast Pacific Ocean: Cenozoic history of atmospheric circulation. *Geolog. Soc. America Bull.*, 94:730–8.

Johnson, G. D., P. Zeitler, C. W. Naeser, N. M. Johnson, D. M. Summers, C. D. Frost, N. D. Opdyke, and R. A. K. Tahirkheli. 1982. The occurrence and fission-track ages of late Neogene and Quarternary volcanic sediments, Siwalik Group, northern Pakistan. *Palaeogeogr. Palaeoclimatol. Palaeoecol.* 37:63–93.

Johnson, N. M., N. D. Opdyke, G. D. Johnson, E. H. Lindsay, and R. A. K. Tahirkheli. 1982. Magnetic polarity stratigraphy and ages of Siwalik Group rocks of the Potwar Plateau, Pakistan. *Palaeogeogr. Palaeoclimatol. Palaeoecol.* 37:17–42.

Johnson, N. M., J. Stix, L. Tauxe, P. F. Cerveny, and R. A. K. Tahirkheli. 1985. Paleomagnetic chronology, fluvial processes and tectonic implications of the Siwalik deposits near Chinji Village, Pakistan. *J. Geol.*, 93:27–40.

Kappelman, J. W. 1986. The paleoecology and chronology of the middle Miocene hominoids from the Chinji Formation of Pakistan. Ph.D. diss., Harvard University, Cambridge, Mass.

Keller, G., and J. A. Barron. 1983. Paleoceanographic implications of Miocene deep-sea hiatuses. *Geolog. Soc. America Bull.*, 94:590–613.

Keller, H. M., R. A. K. Tahirkheli, M. A. Mirza, G. D. Johnson, N. M. Johnson, and N. D. Opdyke. 1977. Magnetic polarity stratigraphy of the upper Siwalik deposits, Pabbi Hills, Pakistan. *Earth Planetary Sci. Letters* 36:187–201.

Kennett, J. P. 1986. Miocene to early Pliocene oxygen and carbon isotope stratigraphy in the Southwest Pacific, Deep Sea Drilling Project Leg 90. *Initial Rep. Deep Sea Drilling Prog.*, 90:1383–411.

Kennett, J. P., G. Keller, and M. S. Srinivansan. 1985. Miocene planktonic foraminiferal biogeography and paleoceanographic development of the Indo-Pacific region. In *The Miocene ocean: Paleoceanography and biogeography,* ed. J. P. Kennett, 197–236. *Geolog. Soc. America Memoir* 163.

Kitchener, D. J., Y. Wang, A. Bradley, R. A. How, and J. Dell. 1987. Small mammals and habitat disturbance near Kunming, south-west China. *Indo-Malayan Zool.* 4:161–86.

Koch, C. F. 1987. Prediction of sample size effects on the measured temporal and geographic distribution patterns of species. *Paleobiology* 13:100–107.

Kurten, B. 1972. The half-life concept in evolution illustrated from various mammalian groups. In *Calibration of hominoid evolution*, ed. W. W. Bishop and J. A. Miller, 187–94. Edinburgh: Scottish Academic Press.

Lindsay, E. H. 1987. Cricetid rodents of lower Siwalik deposits, Potwar Plateau, Pakistan, and Miocene mammal dispersal events. *Proc. Regional Comm. Mediterranean Neogene Stratig. Annales Instituti Geologici Publici Hungarici* 70:483–8.

Lindsay, E. H. 1988. Cricetid rodents from Siwalik deposits near Chinji village. 1. Megacricetodontines, Myocricetodontines, Dendromurines. *Palaeovertebrata* 18:95–154.

Lowrie, W., and W. Alvarez. 1981. One hundred million years of geomagnetic polarity history. *Geology* 9:392–7.

McKinney, M. L. 1986. Biostratigraphic gap analysis. *Geology* 14:36–38.

Mankinen, E. A., and G. B. Dalrymple. 1979. Revised geomagnetic polarity time scale for the interval 0–5 m.y. B.P. *J. Geophysical Res.* 84:615–26.

Mayer, L. A., T. H. Shipley, and E. L. Winterer. 1986. Equatorial Pacific seismic reflectors as indicators of global oceanographic events. *Science* 233:761–4.

Medway, G. G.-H. 1965. Mammals of Borneo. *Monogr. Malaysian Branch R. Asiatic Soc.* 7:1–172.

Nichols. J. D., and K. H. Pollock. 1983. Estimating taxonomic diversity, extinction rates, and speciation rates from fossil data using capture-recapture models. *Paleobiology* 9:150–63.

Olson, E. C. 1966. Community evolution and the origin of mammals. *Ecology* 47:291–302.

Olson, E. C. 1980. Taphonomy: Its history and role in community evolution. In *Fossils in the making, vertebrate taphonomy and ecology* ed. A. K. Behrensmeyer and A. P. Hill, 5–19. Chicago: University of Chicago Press.

Opdyke, N. D., E. Lindsay, G. D. Johnson, N. Johnson, R. A. K. Tahirkheli, and M. A. Mirza. 1979. Magnetic polarity stratigraphy and vertebrate paleontology of the Upper Siwalik Subgroup of northern Pakistan. *Palaeogeogr. Palaeoclimatol. Palaeoecol.* 27:1–34.

Quade, J., T. E. Cerling, and J. R. Bowman. 1989. Development of Asian monsoon revealed by marked ecological shift during the latest Miocene in northern Pakistan. *Nature* 342:163–66.

Raza, S. M. 1983. Taphonomy and paleoecology of middle Miocene vertebrate assemblages, southern Potwar Plateau, Pakistan. Ph.D. diss. Yale University, New Haven, Conn.

Rijksen, H. D. 1978. *A field study on Sumatran orangutans* (Pongo pygmaeus abelli *Lesson, 1827*): *Ecology, behavior and conservation.* Wageningen, Netherlands. H. Veenman and Zonen B. V.

Roberts, T. J. 1977. *The mammals of Pakistan.* London: Ernest Benn.

Savin, S. M., L. Abel, E. Barrera, D. Hodell, G. Keller, J. P. Kennett, J. Killingley, M. Murphy, and E. Vincent. 1985. The evolution of Miocene surface and near-surface marine temperatures: Oxygen isotopic evidence. In *The Miocene ocean: Paleoceanography and biogeography,* ed. J. P. Kennett, 197–236, *Geolog. Soc. America Memoir* 163.

Schaller, G. B. 1967. *The deer and the tiger.* Chicago: The University of Chicago Press.

Schankler, D. M. 1980. Faunal zonation of the Willwood Formation in the central Bighorn Basin, Wyoming. In *Early cenozoic paleontology and stratigraphy of the Bighorn Basin, Wyoming,* ed. P. D. Gingerich, 99–114. *Univ. Michigan Papers Paleontol.* 24.

Stein, R. 1985. The post-Eocene sediment record of DSDP Site 366: Implications for African climate and plate tectonic drift. In *The Miocene ocean: Paleoceanography and biogeography,* ed. J. P. Kennett, 305–15, *Geolog. Soc. America Memoir* 163.

Stenseth, N. C., and J. Maynard Smith, 1984. Coevolution in ecosystems: Red Queen evolution or stasis? *Evolution* 38:870–80.

Tassy, P. 1983. Les Elephantoidea miocènes du Plateau du Potwar, Groupe de Siwalik, Pakistan, Parties I-III. *Annales de Paléontologie* 69:99–136, 235–97, 317–54.

Tauxe, L., and N. D. Opdyke. 1982. A time framework based on Magnetostratigraphy for the Siwalik sediments of the Khaur area, northern Pakistan. *Palaeogeogr. Palaeoclimatol. Palaeoecol.*, 37:43–61.

Thomas, H. 1983. Les Bovidae (Artiodactyla, Mammalia) miocènes du sous-continent indien, de la peninsule arabique et de l'Afrique. Thèse de Doctorat d'Etat, Université de Paris VI.

Thomas, H. 1985. The early and middle Miocene land connection of the Afro-Arabian plate and Asia: a major event for hominoid dispersal? In *Ancestors: the hard evidence,* ed. E. Delson, 42–50. New York: Alan R. Liss.

Van Couvering, J. A. H. 1980. Community evolution in East Africa during the Late Cenozoic, In *Fossils in the making, vertebrate taphonomy and paleoecology,* ed. A. K. Behrensmeyer, and A. P. Hill, 272–98. Chicago: University of Chicago Press.

Van Couvering, J. A. H., and J. A. Van Couvering. 1976. Early Miocene mammal fossils from East Africa: Aspects of geology, faunistics and paleoecology. In *Human origins:* Louis Leakey and the East African evidence, ed. G. L. Isaac and E. R. McCown, 155–207. Menlo Park, Calif.: W. A. Benjamin.

Van Zinderen Bakker, E. M., and J. H. Mercer. 1986. Major late Cainozoic climatic events and palaeoenvironmental changes in Africa viewed in a world wide context. *Palaeogeogr. Palaeoclimatol. Palaeoecol.* 56:217–35.

Vincent, E., J. S. Killingley, and W. H. Berger. 1985. Miocene oxygen and carbon isotope stratigraphy of the tropical Indian Ocean. In *The Miocene Ocean: Paleoceanography and biogeography,* ed. J. P. Kennett, 103–30, Geolog. Soc. America Memoir 163.

Vrba, E. 1980. Evolution, species and fossils: How does life evolve? *South Afr. J. Sci.* 76:61–84.

Vrba, E. 1985. Environment and evolution: Alternative causes of the temporal distribution of evolutionary events. *South Afr. J. Sci.* 81:229–36.

Webb, S. D. 1977. A history of savanna vertebrates in the New World. I. North America. *Ann. Rev. Ecol. Systematics 1977* 8:355–80.

Webb, S. D. 1984. On two kinds of rapid faunal turnover. In *Catastrophes and earth history: The new uniformitarianism,* ed. W. A. Berggren and J. A. Van Couvering, 417–36. Princeton, N.J.: Princeton University Press.

Woodruff, F. 1985. Changes in Miocene deep-sea benthic foraminiferal distribution in the Pacific Ocean: Relationship to paleoceanography. In *The*

Miocene ocean: Paleoceanography and biogeography, ed J. P. Kennett, 131–76, Geolog. Soc. America Memoir 163.

Woodruff, F., S. M. Savin, and R. G. Douglas. 1981. Miocene stable isotope record: a detailed deep Pacific Ocean study and its paleoclimatic implications. *Science* 212:665–8.

15

Quaternary Mammals of the Great Basin: Extinct Giants, Pleistocene Relicts, and Recent Immigrants

Timothy H. Heaton

The Wisconsin glacial, which lasted from about 35,000 to 10,000 years B.P., was the last of a series of ice ages that significantly affected many parts of the earth. These ice ages were ultimately caused by minor temperature fluctuations resulting from cyclic variations in the earth's orbit. This simple cause precipitated a complex web of events with important biogeographical and evolutionary implications, particularly in regions that were not completely deranged by continental ice sheets but were at high enough latitude to experience significant changes in temperature and precipitation. Such a region is the Great Basin. Its peculiar heterogeneous physiography, characterized by large valleys without external drainage and a scattering of tall mountain ranges, makes its Ice Age history of particular interest.

This chapter addresses the effects that the onset and termination of the Wisconsin glacial had on mammal communities of the Great Basin. The following introductory section describes present and past conditions in the region. The mammalian biogeography is then presented, along with current theories for the occurrence of numerous isolated populations of boreal species. The next section covers the Quaternary fossil record of the Great Basin, which helps explain the present biogeography and documents the former existence of many extinct and extirpated mammalian species. This section focuses on Crystal Ball Cave, where I found a dramatic contrast between Pleistocene fossils and the living assemblage. The final sections discuss the evolutionary implications of the changes caused by the Ice Age and the abiotic and biotic pathways by which all these changes took place.

The Great Basin

The Great Basin is defined as that area between the Sierra Nevada and the Rocky Mountains where no external drainage exists (Hunt 1967); it includes nearly all the state of Nevada, the western half of Utah, and

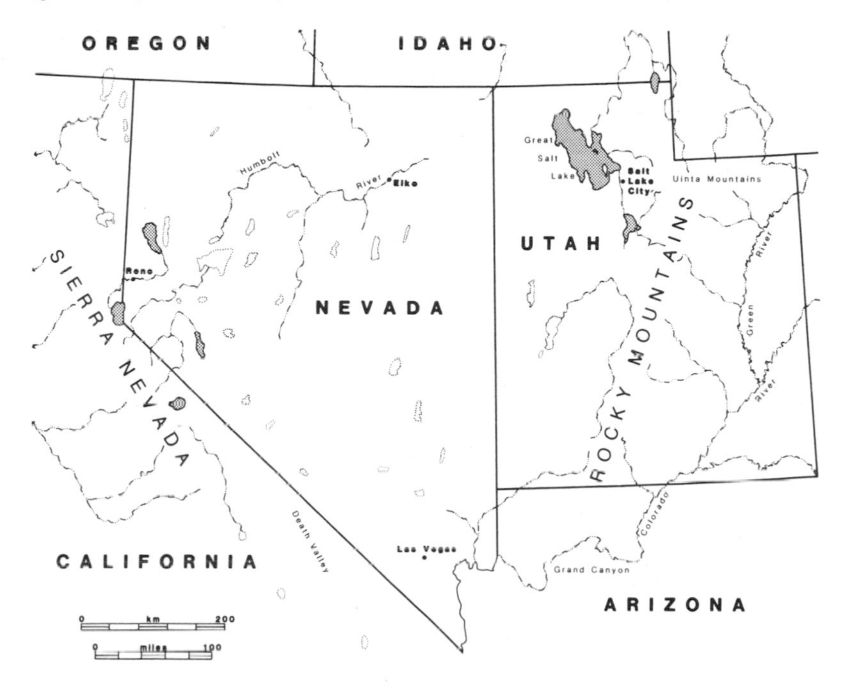

Figure 15.1. Map of the Great Basin showing state boundaries, major cities, and current lakes, rivers, and other physiographic features. Perennial lakes are shaded; intermittent lakes have stippled outlines.

adjacent parts of California, Oregon, and Idaho (fig. 15.1). Geologically it is characterized by active regional extension that has produced north-south trending horsts and grabens. Because of the youthfulness of mountain building, the mountain ranges tend to rise high above the valleys, and 19 of these ranges have peaks over 3,000 m in elevation. In these localized high regions the climate is similar to the Sierra Nevada and Rocky Mountains, but the valleys between them, characterized by desert conditions, are so dry that the water they collect from the mountains evaporates within them. The geomorphology is typical of areas that are dry and tectonically active: massive alluvial fans extend downward from the mountain fronts, and many of the valleys contain pluvial lakes (or lake beds) that expand and shrink as precipitation varies.

The temperature extremes and low precipitation of the Great Basin valleys permit the survival of only the hardiest xeric vegetation. Valleys contain mostly annual grasses, sagebrush, and other small desert plants. Trees occur only locally around water sources. The ground tends to be rocky near the mountain fronts and hard, dry, and alkaline in the valleys. Many mammalian species that are adapted to a desert habitat live in the interconnecting valleys of the Great Basin, and most of these species have ranges that extend southward into the Mojave Desert.

Table 15.1 lists a series of floral belts and life zones as given by Hall (1946) for Nevada. The Upper Sonoran life zone covers nearly all of the Great Basin and includes the pinyon-juniper and sagebrush floral belts. The southernmost part of the Great Basin, the Mojave Desert, is in the Lower Sonoran life zone. All the floral belts above the Sonoran are restricted to the high mountain ranges and occupy only a tiny fraction of the total area (5%), and the steep elevation rise causes the sharply contrasting Sonoran and boreal floras to stand in abrupt contact. Table 15.2 lists the mammal species currently found in the Great Basin, both living and late Pleistocene fossils, and their sizes, their distributions, and the life zones they now occupy.

Geomorphological evidence suggests that the Great Basin was very different as recently as a few tens of thousands of years ago. Terraces attest to the former presence of sizable lakes in most of the intermontane valleys (fig. 15.2). The largest of these was Lake Bonneville, which covered most of western Utah and adjacent parts of Nevada and Idaho and which drained northward into the Snake River. Lake Lahontan and more than a hundred smaller lakes filled the closed basins of Nevada, Oregon, and California. Most of these lakes lacked outlets, but a series of rivers connected lakes in the southernmost Great Basin and carried water to Lake Manly in Death Valley (Smith and Street-Perrott 1983). Glacial moraines in the mountain ranges show that valley glaciers existed on many of the highest peaks and came as low as 2,800 m elevation (Porter, Pierce, and Hamilton 1983), but they never covered enough area to affect mammal ranges significantly.

Climatic reconstruction of the late Pleistocene has been the subject of intense study. Presently the mean temperature in the Great Basin

Table 15.1. Floral belts and life zones of the western United States with species characteristic of the Great Basin (after Hall 1946)

Floral Belts and Their Most Dominant Plants	Life Zones and Their Characteristic Mammals
8. **Alpine Floral Belt** Low-growing species of subalpine belt	**Alpine-Arctic Life Zone** Few mammals range this high
7. **Subalpine Floral Belt** *Pinus longaeva*, Bristlecone Pine *Pinus flexilis*, Limber Pine *Picea engelmannii*, Engelmann Spruce *Juniperus communis*, Common Juniper	**Hudsonian Life Zone** Higher-ranging Canadian species

Table 15.1. (*continued*)

Floral Belts and Their Most Dominant Plants	Life Zones and Their Characteristic Mammals
6. Spruce-Fir Floral Belt *Abies lasiocarpa*, Subalpine Fir *Abies concolor*, White Fir *Pseudotsuga menziesii*, Douglas Fir	**Canadian Life Zone** *Lepus townsendii*, White-tailed Jackrabbit *Eutamias amoenus*, Yellow-pine Chipmunk *Eutamias palmeri*, Palmer's Chipmunk *Neotoma cinerea*, Bushy-tailed Woodrat *Microtus longicaudus*, Long-tailed Vole *Zapus princeps*, Western Jumping Mouse *Mustela erminea*, Ermine
5. Aspen Floral Belt *Populus tremuloides*, aspen	**Canadian Life Zone** Subset of species from list above
4. Yellow Pine Floral Belt *Pinus ponderosa*, Yellow (Ponderosa) Pine *Juniperus scopulorum*, Rocky Mountain Juniper *Cercocarpus ledifolius*, Mountain Mahogany *Quercus* spp., Oak	**Transition Life Zone** (upper section) Mixture of some Canadian and some Upper Sonoran species
3. Pinyon-Juniper Floral Belt *Pinus edulis*, Pinyon Pine (east) *Pinus monophylla*, Single-leaf Pinyon (west) *Juniperus osteosperma*, Utah Juniper	**Upper Sonoran Life Zone** (upper section) *Eutamias Dorsalis*, Cliff Chipmunk *Eutamias panamintinus*, Panamint Chipmunk *Peromyscus truei*, Pinyon Mouse
2. Sagebrush Floral Belt *Artemisia tridentata*, Sagebrush *Atriplex confertifolia*, Shadscale *Sarcobatus vermiculatus*, Greasewwod *Chrysothamnus* spp., Rabbitbrush *Tetradymia* spp., Horsebrush *Ceratoides lanata*, Winterfat *Grayia spinosa*, Hopsage	**Upper Sonoran Life Zone** (main section) *Sylvilagus idahoensis*, Pygmy Rabbit *Eutamias minimus*, Least Chipmunk *Spermophilus richardsonii*, Richardon's Ground Squirrel *Microdipodops megacephalus*, Dark Kangaroo Mouse *Microdipodops pallidus*, Pale Kangaroo Mouse *Dipodomys ordii*, Ord's Kangaroo Rat *Onychomys leucogaster*, Northern Grasshopper Mouse *Lagurus curtatus*, Sagebrush Vole
1. Creosote Bush Floral Belt *Larrea tridentata*, Creosote Bush *Atriplex polycarpa*, Desert Saltbrush *Ambrosia dumosa*, Bursage *Encelia farinosa*, Brittlebush	**Lower Sonoran Life Zone** *Macrotus californicus*, California Leaf-nosed Bat *Spermophilus tereticaudus*, Round-tailed Ground Squirrel *Perognathus penicillatus*, Desert Pocket Mouse *Peromyscus eremicus*, Cactus Mouse *Sigmodon hispidus*, Hispid Cotton Rat

Note: Life zone ranges for all Great Basin mammals can be found in table 15.2.

Table 15.2. Quaternary mammal species of the Great Basin including extinct and extirpated species (after Hall 1946, 1981; Kurten and Anderson 1980)

Scientific Name	Common Name	Size (cm)	Range in G.B.	Life Zone
Class Mammalia				
Order Insectivora				
Family Soricidae				
Sorex vagrans	Vagrant Shrew	14	NW ⅔	2–7
Sorex tenellus	Inyo Shrew	10	SW ⅛	2–4
Sorex palustris	Water Shrew	15	N ¾	2–7
Sorex merriami	Merriam's Shrew	10	NE ⅞	2–3
Notiosorex crawfordi	Crawford's Desert Shrew	10	S ⅛	1–2
Order Chiroptera				
Family Phyllostomidae				
Macrotus californicus	California Leaf-nosed Bat	9	S ⅛	1
Family Vespertilionidae				
Myotis californicus	California Myotis	7	W ½	1–2
Myotis subulatus	Small-footed Myotis	7	All	2–4
Myotis yumanensis	Yuma Myotis	7	SW ¼	1–5
Myotis lucifugus	Little Brown Myotis	8	All	2–7
Myotis volans	Long-legged Myotis	8	All	2–7
Myotis thysanodes	Fringed Myotis	8	All	1–2
Myotis evotis	Long-eared Myotis	8	N ⅞	3–4
Lasionycteris noctivagans	Silver-haired Bat	8	All	1–6
Pipistrellus hesperus	Western Pipistrelle	8	SW ¾	1–2
Eptesicus fuscus	Big Brown Bat	12	All	1–5
Nycteris borealis	Red Bat	11	SW ⅓	1–2
Nycteris cinerea	Hoary Bat	14	All	2–7
Euderma maculatum	Spotted Bat	11	SE ⅞	1–2
Plecotus townsendii	Townsend's Big-eared Bat	10	All	1–4
Antrozous pallidus	Pallid Bat	13	SW ⅔	1–2
Family Molossidae				
Tadarida brasiliensis	Brazilian Free-tailed Bat	10	S ⅔	1–2
Tadarida macrotis	Big Free-tailed Bat	13	SE ⅛	1–2
Order Primates				
Family Hominidae				
Homo sapiens	Modern Man	200	All	
Order Edentata				
Family Megalonychidae				
Megalonyx sp.	Ground Sloth	300	Extinct	
Family Megatheriidae				
Nothrotheriops shastensis	Shasta Ground Sloth	250	Extinct	
Order Lagomorpha				
Family Ochotonidae				
Ochotona princeps	Pika	20	Scat.	3–8
Family Leporidae				
Sylvilagus idahoensis	Pygmy Rabbit	28	N ⅔	2–3
Sylvilagus nuttallii	Nuttall's Cottontail	38	N ⅞	2–6

Introduced species not found as fossils and species inhabiting only the mountainous or Mojave Desert periphery of the physiographic province are not included. Size represents length in centimeters from snout to end of tail of large individuals. Range represents areas of the Great Basin where the species is currently found with no respect to life zones occupied or population density. Life zones occupied correspond to table 15.1.

Table 15.2. (*continued*)

Scientific Name	Common Name	Size (cm)	Range in G.B.	Life Zone
Sylvilagus audubonii	Desert Cottontail	41	S ⅓	1–2
Lepus americanus	Snowshoe Rabbit	50	Extirp.	5–6
Lepus townsendii	White-tailed Jackrabbit	63	N ½	4–7
Lepus californicus	Black-tailed Jackrabbit	60	All	1–3
Order Rodentia				
Family Sciuridae				
Eutamias minimus	Least Chipmunk	21	N ⅔	2
Eutamias amoenus	Yellow-pine Chipmunk	23	N ⅛	4–6
Eutamias dorsalis	Cliff chipmunk	26	E ⅓	3
Eutamias panamintinus	Panamint Chipmunk	21	SW edge	3
Eutamias umbrinus	Uinta Chipmunk	23	Scat.	3–7
Eutamias palmeri	Palmer's Chipmunk	22	S ⅛	4–7
Marmota flaviventris	Yellow-bellied Marmot	70	N ⅔	2–7
Ammospermophilus leucurus	White-tailed Antelope Squirrel	23	S ⅔	1–4
Spermophilus townsendii	Townsend's Ground Squirrel	26	N ⅞	1–3
Spermophilus richardsonii	Richardson's Ground Squirrel	33	N ¼	2–3
Spermophilus armatus	Uinta Ground Squirrel	30	NE edge	
Spermophilus beldingi	Belding's Ground Squirrel	29	N ½	2–7
Spermophilus variegatus	Rock Squirrel	50	SE ⅓	1–4
Spermophilus tereticaudus	Round-tailed Ground Squirrel	26	S ⅛	1
Spermophilus lateralis	Golden-mantled Ground Squirrel	30	N ⅔	3–7
Tamiasciurus douglasii	Douglas' Squirrel	38	W edge	6–8
Family Geomyidae				
Thomomys talpoides	Northern Pocket Gopher	24	N ½	5–8
Thomomys monticola	Mountain Pocket Gopher	27	W edge	6–8
Tomomys umbrinus	Southern Pocket Gopher	27	S ⅔	1–8
Tomomys townsendii	Townsend's Pocket Gopher	30	N ½	3–4
Family Heteromyidae				
Perognathus parvus	Great Basin Pocket Mouse	19	N ⅞	2–3
Perognathus longimembris	Little Pocket Mouse	14	SW ¾	1–2
Perognathus formosus	Long-tailed Pocket Mouse	20	S ½	1–2
Microdipodops megacephalus	Dark Kangaroo Mouse	17	Center	2
Microdipodops pallidus	Pale Kangaroo Mouse	17	SW ⅛	2
Dipodomys ordii	Ord's Kangaroo Rat	27	N⅞	2–3
Dipodomys microps	Chisel-toothed Kangaroo Rat	29	SW ⅞	1–3
Dipodomys panamintinus	Panamint Kangaroo Rat	32	SW edge	1–3
Dipodomys merriami	Merriam's Kangaroo Rat	25	SW ⅓	1–2
Dipodomys deserti	Desert Kangaroo Rat	37	SW ⅛	1–2
Family Castoridae				
Castor canadensis	Beaver	140	N ⅓	1–6
Family Muridae				
Reithrodontomys megalotis	Western Harvest Mouse	16	All	1–4
Peromyscus maniculatus	Deer Mouse	21	All	1–8
Peromyscus crinitus	Canyon Mouse	18	All	1–4
Peromyscus boylii	Brush Mouse	24	S ⅛	1–4
Peromyscus truei	Pinyon Mouse	22	S ¾	3–4
Onychomys leucogaster	Northern Grasshopper Mouse	18	N ⅞	2
Onychomys torridus	Southern Grasshopper Mouse	16	SW ¼	1–2
Neotoma lepida	Desert Woodrat	37	All	1–3
Neotoma fuscipes	Dusty-footed Woodrat	45	Extirp.	
Neotoma cinerea	Bushy-tailed Woodrat	46	N ¾	4–7

Table 15.2. (*continued*)

Scientific Name	Common Name	Size (cm)	Range in G.B.	Life Zone
Phenacomys intermedius	Heather Vole	15	Extirp.	
Microtus californicus	California Vole	20	SW edge	
Microtus pennsylvanicus	Meadow Vole	19	NE edge	
Microtus montanus	Montane Vole	21	N ¾	1–6
Microtus longicaudus	Long-tailed Vole	21	N ⅞	4–7
Lagurus curtatus	Sagebrush Vole	14	All	2–4
Ondatra zibethicus	Muskrat	60	N ½	1–5
Family Zapodidae				
Zapus princeps	Western Jumping Mouse	24	N ½	4–6
Family Erethizontidae				
Erethizon dorsatum	Porcupine	100	All	2–7
Order Carnivora				
Family Canidae				
Canis familiaris	Domestic Dog	130	Domestic	
Canis latrans	Coyote	130	All	1–8
Canis lupus	Gray Wolf	200	Rare	1–7
Canis dirus	Dire Wolf	220	Extinct	
Vulpes vulpes	Red Fox	100	Scat.	2–7
Vulpes velox	Kit Fox	80	SW ⅞	1–2
Urocyon cinereoargenteus	Gray Fox	110	S ½	1–2
Family Ursidae				
Arctodus simus	Giant Short-faced Bear	300	Extinct	
Ursus americanus	Black Bear	200	NW edge	5–7
Ursus horribilis	Brown Bear	300	Extirp.	
Family Procyonidae				
Bassariscus astutus	Ringtail	80	SE ⅓	1–2
Procyon lotor	Raccoon	95	N&S ¼	1–7
Family Mustelidae				
Martes nobilis	Noble Marten	70	Extinct	
Martes americana	Pine Marten	67	Extirp.	5–7
Mustela erminea	Ermine	33	N ⅔	4–7
Mustela nivalis	Least Weasel	20	Extirp.	
Mustela frenata	Long-tailed Weasel	50	N ⅞	1–7
Mustela nigripes	Black-footed Ferret	55	Extirp.	
Mustela vison	Mink	70	N ¼	2–5
Gulo luscus	Wolverine	110	Extirp.	
Taxidea taxus	Badger	85	All	1–7
Spilogale putorius	Spotted Skunk	60	All	1–3
Brachyprotoma brevimala	Western Short-faced Skunk	60	Extinct	
Mephitis mephitis	Striped Skunk	75	N ¾	1–4
Lutra canadensis	River Otter	120	NW ⅓	1–7
Family Felidae				
Smilodon fatalis	Saber-toothed Cat	250	Extinct	
Panthera atrox	American Lion	300	Extinct	
Panthera onca	Jaguar	230	Extirp.	
Acinonyx trumani	American Cheetah	240	Extinct	
Felis concolor	Mountain Lion	240	All	1–5
Lynx canadensis	Canada Lynx	90	N ⅛	
Lynx rufus	Bobcat	120	All	1–5

Table 15.2. (*continued*)

Scientific Name	Common Name	Size (cm)	Range in G.B.	Life Zone
Order Proboscidea				
Family Mammutidae				
Mammut sp.	American Mastodont	500	Extinct	
Family Elephantidae				
Mammuthus columbi	Columbian Mammoth	500	Extinct	
Order Perissodactyla				
Family Equidae				
Equus (large)	Large Horse	300	Extinct	
Equus (small)	Small Horse	250	Extinct	
Order Artiodactyla				
Family Tayassuidae				
Platygonus compressus	Peccary	170	Extinct	
Family Camelidae				
Camelops hesternus	Yesterday's Camel	300	Extinct	
Hemiauchenia macrocephala	Large-headed Llama	260	Extinct	
Family Cervidae				
Cervus elaphus	Wapiti	290	NE ¼	3–4
Odocoileus hemionus	Mule Deer	180	All	2–7
Family Antilocapridae				
Capromeryx minor	Diminutive Pronghorn	80	Extinct	
?*Tetrameryx*	Pronghorn	150	Extinct	
Antilocapra americana	Pronghorn	145	All	1–4
Family Bovidae				
Bison antiquus	Bison	450	Extinct	
Bison bison	Bison	350	N ⅛	
Oreamnos harringtoni	Harrington's Mountain Goat	140	Extinct	
Ovis canadensis	Mountain Sheep	190	All	1–8
Eucerotherium sp.	Shrub Ox	280	Extinct	
Bootherium bombifrons	Woodland Muskox	240	Extinct	

region decreases about 6° C per 1,000 m increase in elevation and 0.8 °C per degree increase in latitude (Porter et al. 1983). Pleistocene snow lines reached 1,000 m lower than they do today, suggesting the climate was several degrees cooler during that time. This would have increased precipitation and decreased evaporation, thus causing water to fill the closed basins. Smith and Street-Perrott (1983) studied the factors that led to the development of pluvial lakes in the Great Basin during the Pleistocene, and they estimated that the streams that feed the basins must have carried 5 to 10 times more water than they do now. From this they estimated that Pleistocene precipitation reached one and one-half to two times its present value. Mifflin and Wheat (1979) estimated that an annual mean temperature of 3° C below present could account for the pluvial Pleistocene lakes.

Plant fossils from late Pleistocene wood rat middens show enormous differences from the vegetation of today. The following patterns have

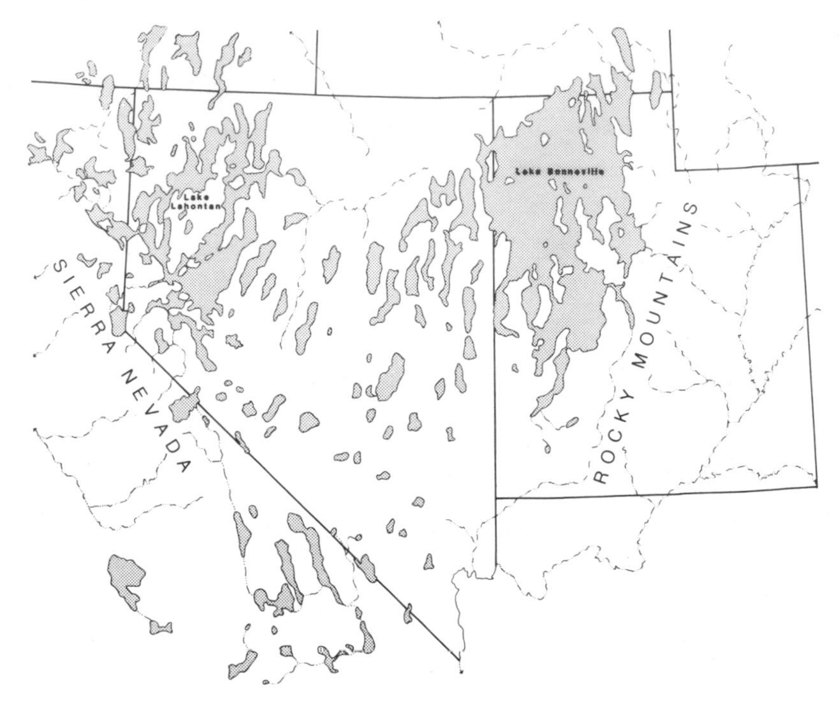

Figure 15.2. Map showing extent of Great Basin lakes during the late Pleistocene (modified from Smith and Street-Perrott 1983).

been documented by Thompson and Mead (1982), Spaulding, Leopold, and Van Devender (1983), and Wells (1983). During the Holocene the creosote-bush community of the Mojave Desert has been expanding into the southern Great Basin and replacing pinyon-juniper woodland. The record also shows that pinyon-juniper woodland was restricted to the Mojave Desert region during the late Wisconsin but has recently expanded more than 500 km northward into the central and northern Great Basin, where it now constitutes a major floral belt. Valleys of the central Great Basin, which are now occupied primarily by sagebrush and shadscale, were dominated by subalpine conifers. Bristlecone pine, Engelmann spruce, and common juniper reached as low as 1,660 m in the central Great Basin. Connecting divides higher than 1,800 m exist across the central Great Basin, which allows a continuous coniferous forest between ranges. In southern Nevada bristlecone pine only reached as low as 1,850 m, but the lower elevations were dominated by limber pine and Douglas fir which advanced from the mountains of southern Utah. All of these species had ranges extending 500 to 1,000 m lower than today. The presently dominant montane conifers of the middle elevations (white fir and Douglas fir) were limited to the southernmost part of the Great Basin during the last glacial.

Timing of the events outlined previously has been worked out reasonably well using a variety of methods. The Wisconsin glacial included multiple periods of glaciation interspersed with periods more like today, but earlier events have been obscured by later events. Based on studies of continental glaciation and sea level changes, the last pulse of glaciation (late Wisconsin) began about 35,000 years B.P. and peaked about 18,000 years B.P. (Bloom 1983). The last rise and fall of Lake Bonneville followed a similar but slightly delayed time frame. According to Scott et al. (1983) the lake started its last filling cycle before 26,000 years B.P. and was at its highest level from about 16,000 to 14,500 years B.P., at which time erosion at the outlet catastrophically lowered the lake level 100 m. About 13,000 years B.P. the lake evaporated below its outlet, and by 11,000 years B.P. a lake configuration similar to today was achieved (Currey 1982; Scott et al. 1983). Lake Lahontan had a more complicated history. It was dry 36,000 years B.P. and then reached its highest level at two different periods about 22,000 and 13,000 years B.P. (Benson 1978; Davis 1978). Plant fossil evidence indicates that the current desert conditions in the Great Basin prevailed in the valleys by 8,000 years B.P. (Wells 1983).

The Mammalian Biogeographical Puzzle: Brown's Nonequilibrium Model

The higher elevations of the mountain ranges in the Great Basin form montane islands with habitats similar to the more extensive high-elevation regions of the Sierra Nevada and Rocky Mountains. Brown (1971, 1978) compared the current distribution of birds and mammals on these montane islands to the distribution of species on oceanic islands studied by MacArthur and Wilson (1963, 1967) using the Sierra Nevada and Rocky Mountains as the equivalent of mainland.

MacArthur and Wilson (1963, 1967) showed that the probability of an oceanic island having a colony of mainland species was related to its distance from the mainland, distance from other islands having such colonies, and the size of the island. Island size not only affects the probability of colonization, but also the probability of extinction once colonization has occurred. Brown (1971, 1978) defined montane islands as ranges having peaks over 3,000 m in elevation and being separated from their closest neighboring island or mainland by at least 8 km with elevation below 2,300 m. The elevation of 2,300 m corresponds to the lower limit of continuous pinyon-juniper woodland in the Great Basin. Brown did not consider the northern part of the Great Basin in

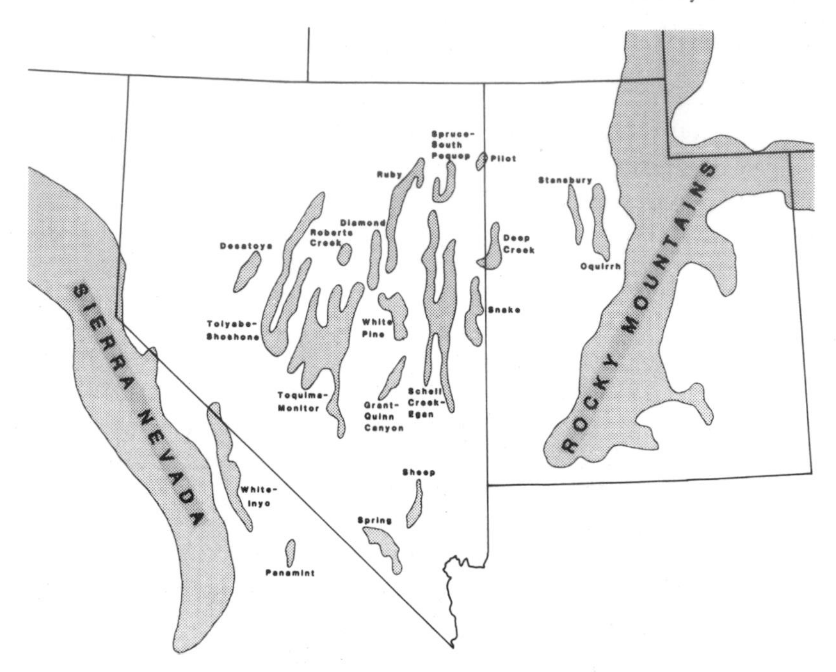

Figure 15.3. Map showing mountain ranges in the Great Basin that act as montane islands. See table 15.3 for pertinent data and table 15.4 for a list of boreal mammal species living on each mountain range.

lacking and because pinyon-juniper woodlands are not well developed there. With these criteria Brown (1978) identified 19 montane islands in the Great Basin (fig. 15.3), and table 15.3 gives data on these mountain ranges.

Brown defined boreal species as those that live at high elevations in the Sierra Nevada, the Rocky Mountains, or both and that do not range lower than the pinyon-juniper floral belt. The mountain ranges of the Great Basin are truly islands to species thus defined. Brown found that the distribution of boreal birds and large mammals fits the colonization-equilibrium model outlined by MacArthur and Wilson, but the distribution of small mammals does not. Table 15.4 shows which small boreal mammals live on each island. Mountain ranges containing the highest numbers of these species are not particularly close to a mainland or to each other, neither do they have a higher-elevation pass to a mainland or another island. The number of boreal species inhabiting an island is strongly related to its size, however, and this relationship is even more pronounced than on oceanic islands. This suggests a very low rate of colonization relative to extinction. Such a system is not in equilibrium and can only be explained by historical factors.

Hall (1946), in an early attempt to explain the discontinuous range of small boreal mammals in the Great Basin, speculated that these animals might disperse from one range to another during the winter, but no evidence has been found to support this. Since they do not fit the colonization-extinction equilibrium model, Brown (1971, 1978) concluded that these small mammals reached the montane islands during the Pleistocene when boreal conditions were thought to have reached lower elevations in the Great Basin and that there have been extinctions but no colonizations since then. Brown then considered factors that might affect the extinction rate. In addition to the correlation with island size, Brown showed that extinction is least likely in mammals with small body size, low trophic level, and high habitat diversity. The Great Basin islands show this very dramatically; species without these qualities occur only on the largest islands. The distribution of some species can be explained on more individual grounds. For example, the Ruby Mountains, one of the larger ranges, does not sup-

Table 15.3. List of mountain ranges acting as montane islands in the Great Basin with pertinent data (after Brown 1971, 1978)

Mountain Range	Area above 2,300 m (km^2)	Highest Peak (m)	Nearest Continent (km)	Current Number of Small Boreal Mammal Species	Current + Fossil Small Boreal Mammal Species
1. White-Inyo	1,910	4,341	16	14	14
2. Panamint	120	3,367	84	3	3
3. Desatoya	210	2,991	134	9	9
4. Toyiabe-Shoshone	1,770	3,593	177	13	13
5. Toquima-Monitor	3,040	3,642	183	12	14
6. Roberts Creek	130	3,089	348	4	4
7. Diamond	410	3,235	306	4	4
8. Ruby	940	3,471	278	12	13
9. Spring	320	3,633	201	7	7–9
10. Sheep	130	3,021	138	3	4–9
11. Grant-Quinn Canyon	380	3,443	222	5	5
12. White Pine	670	3,410	241	8	8
13. Shell Creek-Egan	2,640	3,622	183	8	8
14. Spruce-South Pequop	120	3,128	251	4	4
15. Snake	1,070	3,982	143	10	13–17
16. Deep Creek	570	3,688	167	9	9
17. Pilot	30	3,263	183	3	3
18. Stansbury	140	3,362	63	8	8
19. Oquirrh	210	3,239	31	10	10

Note: Table 15.4 lists boreal mammal species found on these mountain ranges.

Table 15.4. Boreal species of small mammals occupying the Sierra Nevada (SN), Great Basin montane islands (1 to 19; see table 15.3), and/or the Rocky Mountains (RM)

Boreal	West									Mountain Ranges Occupied											East	Living
Mammal Species	SN	1	2	3	4	5	6	7	8	9	10	11	12	13	14	15	16	17	18	19	RM	Total
Sorex cinereus																					x	0
Sorex lyelli[b]	x																					0
Sorex vagrans	x				x	x			x					x		x	x		x	x	x	8
Sorex tenellus[b]		x								x												2
Sorex palustris	x	x			x	x	x		x							x			x	x	x	8
Sorex trowbridgii	x																					0
Ochotona princeps	x	x		x	x	x			x	t	t					t					x	5
Sylvilagus nuttallii	x	x	x	x	x	x			x	x	t	x	x	x		x	x				x	12
Lepus americanus[a]	x															f					x	0
Lepus townsendii	x			x		f			x	x			x			t				x	x	5
Aplodontia rufa[a]	x																					0
Eutamias alpinus[a,b]	x																					0
Eutamias amoenus	x	x									t											1
Eutamias townsendii[a]	x																					0
Eutamias dorsalis		x		x	x	x		x			x	x	x	x	x	x	x	x	x	x	x	15
Eutamias quadrimaculatus[b]	x																					0
Eutamias speciosus[a]	x																					0
Eutamias panamintinus[b]		x	x							x												3

Species																						Total
Eutamias umbrinus		x		x	x	x	x	x	x	x	x	x	x	x	x	x	x		x	x	x	17
Marmota flaviventris	x	x		x	x	x			x	t	t		x	x		x	x		x	x	x	11
Spermophilus armatus[b]																					x	0
Spermophilus beldingi	x				x	x			x							f						3
Spermophilus lateralis	x	x		x	x	x		x	x	x	t	x	x	x	x	x	x	x			x	14
Tamiasciurus hudsonicus[a]																					x	0
Tamiasciurus douglasii[a]	x								f													0
Glaucomys sabrinus[a]	x																				x	0
Neotoma cinerea	x	x	x	x	x	x		x	x	x	x		x	x	x	x	x	x	x	x	x	17
Clethrionomys gapperi[a]																					x	0
Phenacomys intermedius	x					f										f					x	0
Microtus longicaudus	x	x		x	x	x	x		x		f	x	x	x		x	x		x	x	x	13
Microtus richardsoni																					x	0
Zapus princeps	x				x	x	x		x											x	x	5
Martes americana[a]	x															t					x	0
Mustela erminea	x	x			x											x	x		x	x	x	6
Gulo luscus[a]	x	x																			x	1

Note: x = current occupation; t = Quaternary fossils recovered in or near the range; f = possible but uncertain fossil recoveries. After Brown (1971, 1978) and Grayson (1981, 1987).

[a]These species occur only as high in elevation as the yellow pine floral belt.

[b]These species have very restricted ranges and are therefore probably not good habitat indicators. *Sorex tenellus* and *Eutamias panamintinus* have their entire range within the southwestern Great Basin; the others range only outside the Great Basin.

port a pinyon-juniper chipmunk, probably because pinyon-juniper woodland is poorly developed there (Brown 1978).

Brown (1971, 1978) noted the conspicuous absence in the Great Basin of small mammals that are restricted to dense yellow pine forests of the surrounding mainland, in spite of the fact that suitable habitats are available. He listed 11 species in this category: snowshoe rabbit, mountain beaver, alpine chipmunk, Townsend's chipmunk, lodgepole chipmunk, red squirrel, Douglas' squirrel, northern flying squirrel, red-backed mouse, heather vole, and pine marten. Since this suite of species is not found living on any of the montane islands of the Great Basin, Brown concluded that their absence was due to their never having colonized these islands rather than colonization and later extinction. He proposed that climatic conditions never allowed habitat bridges of this higher-elevation suite of plants to exist across the Great Basin. Although the mammals of this habitat are absent, Brown noted that islands of yellow pine forest have many avian species characteristic of such habitats.

Brown did not consider the fossil record in any detail, but Grayson (1981, 1987) derived three implications from Brown's model that could be tested in the fossil record: (1) boreal mammals living on isolated montane islands must once have occupied the intervening lowlands, (2) boreal mammals found on only some of the mountains today should have been present on other ranges in the past, and (3) there may have been species of boreal mammals on isolated mountains in the past that live on none of the mountains today. A fourth prediction is implied by Brown's observation that Great Basin mountain ranges lack boreal species inhabiting only the highest elevations of the mainlands: boreal mammals inhabiting only the yellow pine and higher life zones in the Sierra Nevada and Rocky Mountains never inhabited the Great Basin.

Brown's explanation for the current distribution of boreal mammals in the Great Basin is impressive, and the Quaternary fossil record has supported nearly all of its expectations. But the restricted subset of current species and localities studied by Brown tells only a small part of a much larger story of the changes in mammals inhabiting the Great Basin during the Quaternary period, and it is this larger story that is presented here.

The Quaternary Fossil Record of the Great Basin

Nearly all the Quaternary mammalian fossils from the Great Basin are younger than 30,000 years B.P. (Webb 1986) and are original bone preserved in dry caves, woodrat middens, and lake deposits. Although this fossil record is limited, it supports and augments Brown's (1971,

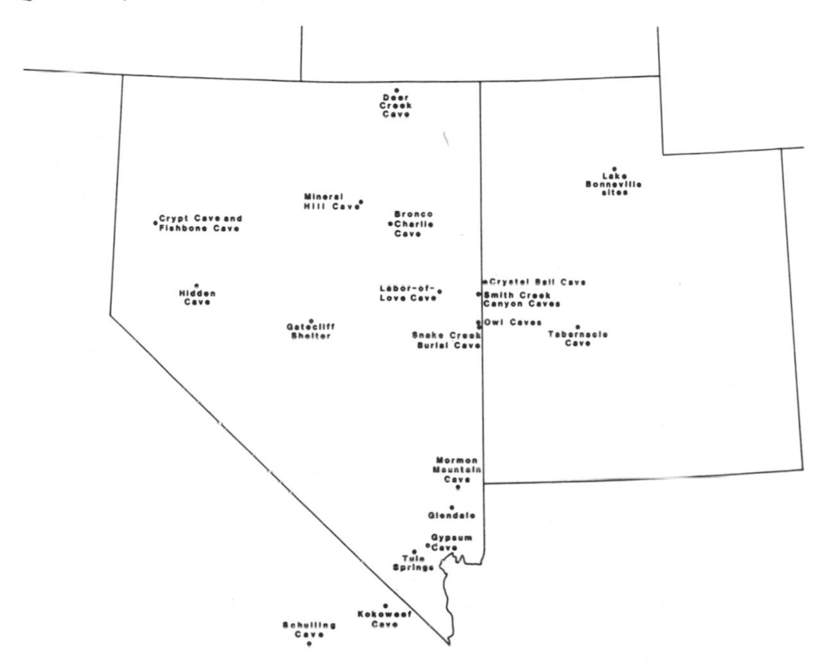

Figure 15.4. Map showing location of Quarternary fossil deposits in the Great Basin. See table 15.5 for pertinent data and tables 15.6 and 15.7 for lists of fossils recovered from each site.

1978) nonequilibrium model for extant species and also records the presence of many species now extinct or extirpated from the Great Basin, including many large mammals. Figure 15.4 shows the more significant fossil sites, and table 15.5 gives pertinent data. Table 15.6 lists mammalian taxa found at the most productive localities.

I conducted an extensive study of Quaternary fossils from Crystal Ball Cave (Heaton 1984, 1985), and this discussion focuses on my findings. The cave is located in a hill in Snake Valley 10 km northeast of the Snake Range, and it was only 1.7 km from Lake Bonneville at its highest level. Dry dust deposits deep within the cave contain abundant fossil bone, accumulated mainly by woodrats. The deposits are unstratified and represent a mixture of Pleistocene and Holocene faunas, but four radiocarbon dates all gave late Pleistocene ages (Heaton 1985). Crystal Ball Cave has provided one of the largest late Pleistocene mammal faunas from the Great Basin, and the contrast between the fossil assemblage and the mammals now living in the area is very dramatic. Species recovered from Crystal Ball Cave can be separated into four categories: (1) desert species that currently live in Snake Valley, (2) boreal species now restricted to high elevations or extirpated

Table 15.5. List of Quaternary fossil localities in the Great Basin with pertinent data

Locality	County and State	Elevation	Deposit and Mechanism	Age (Y.B.P.) or Epoch	Reference
Deer Creek Cave Jarbridge Mtns.	Elko Co., Nevada	1,770 m	Cave Humans	10,485–9,370	Zeigler 1963
Bronco Charlie Cave Ruby Mtns.	Elko Co., Nevada	2,134 m	Cave Humans	Early Holocene	Spiess 1974
Mineral Hill Cave Sulphur Spring Range	Elko Co., Nevada	1,800 m (est.)	Cave Natural	Late Pleistocene, Early Holocene	McGuire 1980
Gatecliff Shelter Monitor Valley	Nye Co., Nevada	2,320 m	Cave Humans and natural	5,970–4,995	Grayson 1983
Hidden Cave Eetza Mtn.	Churchill Co., Nevada	1,251 m	Cave Humans and natural	21,000–<1,500	Grayson 1985
Crypt Cave Lake Winnemucca	Pershing Co., Nevada	1,240 m	Cave Humans	20,400–19,100	Orr 1952, 1969, 1972
Fishbone Cave Lake Winnemucca	Pershing Co., Nevada	1,238 m	Cave Humans	11,650–10,600	Orr 1952, 1956
Lake Bonneville Wasatch Front (many sites)	Salt Lake Co., etc., Utah	1,424–1,515 m	Lake gravels Natural	12,720–12,580, Late Pleistocene	Stokes and Hansen 1937 Stokes and Condie 1961 Stock and Stokes 1969 Nelson and Madsen 1978, 1980, 1983
Tabernacle Cave Black Rock Desert	Millard Co., Utah	1,463 m	Lava tube Natural	11,330–10,820	Romer 1928 Nelson and Madsen 1979
Crystal Ball Cave Snake Valley	Millard Co., Utah	1,760 m	Cave Rodents	>23,000–10,980	Heaton 1984, 1985

Smith Creek Canyon Caves Snake Range	White Pine Co., Nevada	1,950 m	Caves Rodents, birds, humans	29,410–4,140	Miller 1979 Mead et al. 1982
Snake Creek Burial Cave Snake Valley	White Pine Co., Nevada	1,731 m	Cave Natural trap	Late Pleistocene Early Holocene	Barker and Best 1976 Heaton 1987 Mead and Mead 1985, 1989
Owl Cave One and Two Snake Valley	White Pine Co., Nevada	1,707 m	Cave Birds	Late Pleistocene, Holocene	Turnmire 1987
Labor-of-Love Cave Schell Creek Range	White Pine Co., Nevada	2,050 m	Cave Den	Late Pleistocene	Emslie and Czaplewski 1985
Mormon Mountain Cave Mormon Mtn.	Lincoln Co., Nevada	1,372 m	Cave Natural	Late Pleistocene	Jefferson 1982
Glendale Muddy River Basin	Clark Co., Nevada	555 m	Ponds Beavers	Prob. <33,000	Van Devender and Tessman 1975
Gypsum Cave Frenchman Mtns.	Clark Co., Nevada	454 m	Cave Den	11,840–11,100	Harrington 1933 Stock 1931 Mehringer 1967
Tule Springs Las Vegas Valley	Clark Co., Nevada	703 m	Fluvial Natural	>40,000–13,000	Spurr 1903 Mawby 1967 Shutler 1968
Kokoweef Cave Kokoweef Peak	San Bernardino Co., Calif.	1,770 m	Cave Natural trap	Late Pleistocene Early Holocene	Reynolds 1980 Kurten and Anderson 1980 Goodwin and Reynolds 1989
Schuiling Cave Newberry Mtns	San Bernardino Co., Calif.	658 m	Cave Natural and humans	Late Pleistocene	Downs et al. 1959

Note: See figure 15.4 for localities. In addition to the references cited, some information was obtained from Kurten and Anderson (1980), Lundelius et al. (1983), and Harris (1985).

Table 15.6. Mammalian species recovered from the 13 most productive late Pleistocene fossil sites in the Great Basin (Genus entries refer to all occurrences of a genus, including material not identified to species level)

1. Deer Creek Cave
2. Bronco Charlie Cave
3. Mineral Hill Cave
4. Gatecliff Shelter
5. Hidden Cave
6. Crystal Ball Cave
7. Smith Creek Canyon Caves
8. Owl Caves
9. Mormon Mountain Cave
10. Glendale
11. Gypsum Cave
12. Tule Springs
13. Kokoweef Cave

Taxa	Localities: 1	2	3	4	5	6	7	8	9	10	11	12	13
Sorex		X		X	X	X	?	X					
S. vagrans				X									
S. palustris					X								
S. merriami		?											
Notiosorex crawfordi									X				X
Macrotus		X											
M. californicus		?											
Myotis sp.						X		X	X				X
M. yumanensis					X								
Pipistrellus													X
Eptesicus fuscus									X				
Plecotus townsendii					X	?			X				
Antrozous pallidus		X			X		cf.		X				
Megalonyx												X	
Nothrotherium shastensis											X	X	
Ochotona princeps	X	X	X	X		X	X	X	X				X
Sylvilagus	X	X	X	X	X	X	X	X	X		X	X	X
S. idahoensis	X	X		X	X	X	X	X				cf.	cf.
S. nuttallii				X	X	X		X	X				
S. audubonii								cf.					cf.
Lepus	X	X	X	X	X	X	X	X			X	X	X
L. americanus						cf.							?
L. townsendii		X		cf.		X		X			X		
L. californicus				cf.	X	X	cf.	X				X	

Eutamias		X	X	X		X	X	X	X			X
E. minimus				X		X	X	X				X
E. amoenus									X			
E. dorsalis				X		X			cf.			
E. umbrinus				X			cf.					
Marmota sp.	X	X	X	X	X	X	X	X	X			X
M. flaviventris	X		X	X	X	X	X		X			X
Ammospermophilus sp.				X	X	X		X				X
A. leucurus				X	X	X		X				cf.
Spermophilus	X	X	X	X	X	X	X	X	X			X
S. townsendii				X	X	X	cf.	X				X
S. richardsonii		X					cf.	X				
S. beldingi				X			cf.					
S. variegatus		X					X		X			X
S. tereticaudus												cf.
S. lateralis				X			cf.	X	X			X
Tamiasciurus ? douglasii		X										
Thomomys	X	X	X	X	X	X	X	X		X	X	X
T. talpoides							cf.	X				
T. monticola		X										
T. umbrinus		?		X	X	X		X		X	?	
T. townsendii								cf.				
Perognathus				X	X	X	X	X	X	X		X
P. parvus				X	X		cf.	X	X			
P. longimembris					X					X		
P. formosus					X	cf.						
Microdipodops				X	X	X	X					
M. megacephalus				X		X	cf.					
M. pallidus					X							
Dipodomys				X	X	X	X	X		X	X	X
D. ordii					X		cf.					
D. microps					X	X		X				
D. merriami					X							
D. deserti					X							
Castor canadensis	X									X		
Reithrodontomys megalotis					X			X	X			X
Peromyscus		X	X	X	X	X	X	X	X			X
P. maniculatus					X	X		cf.	X			

Table 15.6. (*continued*)

Taxa	Localities:	1	2	3	4	5	6	7	8	9	10	11	12	13
P. crinitus					X	X	cf.							
P. truei						X	cf.		cf.	X				
Onychomys			X		X	X					X			
O. leucogaster						X								
Neotoma		X	X	X	X	X	X	X	X		X			X
N. lepida					X	X	X	X	X		X			cf.
N. fuscipes														X
N. cinerea		X	X		X	X	X	X	X					
Phenacomys cf. *intermedius*					X			X						
Microtus			?	X	X	X	X	X	X	X	X		X	X
M. californicus											?		cf.	X
M. pennsylvanicus							cf.							
M. montanus					cf.	X	cf.	cf.			X			
M. longicaudus					cf.		cf.	cf.	X	cf.				
Lagurus curtatus			?				X		X					X
Ondatra zibethicus						X	X				X		X	
Zapus princeps					cf.									
Erethizon dorsatum		X			X			X						
Canis		X		X	X	X	X	X	X			X	X	X
C. familiaris		X												
C. latrans				X	X	X	cf.	cf.	X				X	X
C. lupus		X				X	cf.	cf.	cf.			?		
C. dirus												?		cf.
Vulpes				X	X	X	X	X				X		X
V. vulpes					X	X	X	X						
V. velox							X	X				X		cf.
Urocyon cinereoargenteus				X										X
Ursus sp.		X				X		X						
U. arctos		X				X								
Bassariscus astutus								X					X	X
Martes		X	X			X	X	X						
M. nobilis			X			X		X						
M. americana		X	?				X							

Mustela		X	X		X	X	X	X			X		X
M. erminea							X	X					
M. frenata		X			X	cf.	X						
M. vison					X	cf.	X						
Taxidea taxus			X		X		X	X					X
Spilogale sp.		X	X	X	X		X						X
S. putorius		X		X	X								X
Brachyprotoma brevimala						X							
Mephitis mephitis				X	X								
Smilodon fatalis						cf.							
Panthera atrox							?					X	
P. onca							X						
Felis concolor						X	X	X	X			X	
Lynx sp.	X		X	X	X	X	X					?	X
L. rufus	X		X	X	X	cf.	X				X		cf.
Mammut sp.										X		X	
M. columbi												X	
Equus sp.			X		X	X	X	X		X	X	X	X
E. sp. (large)			X			X	X	X			X	X	
E. sp. (small)						X	X	X			X	X	X
Camelops sp.					X	X	X	X		X	X	X	X
C. hesternus					cf.	cf.	X	X		X		X	
Hemiauchenia sp.			X			X	?				X		X
Cervus elaphus	?	X		X		cf.	X						
Odocoileus sp.	X	X			X	X	X			X	X	X	X
O. hemionus		X			cf.	X				X	X		
Capromeryx minor							?						
Tetrameryx sp.												?	
Antilocapra americana	X		X	X	X	X	X	X					X
Bison sp.	?	X		X			?	X				X	
B. antiquus								X					
B. bison		X		X									
Oreamnos harringtoni							X						
Ovis canadensis	X	X	X	X	X	X	X	X	X	X	X		X
O. aries						cf.	cf.						
Eucerotherium sp.			X										X
Bootherium bombifrons							cf.						

For locations of these sites see figure 15.4; for pertinent data see table 15.5.

from the Great Basin, (3) species that require perennial water, and (4) extinct species. Each of these groups are discussed from the perspective of the Crystal Ball Cave fauna specifically and the Great Basin fossil record in general.

DESERT SPECIES: RANGE EXTENSIONS AND ECOLOGICAL REPLACEMENTS

Crystal Ball Cave contains many species of mammals that are restricted to the Sonoran life zone in which the cave is now located. Among these are pygmy rabbit, black-tailed jackrabbit, least chipmunk, white-tailed antelope squirrel, Townsend's ground squirrel, long-tailed pocket mouse, dark kangaroo mouse, canyon mouse, desert woodrat, sagebrush vole, kit fox, and pronghorn. However, most of these desert mammal species are greatly outnumbered in the assemblage by more boreal ecological counterparts. Other local species of the Sonoran life zone are entirely absent, for example, desert cottontail, rock squirrel, Great Basin pocket mouse, little pocket mouse, Ord's kangaroo rat, grasshopper mouse, gray fox, and spotted skunk.

One drawback of the Crystal Ball Cave assemblage is that fossil and recent bones are indistinguishable, so the Sonoran life zone species represented in the assemblage may be recent additions to the cave deposits. Yet if bone deposition has occurred at a constant rate for the last 25,000 years and the cave has been in its present desert environment for the last 8,000 years, species restricted to the Sonoran life zone should make up a more sizable part of the assemblage than they do. A decrease in the rate of fossil deposition may have occurred at the end of the Pleistocene, since deserts support a lower density of mammals than do wetter biomes. Another possible explanation is that boreal conditions remained in the area longer than is currently believed. Wells (1983) reported that upward withdrawal of subalpine conifers was tardiest in the Snake Range, and Mead, Thompson, and Van Devender (1982) reported bristlecone pine and common juniper associated with desert species in 7,000-year-old middens from Streamview Rockshelter in the lower part of Smith Creek Canyon, just 100 m higher in elevation than Crystal Ball Cave. At Hidden Cave, in strata less than 1,500 years old, Grayson (1985) found remains of mesic mammals that no longer inhabit that area.

Two desert species were too prominent in the Crystal Ball Cave fauna to be accounted for by recent invasion. By far the most abundant species is Townsend's ground squirrel, and squirrels are not common in the area today. Of four species of ground squirrels currently living in the region (Hall 1981), and possibly two others reported from the nearby Smith Creek Cave sediments (Mead, Thompson, and Van Devender 1982), Townsend's ground squirrel is the only species represented at Crystal Ball Cave. This high density is even more remarkable when one considers that ground squirrels are not cave-dwelling species, yet there are twice as many bones of ground squirrels as woodrats. When conditions are favorable, Townsend's ground squirrel can exist in incredibly high densities (Long 1940; Smith and Johnson 1985), and the area around Crystal Ball Cave must have presented such favorable circumstances at some time in the past. Durrant (1952) found that although Townsend's ground squirrel occurs throughout the Great Basin Desert, "it shows a marked increase in numbers in irrigated land and at desert springs." Snake Valley was much wetter during the Pleistocene, but this does not explain why a desert species like Townsend's ground squirrel was favored over boreal ground squirrels or tree squirrels. Apparently Gandy Mountain, the hill containing Crystal Ball Cave, remained free of subalpine conifers during the Pleistocene.

The other well-represented desert species in the Crystal Ball Cave assemblage is sagebrush vole. According to Hall (1946), this species can be found at almost any elevation but only in association with sagebrush. This species is about twice as abundant in the fauna as the montane species of voles, which suggests that Gandy Mountain retained sagebrush and other woody plants during the last glacial. This would also explain why no montane plant fragments were found in the cave. The vast plains surrounding Gandy Mountain have deep fertile soil that probably supported forest, grassland, and/or meadows during the late Wisconsin. But Gandy Mountain proper, being more exposed and isolated from streams, may have supported a more xeric plant community, as it does today.

The absence or scarcity in the Crystal Ball Cave fauna of desert species now abundant in the area demonstrates that they extended their ranges following the Wisconsin glacial. Of particular interest are cases where a species has been replaced by a closely related ecological counterpart, possibly due to interspecific competition. For example, the bushy-tailed woodrat is better represented in the assemblage than the desert woodrat by a ratio of at least 20 to 1, yet desert woodrat is the only woodrat living in or near the cave today. Similarly, the white-tailed jackrabbit was replaced by blacktailed jackrabbit, Nuttall's cottontail by desert cottontail, and the red fox by gray fox. In each case the species dominating the cave fauna is now found mostly at higher elevations or latitudes or both, and the desert species now found living around the cave is rare as a fossil (or completely absent in the case of gray fox). Currently the species in each pair listed previously have adjoining ranges with little overlap (Hall 1981), and a threshold of temperature, moisture, or both appears to determine where the boundary between them lies. The fossil record demonstrates that these boundaries have changed over time.

Another pair of species that have adjoining ranges with very little overlap are the two living species of grasshopper mouse (Hall 1981). Carleton and Eshelman (1979) found that the northern grasshopper mouse in the Great Basin is very distinct from populations elsewhere. Riddle and Choate (1986) investigated this further and stated: "We hypothesize that [northern grasshopper mouse] was a component of the fauna in a 'cold' desert refuge in the location of the present Mohave 'hot' desert during [full glacial] time. This would place geographic isolation of populations of northern grasshopper mice in the Great Basin from the remainder of the species sometime prior to 20,000 YBP, and would account for the existence at the close of the Pleistocene of size differences among populations of northern Grasshopper Mouse comparable to those among extant populations." The southern grasshopper mouse now inhabits the southern third of the Great Basin and may be extending its range northward. The northern grasshopper mouse has been recovered from Hidden Cave, where the ranges of the two species now meet (Grayson 1985), but no additional paleontological evidence confirms this replacement.

Range extensions in many species can be attributed to an expansion of their preferred resources, particularly the plants that they rely on. The pygmy rabbit and sagebrush vole eat only sagebrush and are only found living where this plant is present (Hall 1946). Late Pleistocene fossils of these species have been found south of their present range, where it is currently too hot for sagebrush (tables 15.1 and 15.6), but these species are presently expanding their ranges eastward in the Great Basin coincident with the expansion of sagebrush (Green and Flinders 1980).

Some rodents require specific substrates. The long-tailed pocket mouse strongly prefers stony ground, the dark kangaroo mouse prefers coarse soils, and the pale kangaroo mouse and desert kangaroo rat are restricted to fine, wind-blown sand containing some plants (Hall 1946). The long-tailed pocket mouse, pale kangaroo mouse, Merriam's kangaroo rat, and the desert kangaroo rat all have similar northern extensions of their ranges into the alkali flats of the Great Basin that were occupied by lakes during the late Pleistocene (Hall 1946).

Pronghorn is rarer in fossil deposits than bighorn sheep and mule deer, but it is presently the dominant artiodactyl in Great Basin valleys due to its preference for open rangeland. This and other desert species must have been more restricted during the last glacial when lakes, forests, and meadows covered the valleys.

In spite of these examples, range extensions at the end of the Pleistocene are few and unimpressive in comparison with range reductions and extinctions. Clearly the Great Basin supported a much larger biomass during the late Pleistocene than it does now. This is documented by the presence of fewer species and by decreased fossil deposition at sites such as Crystal Ball Cave.

BOREAL SPECIES: EXTIRPATIONS AND RANGE REDUCTIONS

Many fossils from Crystal Ball Cave are of extant species that do not live in Snake Valley at present. Included in this list of species are two small rodents that are restricted to the pinyon-juniper floral belt: the cliff chipmunk and the pinyon mouse. Although these are desert species, their absence from the sagebrush floral belt makes their presence in the cave unexpected. Neither pinyon nor juniper trees occur on or near Gandy Mountain today, and their presence in the lower elevations of the Snake Range is thought to be a recent invasion from southern Nevada (Wells 1983). It is possible that pinyon-juniper woodland extended closer to the cave at some time during the early Holocene, but the scarcity of these rodents in the cave fauna suggests that such woodland never dominated the area.

Many species in the Crystal Ball Cave assemblage occur no lower than the transition life zone, so it seems very unlikely that they could have reached the cave in recent times. Some of these are still found at higher elevations in the nearby Snake Range: white-tailed jackrabbit, marmot, bushy-tailed woodrat, long-tailed vole, and wapiti (extirpated in recent times). Others are extirpated from the Snake Range: pika, snowshoe rabbit, meadow vole, red fox, and pine marten. All Ice Age fossil localities in the Great Basin show similar patterns, and in many cases alpine species such as marmot and pika are the most abundant fossils (Grayson 1987; Mead 1987; Zimina and Gerasimov 1969). Extinct species considered to be boreal are discussed subsequently.

Two main factors restrict boreal species from inhabiting deserts: (1) inability to survive under hot and/or dry conditions and (2) lack of preferred food items. Interspecific competition with desert ecological counterparts may also have been an important factor in some cases as discussed previously, but other boreal species were not replaced. When conditions were wetter and cooler, boreal mammals extended their ranges and increased in numbers. When conditions became hotter and drier and conifers receded to higher elevations, desert mammals extended their ranges at the expense of boreal mammals. Crystal Ball Cave is located in a place where this replacement shows up very clearly.

Sixteen species of small boreal mammals are restricted to isolated mountaintops in the Great Basin but have extensive ranges in the Sierra Nevada, Rocky Mountains, or both: vagrant shrew, water shrew, pika, Nuttall's cottontailed, white-tailed jackrabbit, yellow pine chipmunk, cliff chipmunk, Uinta chipmunk, marmot, Belding's ground squirrel, Golden-mantled ground squirrel, Bushy-tailed woodrat, long-tailed vole, western jumping mouse, ermine, and wolverine (see table 15.4). Heather vole, pine marten, and probably snowshoe rabbit and Douglas' squirrel can be added to this list based on fossil occurrences. Some species have a continuous range in the northern Great Basin but a discontinuous range farther south. Panamint chipmunk, northern pocket gopher, montane vole, long-tailed weasel, mule deer, and most of the species listed previously have at least one isolated mountaintop population far south of their main ranges (Hall 1946, 1981). Brown (1971, 1978) and Grayson (1981, 1987) argued that these discontinuous ranges could only be explained by the reduction of a continuous late Pleistocene distribution over the Great Basin. The fossil record confirms this scenario, since many of the sites are located at low elevations and at latitudes far south of the present ranges of many species (fig. 15.4, tables 15.5 and 15.6). In a few cases isolation has led to speciation, such as the evolution of Palmer's chipmunk from uinta chipmunk (discussed subsequently), further supporting the relectual interpretation.

The following species now extirpated from the Great Basin have been found at fossil sites far from their current ranges: snowshoe rabbit, meadow vole, and pine marten from Crystal Ball Cave (Heaton 1985); Douglas' squirrel and mountain pocket gopher from Bronco Charlie Cave (Spiess 1974); dusky-footed woodrat from Kokoweef Cave (Reynolds 1980); heather vole from Gatecliff Shelter and Smith Creek Cave (Grayson 1981; Thompson and Mead 1982); brown bear from Deer Creek, Hidden, and Labor-of-Love caves (Ziegler 1963; Grayson 1985; Emslie and Czaplewski 1985); least weasel, black-footed ferret, and wolverine from Snake Creek Burial Cave (Barker and Best 1976; Mead and Mead 1989); and jaguar from Smith Creek Cave (Miller 1979). Fossils also document the former presence of species that have been eliminated from the Great Basin in recent times: black bear (Labor-of-Love Cave; Emslie and Czaplewski 1985), gray wolf, wapiti, and bison (table 15.6).

SPECIES REQUIRING PERENNIAL WATER

The Crystal Ball Cave assemblage includes bones of several kinds of fish and of two semiaquatic mammals: muskrat and mink. Other deposits contain fossils of water shrew and beaver (table 15.6). These animals live in a wide variety of life zones, but only where perennial streams, lakes, or both occur. There is a large warm spring at the south end of Gandy Mountain, but upon its discovery it contained only small minnows. Muskrat and mink do not presently occur in or near the Snake Range, but they do range into the northern Great Basin. Since present conditions cannot support these species, conditions must have been wetter in the past.

Lake Bonneville at its highest level included a shallow arm that extended southward into Snake Valley (fig. 15.2); during that time the lake was only 1.7 km east of and 195 m lower than Crystal Ball Cave. The lake was within 8 km of the cave for about half the time that fossils were accumulating. Washes near Gandy Mountain that drain the northern Snake Range may also have contained perennial streams during the late Wisconsin and provided suitable habitats for semiaquatic animals. These species are rare in Quaternary fossil assemblages of the Great Basin, so Crystal Ball Cave probably contains them only because of its proximity to Lake Bonneville.

EXTINCT SPECIES

The discovery of over 20 species of extinct mammals that formerly lived in the Great Basin (table 15.7) has greatly contributed to the study of biotic changes occurring since the last glacial. The extinct species include the largest Pleistocene herbivores (mammoths, mastodons, ground sloths, bison, horses, peccaries, camels, and oxen) and carnivores (short-faced bears, saber-toothed cats, lions, and cheetahs). The only smaller mammals to go extinct were diminutive pronghorn, noble marten, and short-faced skunk. None of these extinct mammals were restricted to the Great Basin; they ranged over much of North America.

Since mainly large mammals went extinct, it is interesting to note the status of extant large mammals. Of seven species extirpated from the Great Basin at the end of the Pleistocene, four were large: brown bear, wolverine, jaguar, and mountain goat. This is even more striking when one considers that small species greatly outnumber large species. Of the large mammals reported living in the Great Basin in recent times, most are rare and possibly extirpated: beaver, wolf, black bear, otter, lynx, wapiti, and bison. Bighorn sheep have also become scarce. The only large mammals found today in any abundance in the Great

Table 15.7. Extinct mammals that lived in Great Basin during the late Quaternary period

Scientific Name	Locality	Reference
Megalonyx sp. Ground Sloth	Tule Springs	Mawby 1967
Nothrotheriops shastensis Shasta Ground Sloth	Gypsum Cave Tule Springs	Harrington 1933; Stock 1931 Mawby 1967
Canis cf. *dirus* Dire Wolf	Kokoweef Cave	Reynolds 1980
Arctodus simus Giant Short-faced Bear	Labor-of-Love Cave Lake Bonneville	Emslie and Czaplewski 1985 Nelson and Madsen 1980, 1983
Martes nobilis Noble Marten	Bronco Charlie Cave Hidden Cave Smith Creek Cave	Grayson 1987; Spiess 1974 Grayson 1985 Miller 1979
Brachyprotoma brevimala Western Short-faced Skunk	Crystal Ball Cave	Heaton 1985
Smilodon cf. *fatalis* Saber-toothed Cat	Crystal Ball Cave	Heaton 1985
Panthera atrox American Lion	?Smith Creek Cave Tule Springs	Miller 1979 Mawby 1967
Acinonyx trumani American Cheetah	Crypt Cave	Adams 1979; Orr 1969
Mammut sp. American Mastodont	near Tule Springs	Spurr 1903
Mammuthus columbi Columbian Mammoth	Lake Bonneville Tule Springs	Nelson and Madsen 1980 Mawby 1967
Equus (large) Large Horse	Crypt Cave Crystal Ball Cave Fishbone Cave Gypsum Cave Owl Cave Two Schuiling Cave Smith Creek Cave Tule Springs	Grayson 1987; Orr 1952 Heaton 1985 Grayson 1987; Orr 1956 Harrington 1933; Stock 1931 Turnmire 1987 Downs et al. 1959 Miller 1979 Mawby 1967
Equus (small) Small Horse	Crystal Ball Cave Kokoweef Cave Owl Cave Two Schuiling Cave Smith Creek Cave Tule Springs	Heaton 1985 Reynolds 1980 Turnmire 1987 Downs et al. 1959 Miller 1979 Mawby 1967

Table 15.7. (*continued*)

Scientific Name	Locality	Reference
Camelops hesternus	Crystal Ball Cave	Heaton 1985
Yesterday's Camel	Fishbone Cave	Grayson 1987; Orr 1956
	Gypsum Cave	Harrington 1933; Stock 1931
	Hidden Cave	Grayson 1985
	Kokoweef Cave	Reynolds 1980
	Owl Cave Two	Turnmire 1987
	Smith Creek Cave	Miller 1979
	Tabernacle Cave	Nelson and Madsen 1979
	Tule Springs	Mawby 1967
Hemiauchenia	Crystal Ball Cave	Heaton 1985
macrocephala	Gypsum Cave	Harrington 1933; Stock 1931
Large-headed Llama	Kokoweef Cave	Reynolds 1980
	Mineral Hill Cave	McGuire 1980
	?Schuiling Cave	Downs et al. 1959
	?Smith Creek Cave	Miller 1979
Capromeryx minor	Schuiling Cave	Downs et al. 1959
Dimunitive Pronghorn	?Smith Creek Cave	Miller 1979
?*Tetrameryx* sp.	Tule Springs	Mawby 1967
Pronghorn		
Bison antiquus	Owl Cave Two	Turnmire 1987
Bison		
Oreamnos harringtoni	Smith Creek Cave	Miller 1979
Harrington's Mountain Goat		
Eucerotherium sp.	Kokoweef Cave	Reynolds 1980
Shrub Ox	Mineral Hill Cave	McGuire 1980
Bootherium bombifrons[a]	Crystal Ball Cave	Heaton 1985
Woodland Ox	Lake Bonneville	Nelson and Madsen 1978

Note: See table 15.5 for further information and references on localities. The list of localities is not exhaustive.

[a]*Symbos cavifrons* is now considered a synonym of *Bootherium bombifrons* (McDonald and Ray 1989).

Basin are coyote, fox, mountain lion, bobcat, mule deer, and pronghorn. Mountain lions feed primarily on deer, whereas the other carnivores feed mainly on small mammals (Hall 1946).

Although the largest mammals, such as mammoths, mastodons, sloths, bison, and oxen, seem to have never been abundant in the Great Basin, some large mammals are found in great numbers at most fossil sites, especially horses (several species), camels, llamas, and bighorn sheep. It appears that the Great Basin is no longer able to support such a diverse fauna of large mammals.

A few cases of extinction may be attributable to the invasion of a new competitor. The short-faced bear was the largest native bear of North America when brown bears immigrated from Asia during the last glacial. These two bears are of similar size and have been found associated at Labor-of-Love Cave in Nevada (Emslie and Czaplewski 1985) and Little Box Elder Cave in Wyoming (Kurten and Anderson 1974), so the introduction of the brown bear may have promoted the extinction of the short-faced bear. Noble marten fossils have been found throughout the Great Basin, whereas pine marten fossils have been found only in northern and eastern sites. During the late Pleistocene, the pine marten inhabited primarily the eastern part of the continent and the noble marten the western part; the pine marten has replaced noble marten in the west, probably by competition (Kurten and Anderson 1980). The only Holocene records of noble marten are from the Great Basin, where it survived until about 3300 years B.P. (Grayson 1987); desert barriers preventing pine martens from invading the central Great Basin may account for the noble marten's extended existence there. The extinct short-faced skunk may have competed with the spotted skunk, but no evidence substantiates this. Most large extinct mammals were not replaced by ecological equivalents, so competition could not have been a factor in their extinctions.

Martin (1984) has proposed that the primary cause of the late Pleistocene megafauna extinctions was overkill by human hunters. He envisioned a sudden human invasion across the Bering land bridge about 12,000 years B.P. that encountered a megafauna having no fear of humans. Mosimann and Martin (1975) postulated that the human population grew rapidly in this plentiful environment and swept across the continent, driving the megafauna into a smaller and smaller area and finally to extinction. Evidence cited for this model includes the rapidity of the extinctions, the fact that mainly large species were affected, and coincidence with abundant human archaeological sites. Although most workers allow human activities some role in the extinctions, Martin's theory that humans were the exclusive cause has not been well accepted. Evidence cited against the overkill hypothesis has included the similarity of the climatic backdrop to many prehuman extinctions (Webb 1984) and the archaeological distinctness of long-term human cultures in different parts of North America (MacNeish 1976). Since the extinctions are only one component of a dramatic restructuring of the mammalian community, human intervention is an unnecessary explanation. Humans were clearly contemporary with many now extinct species of large mammals in the Great Basin, but little evidence has been found to demonstrate association between them (Grayson 1982).

Evolutionary Implications

Because the Great Basin has been a region of much ecological discontinuity and reorganization during the Quaternary period, it has afforded many opportunities for evolutionary change. This paper covers a time frame of only 30,000 years, and it is therefore not surprising that no major evolutionary transformations have taken place. However, evolutionary modifications with respect to physiological tolerance, body size, and various minor characters have occurred. These short-term changes, which are well documented because of their recency and because they involve living species, show patterns that may have been characteristic of transitional environments in past ages.

Several cases of physiological evolution have been documented in species with ecological opportunity. Northern pocket gophers inhabit the northern latitudes and higher elevations of the Great Basin and compete with southern pocket gophers, which inhabit the southern latitudes and lower elevations. Only the southern pocket gopher was recovered from Crystal Ball Cave, and this is the only pocket gopher found living in the Snake Range, where it reaches exceptionally high elevations (Heaton 1985). Hall (1946) proposed that only the southern pocket gopher was in the area when the Snake Range became available for colonization, and in the absence of the northern species it adapted to the higher elevations (although it does not occur in as great of numbers at high elevations as northern pocket gopher does in other ranges). The same situation apparently exists in the Toquima Range (Hall 1946; Grayson 1987). In the Pilot Range, in the absence of a high-elevation chipmunk such as Uinta Chipmunk, the cliff chipmunk, which is usually restricted to the pinyon-juniper floral belt, has extended its elevational range upward into the mixed conifers on the single peak (Brown 1978; Harper et al. 1978). Such cases show that restrictions are imposed on species by congeneric competition. When these restrictions are lifted, physiological traits can evolve rapidly to allow the invasion of new habitats.

Ecological fluctuations have been mirrored by changes in body size in some species. Stokes and Condie (1961) described enormous individuals of Pleistocene bighorn sheep from lake deposits throughout the Great Basin. Because of their large size, great orbital width, wide rostrum, and massive horn cores, they were first assigned to a species distinct from the living form. Further investigation by Stock and Stokes (1969) and Harris and Mundel (1974) showed that this large form evolved over about 2,000 years into the smaller modern form, so they are now considered conspecific. Black bear, wolverine, and bison also increased in size during the Wisconsin, then decreased afterward (Kurten and Anderson 1980).

Geist (1971, 1983) proposed that during the glacial period when there was an abundance of forage, large mammals altered their morphology because of selection for males with generalized feeding structures but exceptional sexual structures (large body size and enlarged ornate characters) and high intelligence (larger brains). Geist sees Irish elk, moose, muskoxen, polar bears, and humans as examples of this pattern. When the exceptional environmental conditions deteriorated, these species were forced to reduce their investment in ornate characters or be driven to extinction.

SPECIATION EVENTS

Mayr (1954, 1963) proposed that most new species arise from peripheral isolated populations, and Eldredge and Gould (1972) used this model to explain discontinuities in the fossil record. The remnant populations on habitat islands discussed by Brown (1971, 1978) fit this model perfectly, and a number of new species appear to have arisen in the Great Basin during the Quaternary period as a result of isolation. Few fossil data are available to document when these species originated (most of them are very localized and distant from fossil sites), but their present distributions suggest a recent origin. None of these evolutionary events represent major morphological change, but some are worthy of species recognition.

The Mt. Lyell shrew currently inhabits a small area in central California and westernmost Nevada, and the Great Basin separates it from its parent species, the masked shrew, which is found throughout northern North America (Hall 1946, 1981). Boreal habitat bridges across the northwestern Great Basin during the Pleistocene appear to have allowed the invasion of this shrew into California, where it speciated following isolation. The Inyo shrew lives in a few isolated mountainous environments in the southwestern Great Basin, and it clearly evolved from the ornate shrew of the Sierra Nevada and coastal ranges of California (Hall 1946, 1981). The development and subsequent deterioration of boreal habitat bridges between mountain ranges in the Great

Basin could have led to the dispersal and subsequent isolation of the ornate shrew and, finally, the origin of Inyo shrew. The Panamint chipmunk has a similar distribution and probably developed in the same way from yellow pine chipmunk.

Palmer's chipmunk is restricted to Charleston Peak in the Spring Mountains of southern Nevada. Its parent species, the Uinta chipmunk, is found on mountain peaks across the central Great Basin, with the closest population, a distinct subspecies, being restricted to the Sheep Mountains just 50 km northwest of Charleston Peak (Hall 1946, 1981). The morphological distinctness of each isolated population appears correlated with the length of time that it has been isolated. The development of the yellow-eared pocket mouse and white-eared pocket mouse from the Great Basin pocket mouse on isolated mountain ranges in southern California follows an identical pattern (Hall 1981). As thermal thresholds for boreal species migrated northward at the end of the Pleistocene, populations first became isolated on the most southern mountaintops, and there they have had the most time to diverge from their parent species.

None of the species discussed previously are known from the fossil record of the Great Basin, partly because of their limited ranges and the sparsity of fossil localities. Another reason is that most fossils occur at low elevations in lake deposits and dry caves, and moist erosional conditions in the high mountains are less favorable for fossil preservation. Biases such as this may account for seeming gaps in the fossil record as discussed by Eldredge and Gould (1972).

Discussion: Causal Pathways of Change

The current distribution and many other features of Great Basin mammals cannot be explained by equilibrium conditions but only by historic considerations. In the past 10,000 years, the broad valleys that comprise almost all of the land area have changed from lush savannas, forests, and lakes to dry, desolate flats. The major climatic and ecological changes described in this paper can be attributed to temperature fluctuations caused by periodic variations in the earth's orbit termed Milankovitch cycles (Hays, Imbrie, and Shackleton 1976; Imbrie and Imbrie 1980; Kutzback 1983). But while the ultimate cause of these changes may be simple, the pathways by which they occurred and interacted are not. Few mammals would have been affected appreciably by the temperature fluctuations alone, but the indirect effects included large changes in range as well as many extinctions. Some of the pathways of change were abiotic: increased precipitation, the development of lakes and glaciers, and the associated lowering of sea level. Other pathways were biotic, for example, mammalian exchanges between continents, changes in the abundance and types of vegetation, development and destruction of habitat bridges, and interspecific competition. Figures 15.5 and 15.6 diagram some of the major pathways by which mammals were affected by the onset and the termination of the last Wisconsin glacial.

ABIOTIC CHANGES

As discussed previously, a climatic cooling caused the Ice Age. Cooler air leads to increased precipitation, although the exact function of this

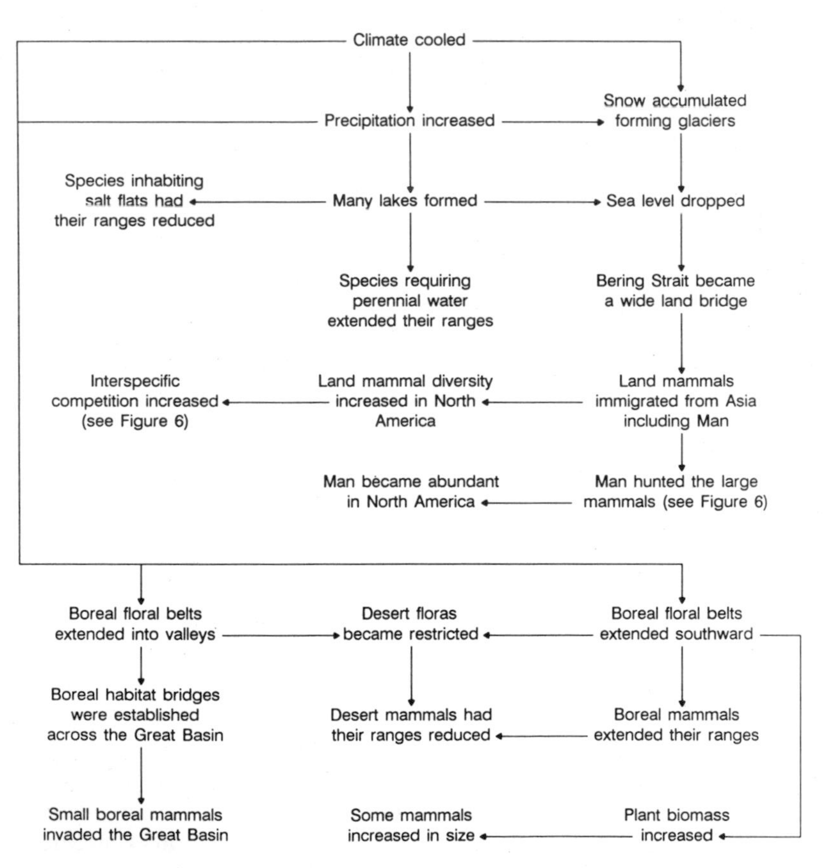

Figure 15.5. Schematic diagram showing pathways of change caused by the onset of the last Ice Age.

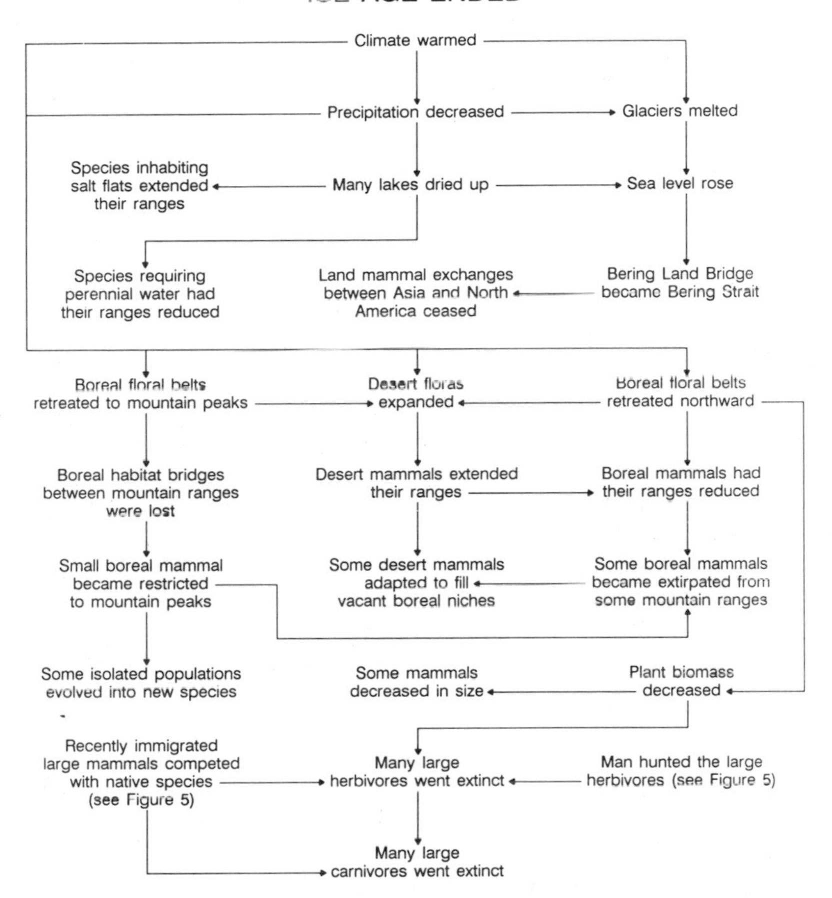

Figure 15.6 Schematic diagram showing pathways of change caused by the termination of the last Ice Age.

relationship is complex. A cooler climate also decreases the melting of snow and evaporation of lake water. These effects led to the formation of continental glaciers in northern North America and valley glaciers on many mountain peaks during the late Pleistocene. A more significant effect in the Great Basin was the formation and expansion of pluvial lakes in the numerous closed valleys. Although these lakes consumed the range of many desert mammals that live on salt flats, they provided an increased range for mammals that require perennial water. At the end of the Wisconsin glacial all of these trends were reversed. The fossil record shows that as the lakes and glaciers dried up, desert mammals proliferated and semiaquatic mammals could not survive.

Correlated with the formation of glaciers and lakes was a drop in sea level, and the shallow Bering Strait became a wide land bridge. This allowed faunal exchanges between Asia and North America that greatly increased mammalian diversity, further disrupting the biological equilibrium, especially among large species because of their greater mobility. Humans also entered North America at this time, and their role in the megafaunal extinction is a matter of some controversy. When the Ice Age ended and sea level rose, the intercontinental land bridge disappeared.

BIOTIC CHANGES

The increase in precipitation and decrease in temperature at the beginning of the last glacial had an enormous effect on vegetation. A profusion of vegetation provided more food for herbivorous mammals, which increased their population density and diversity and in some cases even their body size. This trend extended up the food web to carnivores as well. At the end of the Ice Age the vegetational biomass decreased again, reducing the diversity of large mammals through extinctions, extirpations, and range reductions.

Because of the elevational configuration of the Great Basin, the migration of vegetational thresholds had a unique impact. Boreal conditions are now restricted to isolated mountain islands between which small boreal mammals cannot disperse. But during the Ice Age boreal conditions became continuous across the Great Basin, which allowed for dispersal of boreal mammals between ranges. Now that these habitat bridges have disappeared, many species are isolated in mountaintop populations, and in a few cases these populations have undergone allopatric speciation. Some mountain populations have died out and are only known from the fossil record.

The diets of some species of mammals are restricted to particular plant foods, and the distributions of these species closely track the distributions of their favored plants. This occurs most often in desert life zones that contain one or two principal plant species. Certain favored plants, such as sagebrush and juniper, now cover vast areas of the Great Basin because they are well adapted to the hot valleys and lower mountain slopes. During the last glacial, however, these life zones had been driven southward by the expanding boreal flora or were restricted to small, isolated areas, and the number of mammals depending on these plants decreased. When the climate became warmer and drier after the

Ice Age, mesic vegetation could no longer survive in the valleys of the Great Basin, so the ranges of desert plants and their associated mammals expanded. Range expansions for some such mammals have even been documented in historic times.

Interspecific competition often leads to ecological specialization, and some species would inhabit much wider ranges of environments if those habitats were not occupied by other closely related species. Many of the range shifts that occurred in the Great Basin during the Quaternary period appear to have resulted primarily from competition and only secondarily from climatic and vegetational changes. This accounts for replacement of one species by another among rabbits, woodrats, and foxes, for example. As the climate cooled at the beginning of the Ice Age, mesic species were able to utilize the environment slightly better and thus outcompete xeric species. When the climate warmed again, the reverse occurred. In several cases the habitat islands of the Great Basin provided ecological opportunity. On some peaks where a boreal competitor was absent, desert species of pocket gopher and chipmunk were able to invade the high elevations and adapt to the cooler conditions.

During the last glacial, when vegetation was abundant, the northern hemisphere supported many large-bodied mammals. Continental exchanges across the Bering land bridge added greatly to this diversity. It appears that these species coexisted with ease during the Ice Age under favorable conditions and diverse habitats, and interspecific competition was at a minimum. Instead, competition was more intraspecific, and sexual selection led to the development of large body size and bizarre external features at high metabolic cost. When the Ice Age ended, many habitats deteriorated, and therefore interspecific competition increased again. With lush vegetation restricted to the mountain peaks, the Great Basin was no longer able to support a diverse population of large mammals, and only the large mammals adapted to desert environments or mountainous terrain have survived.

Conclusion

Temperature fluctuations during the Quaternary Period had far-reaching abiotic and biotic implications in the Great Basin. At the beginning of the last Ice Age, lakes formed in the numerous closed valleys, and boreal floras developed a continuous distribution over much of the region. These conditions supported a diverse mammalian community, including many large species that are now extinct or extirpated. Boreal mammals currently living in the Great Basin are restricted to scattered mountaintops, and their distributions can only be understood as a relict of formerly continuous habitats. This is demonstrated by both the nonequilibrium pattern of their distributions and fossil occurrences at low elevations. In a few cases, isolated mountaintop populations have undergone allopatric speciation.

At the end of the Ice Age, as lakes dried up and boreal floras diminished, desert mammals became the dominant species of the Great Basin. The recency of their expansion is demonstrated by their poor representation in fossil deposits compared to boreal species. Factors that influenced overall diversity reduction and desert species proliferation included a warming and drying of the climate, the replacement of lush mesic floras by sparse xeric floras, and interspecific competition for limited resources. The pathways by which these changes took place are very complex and can only be understood by studying diets, tolerances, and habits of living species in connection with fossil distributions and associations.

Acknowledgments

My work at Crystal Ball Cave was funded by Brigham Young University, the National Speleological Society, and Herbert H. Gerisch (an early collector at the cave). I thank Jim I. Mead, Wade E. Miller, Donald K. Grayson, Julia S. Heaton, and the editors for helpful comments on the manuscript.

Literature Cited

Adams, D. B. 1979. The cheetah: Native American. *Science* 205:1155–8.

Barker, M. S., Jr., and T. L. Best. 1976. The wolverine (*Gulo luscus*) in Nevada. *Southwest Nat.* 21(1):133.

Benson, L. V. 1978. Fluctuation in the level of pluvial Lake Lahontan during the last 40,000 years. *Quat. Res.* 9:300–318.

Bloom, A. L. 1983. Sea level and coastal morphology of the United States through the late Wisconsin glacial maximum. In *Late-Quaternary environments of the United States, vol. 1, The late Pleistocene,* ed. S. C. Porter, 215–29. Minneapolis: University of Minnesota Press.

Brown, J. H. 1971. Mammals on mountaintops: Nonequilibrium insular biogeography. *Am. Nat.* 105(945):467–78.

Brown, J. H. 1978. The theory of insular biogeography and the distribution of boreal birds and mammals. In *Intermountain biogeography: a symposium, Great Basin Nat. Mem.* 2, ed. K. T. Harper and J. L. Reveal, 209–27.

Carleton, M. D., and R. E. Eshelman. 1979. A synopsis of fossil grasshopper mice, genus *Onychomys*, and their relationships to Recent species. *Pap. Paleontol. Univ. Michigan Mus. Paleontol.* 21:1–63.

Currey, D. R. 1982. Lake Bonneville: Selected features of relevance to neotectonic analysis. *U.S. Geol. Surv. Open-File Rep.*, 82–1070.

Davis, J. O. 1978. Quaternary tephrochronology of the Lake Lahontan Area, Nevada and California, *Nevada Archaeol. Surv. Res. Pap.* 7.

Downs, T., H. Howard, T. Clements, and G. A. Smith. 1959. Quaternary animals from Schuiling Cave in the Mojave Desert, California. *Los Angeles Co. Mus. Contrib. Sci.*, no. 29, pp. 1–21.

Durrant, S. D. 1952. *Mammals of Utah: Taxonomy and distribution.* Lawrence: University of Kansas.

Eldredge, N., and S. J. Gould. 1972. Punctuated equilibria: an alternative to phyletic gradualism. In *Models in paleobiology,* ed. T. J. M. Schoph, 82–115. San Francisco: Freeman, Cooper and Co.

Emslie, S. D., and N. J. Czaplewski. 1985. A new record of giant short-faced bear, *Arctodus simus*, from western North America with a re-evaluation of its paleobiology. *Contrib. Sci.*, no. 371, pp. 1–12.

Geist, V. 1971. *Mountain sheep: a study in behavior and evolution.* Chicago: University of Chicago Press.

Geist, V. 1983. On the evolution of Ice Age mammals and its significance to an understanding of speciation. *Am. Soc. Biol. Bull.* 30(3):110–33.

Goodwin, H. T., and R. E. Reynolds. 1989. Sciuridae from Kokoweef Cave, San Bernardino County, California. *Bull. South. Calif. Acad. Sci.* 88(1):21–32.

Grayson, D. K. 1981. A mid-Holocene record of the heather vole, *Phenacomys* cf. *intermedius*, in the central Great Basin and its biogeographic significance. *J. Mamm.* 62:115–21.

Grayson, D. K. 1982. Toward a history of Great Basin Mammals during the past 15,000 years. In *Man and environment in the Great Basin, Soc. Am. Archaeol. Pap.* 2, ed. D. B. Madsen and J. F. O'Connell, 82–101.

Grayson, D. K. 1983. Small mammals. In *The archaeology of Monitor Valley. 2. Gatecliff Shelter, Am. Mus. Nat. Hist. Anthropol. Pap.* 59 ed. D. H. Thomas. 99–126.

Grayson, D. K. 1985. The paleontology of Hidden Cave: Birds and mammals. In *The archaeology of Hidden Cave, Nevada, Am. Mus. Nat. Hist. 61, Anthropol. Pap.* ed. D. H. Thomas, 125–61.

Grayson, D. K. 1987. The biogeographic history of small mammals in the Great Basin: observations on the last 20,000 years. *J. Mamm.* 68(2):359–75.

Green, J. S. and J. T. Flinders. 1980. *Brachylagus idahoensis. Mamm. Species* 125:1–4.

Hall, E. R. 1946. *Mammals of Nevada.* Berkeley and Los Angeles: University of California Press.

Hall, E. R. 1981. *The mammals of North America,* 2d ed. New York: Ronald Press.

Harper, K. T., D. C. Freeman, W. K. Ostler, and L. G. Klikoff. 1978. The flora of Great Basin Mountain Ranges: Diversity, sources, and dispersal ecology. In *Intermountain biogeography: a symposium, Great Basin Nat. Mem.* 2 ed. K. T. Harper and J. L. Reveal, 81–103.

Harrington, M. R. 1933. Gypsum Cave, Nevada. *Southwest Mus. Pap.* 8:1–197.

Harris. A. H. 1985. *Late Pleistocene vertebrate paleoecology of the West*. Austin: University of Texas Press.

Harris, A. H., and P. Mundel. 1974. Size reduction in bighorn sheep (*Ovis canadensis*) at the close of the Pleistocene. *J. Mamm.* 53(3), 678–80.

Hays, J. D., J. Imbrie, and N. J. Shackleton. 1976. Variations in the Earth's orbit: Pacemaker of the ice ages. *Science* 194:1121–32.

Heaton, T. H. 1984. Preliminary report on the Quaternary vertebrate fossils from Crystal Ball Cave, Millard County, Utah. *Curr. Res. Pleist.* 1:65–67.

Heaton, T. H. 1985. Quaternary paleontology and paleoecology of Crystal Ball Cave, Millard County, Utah: with emphasis on mammals and description of a new species of fossil skunk. *Great Basin Nat.* 45(3):337–90.

Heaton, T. H. 1987. Initial investigation of vertebrate remains from Snake Creek Burial Cave, White Pine County, Nevada. *Curr. Res. Pleist.* 4:107–9.

Hunt, C. B. 1967. *Physiography of the United States*. San Francisco: W. H. Freeman and Co.

Imbrie, J., and J. Z. Imbrie. 1980. Modeling the climatic response to orbital variations. *Science* 207:943–53.

Jefferson, G. T. 1982. Late Pleistocene vertebrates from a Mormon Mountain cave in southern Nevada. *Bull. South. Calif. Acad. Sci.* 81(3):121–27.

Kurten, B., and E. Anderson. 1974. Association of *Ursus arctos* and *Arctodus simus* (Mammalia: Ursidae) in the late Pleistocene of Wyoming. *Breviora* no. 426, pp. 1–6.

Kurten, B., and E. Anderson. 1980. *Pleistocene mammals of North America*. New York: Columbia University Press.

Kutzback, J. E. 1983. Modeling of Holocene climates. In *Late-Quaternary Environments of the United States, vol. 2, The Holocene*, ed. H. E. Wright, Jr., 271–7. Minneapolis: University of Minnesota Press.

Long. W. S. 1940. Notes on the life histories of some Utah mammals. *J. Mamm.* 21(2):170–80.

Lundelius, E. L., Jr., R. W. Graham, E. Anderson, J. Guilday, J. A. Holman, D. W. Steadman, S. D. Webb. 1983. Terrestrial vertebrate faunas. In *Late-Quaternary environments of the United States, vol. I, The Late Pleistocene*, ed. S. C. Porter, 311–53. Minneapolis: University of Minnesota Press.

MacArthur, R. W., and E. O. Wilson. 1963. An equilibrium theory of insular biogeography. *Evolution* 17:373–87.

MacArthur, R. W., and E. O. Wilson. 1967. *The theory of island biogeography*, Princeton, N.J.: Princeton University Press.

MacNeish, R. S. 1976. Early man in the New World. *Am. Sci.* 64:316–27.

Martin, P. S. 1984. Prehistoric overkill: the global model. In *Quaternary ex-*

tinctions: a prehistoric revolution, ed. P. S. Martin and R. G. Klein, 354–403. Tucson: University of Arizona Press.

Mawby, J. E. 1967. Fossil vertebrates of the Tule Springs Site, Nevada. *Nevada State Mus. Anthrop. Pap.* 13:105–28.

Mayr, E. 1954. Change of genetic environment and evolution. In *Evolution as a process*, ed. J. Huxley, A. C. Hardy, and E. B. Ford, 157–80. London: Allen and Unwin.

Mayr, E. 1963. *Animal speciation and evolution*. Cambridge, Mass.: Harvard University Press.

McDonald, J. N., and C. E. Ray. 1989. The autochthonous North American musk oxen *Bootherium*, *Symbos*, and *Gidleya* (Mammalia: Artiodactyla: Bovidae). *Smithsonian Contrib. Paleobiol.* 66:1–77.

McGuire, K. R. 1980. Cave sites, faunal analysis, and big-game hunters of the Great Basin: a caution. *Quat. Res.* 14:263–8.

Mead, E. M., and J. I. Mead 1989. Snake Creek Burial Cave and a review of the Quaternary mustelids of the Great Basin. *Great Basin Nat.* 49:143–54.

Mead, J. I. 1987. Quaternary records of pika, *Ochotona*, in North America. *Boreas* 16:165–71.

Mead, J. I., and E. M. Mead. 1985. A natural trap for Pleistocene animals in Snake Valley, eastern Nevada. *Curr. Res. Pleist.* 2:105–6.

Mead, J. I., R. S. Thompson, and T. R. Van Devender. 1982. Late Wisconsinan and Holocene fauna from Smith Creek Canyon, Snake Range, Nevada. *Trans. San Diego Soc. Nat. Hist.* 20(1):1–26.

Mehringer, P. J., Jr. 1967. The environment of extinction of the late-Pleistocene megafauna in the arid southwestern United States. In *Pleistocene extinctions: the search for a cause*, ed. P. S. Martin and H. E. Wright, Jr., 247–66. New Haven, Conn.: Yale University Press.

Mifflin, M. D., and M. M. Wheat. 1979. Pluvial lakes and estimated pluvial climates of Nevada. *Nev. Bureau Mines Geol. Bull.* 94:1–57.

Miller, S. J. 1979. The archaeological fauna of four sites in Smith Creek Canyon. In *The archaeology of Smith Creek Canyon, Eastern Nevada*, *Nevada State Mus. Anthrop. Pap.* 17, ed. D. R. Tuohy and D. L. Rendall, 271–329.

Mosimann, J. E., and P. S. Martin, 1975. Simulating overkill by paleoindians. *Am. Sci.* 63:304–13.

Nelson, M. E., and J. H. Madsen, Jr. 1978. Late Pleistocene musk oxen from Utah. *Trans. Kan. Acad. Sci.* 81(4):277–95.

Nelson, M. E., and J. H. Madsen, Jr. 1979. The Hay-Romer debate: Fifty years later. *Univ. Wyoming Contrib. Geol.* 18:47–50.

Nelson, M. E., and J. H. Madsen, Jr. 1980. Paleoecology of the late Pleistocene, large mammal community in the northern Bonneville Basin, *Utah*, *Geol. Soc. Am. Abs. Prog.* 12(6):299.

Nelson, M. E., and J. H. Madsen, Jr. 1983. A giant short-faced bear (*Arctodus simus*) from the Pleistocene of northern Utah. *Trans. Kan. Acad. Sci.* 86(1):1–9.

Orr, P. C. 1952. Preliminary excavations of Pershing County caves. *Nev. State Mus. Dept. Archaeol. Bull. 1.*

Orr, P. C. 1956. Pleistocene man in Fishbone Cave, Pershing County, Nevada. *Nev. State Mus. Dept. Archaeol. Bull* 2.

Orr, P. C. 1969. *Felis trumani*, a new radiocarbon dated cat skull from Crypt Cave, Nevada. *Bull. Santa Barbara Mus. Nat. Hist. Dept. Geol.* 2:1–8.

Orr, P. C. 1972. The eighth Lake Lahontan (Nevada) expedition, 1957. *Natl. Geog. Res. Rept. 1955–1960 Proj.*, 123–6.

Porter, S. C., K. L. Pierce, and T. D. Hamilton. 1983. Late Wisconsin mountain glaciation in the western United States. In *Late-Quaternary environments of the United States, vol. I, The Late Pleistocene,* ed. S. C. Porter, 71–111. Minneapolis: University of Minnesota Press.

Reynolds, R. E. 1980. Kokoweef Cave faunal list. Unpublished.

Riddle, B. R., and J. R. Choate. 1986. Systematics and biogeography of *Onychomys leucogaster* in Western North America. *J. Mamm.* 67(2):233–55.

Romer, A. S. 1928. A "fossil" camel recently living in Utah. *Science* 68:19–20.

Scott, W. E., W. D. McCoy, R. R. Shroba, and M. Rubin. 1983. Reinterpretation of the exposed record of the last two cycles of Lake Bonneville, Western United States. *Quat. Res.* 20:261–85.

Shutler, R. 1968. Tule Springs: Its implications to early man studies in North America. *Contrib. Anthropol. East. N.M. Univ.* 1(4):19–26.

Smith, G. W., and D. R. Johnson. 1985. Demography of a Townsend ground squirrel population in southwestern Idaho. *Ecology* 66(1):171–8.

Smith, G. I., and F. A. Street-Perrott. 1983. Pluvial lakes in the western United States. In *Late-Quaternary environments of the United States, vol. I, The Late Pleistocene*, ed. S. C. Porter, 190–212. Minneapolis, University of Minnesota Press.

Spaulding, W. G., E. B. Leopold, and T. R. Van Devender. 1983. Late Wisconsin paleoecology of the American Southwest. In *Late-Quaternary environments of the United States, vol. I, the Late Pleistocene,* ed. S. C. Porter, 259–93. Minneapolis: University of Minnesota Press.

Spiess, A. 1974. Faunal remains from Bronco Charlie Cave (26EK801), Elko County, Nevada. In *The prehistoric human ecology of southern Ruby Valley, Nevada*, ed. L. Casjens, 452–86. Ph.D. diss. Harvard University, Cambridge, Mass.

Spurr, J. E. 1903. Descriptive geology of Nevada south of the fortieth parallel and adjacent portions of California. *U.S. Geol. Surv. Bull. 208.*

Stock, A. D., and W. L. Stokes. 1969. A re-evaluation of Pleistocene bighorn sheep from the Great Basin and their relationship to living members of the genus *Ovis. J. Mamm.* 50:805–7.

Stock, C. 1931. Problems of antiquity presented in Gypsum Cave, Nevada. *Scientific Monthly* 32:22–32.

Stokes, W. L., and K. C. Condie. 1961. Pleistocene bighorn sheep from the Great Basin. *J. Paleontol.* 35(3):598–609.

Stokes, W. L., and G. H. Hansen. 1937. Two Pleistocene musk-oxen from Utah. *Utah Acad. Sci. Arts Let.* 14:63–65.

Thompson, R. S., and J. I. Mead. 1982. Late Quaternary environments and biogeography of the Great Basin. *Quat. Res.* 17:39–55.

Turnmire, K. L. 1987. An analysis of the mammalian fauna from Owl Cave One and Two, Snake Range, east-central Nevada. M.S. thesis, University of Maine, Orono.

Van Devender, T. R., and N. T. Tessman. 1975. Late Pleistocene snapping turtles (*Chelydra serpentina*) from southern Nevada. *Copeia* 1975, 249–53.

Webb, R. H. 1986. Spatial and temporal distribution of radiocarbon ages on rodent middens from the southwestern United States. *Radiocarbon* 28:1–8.

Webb, S. D. 1984. Ten million years of mammal extinctions in North America. In *Quaternary extinctions: a prehistoric revolution*, ed. P. S. Martin and R. G. Klein, 189–210. Tucson: University of Arizona Press.

Wells, P. V. 1983. Paleobiogeography of montane islands in the Great Basin since the last glaciopluvial. *Ecolog. Mongr.* 53(4):341–82.

Ziegler, A. C. 1963. Unmodified mammal and bird remains from Deer Creek Cave, Elko County, Nevada. In *Deer Creek Cave*, Nevada State Mus. Anthrolpol. Pap. 11, ed. M. E. Shutler and R. Shutler, Jr., 15–24.

Zimina, R. P., and I. P. Gerasimov. 1969. The periglacial expansion of marmots (*Marmota*) in middle Europe during the Upper Pleistocene. In *Etudes sur le Quaternaire dans le Monde*, ed. M. Ters, 465–72. Paris: CNRS.

Author Index

Index excludes authors cited only in tables or appendixes or as taxonomic authorities.

General Index